# Global Change and River Ecosystems—Implications for Structure, Function and Ecosystem Services

# Developments in Hydrobiology 215

*Series editor*
K. Martens

For other titles published in this series, go to
www.springer.com/series/5842

# Global Change and River Ecosystems—Implications for Structure, Function and Ecosystem Services

*Editors*

R. Jan Stevenson[1], Sergi Sabater[2]

[1]*Center for Water Sciences, Department of Zoology, University of Michigan, East Lansing, USA*
[2]*Institute of Aquatic Ecology at the University of Girona, Girona, Spain*

Previously published in *Hydrobiologia,* Volume 657, 2010

*Editors*
R. Jan Stevenson
Center for Water Sciences
Department of Zoology
University of Michigan
East Lansing
USA

Sergi Sabater
Institute of Aquatic Ecology at the
University of Girona
Girona, Spain

QH
541.5
.S7
G58
2011

ISBN 978-94-007-0607-1

Springer Dordrecht Heidelberg London New York

© Springer Science+Business Media B.V. 2011
No part of this work may be reproduced, stored in a retrieval system, or transmitted in any form or by any means, electronic, mechanical, photocopying, microfilming, recording or otherwise, without written permission from the Publisher, with the exception of any material supplied specifically for the purpose of being entered and executed on a computer system, for exclusive use by the purchaser of the work.

Printed on acid-free paper.

Springer is part of Springer Science+Business Media (www.springer.com)

# Contents

**Foreword: Global change and river ecosystems—implications for structure, function, and ecosystem services**
S. Sabater · R.J. Stevenson  **1**

**Understanding effects of global change on river ecosystems: science to support policy in a changing world**
R.J. Stevenson · S. Sabater  **3**

**Biogeochemical implications of climate change for tropical rivers and floodplains**
S.K. Hamilton  **19**

**Dynamics of a benthic microbial community in a riverine environment subject to hydrological fluctuations (Mulargia River, Italy)**
A. Zoppini · S. Amalfitano · S. Fazi · A. Puddu  **37**

**Global changes in pampean lowland streams (Argentina): implications for biodiversity and functioning**
A. Rodrigues Capítulo · N. Gómez · A. Giorgi · C. Feijoó  **53**

**Factors regulating epilithic biofilm carbon cycling and release with nutrient enrichment in headwater streams**
S.E. Ziegler · D.R. Lyon  **71**

**Periphyton biomass and ecological stoichiometry in streams within an urban to rural land-use gradient**
P.J. O'Brien · J.D. Wehr  **89**

**The physico-chemical habitat template for periphyton in alpine glacial streams under a changing climate**
U. Uehlinger · C.T. Robinson · M. Hieber · R. Zah  **107**

**The periphyton as a multimetric bioindicator for assessing the impact of land use on rivers: an overview of the Ardières-Morcille experimental watershed (France)**
B. Montuelle · U. Dorigo · A. Bérard · B. Volat · A. Bouchez · A. Tlili · V. Gouy · S. Pesce  **123**

**Discharge and the response of biofilms to metal exposure in Mediterranean rivers**
H. Guasch · G. Atli · B. Bonet · N. Corcoll · M. Leira · A. Serra  **143**

**Effects of eutrophication on the interaction between algae and grazers in an Andean stream**
J.Ch. Donato-Rondón · S.J. Morales-Duarte · M.I. Castro-Rebolledo  **159**

**Comparing fish assemblages and trophic ecology of permanent and intermittent reaches in a Mediterranean stream**
E. Mas-Martí · E. García-Berthou · S. Sabater · S. Tomanova · I. Muñoz  **167**

**Global change and food webs in running waters**
D.M. Perkins · J. Reiss · G. Yvon-Durocher · G. Woodward  **181**

**Effects of hydromorphological integrity on biodiversity and functioning of river ecosystems**
A. Elosegi · J. Díez · M. Mutz  **199**

**Organic matter availability during pre- and post-drought periods in a Mediterranean stream**
I. Ylla · I. Sanpera-Calbet · E. Vázquez · A.M. Romaní · I. Muñoz · A. Butturini · S. Sabater  **217**

**Flow regime alteration effects on the organic C dynamics in semiarid stream ecosystems**
V. Acuña   233

**A multi-modeling approach to evaluating climate and land use change impacts in a Great Lakes River Basin**
M.J. Wiley · D.W. Hyndman · B.C. Pijanowski · A.D. Kendall · C. Riseng · E.S. Rutherford · S.T. Cheng · M.L. Carlson · J.A. Tyler · R.J. Stevenson · P.J. Steen · P.L. Richards · P.W. Seelbach · J.M. Koches · R.R. Rediske   243

**Implications of global change for the maintenance of water quality and ecological integrity in the context of current water laws and environmental policies**
A.T. Hamilton · M.T. Barbour · B.G. Bierwagen   263

GLOBAL CHANGE AND RIVER ECOSYSTEMS

# Foreword: Global change and river ecosystems—implications for structure, function, and ecosystem services

Sergi Sabater · R. Jan Stevenson

Published online: 22 September 2010
© Springer Science+Business Media B.V. 2010

Rivers around the world are threatened by changes in land use, climate, hydrologic cycles, and biodiversity. These changes originate from economic and social processes, which can be collectively considered as drivers of global change. Among the most relevant, global changes in rivers are expressed as water flow interruptions, elevated temperatures, loss of hydrological connectivity, higher water residence times, increased nutrient and sediment loads, exposure to new chemicals, simplified physical structure, greater exposure to invasive species, and biodiversity losses. Most of these are not occurring independently, but mostly as combined or multiple interacting factors.

---

Guest editors: R.J. Stevenson, S. Sabater / Global Change and River Ecosystems – Implications for Structure, Function and Ecosystem Services

S. Sabater
Department of Environmental Sciences,
Institute of Aquatic Ecology, Faculty of Sciences,
University of Girona, 17071 Girona, Spain

S. Sabater
Catalan Institute for Water Research (ICRA), Scientific and Technological Park of the UdG, Emili Grahit 101, 17003 Girona, Spain

R. J. Stevenson (✉)
Department of Zoology, Michigan State University,
203 Natural Science Building, East Lansing, MI 48824, USA
e-mail: rjstev@msu.edu

All of them affect structure (e.g., number and identity of taxa or biomass) and functioning of the river ecosystems (e.g., productivity, use of organic matter, processing efficiency). Hence, these effects result in a loss or malfunction of the ecosystem services that rivers provide.

Understanding the responses of river ecosystems and the services they provide is essential for human well being in all regions of the planet. Rivers provide benefits to human well being with food from fisheries and irrigation, by regulating biogeochemical balances and enriching our esthetic and cultural experience. Predicting responses of rivers to global change is challenged by the complexity of interactions among these man-made drivers across a mosaic of natural hydrogeomorphic and climatic settings.

This special issue will explore the broad range of determinants defining global change and their effects on river ecosystems. We used a simple conceptual model of global change to delineate the breadth of topics for this special issue for river ecosystems and to provide variables around which the papers can be organized. The growing human populations as well as political, economic, and social processes are changing human activities at global scales. The resulting pollutants, habitat alterations, and climate change affect the structure, function, and services of river ecosystems in ways that vary greatly among naturally defined hydrogeomorphic and climatic regimes around the world. We do not intend to have an encyclopedic coverage of all topics possible. Instead, we asked authors for insightful treatments of specific topics that relate to the broader theme of global change regulation of river ecosystems.

We invited authors from different ecoregions around the world to submit contributions on the list of hypotheses that emerged from our conceptual model. We asked for overarching papers that related human drivers of land use and climate change to the structure and/or functioning of river ecosystems and also asked for papers dealing with specific pollutant, habitat alteration, and climate change effects in rivers. We asked for studies from biomes around the world and for different groups of organisms. We also sought papers that dealt directly with application of science in developing environmental policy. With this special issue of Hydrobiologia, we expect to advance frameworks for generating, synthesizing, and applying river science, as well as examples of that science, which will guide future river research and thereby preserve ecosystem services.

GLOBAL CHANGE AND RIVER ECOSYSTEMS

# Understanding effects of global change on river ecosystems: science to support policy in a changing world

R. Jan Stevenson · Sergi Sabater

Received: 11 June 2010 / Accepted: 18 July 2010 / Published online: 17 August 2010
© Springer Science+Business Media B.V. 2010

**Abstract** The generation of scientific knowledge to inform environmental management is crucial with current rates of global change. Although ecology and river science in particular have advanced greatly in the last 40 years, gaps remain between what we know and what environmental managers need to know to protect and restore aquatic resources. We argue that detailed quantitative relationships among human activities, contaminants, habitat alterations, and ecosystem services are needed to fill many of these gaps. Given that detailed research efforts cannot be conducted on all water bodies of the planet, scientists need to develop methods for transferring these global change relationships (models) from one system and region to another. Complexity in global change relationships is caused by natural variation among rivers and variation among responses to human activities. We propose resolving this complexity with a set of guiding principles intended to facilitate transfer of knowledge learned in one river or region to another. The ecology of disturbance provides the theoretical framework for predicting effects of human activities on rivers as well as management activities. Predicting river responses to human activities is challenged by the diversity of contaminants and habitat alterations associated with these activities, but predicting effects of human activities can be improved by recognizing: similarities in sets of stressors within classes of human activities; similarities in how different stressors affect rivers; and distinguishing effects of stressors having direct versus indirect regulation of ecosystem services. Geology and climate are key variables for predicting ecological response to human activities because they regulate the natural variation in river structure and function as well as the human activities and corresponding sets of stressors in watersheds. Transferring relationships among systems can be facilitated by emphasis on direct rather than indirect relationships and developing predictions of how geology and climate regulate direct relationships in global change ecology. These guiding principles for predicting effects of human activities should be tested and refined to resolve complexity and to manage ecosystem services, which will emerge as an important currency for global assessment of ecosystems.

Guest editors: R. J. Stevenson, S. Sabater / Global Change and River Ecosystems – Implications for Structure, Function and Ecosystem Services

R. J. Stevenson (✉)
Department of Zoology, Center for Water Sciences, Michigan State University, East Lansing, MI 44864, USA
e-mail: rjstev@msu.edu

S. Sabater
Institute of Aquatic Ecology, Faculty of Sciences, University of Girona and Catalan Institute for Water Research (ICRA), Girona, Spain
e-mail: sergi.sabater@udg.edu

**Keywords** Land use · Environment · Geology · Climate · Disturbance · Stressor

**Introduction**

Rivers around the world are threatened by socioeconomic drivers that degrade environmental conditions by altering land use and climate, thereby affecting hydrology and water quality. Approximately 40% of the earth's surface is occupied by croplands and pastures (Ramankutty & Foley, 1999; Asner et al., 2004; Foley et al., 2005). Phosphorus and nitrogen fertilizer use have increased two- and seven-fold, respectively and irrigated cropland has increased 100% during the last 40 years (Tilman et al., 2001). The continued increase in population, global development of economies, and migration of people from rural to urban locations will increase the demand for food production and transportation as well as the intensity of alterations of urban ecosystems (Grimm et al., 2008). Global pressures on water resources are on the rise (Oki & Kanae, 2006) for hydropower, flood control, and water supply, which today traps 15% of the world's total runoff (40,000 km$^3$ year$^{-1}$) in 45,000 large dams (Nilsson et al., 2005). These so called land use changes have and will continue to alter hydrology, habitat availability, nutrient and sediment inputs to rivers (Sabater, 2008) as well as toxic chemicals that have unknown effects on tens of thousands of species (Schwarzenbach et al., 2006; Sumpter, 2009).

Climate change is predicted to increase air temperature and extreme weather events as well as alter spatial patterns in precipitation and runoff (Milly et al., 2005). The Intergovernmental Panel on Climate Change (IPCC) predicts continued increases in greenhouse gases will push global temperatures higher by 2–4.5°C in the next 50 years (Solomon et al., 2007). Rates of river warming in the United States (US) range from 0.009 to 0.077°C year$^{-1}$, with greatest increases in urban areas (Kaushal et al., 2010). The IPCC (Bates et al., 2008) also indicated that annual average river runoff might increase by 10–40% at higher latitudes and decrease by 10–30% over some dry regions. Thus, climate change will increase flood and drought frequencies, alter stream geomorphology and habitat availability, as well as increase water temperature, sediments, and nutrient concentrations (Baron et al., 2002; Sabater & Tockner, 2010).

Land use and climate, in conjunction with geology, are the ultimate determinants of hydrology and water quality, thereby acting as the primary drivers of change in structure and function of rivers (Leopold et al., 1964; Dunne & Leopold, 1978; Stevenson, 1997a; Alcamo et al., 2007). Rivers provide direct benefits to human well being (sensu Millennium Assessment, 2005) by regulating biogeochemical cycles, supplying water for drinking, industrial, and irrigation purposes, fisheries production, and enriching our aesthetic and cultural experience (Postel & Carpenter, 1997; Palmer et al., 2009). Thus, predicting response of rivers to global change, i.e., the combination of land use and climate change, is critical for managing aquatic resources, ecosystem services, and human well being.

Predicting responses of rivers to global change is challenged by the complexity of interactions among the diversity of man-made stressors across a worldwide mosaic of natural hydrogeomorphic and climatic settings. In addition, management targets vary across cultures, socioeconomic gradients, and geoclimatic regions. However, basic theory and ecological relationships may be relatively transferable among regions. By combining first principles with a place-based approach, complexity of the global change problems in rivers could be managed to develop an understanding of the major threats to rivers and to design strategies for protection or restoration.

The overarching goal of this paper is to facilitate the generation and transfer of the scientific information needed to manage rivers in a changing world. This paper reviews the major issues of global change in river ecosystems and the types of science needed to support environmental policy. We start with a review of the tools needed by aquatic resource managers and the kinds of information used to develop these tools. We then relate the information that most ecological scientists generate to results needed to inform management tools. In addition, we propose a set of generalities that enable broad application of scientific information, including ecological concepts and theory as well as rules for transferring information learned in one ecosystem to another. The papers in this special volume are invited contributions that address specific sets of issues and advance our understanding of global change effects on river ecosystems. These papers will be referenced throughout our paper.

## Integration of science into water policy

Rivers are managed throughout the world, but often with different goals and for immediate benefits. Aquatic resource managers should ask three basic questions (Stevenson et al., 2004). Does a problem exist in the river? What is causing the problem? How can we fix the problem? Problems are generally defined as rivers not meeting expectations of their management (often referred to as uses in government regulations), whether within the river system itself or downstream, where effects of a poorly managed river can be manifested. Expectations for management may include water supply for drinking, industry, and irrigation, recreation, fisheries production, and support of biodiversity. In the US, observed ecological conditions are compared to water quality standards to determine whether a water body is meeting its use. Statutory requirements for European waters circumscribed by the European Union (EU) Water Framework Directive (European Commission, 2000) as well as the US demand the cause of the problem to be diagnosed. In most cases, when the cause of problems is determined, management strategies must be developed that will reduce contaminants or habitat alterations (the two herein referred to as stressors) to levels that will allow restoration of ecological conditions so uses are met. When possible, managers strive to avoid degradation, not just restore impaired ecosystems. So predicting effects of global change on ecological systems is critical.

Whether developing water quality criteria, diagnosing stressors, and forecasting effects of different management scenarios, detailed quantitative understandings are needed for relationships among human activities, stressors, biological condition, and ecosystem services in rivers (Wiley et al., 2010). We can call this *global change ecology*. Historically, water quality criteria have been used to protect minimally disturbed condition (sensu Stoddard et al., 2006). The assumption has been that minimal disturbance of ecosystems will provide other ecosystem services as well, i.e., ecosystem services and condition were positively correlated. However, we know tradeoffs exist among services of agricultural, urban, and aquatic systems within a watershed (Fig. 1; Robertson & Swinton, 2005), and they commonly result in the impairment of aquatic biodiversity as human uses increase (Hilton et al., 2006). Given these tradeoffs, aquatic resource managers in the EU and many states of the US defined different levels of resource protection (Davies & Jackson, 2006) in different waters, largely based on the land use in the watershed. These so-called tiered uses (sensu Davies & Jackson, 2006) typically provide higher levels of protection for biodiversity in minimally disturbed waters than if the same criterion was used for all waters. In addition, tiered criteria for tiered uses provide incremental restoration goals for impaired waters and appropriate management goals for watersheds altered extensively by humans.

Quantitative relationships among ecosystem condition, services, and stressors are important for establishing stressor criteria. Thresholds in ecological response along stressor gradients are increasingly being used for developing stressor criteria (Dodds et al., 1997; Soranno et al., 2008; Stevenson et al., 2008; Fig. 1). These dose–response-like approaches are common in ecotoxicology (Suter, 1993), but have only recently been adopted for the criteria development of nontoxic stressors, such as nutrient concentrations (Stevenson et al., 2004, 2008) and flow

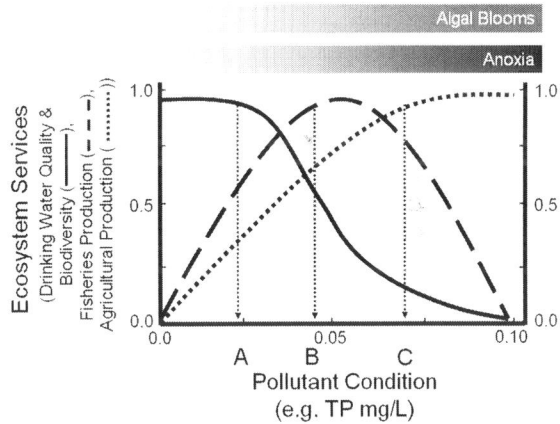

Fig. 1 Tradeoffs among uses of rivers indicated by hypothetical relationships between a resource stressor (e.g. nutrient concentrations) and a suite of ecosystem services of watersheds: drinking water quality; algal, invertebrate, and fish biodiversity; fisheries production; and agricultural production. The *vertical lines* indicate nutrient criteria that could be used to protect different uses in different waters

regimes to protect biodiversity (Poff et al., 2010). Thresholds in ecological responses along stressor gradients delineate specific stressor levels at which risk of undesired changes dramatically increase. Such threshold responses are important for developing stakeholder consensus and support of water policy (Muradian, 2001).

Quantitative relationships among human activities, stressors, biological condition, and ecosystem services also enable rigorous evaluation of tradeoffs among ecosystem services and comparisons among countries in levels of protection. Managing rivers for their ecosystem services is practiced throughout the world, but many rivers are managed to generate energy with hydroelectric dams, flood control, and water supply rather than their support of biodiversity and nutrient retention. In part, this is due to great differences among countries with different cultures, economic status, and natural resource availability which affect the valuation of ecosystem services and relationships to human well-being (Schulze, 2007). Some tradeoffs among river ecosystem services allow meeting human needs and substantial ecosystem conservation. For example, greater fisheries production is possible with nutrient concentrations moderately elevated above the natural concentrations needed to protect biodiversity (Fig. 1). However, sustainable management of resources for support of human well-being rather than biodiversity has been and will likely continue to be the management goal for many governments. A full accounting of ecosystems services will be important for evaluating fair trade policies and incorporating ecosystem services into ecological assessments.

## Application of ecological theories and concepts to global change of rivers

Ecology, the study of interactions among organisms and their environment, provides theories and concepts that should facilitate application of science to manage global change (National Research Council, 1986; Jørgensen & Müller, 2000; Collins et al., 2007). Predictive models of ecological response to environmental change developed rapidly during the philosophical transition from ecology largely being the study of equilibrium (steady-state) systems to dynamic systems. Models such as the subsidy-stress perturbation theory (Odum et al., 1979), the intermediate disturbance hypothesis (Grime, 1973; Connell, 1978; Reynolds et al., 1993), and pulse-press disturbance models (Bender et al., 1984; Collins et al., 2007; Smith et al., 2009) provided theories to test across ecological systems. The succession model of Pickett et al. (1987) provided a list of factors to consider as communities reassembled following disturbance.

Fundamentally, we can think of global changes as disturbances of ecological systems (Odum et al., 1979; Collins et al., 2007; Smith et al., 2009). Although rivers require natural disturbance to maintain natural structure and function, the disturbances introduced by humans are often novel in composition, such as pesticides and pharmaceutical products (Ricart et al., 2010), or occur at spatial and temporal scales to which river systems are not adapted, such as channelization and damming (Nilsson et al., 2005). In addition, the energy of disturbances can be increased by humans through combinations of stressors. The energy associated with disturbances is related to their intensity and frequency, which are inversely related (Vitousek, 1994; Margalef, 1997). Thus, high intensity disturbances like earthquakes or huge landslides are relatively rare. The extreme drought of cold deserts of some polar regions, intense waves in some littoral habitats, or siltation in some streams are examples of high frequency disturbances of medium magnitude that severely limit biological activity. Diurnal variations in temperature and light conditions are examples of disturbances with high frequency and low intensity to which organisms would be adapted (Fig. 2). Flood disturbances are examples of relatively high intensity but low frequency disturbances to which biota are adapted.

Effects of disturbance depend both on the associated energy as well as on its spatial and temporal scale (O'Connor & Lake, 1994; Richards et al., 1996; Stevenson, 1997b) (Fig. 2). Global changes are generally considered to be press (i.e. chronic stresses sensu Smith et al., 2009) or ramp (increasing and chronic; Lake, 2000) disturbances in which exogenous forces cause changes in environmental conditions that last many times the generation times of organisms in the ecosystem. Agricultural and urban development, hydrologic modifications, invasive species introductions, and temperature increases are good examples of such long term changes in land uses and stressor that affect river ecosystems, though

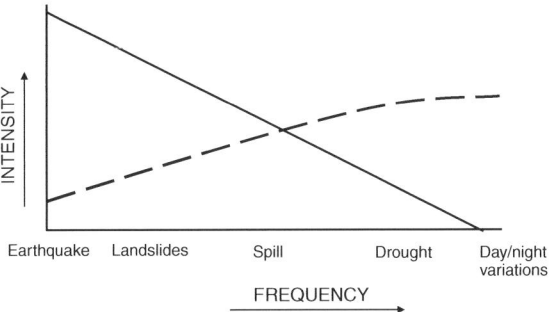

**Fig. 2** The range of disturbance regimes in which biota are adapted to the intensity and frequency of disturbance. The inverse relationship between the intensity and frequency of disturbance (plotted on logarithmic scales) affecting ecosystems is indicated by the *solid straight line*, which delineates the range of possible disturbance conditions. Some examples of disturbances with rising energy and decreasing frequency (from left to right) are given along the *x*-axis. The area below the curve (*dashed line*) represents the disturbance regime in which biota are able to adapt to disturbance. Biota are not resilient or resistant to high-energy disturbances as a result of either high or medium intensity, with low or medium frequencies

their associated energy may differ. Thus, river biota may be resilient to short-term anthropogenic disturbances of low magnitude with stressors to which they have previously been exposed, such as slight nutrient enrichment with inappropriate agricultural practices that might occur naturally with flood disturbances. However, chronic or persistent disturbances by humans when pressure is continuously maintained on the ecosystem or when intensity of disturbances is high will cause significant structural simplification of the community and loss of resistance as well as function. With short-term and low intensity anthropogenic disturbance we can expect species adapted to the new conditions to replace existing species with the potential that ecosystem functions and some services can be maintained (Bender et al., 1984; Stevenson, 1997b). However, high energy disturbances undoubtedly cause environmental alterations to which the biological system cannot respond, adapt, or recover—thereby resulting in collapse of ecosystem services.

Rivers are relatively unique because they are asymmetrically linked ecosystems with transport of energy and materials in a downstream direction, which forms a predictable succession of ecological conditions (Vannote et al., 1980). This succession in conditions differs at scales ranging from the habitat, reach, and entire watershed (Frissell et al., 1986). Theories of dissipative structure and stratified stability predict hierarchies in ecological systems are more stable than a random grouping of assemblages (D'Angelo et al., 1997). The resulting complexity of the hierarchical structure of the river network may provide a buffer to disturbances, since refuges for organisms may be more numerous than in a simple linear channel system. Refuges and nurseries in riffle-pool scale habitats or interconnected lakes and wetlands provide sources of immigrants for recolonization of habitats. Higher productivity and stability of river ecosystems has been proposed as emergent properties of this heterogeneity (D'Angelo et al., 1997; Stevenson, 1997b). Further, hyporheic and riparian-floodplain compartments are also hierarchically related to the river channel, since their relevance increase progressively with river size (Burt et al., 2010). Thus, properties of connectivity, heterogeneity, and stability may increase in a downstream direction and may make rivers more resistant and resilient to global change than other types of ecosystems.

**Stressor classification**

Responses of ecological systems to global change are also highly dependent upon the type of stressor and combination of stressors (Rapport et al., 1985). Ecosystem-level responses are highly dependent upon biological responses, which also vary with spatial and temporal extent and intensity of the disturbance (Pickett et al., 1987; Suter, 1993; Stevenson, 1997b; Smith et al., 2009). Based on these principles, it is possible to advance a classification of stressors (Table 1). Initially, we consider whether a stressor is a *contaminant* (in the broadest sense) or *habitat alteration* (Level I). The scale of disturbance both in spatial, temporal, and energetic dimensions are greater for habitat alterations than most contaminants. Then, we classify stressors by whether effects are *direct* or *indirect* on the physiology or habitat suitability of the individual (Level II). In general, contaminants have direct effects, and habitat alterations have indirect effects of ecological response. We also classify stressors as *resource*, *toxic*, or *complex* depending upon the direct effects of a stressor on an individual organism (Level III).

Table 1 Stressor arranged from top to bottom of the table in relation to their associated energy in the systems, and therefore relation to disturbances ranging from more transient to more irreversible

| Stressor | Stressor classification | | | Degree of reversibility |
|---|---|---|---|---|
| | Level I | Level II | Level III | |
| Nutrients | C | Direct | Resource | BC |
| Light | C | Direct | Complex | BC |
| Human and animal organic waste | C | Direct | Resource | BC |
| Temperature | C | Direct | Complex | BC, WE |
| Salinity | C | Direct | Complex | WE |
| Acidification | C | Direct | Complex | WE |
| Fine sediments | C | Direct | Toxic | WE |
| Heavy metals | C | Direct | Toxic | BC |
| Pesticides | C | Diverse | Toxic | BC |
| Emergent pollutants (nanoparticles, pharmaceuticals) | C | Diverse | Complex | BC |
| Invasive species | C | Indirect | Complex | BC |
| Channel filling | HA | Direct | Complex | WE |
| Current alteration | HA | Indirect | Complex | WE |
| Flow variability | HA | Indirect | Complex | WE |
| Water abstraction | HA | Indirect | Complex | WE |
| Interbasin transfer | HA | Indirect | Complex | WE |
| Impervious surface | HA | Indirect | Complex | WE |
| Channel simplification | HA | Indirect | Complex | WE |
| Dams | HA | Indirect | Complex | WE |

Level I refers to the scale of disturbance, as whether it is a *C* contaminant or *HA* habitat alteration. Level II refers to the *direct* or *indirect* effects being produced by the stressor. Level III classify stressors as *resource*, *complex* or *toxic*. Degree of reversibility defines the potential effects if limited at one or several biological class/es (*BC*), instead ofs the whole ecosystem (*WE*), in a scale of rising irreversibility

Resource stressors (Smith et al., 2009), such as nutrients, stimulate response of individual metabolism with little negative direct effect; however, indirect effects as a result of interactions with other organisms may produce strong negative responses. Increases in toxic stressors (e.g. heavy metal pollution, or pesticides) have mostly negative effects on performance. Complex stressors, such as temperature and current velocity, may stimulate performance of organisms at low levels and then, typically due to other processes, negatively affect performance of organisms at high levels. Note, few of these relationships are predicted to be linear, which can increase complexity of system responses to disturbance. Finally, the energy associated with the stressor can define a stressor to be reversible or irreversible. In some cases, the effects of the stressors are limited to one or several biological classes and this might also regulate reversibility up to a certain threshold. As disturbance increases, either in frequency, duration, or intensity, responses will be expressed at individual, population, guild, and community levels. Different stressors potentially have different effects on biological response at each of these levels of organization. When effects of stressors are extended to the whole ecosystem, mostly as a result of physical as well as biological alterations, likelihood increases for irreversible effects in river structure and function.

While it is beyond the scope of this paper to review all that we know about these stressors, we illustrate how classification of stressors can help predict response of ecosystems to global change by reviewing stressor effects and explain their classification. Even though a resource stressor has a positive effect on physiology of organisms, ultimate effects may be negative for the same organism and for the whole ecosystem. Nutrients (or food availability for consumers) are the best example of a resource stressor because they have a positive effect on metabolism (e.g. of algae and bacteria, or in consumers), but their increase in concentration may make the habitat available (sensu Pickett et al., 1987) for species that require high nutrient concentration and often have high growth rates (Leira & Sabater, 2005; Manoylov & Stevenson, 2006; Stevenson et al., 2008). Evidence suggests that these high nutrient species may competitively displace low nutrient species (Stevenson et al., 2008). High levels of nutrient enrichment can stimulate algal and bacterial biomass accrual to cause habitat alterations, low dissolved oxygen, and high pH that affects all organisms, and even impair aesthetics for humans (Sabater et al., 2003; Suplee et al., 2008; Stevenson et al., submitted).

Organic wastes can also be considered a resource stressor because they stimulate productivity of some bacteria, fungi, and a few algae, directly, as well as the performance of some invertebrates, indirectly. Organic wastes cause accumulation of high heterotroph biomass but produce great impacts in the ecosystem because of their role in depleting dissolved oxygen concentration. Light is a factor that has a positive effect on photosynthetic rates of primary producers over a substantial range of intensities, though should be classified as a complex stressor because high light can have negative effects on performance (Hill et al., 1995). Further, light availability may have implications for the ecosystem functioning by causing a shift from heterotrophy to autotrophy when occurring together with higher nutrient availability (Sabater et al., submitted).

Temperature, salinity, and acidification are direct stressors on many aquatic organisms, upon which they have complex effects. All organisms have optimal performance at some intermediate tipping point along temperature, conductivity, and pH gradients. Increases in temperature below optima increase performance by increasing kinetic energy and metabolic rates, but high temperatures denature enzymes, reduce enzymatic function, slow metabolic rates, and alter biogeochemical rates (Hamilton, 2010). Increases in water temperature will have profound effects on temperature sensitive organisms, such as some macroinvertebrates and fish. Warmer waters and other stresses may increase the possibility of fungal, viral, and bacterial infections on fish. Increasing temperatures may support eurythermic species and generalists, resulting in less specialized communities. Higher temperature may enhance respiration more than photosynthesis and therefore have implications for ecosystem processing. Altered thermal regimes in lakes and rivers, because of climate change or human action, may have deep impact in food web structure, particularly in fragile ecosystems (Perkins et al., 2010). Similarly, optimal metabolic rates are at very specific conductivity and pH conditions. An ionic gradient, defined by conductivity and pH, is often the most important factor regulating variability in diatom species composition among rivers (Pan et al., 1996).

Sediment transport has changed significantly as a result of hydrologic alterations and land use changes (Syvitski et al., 2005). Increases in fine sediment accumulation are classified as a toxic contaminant because performance of most organisms are directly impaired from increased levels fine sediments on substrata, even though a few rare taxa may benefit. For example, a small number of diatom and macroinvertebrate taxa may be adapted to tolerate fine sediments, but their success in fine sediment habitats is probably due to the reduction in competition from species that are not adapted to fine sediments. The end result is an impairment of biodiversity and many ecosystem functions with elevated sediments in most habitats. In contrast, bank stabilization and dams can reduce downstream sediment availability (i.e. sediment starvation), thereby altering natural substratum conditions or increasing stream power (Bridge, 2003). Elevated stream power causes downstream channel erosion, bank destabilization, or potentially unnatural increases in river sediment loads. Thus, alterations in sediment transport, either increases or decreases from natural channel requirements, can negatively affect river structure and function. Alterations in sediment provide an excellent example of how deviations from natural conditions, either increases or decreases in an environmental factor, can be considered a stressor on ecosystem structure and function.

A diverse group of chemical contaminants, heavy metals, pesticides, pharmaceuticals, and nanoparticles have wide ranging and often complex effects on biota, which in many cases are still undetermined but are mostly toxic. Each of these chemicals could have direct or indirect effects on different groups of organisms in a habitat. For instance, heavy metal effects are cumulative and increase their toxicity after repetitive pulses (Guasch et al., 2010). Several contaminants in this list may also be considered complex because they have different effects on different organisms. For example, some herbicides affect algal growth but indirectly inhibit invertebrates (López-Doval et al., 2010). Considerable research is needed to better understand the roles of this diverse group of contaminants (Sumpter, 2009).

We include invasive species as a biological contaminant of a habitat, with mostly indirect and complex effects. Similarly, we could think of stocked fish and shellfish in a habitat as biological contaminants. These organisms seldom have positive benefits on other organisms that naturally occur in the habitat. Their effects are indirect through competitive exclusion, habitat alteration, or potentially other negative biotic interactions. Nutrient enrichment, temperature change, and habitat alteration are factors that may facilitate invasion of species (Perkins et al., 2010) and cause further habitat impairment. If invading species are keystone species or ecological engineers, for example *Didymosphaenia*, *Cladophora*, *Potamopyrgus*, or stocked fish (e.g. brown trout), dramatic changes in structure and function may occur with relative small incremental changes in stressors (Sabater et al., 2005; Stevenson et al., submitted).

Habitat alterations range from filling and straightening river channels to dam building. In some mountainous mining regions of the US, mined rock and soils are placed in headwater stream channels with dramatic direct impacts of biota and downstream biogeochemical processes (Palmer et al., 2009). Throughout agricultural landscapes, small headwater streams are integrated into agricultural fields by filling and plowing. Reduction and alteration of riparian vegetation affects the stream hydromorphological integrity and its biodiversity and ecosystem functioning (Elosegi et al., 2010). Headwater streams, comprising a great proportion of drainage networks, are critical in biogeochemical transformation and retention of inorganic and organic nutrients draining from surrounding landscapes (Alexander et al., 2007). Alteration of natural flows regimes (affecting patterns of current velocity and temporal variability in discharge) by dams, channel straightening and dredging, surface and ground water abstraction, interbasin water transfers, and impervious surfaces are one of the most widespread and intense (high energy) disturbances of river ecosystems (Sabater & Tockner, 2010). Seasonal flows fall virtually to zero along extensive sections and over long periods of the year in such major rivers as the Yellow River in China (Fu et al., 2004) and the Colorado and Grande rivers in the southwestern US (Molles et al., 1998). Climate-related increases in drought frequency and duration will reinforce the ever-increasing water abstraction from rivers and groundwaters by humans, thereby causing intensification of drought effects on rivers. Decreases in river discharge are predicted to have substantial effects on fish biodiversity (Xenopoulos & Lodge, 2006). Drought constrains species diversity to short-generation or highly mobile organisms and limits ecosystem function to shorter periods (Sabater & Tockner, 2010; Mas-Martí et al., 2010).

Loss of hydrological variability resulting from regulation and abstraction directly affects diversity (Margalef, 1997). Habitat disturbance, alteration and loss by channel modification, dredging, and filling significantly affect biodiversity and ecosystem functioning (Strayer, 2006). In the vast majority of cases, altering the magnitude, timing, duration, and frequency of flows reduced measures of biological condition, especially for fish; but in some cases invertebrates and riparian vegetation responded positively (Poff & Zimmerman, 2010). Flow regime regulates algal productivity and food web structure (Power et al., 1995; Riseng et al., 2004; Stevenson et al., 2006). Greatest algal biomass can accumulate in intermediate flow variability, with macroalgae scoured at high flow disturbance regimes and microalgae grazed at low disturbance regimes when primary consumers accumulate (Stevenson et al., 2006).

A trait-based assessment of species shifts with global change could improve predictions of responses of ecological systems and biodiversity to global change as well as restoration and protection strategies. While the biota in Mediterranean rivers are better adapted to naturally high seasonal and inter-annual flow variability (Mas-Martí et al., 2010; Zoppini et al., 2010), biota in more temperate rivers may be more sensitive to flow reduction. Xenopoulos et al. (2005) predict loss of fish in rivers based on discharge requirements by species. Fisheries managers develop water release strategies for dams based on species traits which optimize habitat conditions for as great a diversity of fish as possible. This approach could be extended to predict changes in species composition along many stressor gradients and forecast how global changes and management strategies will affect species composition of assemblages and ecosystem services in rivers.

Predicting responses of ecological systems can be facilitated by a more thorough understanding of individual stressor effects, as advanced in our classification of stressors, in two ways. First, within regions with similar land uses, climate, and geology, we can expect similar suites of stressors to affect river systems with global change. By knowing generalized responses of stressors, we can predict responses of ecosystems to environmental change. In addition, classifying stressors can help predict responses to multistressor disturbances, which is the form of most human alterations of aquatic resources. Predicting multistressor responses is challenging, but some general principles provide a conceptual starting point for predicting response. Many of these concepts are rooted in ecotoxicology (e.g. Suter, 1993). Stressor effects can be aggregated by similarity in effects and evaluated for synergistic interactions. Combinations of stressors can be synergistic or positive, antagonistic or negative, and indifferent. For example, elevated light and nutrient concentrations are more likely to stimulate algal accumulations than either factor alone (Ylla et al., 2007). Similarly, antagonistic effects would be predicted for heavy metals and phosphate (Guasch et al., 2004). As in the latter case, more rigorous challenges in multistressor analysis occur with stressors that have opposing effects, and this requires using innovative combinations of biomarkers that can produce reliable assessments (Montuelle et al., 2010). Considerable research is needed to refine our understanding of individual and interactive stressor effects in multistressor situations.

## River classification

As ultimate determinants of river ecosystems, geology and climate determine hydrology, topography, and geomorphology (Leopold et al., 1964; Biggs, 1996; Stevenson, 1997b), and consequently, flow regime, biogeochemistry (e.g. Meybeck & Helmer, 1989), and biological structure and function (Margalef, 1983). These largely exogenous abiotic determinants also regulate water chemistry, productivity, trophic structure, and ecosystem processes (Biggs 1996, Stevenson, 1997a; Riseng et al., 2004). The theory, that climate and geology are ultimate determinants of the structure and function of rivers, led to the application of ecoregions (sensu Omernik, 1987) as organizing units for environmental management. Thus, we can expect different responses of rivers to global change across climate and geologic regimes because the natural determinants of river ecology differ among geoclimatic regimes (Fig. 3).

Although we illustrate climate, geology, and anthropogenic gradients as a relatively simple multivariate axes in three-dimensional space (Fig. 3), we recognize the complexity of these gradients. We could simplify natural gradients to a few dominant factors, so we can develop generalities to predict directions of global changes that would hold for a substantial proportion of river ecosystems. Climate involves temperature and precipitation which, respectively, directly and indirectly affect structure and function of river ecosystems. When precipitation is linked to geology, hydrologic variability with altered flood and drought regime becomes an important factor along geologic gradients. While the prediction is that systems will get warmer and subject to greater hydrologic variability, ecoregional differences imply that some will get wetter and others drier with changes in mean annual precipitation (Milly et al., 2005).

In addition, many different combinations of stressors may be associated with the human disturbance

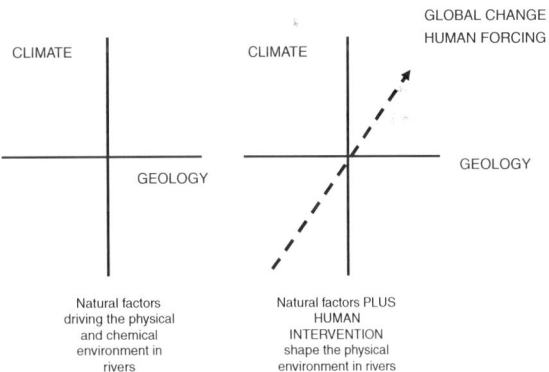

Fig. 3 Climate and geology define the two-dimensional space of the physical environment in river ecosystems. Human forcing makes up a third dimension that configures a new, more complex space in which the physical and biological compartments of the river are regulated

gradient depending upon the types of human activities in the watershed as well as economic condition and cultural practices. Ellis & Ramankutty (2008) have developed the concept of anthropogenic biomes, which is a biome classification that includes human interactions with ecosystems. Human activities in watersheds are greatly regulated by climate and geology (Ellis & Ramankutty, 2008). Fewer people live in mountainous regions than along larger rivers and coastlines because historic and current routes of transportation are more limited in mountainous regions. Fewer people also live at higher latitudes, largely because of historic limitations on transportation of food and harsh conditions. Agriculture becomes more sustainable with increasing precipitation, water storage potential, and lower slope in watersheds. Such a classification could be used to develop subgroups of anthropogenic ecoregions in which gradients of human disturbance could be defined with similar ratios of human activities, and presumably similar multistressor gradients. The conceptual habitat template outlined in Fig. 3 could provide a foundation upon which predictions of global change ecology can be organized by class of river.

**Transferability of relationships among regions**

Many relationships among human activities, natural disturbances, stressors, and ecosystem services may be transferable among classes of stream ecosystems. Many stressor responses and predictions of ecological effects are relatively consistent among regions, but vary in magnitude of effect. Basic ecological principles with associated predications should be transferable among stream systems. Timing of extreme events rather than their magnitude can be relevant to river functioning (Acuña, 2010). Direct effects on organisms are more likely transferable across ecoregions than indirect effects because of added interactive effects. The mechanistic responses to differing nutrient pollution that occur from microbes (Ziegler & Lyon, 2010) to invertebrates (Donato-Rondón et al., 2010) and fish (Miltner & Rankin, 1998; Wang et al., 2007) are similar among regions and are even expressed in the stoichiometric composition of the organisms (O'Brien & Wehr, 2010). Multiple investigations indicate direct effect of phosphorus on algal biomass is greatest when phosphorus concentrations are less than 30 µg TP $l^{-1}$ (Bothwell, 1988; Dodds et al., 1997; Stevenson et al., 2008, submitted).

Our greatest challenge may be transferring predictions of global change ecology across classes of rivers, as defined by river typologies in different geoclimatic regions. What are the similarities and differences in ecological responses to stressors in different classes of rivers? It could be argued that some areas are particularly sensitive because of their fragility, or simply because the windows of opportunity for protection may narrow because of global change (Uehlinger et al., 2010) and unexpected impacts. Biodiversity hot spots like boreal, tropical, and Mediterranean regions (Sala et al., 2000) may be more sensitive than other areas. However, similarities in responses among river typologies are likely larger than differences, which may largely be associated with system sensitivity and magnitude of responses. Effects of urbanization and agricultural activity produce long-lasting effects in biodiversity and functioning elsewhere, from the tropical regions (Donato-Rondón et al., 2010) to flood plain rivers (Rodrigues Capítulo et al., 2010) and headwater streams (O'Brien & Wehr, 2010).

Transfer coefficients can be developed for account for differences among river typologies. Riseng et al. (2004) empirically derived relationships for algal and invertebrate response to nutrient pollution in a survey of streams in two ecoregions with different disturbance regimes. In one ecoregion, streams were hydrologically flashy with greater flood and drought frequency and intensity than the other region, which had hydrologically stable streams. Riseng et al. (2004) found nutrients regulated algal and consumer biomass, but top–down regulation of algal biomass by primary consumers decreased with increasing flood and drought disturbance. This produced dramatically different food web structure in the two ecoregions, with low biomass of short-generation herbivores (e.g., mayflies) in the high disturbance ecoregion and high biomass of more long-generation caddisflies in the hydrologically stable ecoregion. Although positive relationships were observed in both algal and invertebrate biomass in the two regions, algal biomass was an order magnitude greater in the flashy streams and invertebrate biomass was an order of magnitude higher in the hydrologically stable streams (Fig. 4). The differing responses of algae and invertebrates in hydrologically stable and flashy streams provide an

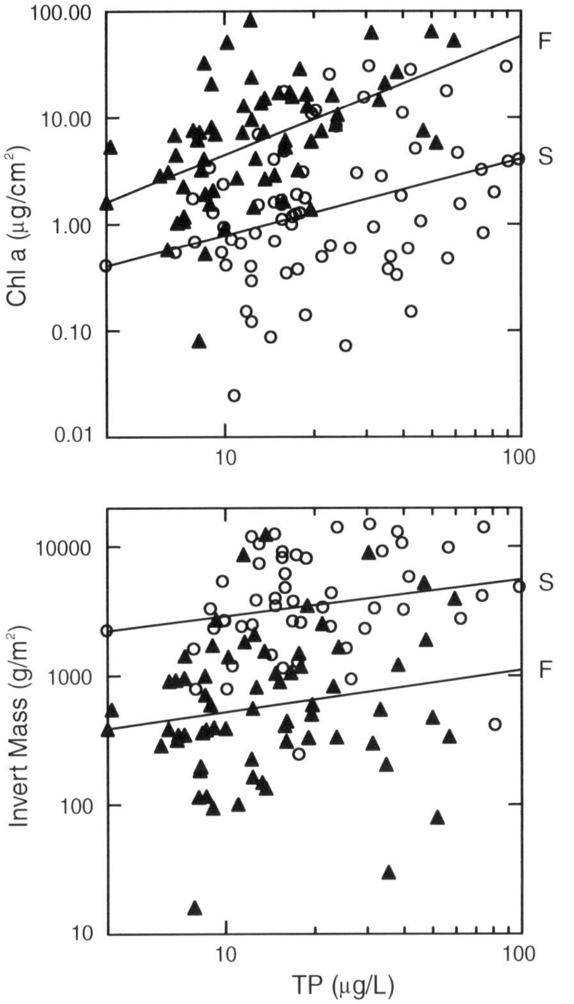

**Fig. 4** Biomass of benthic algae and invertebrates in hydrologically flashy (*F*, *open circles*) and stable (*S*, *filled triangles*) streams with data replotted from Riseng et al. (2004) and Stevenson et al. (2006)

opportunity to expand the model for algal and invertebrate response to nutrients in streams by accounting for hydrologic disturbance regime (Riseng et al., 2004; Stevenson et al., 2006).

## Future challenges of global change for river science

River sciences have advanced greatly in the last four decades, from a concentration in North America to Europe, Australia, New Zealand and other regions of the world (Benke & Cushing, 2005; Allan & Castillo, 2007; Tockner et al., 2009). The early focus on identifying biota and biotic interactions among them rapidly grew to include a quest to understand the abiotic factors that regulated those interactions and the broader interdisciplinary linkages among the biota, ecosystem function, biogeochemistry, and hydrology. Aquatic resource managers have been engaged in river ecology through most of the development of river science in their educational training and continued scientific engagement. Still, significant gaps remain between the science provided by many river ecologists and needed by aquatic resource managers. These gaps are not restricted to recent times or unique to river ecology (e.g. Aumen & Havens, 1997; Carpenter et al., 2009), but the need and opportunities for partnerships between scientists and policy makers has never been as great as it is today (Hamilton et al., 2010).

While emerging stressors (e.g. pharmaceuticals, nanoparticles, and invasive species) and unforeseen climatic conditions (higher drought frequency, large flood episodes, increasing frequency of heat waves) may present new variables for study, great demand exists for more thorough study of existing relationships in global change ecology of rivers. Though exploring new variables or new combinations of variables has great value, challenges exist for comparing relationships with the same variables, in different settings, and more thoroughly. Cross-region or cross-typology studies will increase transferability of relationships among the classes of rivers studied as well as to other rivers, if cross-system patterns can be identified. Predicting differences in global change relationships across river typologies is challenging and important for anticipating inter-regional differences and also for predicting which types of rivers will be most sensitive to global change.

More detailed quantification of relationships among human activities, stressors, structure and function of rivers, as well as ecosystems is needed (Boix et al., 2010; Ricart et al., 2010), versus the classical high-low experimental manipulations. Few relationships are sufficiently well quantified that we can develop predictive models of river ecosystem response under different global change scenarios. Many relationships in river ecology are non-linear, which generates large differences in predicted conditions with small increments of environmental change (Stevenson et al., 2008; Sabater, 2008). Threshold responses resulting from non-linearities are also valuable for setting management targets for both protection and restoration of rivers (Muradian, 2001; Stevenson et al., 2004, 2008). Balancing tradeoffs among different ecosystem services, which are not necessarily concordantly related to environmental gradients, also demands accurate quantification of relationships. Recent investments in large scale ecological assessments in the United States and Europe provide excellent opportunities for gathering sufficient date for both quantifying relationships, but also comparing them across ecoregions and river typologies.

Water policy, like other environmental policies, requires a high level of scientific certainty before widespread support can be developed among stakeholders for a policy. The challenge with development of climate and greenhouse gas regulation policies demonstrates that the scientific certainty required for policy development increases with the demands on society that will result from that policy. A great source of uncertainty in our understanding problems is the diagnosis of the specific pollutants and habitat alterations that are threatening or causing impairment of ecosystem services. As an example, the effects of hydrological disturbances (e.g. abstraction) may directly affect the biota performance on organic matter use, but also indirectly through organic matter transport (Ylla et al., 2010). The amazing adaptability of some biota to human disturbance (Zoppini et al., 2010; Mas-Martí et al., 2010) can provide a false sense of security and generate ambiguity in perceived effects, and therefore provides a further source of uncertainty. However, rules for causal inference are relatively well established in environmental epidemiology (Beyers, 1998; Norton et al., 2009) and should be more broadly applied by basic as well as applied scientists. These rules include significant effects in exposed area, relationship to exposure concentration, the same causes and effects observed elsewhere, ecological plausibility, and concordance of experimental and field survey results. These rules, as well as an understanding of the kind of information that we need to know, can be used to guide our gathering and generation of knowledge to inform water policy.

Highlighted within the rules for confirming causal relationships are the three main scientific approaches used in ecology that should be applied to develop an understanding of interrelationships between human activities, contaminants, habitat alterations, and ecosystem services: observations of patterns in the field, experiments in which specific factors are isolated and manipulated, and process-based modeling. Observations within the field provide evidence for significant effects in exposed areas and for relationships between effects and the magnitude of exposure to stressors, but these relationships could be caused by other variables. Experiments provide the opportunity to isolate specific variables, determine effects without ambiguity about cause, show the possible responses of ecosystems to stressors, and establish ecological plausibilities. Coupled, process-based models (e.g. Wiley et al., 2010) provide syntheses of multiple single models, evaluate interactive effects, and explore ecological plausibilities with a more thorough approach than simple statistical models and experiments. In addition to these three scientific approaches, the fundamental practice of repeating experiments and field studies and comparing relationships among similar studies is also important for building a knowledge foundation for evaluating causality. This multiple lines of evidence are important for understanding the complexities of coupled human and natural systems (CHANS; Collins et al., 2007; Liu et al., 2007; Smith et al., 2009; Stevenson, in press) with sufficient certainty that water policies can be broadly and sustainably supported by stakeholders.

Relating river science and management more completely to ecosystem services will provide a common denominator for river management around the world. Although valuation of ecosystem services varies among regions with different socioeconomic and cultural settings, ecosystem services are the same; and many services have value beyond the region in which they are produced. Given the globalization of economies and need for global assessments of environmental compliance, ecosystem services are internationally recognized endpoints for environmental assessment and management. The need to quantify ecosystem services and relate them to environmental change emphasizes the need for functional as well as structural assessments of river ecosystems, as well as bridging basic and applied science.

**Acknowledgments** Stevenson and Sabater acknowledge Judit Padisak for the invitation to prepare this special issue of Hydrobiologia. Stevenson acknowledges financial support from the Great Lakes Fisheries Trust and the following grants from the United States Environmental Protection Agency,

G2M104070 and R-83059601. Sabater acknowledges the financial support provided by projects CGL2007-65549/BOS, CGL2008-05618-C02-01 and SCARCE (Consolider Ingenio 2010, CSD2009-00065) of the Spanish Ministry of Science and Innovation.

# References

Acuña, V., 2010. Flow regime alteration effects on the organic C dynamics in semiarid stream ecosystems. Hydrobiologia. doi:10.1007/s10750-009-0084-3.

Alcamo, J., M. Flörke & M. Märker, 2007. Future long-term changes in global water resources driven by socioeconomic and climatic change. Hydrological Sciences Journal 52: 247–275.

Alexander, B. N., E. W. Boyer, R. A. Smith, G. E. Schwarz & R. B. Moore, 2007. The role of headwater streams in downstream water quality. Journal of the American Water Resources Association 43: 41–49.

Allan, J. D. & M. A. Castillo, 2007. Stream Ecology: Structure and Function Of Running Waters, 2nd ed. Springer, Dordrecht, The Netherlands.

Asner, G. P., A. J. Elmore, L. P. Olander, R. E. Martin & A. T. Harris, 2004. Grazing systems, ecosystem responses, and global change. Annual Review of Environment and Resources 29: 261–299.

Aumen, N. G. & K. E. Havens, 1997. Needed: a new cadre of applied scientists skilled in basic science, communication, and aquatic resource management. Journal of the North American Benthological Society 16: 710–716.

Baron, J. S., N. L. Poff, P. L. Angermeier, C. N. Dahm, P. H. Gleick, N. G. Hairston, R. B. Jackson, C. A. Johnston, B. D. Richter & A. D. Steinman, 2002. Meeting ecological and societal needs for freshwater. Ecological Applications 12: 1247–1260.

Bates, B. C., Z. W. Kundzewicz, S. Wu & J. P. Palutikof (eds), 2008. Climate Change and Water. Technical Paper of the Intergovernmental Panel on Climate Change, IPCC Secretariat, Geneva: 210.

Bender, E. A., T. J. Case & M. E. Gilpin, 1984. Perturbation experiments in community ecology: theory and practice. Ecology 65: 1–13.

Benke, A. C. & C. E. Cushing, 2005. Rivers of North America. Academic/Elsevier, Amsterdam/Boston.

Beyers, D. W., 1998. Causal inference in environmental impact studies. Journal of the North American Benthological Society 17: 367–373.

Biggs, B. J. F., 1996. Patterns in benthic algae of streams. In Stevenson, R. J., M. L. Bothwell & R. L. Lowe (eds), Algal Ecology: Freshwater Benthic Ecosystems. Academic Press, San Diego, CA: 31–56.

Boix, D., E. García-Berthou, S. Gascón, L. Benejam, E. Tornés, J. Sala, J. Benito, C. Munné & S. Sabater, 2010. Response of community structure to sustained drought in Mediterranean rivers. Journal of Hydrology 383: 135–146.

Bothwell, M. L., 1988. Growth rate responses of lotic periphytic diatoms to experimental phosphorus enrichment: the influence of temperature and light. Canadian Journal of Fisheries and Aquatic Sciences 45: 261–270.

Bridge, J. S., 2003. Rivers and Floodplains: Forms, Processes, and Sedimentary Record. Blackwell Publishing, Oxford, UK.

Burt, T. P., G. Pinay & S. Sabater, 2010. Riparian zone hydrology and biogeochemistry: a review. In Burt, T. P., G. Pinay & S. Sabater (eds), Hydrology and Nitrogen Buffering Capacity of Riparian Zones. International Association of Hydrological Sciences Benchmark Papers in Hydrology, Toronto.

Carpenter, S. R., H. A. Mooney, J. Agard, D. Capistrano, R. S. DeFries, S. Díaz, T. Dietz, A. K. Duraiappah, A. Oteng-Yeboah, H. M. Pereira, C. Perrings, W. V. Reid, J. Sarukhan, R. J. Scholes & A. Whyte, 2009. Science for managing ecosystem services: beyond the Millennium Ecosystem Assessment. Proceedings of the National Academy of Sciences of the United States of America 106: 1305–1312.

Collins, S. L., S. M. Swinton, C. W. Anderson, B. J. Benson, J. Brunt, T. Gragson, N. B. Grimm, M. Grove, D. Henshaw, A. K. Knapp, G. Kofinas, J. J. Magnuson, W. McDowell, J. Melack, J. C. Moore, L. Ogden, J. H. Porter, O. J. Reichman, G. P. Robertson, M. D. Smith, J. V. Castle & A. C. Whitmer, 2007. Integrated science for society and the environment: a strategic research initiative. In Miscellaneous Publications of the LTER Network. Available at http://www.lternet.edu.

Connell, J. H., 1978. Diversity in tropical rain forests and coral reefs—high diversity of trees and corals is maintained only in a non-equilibrium state. Science 199:1302–1310

D'angelo, D. J., S. V. Gregory, L. R. Ashkenas & J. L. Meyer, 1997. Physical and biological linkages within a stream geomorphic hierarchy: a modeling approach. Journal of the North American Benthological Society 16: 480–502.

Davies, S. P. & S. K. Jackson, 2006. The biological condition gradient: a descriptive model for interpreting change in aquatic ecosystems. Ecological Applications 16: 1251–1266.

Dodds, W. K., V. H. Smith & B. Zander, 1997. Developing nutrient targets to control benthic chlorophyll levels in streams: a case study of the Clark Fork River. Water Research 31: 1738–1750.

Donato-Rondón, J. C., S. J. Morales-Duarte & M. I. Castro-Rebolledo, 2010. Effects of eutrophication on the interaction between algae and grazers in an Andean stream. Hydrobiologia. doi:10.1007/s10750-010-0194-y.

Dunne, T. & L. B. Leopold, 1978. Water in Environmental Planning. W.H. Freeman, San Francisco.

Ellis, E. C. & N. Ramankutty, 2008. Putting people in the map: anthropogenic biomes of the world. Frontiers in Ecology and the Environment 6: 439–447.

Elosegi, A., J. Díez & M. Mutz, 2010. Effects of hydromorphological integrity on biodiversity and functioning of river ecosystems. Hydrobiologia. doi:10.1007/s10750-009-0083-4.

European Union Commission., 2000. Directive 2000/60/EC of the European Parliament and of the Council of 23 October 2000 establishing a framework for Community action in the field of water policy. The European Parliament and the Council of the European Union. Official Journal of the European Communities L 327/1, 1-72.

Foley, J. A., R. DeFries, G. P. Asner, C. Barford, G. Bonan, S. R. Carpenter, F. S. Chapin, M. T. Coe, G. C. Daily, H. K. Gibbs, J. H. Helkowski, T. Holloway, E. A. Howard, C. J. Kucharik, C. Monfreda, J. A. Patz, I. C. Prentice, N. Ramankutty & P. K. Snyder, 2005. Global consequences of land use. Science 309: 570–574.

Frissell, C. A., W. J. Liss, C. E. Warren & M. D. Hurley, 1986. A hierarchical framework for stream habitat classification: viewing streams in a watershed context. Environmental Management 10: 199–214.

Fu, G. B., S. L. Chen, C. M. Liu & D. Shepard, 2004. Hydroclimatic trends of the Yellow River basin for the last 50 years. Climatic Change 65: 149–178.

Grime, J. P., 1973. Competitive exclusion in herbaceous vegetation. Nature 242: 344–347.

Grimm, N. B., S. H. Faeth, N. E. Golubiewski, C. L. Redman, J. Wu, X. Bai & J. M. Briggs, 2008. Global change and the ecology of cities. Science 319: 756–760.

Guasch, H., E. Navarro, A. Serra & S. Sabater, 2004. Phosphate limitation influences the sensitivity to copper in periphytic algae. Freshwater Biology 49: 463–473.

Guasch, H., G. Atli, B. Bonet, N. Corcoll, M. Leira & A. Serra, 2010. Discharge and the response of biofilms to metal exposure in Mediterranean rivers. Hydrobiologia. doi: 10.1007/s10750-010-0116-z.

Hamilton, S. K., 2010. Biogeochemical implications of climate change for tropical rivers and floodplains. Hydrobiologia. doi:10.1007/s10750-009-0086-1.

Hamilton, A.T., M. T. Barbour & B. G. Bierwagen, 2010. Implications of global change for the maintenance of water quality and ecological integrity in the context of current water laws and environmental policies. Hydrobiologia. doi:10.1007/s10750-010-0316-6.

Hill, W. R., M. G. Ryon & E. M. Schilling, 1995. Light limitation in a stream ecosystem: responses by primary producers and consumers. Ecology 76: 1297–1309.

Hilton, J., M. O'Hare, M. J. Bowes & I. I. Jones, 2006. How green is my river? A new paradigm of eutrophication in rivers. Science of the Total Environment 365: 66–83.

Jørgensen, S. E. & F. Müller, 2000. Handbook of Ecosystem Theories and Management. Lewis Publishers, Boca Raton, FL, USA.

Kaushal, S. S., G. E. Likens, N. A. Jaworski, M. L. Pace, A. M. Sides, D. Seekell, K. T. Belt, D. H. Secor & R. L. Wingate, 2010. Rising stream and river temperatures in the United States. Frontiers in Ecology and the Environment. (in press).

Lake, P. S., 2000. Disturbance, patchiness, and diversity in streams. Journal of the North American Benthological Society 19: 573–592.

Leira, M. & S. Sabater, 2005. Diatom assemblages distribution in catalan rivers, NE Spain, in relation to chemical and physiographical factors. Water Research 39: 73–82.

Leopold, L. B., M. G. Wolman & J. P. Miller, 1964. Fluvial Processes in Geomorphology. W.H. Freeman, San Francisco.

Liu, J., T. Dietz, S. R. Carpenter, M. Alberti, C. Folke, E. Moran, A. N. Pell, P. Deadman, T. Kratz, J. Lubchenco, E. Ostrom, Z. Ouyang, W. Provencher, C. L. Redman, S. H. Schneider & W. W. Taylor, 2007. Complexity of coupled human and natural systems. Science 317: 1513–1516.

López-Doval, J. C., M. Ricart, H. Guasch, A. M. Romaní, S. Sabater & I. Muñoz, 2010. Does grazing pressure change modify diuron toxicity in a biofilm community? Archives of Environmental Contamination and Toxicology 58: 955–962.

Manoylov, K. M. & R. J. Stevenson, 2006. Density-dependent algal growth along N and P nutrient gradients in artificial streams. In Ognjanova-Rumenova, N. & K. Manoylov (eds), Advances in Phycological Studies. Pensoft Publishers, Moscow, Russia: 333–352.

Margalef, R., 1983. Limnología. Omega, Barcelona.

Margalef, R., 1997. Our Biosphere. Ecology Institute, Oldendorf.

Mas-Martí, E., E. García-Berthou, S. Sabater, S. Tomanova & I. Muñoz, 2010. Comparing fish assemblages and trophic ecology of permanent and intermittent reaches in a Mediterranean stream. Hydrobiologia. doi:10.1007/s10750-010-0292-x.

Meybeck, M. & R. Helmer, 1989. The quality of rivers: from pristine stage to global pollution. Paleogeography, Paleoclimatology, Paleoecology 75: 283–309.

Millennium Ecosystem Assessment, 2005. Ecosystems and Human Well-being: Synthesis. Island Press, Washington, DC.

Milly, P. C. D., K. A. Dunne & A. V. Vecchia, 2005. Global pattern of trends in streamflow and water availability in a changing climate. Nature 438: 347–350.

Miltner, R. J. & E. T. Rankin, 1998. Primary nutrients and the biotic integrity of rivers and streams. Freshwater Biology 40: 145–158.

Molles, M. C. Jr., C. S. Crawford, L. M. Ellis, H. M. Valett & C. N. Dahm, 1998. Managed flooding for riparian ecosystem restoration. Bioscience 48: 749–756.

Montuelle, B., U. Dorigo, A. Bérard, B. Volat, A. Bouchez, A. Tlili, V. Gouy & S. Pesce, 2010. The periphyton as a multimetric bioindicator for assessing the impact of land use on river: an overview on the Ardières-Morcille experimental watershed (France) Hydrobiologia. doi: 10.1007/s10750-010-0105-2.

Muradian, R., 2001. Ecological thresholds: a survey. Ecological Economics 38: 7–24.

National Research Council, 1986. Ecological Knowledge and Environmental Problem-Solving: Concepts and Case Studies. National Academy Press, Washington, DC.

Nilsson, C., C. A. Reidy, M. Dynesius & C. Revenga, 2005. Fragmentation and flow regulation of the world's large river systems. Science 308: 405–408.

Norton, S. B., S. M. Cormier, G. W. Suter II, K. Schofield, L. Yuan, P. Shaw-Allen & C. R. Ziegler, 2009. CADDIS: the causal analysis/diagnosis decision information system. In Marcomini, A., G. W. Suter II & A. Critto (eds), Decision Support Systems for Risk-Based Management of Contaminated Sites. Springer, Berlin: 1–24.

O'Brien, P. J. & J. D. Wehr, 2010. Periphyton biomass and ecological stoichiometry in streams within an urban to rural land-use gradient. Hydrobiologia. doi:10.1007/s10750-009-9984-5.

O'Connor, N. A. & P. S. Lake, 1994. Long-term and seasonal large-scale disturbances of a small lowland stream.

Australian Journal of Marine and Freshwater Research 45: 243–255.

Odum, E. P., J. T. Finn & E. H. Franz, 1979. Perturbation theory and the subsidy-stress gradient. BioScience 29: 349–352.

Oki, T. & S. Kanae, 2006. Global hydrological cycles and world water resources. Science 313: 1068–1072.

Omernik, J. M., 1987. Ecoregions of the conterminous United States. Annals of the Association of American Geographers 77: 118–125.

Palmer, M. A., J. D. P. Lettenmaier, N. L. Poff, S. L. Postel, B. Richter & R. Warner, 2009. Climate change and river ecosystems: protection and adaptation options. Environmental Management 44: 1053–1068.

Pan, Y., R. J. Stevenson, B. H. Hill, A. T. Herlihy & G. B. Collins, 1996. Using diatoms as indicators of ecological conditions in lotic systems: a regional assessment. Journal of the North American Benthological Society 15: 481–495.

Perkins, D. M., J. Reiss, G. Yvon-Durocher & G. Woodward. 2010. Global change and food webs in running waters. Hydrobiologia. doi:10.1007/s10750-009-0080-7.

Pickett, S. T. A., S. L. Collins & J. J. Armesto, 1987. Models, mechanisms, and pathways of succession. Botanical Review 53: 335–371.

Poff, N. L. & J. K. H. Zimmerman, 2010. Ecological responses to altered flow regimes: a literature review to inform environmental flows science and management. Freshwater Biology 55: 194–205.

Poff, N. L., B. Richter, A. H. Arthington, S. E. Bunn, R. J. Naiman, E. Kendy, M. Acreman, C. Apse, B. P. Bledsoe, M. Freeman, J. Henriksen, R. B. Jacobson, J. Kennen, D. M. Merritt, J. O'Keeffe, J. D. Olden, K. Rogers, R. E. Tharme & A. Warner, 2010. The Ecological Limits of Hydrologic Alteration (ELOHA): a new framework for developing regional environmental flow standards. Freshwater Biology 55: 147–170.

Postel, S. L. & S. R. Carpenter, 1997. Freshwater ecosystem services. In Daily, G. C. (ed.), Nature's Services. Island Press, Washington, DC: 195–214.

Power, M. E., A. Sun, G. Parker, W. E. Dietrich & J. T. Wootton, 1995. Hydraulic food chain models. BioScience 45: 159–167.

Ramankutty, N. & J. Foley, 1999. Estimating historical changes in global land cover: croplands from 1700 to 1992. Global Biogeochemical Cycles 13: 997–1028.

Rapport, D. J., H. A. Regier & T. C. Hutchinson, 1985. Ecosystem behaviour under stress. American Naturalist 125: 617–640.

Reynolds, C. S., J. Padisák & U. Sommer, 1993. Intermediate disturbance in the ecology of phytoplankton and the maintenance of species diversity: a synthesis. Hydrobiologia 249: 183–188.

Ricart, M., H. Guasch, D. Barceló, R. Brix, M. H. Conceição, A. Geiszinger, M. López De Alda, J. López-Doval, I. Muñoz, A. M. Romaní, M. Villagrassa & S. Sabater, 2010. Primary and complex stressors in polluted Mediterranean rivers: pesticide effects on biological communities. Journal of Hydrology 383: 52–61.

Richards, C., L. B. Johnson & G. E. Host, 1996. Landscape-scale influences on stream habitats and biota. Canadian Journal of Fisheries and Aquatic Sciences 53: 295–311.

Riseng, C. M., M. J. Wiley & R. J. Stevenson, 2004. Hydrologic disturbance and nutrient effects on benthic community structure in midwestern US streams: a covariance structure analysis. Journal of the North American Benthological Society 23: 309–326.

Robertson, G. P. & S. M. Swinton, 2005. Reconciling agricultural productivity and environmental integrity: a grand challenge for agriculture. Frontiers in Ecology and the Environment 3: 38–46.

Rodrigues Capítulo, A., N. Gómez, A. Giorgi & C. Feijoo, 2010. Global changes in pampean lowland streams (Argentina): implications for biodiversity and functioning. Hydrobiologia. doi:10.1007/s10750-010-0319-3.

Sabater, S., 2008. Alterations of the global water cycle and their effects on river structure, function and services. Freshwater Reviews 1: 75–88.

Sabater, S. & K. Tockner, 2010. Effects of hydrologic alterations on the ecological quality of river ecosystems. In Sabater, S. & D. Barcelo (eds), Water Scarcity in the Mediterranean: Perspectives Under Global Change. Springer Verlag, Berlin: 15–39.

Sabater, S., E. Vilalta, A. Gaudes, H. Guasch, I. Munoz & A. Romani, 2003. Ecological implications of mass growth of benthic cyanobacteria in rivers. Aquatic Microbial Ecology 32: 175–184.

Sabater, S., V. Acuna, A. Giorgi, E. Guerra, I. Munoz & A. M. Romani, 2005. Effects of nutrient inputs in a forested Mediterranean stream under moderate light availability. Archiv Fur Hydrobiologie 163: 479–496.

Sabater, S., J. Artigas, A. Gaudes, I. Muñoz, G. Urrea & A. M. Romaní, 2010. Moderate long-term nutrient input enhances autotrophy in a forested Mediterranean stream. Freshw Biol (submitted).

Sala, O. E., F. S. Chapin, J. J. Armesto, E. Berlow, J. Bloomfield, R. Dirzo, E. Huber Sanwald, L. F. Huenneke, R. B. Jackson, A. Kinzig, R. Leemans, D. M. Lodge, H. A. Mooney, M. Oesterheld, N. L. Poff, M. T. Sykes, B. H. Walker, M. Walker & D. H. Wall, 2000. Biodiversity – global biodiversity scenarios for the year 2100. Science 287: 1770–1774.

Schulze, R. E., 2007. Some foci of integrated water resources management in the "South" which are oft-forgotten by the "North": a perspective from southern Africa. Water Resources Management 21: 269–294.

Schwarzenbach, R. P., B. I. Escher, K. Fenner, T. B. Hofstetter, C. A. Johnson, U. Von Gunten & B. Wehrli, 2006. Global hydrological cycles and world water resources. Science 313: 1072–1077.

Smith, M. D., A. K. Knapp & S. L. Collins, 2009. A framework for assessing ecosystem dynamics in response to chronic resource alterations induced by global change. Ecology 90: 3279–3289.

Solomon, S., D. Qin, M. Manning, R. B. Alley, T. Berntsen, N. L. Bindoff, Z. Chen, A. Chidthaisong, J. M. Gregory, G. C. Hegerl, M. Heimann, B. Hewitson, B. J. Hoskins, F. Joos, J. Jouzel, V. Kattsov, U. Lohmann, T. Matsuno, M. Molina, N. Nicholls, J. Overpeck, G. Raga, V. Ramaswamy, J. Ren, M. Rusticucci, R. Somerville, T. F. Stocker, P. Whetton, R. A. Wood & D. Wratt, 2007. Technical Summary. In Solomon, S., D. Qin, M. Manning, Z. Chen, M. Marquis, K. B. Averyt, M. Tignor & H. L. Miller (eds),

Climate Change 2007: The Physical Science Basis. Contribution of Working Group I to the Fourth Assessment Report of the Intergovernmental Panel on Climate Change. Cambridge University Press, Cambridge, United Kingdom and New York, NY, USA.

Soranno, P. A., K. S. Cheruvelil, R. J. Stevenson, S. L. Rollins, S. W. Holden, S. Heaton & E. Torng, 2008. A framework for developing ecosystem-specific nutrient criteria: integrating biological thresholds with predictive modeling. Limnology and Oceanography 53: 773–787.

Stevenson, R. J., 1997a. Scale-dependent determinants and consequences of benthic algal heterogeneity. Journal of North American Benthological Society 16: 248–262.

Stevenson, R. J., 1997b. Resource thresholds and stream ecosystem sustainability. Journal of the North American Benthological Society 16: 410–424.

Stevenson, R.J., in press. Coupling human and natural systems to solve environmental problems. Physics and Chemistry of the Earth.

Stevenson, R. J., R. C. Bailey, M. C. Harass, C. P. Hawkins, J. Alba-Tercedor, C. Couch, S. Dyer, F. A. Fulk, J. M. Harrington, C. T. Hunsaker & R. K. Johnson, 2004. Interpreting results of ecological assessments. In Barbour, M. T., S. B. Norton, H. R. Preston & K. W. Thornton (eds), Ecological Assessment of Aquatic Resources: Linking Science to Decision-Making. Society of Environmental Toxicology and Contamination Publication, Pensacola, FL: 85–111.

Stevenson, R. J., S. T. Rier, C. M. Riseng, R. E. Schultz & M. J. Wiley, 2006. Comparing effects of nutrients on algal biomass in streams in two regions with different disturbance regimes and with applications for developing nutrient criteria. Hydrobiologia 561: 140–165.

Stevenson, R. J., B. E. Hill, A. T. Herlihy, L. L. Yuan & S. B. Norton, 2008. Algal-P relationships, thresholds, and frequency distributions guide nutrient criterion development. Journal of the North American Benthological Society 27: 783–799.

Stevenson, R. J., B. J. Bennett, D. N. Jordan & R. D. French. Phosphorus regulates stream injury by filamentous green algae, thresholds, DO, and pH. (Hydrobiologia submitted).

Stoddard, J. L., D. P. Larsen, C. P. Hawkins, R. K. Johnson & R. H. Norris, 2006. Setting expectations for the ecological condition of streams: the concept of reference condition. Ecological Applications 16: 1267–1276.

Strayer, D. L., 2006. Challenges for freshwater invertebrate conservation. Journal of the North American Benthological Society 25: 271–287.

Sumpter, J. P., 2009. Protecting aquatic organisms from chemicals: the harsh realities. Philosophical Transactions of the Royal Society A 367: 3877–3894.

Suplee, M. W., V. Watson, M. Teply & H. McKee, 2008. How green is too green? Public opinion of what constitutes undesirable algae levels in streams. Journal of the American Water Works Association 44: 1–18.

Suter, G. W., 1993. Ecological Risk Assessment. Lewis Publishers, Boca Raton, FL.

Syvitski, J. P. M., C. J. Vorosmarty, A. J. Kettner & P. Green, 2005. Impact of humans on the flux of terrestrial sediment to the global coastal ocean. Science 308: 376–380.

Tilman, D., J. Fargione, B. Wolff, C. D'Antonio, A. Dobson, R. Howarth, D. Schindler, W. H. Schlesinger, D. Simberloff & D. Swackhamer, 2001. Forecasting agriculturally driven global environmental change. Science 292: 281–284.

Tockner, K., U. Uehlinger & C. T. Robsinson, 2009. Rivers of Europe. Academic Press, London.

Uehlinger, U., C. T. Robinson, M. Hieber & R. Zah, 2010. The physico-chemical habitat template for periphyton in Alpine glacial streams under a changing climate. Hydrobiologia. doi:10.1007/s10750-009-9963-x.

Vannote, R. L., G. W. Minshall, K. W. Cummins, J. R. Sedell & C. E. Cushing, 1980. The river continuum concept. Canadian Journal of Fisheries and Aquatic Science 37: 130–137.

Vitousek, P. M., 1994. Beyond global warming: ecology and global change. Ecology 75: 1861–1876.

Wang, L., D. M. Robertson & P. L. Garrison, 2007. Linkages between nutrients and assemblages of macroinvertebrates and fish in wadeable streams: implication to nutrient criteria development. Environmental Management 39: 194–212.

Wiley, M. J., D. W. Hyndman, B. C. Pijanowski, A. D. Kendall, C. Riseng, E. S. Rutherford, S. T. Cheng, M. L. Carlson, J. A. Tyler, R. J. Stevenson, P. J. Steen, P. L. Richards, P. W. Seelbach & J. M. Koches, 2010. A multi-modeling approach to evaluating climate and land use change impacts in a great lakes tributary river basin. Hydrobiologia. doi:10.1007/s10750-010-0239-2.

Xenopoulos, M. A. & D. M. Lodge, 2006. Going with the flow: using species-discharge relationships to forecast losses in fish biodiversity. Ecology 87: 1907–1914.

Xenopoulos, M. A., D. M. Lodge, J. Alcamo, M. Märker, K. Schulze & D. P. van Vuuren, 2005. Scenarios of freshwater fish extinctions from climate change and water withdrawal. Global Change Biology 11: 1557–1564.

Ylla, I., A. M. Romaní & S. Sabater, 2007. Differential effects of nutrients and light on the primary production of stream algae and mosses. Fundamental and Applied Limnology/Archiv für Hydrobiologie 170: 1–10.

Ylla, I., I. Sanpera-Calbet, E. Vázquez, A. M. Romaní, I. Muñoz, A. Butturini & S. Sabater, 2010. Organic matter availability during pre and post-drought periods in a Mediterranean stream. Hydrobiologia. doi:10.1007/s10750-010-0193-z.

Ziegler, S. E., D. R. Lyon, 2010. Factors regulating epilithic biofilm carbon cycling and release with nutrient enrichment in headwater streams. Hydrobiologia. doi:10.1007/s10750-010-0296-6.

Zoppini, A., S. Amalfitano, S. Fazi, A. Puddu, 2010. Dynamics of a benthic microbial community in a riverine environment subject to hydrological fluctuations (Mulargia River, Italy). Hydrobiologia. doi:10.1007/s10750-010-0199-6.

GLOBAL CHANGE AND RIVER ECOSYSTEMS

# Biogeochemical implications of climate change for tropical rivers and floodplains

Stephen K. Hamilton

Received: 25 June 2009 / Accepted: 30 December 2009 / Published online: 13 January 2010
© Springer Science+Business Media B.V. 2010

**Abstract** Large rivers of the tropics, many of which have extensive floodplains and deltas, are important in the delivery of nutrients and sediments to marine environments, in methane emission to the atmosphere and in providing ecosystem services associated with their high biological productivity. These ecosystem functions entail biogeochemical processes that will be influenced by climate change. Evidence for recent climate-driven changes in tropical rivers exists, but remains equivocal. Model projections suggest substantial future climate-driven changes, but they also underscore the complex interactions that control landscape water balances, river discharges and biogeochemical processes. The most important changes are likely to involve: (1) aquatic thermal regimes, with implications for thermal optima of plants and animals, rates of microbially mediated biogeochemical transformations, density stratification of water bodies and dissolved oxygen depletion; (2) hydrological regimes of discharge and floodplain inundation, which determine the ecological structure and function of rivers and floodplains and the extent and seasonality of aquatic environments; and (3) freshwater–seawater gradients where rivers meet oceans, affecting the distribution of marine, brackish and freshwater environments and the biogeochemical processing as river water approaches the coastal zone. In all cases, climate change affects biogeochemical processes in concert with other drivers such as deforestation and other land use changes, dams and other hydrological alterations and water withdrawals. Furthermore, changes in riverine hydrology and biogeochemistry produce potential feedbacks to climate involving biogeochemical processes such as decomposition and methane emission. Future research should seek improved understanding of these changes, and long-term monitoring should be extended to shallow waters of wetlands and floodplains in addition to the larger lakes and rivers that are most studied.

**Keywords** Global warming · Temperature · Wetlands · Carbon dioxide · Oxygen · Biogeochemistry

Guest editors: R. J. Stevenson, S. Sabater / Global Change and River Ecosystems – Implications for Structure, Function and Ecosystem Services

S. K. Hamilton (✉)
W.K. Kellogg Biological Station, Michigan State University, 3700 E. Gull Lake Drive, Hickory Corners, MI 49060-9516, USA
e-mail: hamilton@msu.edu; hamilton@kbs.msu.edu

S. K. Hamilton
Department of Zoology, Michigan State University, 3700 E. Gull Lake Drive, Hickory Corners, MI 49060-9516, USA

## Introduction

In the process of conveying water and materials from uplands to the oceans, large tropical rivers and their floodplains are sites of intense biogeochemical activity associated with the passage of river water through highly productive floodplains and estuaries. Rivers and their floodplains provide important ecosystem services, supporting biodiversity, freshwater and marine fisheries, traditional human livelihoods and productive agricultural land (Junk, 1997). Many large tropical rivers have not been regulated by dams (Nilsson et al., 2005), and therefore retain seasonally variable hydrological regimes and periodically inundated floodplains, characteristics that exert important influence on aquatic biogeochemical processes and fluxes (Lewis et al., 2000). Extensive seasonal inundation in many tropical floodplains makes them major natural sources of methane emission to the atmosphere (Melack et al., 2004; Denman et al., 2007). Large tropical rivers are important in the delivery of nutrients and sediments to marine environments, and this material influx supports productive estuarine and coastal marine ecosystems (Cotrim da Cunha et al., 2008). Where rivers meet the oceans, deltas and estuaries are important sites of biogeochemical transformations including biological production, sediment retention and nutrient transformation (Bianchi, 2007).

This article reviews literature on whether climate-driven changes are already taking place in tropical ecosystems, how future climate change is expected to affect large tropical rivers and their floodplains, and the potential biogeochemical implications of these effects. Large tropical river systems drain catchments ranging from rainforest to dryland and from relatively pristine to heavily impacted by human activity. In all cases, climate change will affect biogeochemical processes in concert with other anthropogenic drivers, such as land use change, construction of dams and other river channel alterations and water withdrawals. Also, changes in catchment vegetation and land use likely will produce potential hydrological feedbacks to climate that remain poorly understood. Evidence reviewed here indicates that climate change potentially is expected to profoundly influence the biogeochemical processes that underpin ecosystem functions of large rivers and their floodplains and estuaries, most prominently via hydrological changes in the catchments and rising sea levels at the river mouths, but potentially also via a host of other interactions. Tropical river catchments encompass a broad range of climatic settings, human activity and geomorphology, and thus the hydrological and biogeochemical implications of climate change will be variable across river systems.

## Tropical river hydrology and climatic regimes

Major tropical river systems and their associated floodplains are described by Welcomme (1985) and include systems with largely natural hydrological regimes, such as the Amazon and Orinoco rivers of South America, as well as systems that have been strongly impacted by hydrological modifications including large dams, such as the Paraná River in South America and the Niger River in Africa. All of these river systems are seasonally variable in discharge and many have extensive floodplains and deltas, although large dams can reduce variability in discharge and inundation. Floodplains fringing large rivers typically include areas flooded by riverine overflow as well as areas subject to inundation from local sources that drains through floodplains into a river system (Mertes, 2000; Hamilton, 2008a). Large tropical rivers deliver $\sim 35\%$ of global freshwater discharge to the oceans (Vörösmarty et al., 1997).

Seasonal variation in temperature, precipitation and river discharge occurs over most of the tropics, and is most marked in the 'wet–dry tropics' (tropical latitudes higher than $\sim 10°$), where climatic seasonality is driven by the shifting latitudinal location of the equatorial trough and the inter-tropical convergence zone (Talling & Lemoalle, 1998). Seasonal variation in temperature and river discharge may or may not be coincident in timing. For example, inundation of much of the Pantanal wetland of Brazil takes place in the warmer months (Hamilton et al., 1996), whereas river flooding in the Okavango Delta in Botswana coincides with the cool season (Ramberg et al., 2006). The Amazon River system drains catchments on both sides of the equator with roughly opposite seasonal discharge patterns and as a result the seasonality of discharge and floodplain inundation along the main stem is somewhat attenuated (Hamilton et al., 2002).

The El Niño-Southern Oscillation (ENSO) cycle, which is associated with sea surface temperature anomalies in the tropical Pacific, exerts considerable influence on precipitation patterns in the tropics and subtropics (Foley et al., 2002; Bates et al., 2008). ENSO variability partly explains the interannual variability in river discharges and floods and droughts in regions as distant as southern Africa and Australia. Large tropical rivers with discharges that are particularly influenced by ENSO events include the Nile, Amazon, Congo, Paraná and Ganges (Khan et al., 2006). The erratic behaviour of tropical cyclones also contributes to the interannual variability in the hydrological regimes of tropical rivers in regions where much of the rainfall is associated with cyclonic activity, such as southeast Asia and northern Australia.

Floodplain waters are amongst the warmest surface waters in tropical regions because they tend to be shallow (and thus they concentrate heat accumulation and track short-term fluctuations in air temperatures) and they lie at low elevations. Water depths can be >10 m in the main channels of the largest rivers, but the mean depth on fringing floodplains is typically much less, and often not deep enough to develop persistent hypolimnetic anoxia, even in the permanent floodplain lakes (e.g. Hamilton & Lewis, 1990; Hamilton et al., 1990). Most discussions of thermal regimes in tropical fresh waters have dealt with large lakes, for which there are many published examples, and often these are deep enough to be seasonally stratified and many of the most studied tropical lakes lie well above sea level (Talling & Lemoalle, 1998).

Shallow and quiescent waters on floodplains or in rivers with little or no flow are more subject to diel temperature fluctuation in response to short-term gains and losses of heat. Diurnal stratification is characteristic of these waters, and is a phenomenon subject to change in a warming climate, as discussed later. An example of thermal seasonality in relatively shallow waters of tropical rivers and floodplains is presented in Fig. 1, which shows typical seasonal fluctuations in daytime temperatures in waters of the Brazilian Pantanal. Figure 2 shows a diel cycle of temperature during the late dry-season in a slowly flowing monsoon river of northern Australia.

Similar water temperatures, commonly exceeding 30°C, but not often approaching 40°C during the warm season, have been measured in other shallow tropical waters in the wet–dry tropics. For example, measurements made in a rice field pool of northeastern Thailand (~17.5°N latitude) by Heckman (1979) showed frequent maxima above 35°C. Water temperatures in more equatorial locations usually do not reach such extremes due to attenuation of solar radiation by clouds (Lewis, 1987). Also, floodplains

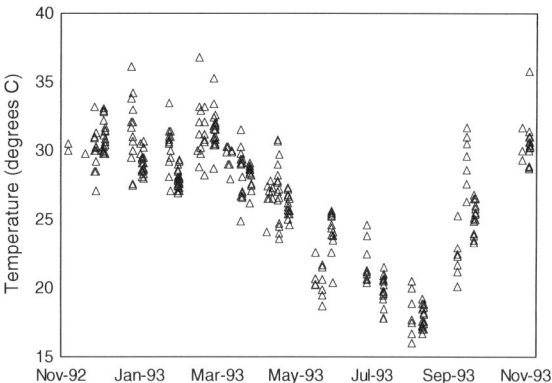

**Fig. 1** Seasonal fluctuations in daytime temperatures in surface waters of the Brazilian Pantanal, collected by the author at numerous river and floodplain sites over the course of biogeochemical investigations (Hamilton et al., 1995). Most of the waters sampled were <3 m and many were <1 m deep, and were located between 17 and 18°S latitude

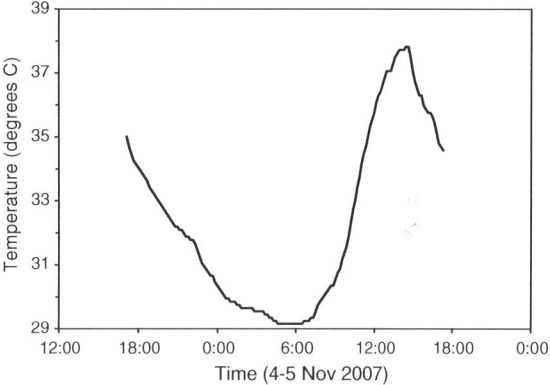

**Fig. 2** Example of a diel temperature cycle in an open (uncanopied) riffle of a slowly flowing river of northern Australia during the late dry season (Mitchell River, 16.5°S latitude, 4–5 November 2007)

where precipitation is more equitably distributed throughout the year tend to have dense vegetation canopies that shade the water. Nonetheless, in shallow and uncanopied water bodies daytime water temperatures can surpass 30°C, as shown for example in data for an Amazon floodplain lake by de Melo & Huszar (2000), for Orinoco floodplain lakes by Hamilton et al. (1990), and in the Paraguay River by Hamilton et al. (1997). Thus, for this review, the water temperature range of interest is approximately 15–40°C, recognising that maximum temperatures above ~35°C are likely only in the shallowest waters.

Water temperatures in lakes and larger rivers are commonly measured although typically published only in summary form, obscuring the minima and maxima that can be important for biogeochemical and ecological processes. Temperatures in shallow waters of wetlands and floodplains are not measured as often, and when done so it is usually for a limited time, serving as ancillary information for studies focused on other topics. Long-term monitoring of water temperatures in a diversity of tropical water bodies, in conjunction with meteorological data collection nearby, would be a wise investment to improve our understanding of the thermal implications of climate change for aquatic ecosystems and could readily be done using inexpensive continuous logging systems, provided that rigorous calibration protocols were in place.

## Climate change in the tropics: past trends and future projections

Current consensus on how climate change is projected to affect tropical regions is drawn mainly from the Intergovernmental Panel on Climate Change (IPCC) Technical Paper on Climate Change and Water (Bates et al., 2008), which is based on the Fourth Assessment Report (IPCC, 2007; Kundzewicz et al., 2008), and references cited therein. Additional information was drawn from CCSP (2008).

Temperatures in the tropics are projected to warm 2–4°C by 2100 (Meehl et al., 2007). The warming temperatures will extend tropical climates poleward and upward in elevation. Evaporation rates and atmospheric moisture fluxes are often projected to increase significantly as the temperatures rise because the saturation water vapour pressure increases non-linearly with temperature. However, evaporation rates are influenced by complex interacting factors associated with local geography and prevailing climatic conditions, and increases in atmospheric humidity and/or wind can counterbalance the temperature effect and even result in decreased evaporation rates (Hobbins et al., 2004, 2008).

Hydrological fluxes including precipitation, evaporation and runoff will intensify as a result of the increase in atmospheric temperature and moisture and consequent changes in atmospheric circulation. A general tendency in model projections for a warming climate is that presently wet climates become wetter and dry climates become drier, and this appears to hold for the tropics as well as globally (IPCC, 2007). Increasing subtropical aridity is also projected, but mainly at the poleward flanks of the subtropics (Cook et al., 2008).

Timing and volume of river discharges respond strongly to changes in precipitation and are usually somewhat less sensitive to changes in evaporation rates and storage reservoirs (e.g. ground water, soil moisture, artificial reservoirs and in high-elevation catchments, snow and ice). Thus altered river hydrological regimes are expected to occur particularly as a result of changes in precipitation regimes. However, certain rivers also may change significantly in response to changes in evaporation rates, particularly in dryland catchments where a larger fraction of the precipitation is lost to evaporation (Goudie, 2006), as well as in high-elevation catchments such as the Himalayas where seasonal water storage as snow is important (Nijssen et al., 2001).

Many studies have searched for global trends in river discharge, precipitation and water quality over recent decades, but a clear consensus has yet to emerge. Recent study has pointed to an overall trend of increasing river discharge over the past century (Labat et al., 2004; Gerten et al., 2008), although the quality of the data set underpinning this conclusion has been questioned (Peel & McMahon, 2006; Milliman et al., 2008), and the increase has yet to be rigorously verified. Precipitation also fails to show discernable trends in total amount in the tropics (Bates et al., 2008), although trend detection is difficult in the face of high interannual variability in precipitation, in part, related to ENSO variation. While there may have been a global increase in river discharges, evidence so far indicates that trends across the tropics have certainly not been homogeneous (Bates et al., 2008), and are likely to be spatially variable in the future. For example, regional decreases in annual runoff have been noted for parts of West Africa and southern South America (Milly et al., 2005; Li et al., 2007). Recent assessments of trends in water quality also fail to show globally consistent changes that can be attributed to climate change (Bates et al., 2008). This lack of conclusive evidence for global trends is not surprising because hydrological changes associated with a changing global climate are expected to be spatially heterogeneous and difficult to discern from natural variability and catchment-specific hydrological alterations by human activities.

Despite the ongoing debates about whether precipitation and river discharges have changed in the recent past, there is good reason to believe that a combination of global climate change and human interferences in land and river hydrology will increasingly alter river hydrological regimes in the future. Specific projections are challenging, however, because river runoff reflects the net result of multiple interacting processes. Several studies have examined the idea that increased runoff could arise from increased plant water use efficiency in response to greater atmospheric $CO_2$ availability, via reduced need for stomatal gas exchange and consequent evapotranspirative water losses (Gedney et al., 2006; Betts et al., 2007). However, another modeling study has suggested that this plant physiological response should be compensated by $CO_2$-stimulated increases in plant biomass, and that in some tropical regions past increases in runoff appear to have been driven largely by changing land use, with climate change of secondary importance (Piao et al., 2007). Another recent modelling study of past discharge records has demonstrated the potential importance of changes in land cover, temperature and irrigation in addition to precipitation and $CO_2$ (Gerten et al., 2008).

Multi-model ensemble projections extending to time spans of 50–100 years provide the best available indication of where future climatic changes are likely to be most important (Nohara et al., 2006). In general, these model ensembles project that by mid-century mean annual precipitation and runoff will decrease by 10–30% in some dry subtropical and tropical regions and increase by 10–40% in some wet tropical regions (Milly et al., 2005; Bates et al., 2008). For example, mean annual runoff is projected to decrease in southern Africa, tropical Mexico, central America and northeast Brazil, to increase in eastern equatorial Africa and the La Plata River catchment in South America, and to change less in much of the Amazon and Orinoco catchments. Regions subject to monsoonal regimes are likely to become wetter, including southern Asia and northern Australia. Evaporation and evapotranspiration may increase with warming temperatures, and as mentioned before this could counteract the effect of increased precipitation on runoff, especially in dryland rivers (Dai et al., 2004; Cai & Cowan, 2008). However, climate warming models that project increased evaporation in a warmer world may overemphasise the role of temperature alone as a driver of evaporation (Hobbins et al., 2008).

Despite the uncertainty of model projections, in many tropical rivers the projected changes in discharge over the next 100 years are similar to or greater than the envelope of natural variability over the past 9,000 years, a period of substantial variability driven mainly by changes in solar insolation arising from orbital variations (Aerts et al., 2006). The comparative analysis by Aerts et al. suggests that the Ganges, Mekong, Volta, Congo and Amazon rivers are in this category, whereas the Nile appears less sensitive to future climate change.

One globally consistent projected trend is towards more intense precipitation events (Cook et al., 2008), which has implications for river hydrological regimes and floodplain inundation even if mean annual precipitation and river discharges are unchanged. The frequency and severity of heavy precipitation events are likely to increase in tropical catchments, as are the extent and occurrence of droughts, with both changes projected to take place in some catchments (Bates et al., 2008; Kundzewicz et al., 2008). For rivers, in general, studies have documented how precipitation intensity strongly influences rates of soil erosion and sediment transport into rivers, peak discharges that control many fluvial geomorphological processes, and timing and duration of floodplain inundation (Goudie, 2006). Tropical rivers would behave similarly in these regards, albeit with the most potential for soil erosion in seasonally dry catchments.

Droughts appear to have become more common in recent years, showing some correspondence with elevated sea surface temperatures in tropical regions, although their attribution to climate change remains uncertain (Bates et al., 2008; cf. Marengo et al., 2008; Sheffield & Wood, 2008). Drier parts of the northern subtropics are particularly likely to experience greater seasonal water limitation (Bates et al., 2008). It is important to recognise that seemingly mild droughts in normally humid regions can cause major changes in vegetation and organic carbon stocks, often via increased fire influence (Marengo et al., 2008). For river systems in populated regions with limited water supplies, stronger or more protracted droughts will increase pressure to withdraw river water for consumptive uses or interbasin diversions (Palmer et al., 2008).

Glacier-fed rivers will temporarily increase in discharge then decrease as the glacial ice becomes diminished. Loss of glacial and snowmelt sources will in some cases impact significant tropical rivers and water supplies downstream, as for example in tropical Andean alpine areas (Bradley et al., 2006) and in rivers arising in the Himalayas (WWF, 2005). Glacial retreat or disappearance may not greatly affect discharge of the larger South American rivers because most of their water originates from more lowland catchments. However, glaciers across the Himalayas and the Qinghai-Tibet Plateau are important sources of water, particularly during low-discharge seasons, to rivers, such as the Yangtze and Yellow in China, the Mekong in Indochina, the Brahmaputra in Bangladesh and the Ganges in India.

Models indicate that tropical cyclones are likely to become more intense in the future (Bates et al., 2008). It remains unclear whether tropical cyclone activity has increased as a consequence of climate change (Goudie, 2006), although there is evidence that tropical cyclones are becoming more severe (Webster et al., 2005). Notably, storm surges associated with coastal storms will increase in impact and area of influence with sea level rise, whether or not the storms become more severe.

Sea level rise is one of the most certain consequences of a warmer climate, but the magnitude and timing are not easily forecast. Projections of sea level rise are being revised in light of the surprisingly accelerated degradation of continental ice sheets, and rises of 40–100 cm over the next century are now being projected (Horton et al., 2008), substantially greater than the 28–42 cm estimates of global models evaluated by the IPCC (2007). Even greater rises cannot be ruled out due to the difficulty in predicting ice sheet behaviour (Hansen, 2007; Steffen et al., 2008; Solomon et al., 2009). In some large river deltas including those of the Mississippi, Nile and Ganges, the effective sea level rise has been considerably exacerbated by land subsidence, which can result from a variety of processes including decreased sediment loads due to upstream impoundments, oxidation of drained organic soils and localised withdrawals of groundwater, oil or gas (Ericson et al., 2006).

Large-scale changes in terrestrial vegetation may accompany climate change, particularly where droughts and seasonal water limitation increase to the point where forests are replaced by grasslands, and where the impacts of fires increase. Vegetation water use and land cover have consequences for groundwater recharge and runoff (Li et al., 2007; Costa-Cabral et al., 2007), and may also affect soil erosion and sediment transport into rivers. A substantial fraction of forest in the Amazon catchment may be vulnerable to increased water stress and fires during the dry season, even though total annual precipitation is not projected to change greatly (Bates et al., 2008; Barlow & Peres, 2008). Deforestation in the eastern Amazon has also been linked to decreased plant water use and increased runoff (Chaves et al., 2008), which in turn explain recent increases in discharge of large rivers (Coe et al., 2009). Extensive deforestation could result in regional decreases in precipitation through this feedback, however, and the net balance of these counteracting changes is likely to vary depending on catchment characteristics (Coe et al., 2009). As discussed above, such changes in vegetation would affect plant water use (evapotranspiration), and the water use efficiency of plants is also affected by increasing $CO_2$ availability. Large-scale changes in vegetation thus represent one of the strongest feedbacks to climate, particularly where changing vegetation produces large changes in carbon sequestration, water use or radiation balance (i.e. albedo), but these feedbacks are complex and incompletely understood.

## Conceptual model of biogeochemical implications of climate change

A conceptual model illustrates the major ways in which climate change is hypothesised to affect riverine and floodplain biogeochemistry (Fig. 3). For simplicity, the main driver of global change is depicted as radiative forcing and the resultant global atmospheric temperature increase. Ecosystem-wide increases in air temperature will result in myriad ecological implications, and some would propagate from the terrestrial catchments through hydrological flow paths to affect streams and rivers, in addition to changing land cover and ecosystem processes on seasonally inundated floodplains. Amongst the most notable catchment-wide, direct ecological effects of rising temperatures are increased thermal stress for

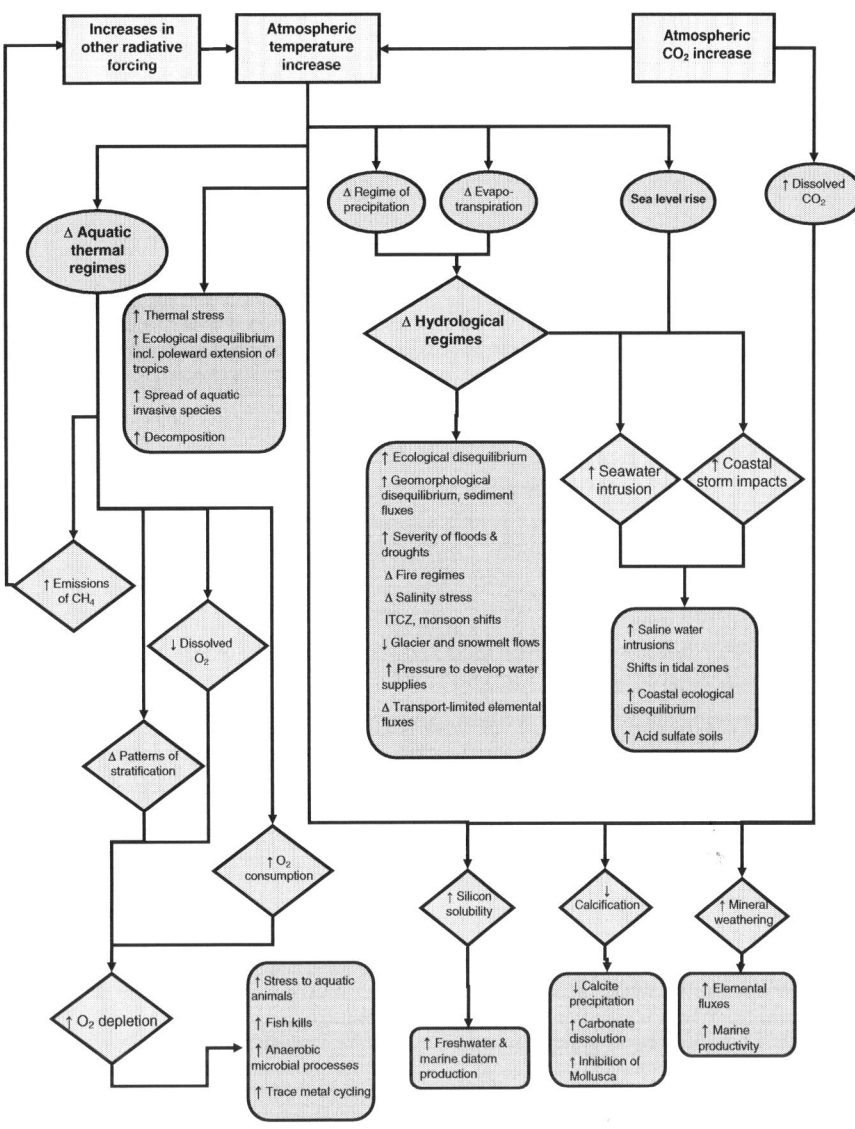

Fig. 3 Conceptual model depicting the major ways in which climate change is hypothesised to affect riverine and floodplain biogeochemistry. First-order effects of the increased atmospheric temperature on aquatic ecosystems are depicted as ovals and higher-order effects as diamonds. These effects have ecological and biogeochemical implications that are listed in the boxes. Hydrological regimes encompass the magnitude, timing, duration, frequency and rate of change of key characteristics including water level, discharge and extent of inundation (Poff et al., 1997)

some organisms, a general ecological disequilibrium producing shifts in ecosystem structure including poleward and altitudinal extension of the ranges of some tropical species, and spread of invasive species that exploit new habitats as thermal limitations are reduced. Changes in terrestrial ecosystems are important, but beyond the scope of Fig. 3, which focuses on aquatic biogeochemistry.

Based on my review of the literature, I conclude that the most broadly applicable and probably the most important effects of atmospheric temperature increases on tropical rivers and floodplains entail changes in aquatic thermal and hydrological regimes and, in low-lying coastal areas, sea level rise, which will interact with the other drivers. Increased dissolved $CO_2$ in response to higher partial pressures in overlying air is a distinct and possibly significant first-order effect that interacts with higher temperatures to affect mineral weathering and solubility, which in turn control water chemistry at the site of mineral weathering and may affect riverine solute fluxes. Each of these first-order effects and the higher-order effects that follow from them are discussed below.

## Changes in thermal regimes

Temperature exerts pervasive effects on many aspects of ecosystem structure and function including biogeochemical processes. Direct impacts of warmer temperatures operate mainly through thermodynamics and the physiological optima of organisms. Thermodynamics drive faster chemical reaction rates at higher temperatures, and many plants and animals increase their growth efficiency (production:biomass ratios) in response to increasing temperature (Talling & Lemoalle, 1998). However, as temperatures climb the point is reached where biotically mediated reactions are subject to biochemical limitations such as enzyme denaturation and loss of membrane functions. Thermal tolerances for organisms are often cited as temperatures at which they no longer function, but their thermal optima are likely well below those tolerances.

The thermal optima of vascular plants, insects and microbes may be of greatest interest in regard to biogeochemical cycles, although biotic changes at all levels of food webs can have significant biogeochemical implications. Thermal optima tend to be narrower for tropical organisms, at least for tropical ectotherms ranging from insects to vertebrates, than they are for equivalent temperate-zone biota, and the thermal response tends to drop sharply from the optimum range to the maximum tolerable temperature (Deutsch et al., 2008). Seasonal minimum temperatures may dictate the ranges of many tropical plants and animals. At high temperatures (e.g. >30°C), differences in thermal optima affect competitive dynamics amongst species, and often species diversity begins to drop around 35°C. Hence seemingly modest temperature increases in tropical environments could produce unexpectedly strong reductions or shifts in biotic composition and diversity, with unknown biogeochemical ramifications.

Heterotrophic microbial activity in natural environments tends to increase with increasing temperature, and experimental work with natural wetland soils and sediments supports this generalisation over the temperature range of interest here (i.e. 15–40°C). Overall decomposition rates increase with temperature, although the net effect for soil or sediment organic matter pools depends on whether soil moisture and redox status also change, and whether organic inputs change as a result of changes in plant production (Davidson & Janssens, 2006). Currently, there is not a consensus as to the importance of temperature increases for net ecosystem carbon balances, and most study has focused on upland soils and on northern wetland soils including permafrost. In tropical floodplains subject to seasonal inundation and desiccation, a significant increase in temperature is likely to be accompanied by changes in landscape water balance, the seasonality of soil moisture and inundation, and fire regimes, all factors which may have overriding influences on organic matter pools and fluxes compared to changes in rates of microbial metabolism.

Primary production rates of algae tend to increase with temperature in the range of 15–30°C, but may fall sharply above about 35°C, while respiration rates continue to increase (Kirk, 1994; Talling & Lemoalle, 1998). However, in most tropical waters, primary production rates often are more constrained by nutrient limitation than by temperature (Lewis, 1987). Vascular plants with leaves above water are more likely to suffer negative impacts of excessive heating, although plant reaction to overheating has been investigated mainly in terrestrial plants (Berry & Björkman, 1980). Grasses with the $C_4$ photosynthetic pathway can be important components of aquatic vegetation in tropical floodplains (e.g. Junk, 1997), and are more tolerant of high temperatures than $C_3$ plants. Thus increasing temperatures may tip the balance of competition in favour of $C_4$ grasses, which may decrease the nutritional quality of organic matter for consumers, particularly because $C_4$ grasses tend to have lower nitrogen contents (Gibson, 2009) and provide little support of aquatic food webs (Hamilton et al., 1992; Clapcott & Bunn, 2003).

Anaerobic microbial processes may be responsive to temperature increases. Denitrification seems to be a more effective sink for nitrate in warmer waters, and may contribute to the tendency for nitrogen limitation of primary production in tropical fresh waters and riverine estuaries (Talling & Lemoalle, 1998; Downing et al., 1999). Methanogenesis and iron reduction are two competing processes of anaerobic decomposition of particular importance in tropical floodplains (Roden & Wetzel, 2002; Weber et al., 2006). Both methanogenesis and iron reduction show temperature optima of 32–41°C in anoxic soils from rice paddies (Yao & Conrad, 2000; Fey et al., 2001). Although total methanogenesis increases with temperature, there is molecular and isotopic evidence for strong shifts in the phylogenetic composition of methanogenic archaea as temperatures climb above 30°C (Fey et al., 2004), and further shifts in methanogen community composition and methanogenic pathways as temperatures climb above 40°C (Fey et al., 2001; Wu et al., 2006; Conrad et al., 2009). Bacterial community structure also changes between 35 and 45°C (Conrad et al., 2009).

Based on this body of study focused especially on methanogenesis, it appears likely that heterotrophic microbial metabolism, including anaerobic processes of key interest in biogeochemistry, will increase in response to higher temperatures. Most estimates of the $Q_{10}$ for aquatic respiration, which refers to the relative increase in rate for a 10°C increase in temperature within the normal temperature range of the environment, and if all else remains unchanged, range between 2 and 3 (Fenchel, 2005). Methanogenesis can have a higher $Q_{10}$ (Roehm, 2005). There may be changes in microbial community composition, but functional redundancy may maintain rates of biogeochemical processes. Thermal stress on the upper end of the temperature range is thus more likely to affect animals and higher plants, while microbes and the processes they mediate seem more resilient within the range of water temperatures that might be experienced.

Tropical wetlands are globally important sources of methane emission to the atmosphere (Denman et al., 2007; Bergamaschi et al., 2007), and hence changes in their contribution can be important as a feedback to climate change. Increased emission of methane by tropical wetlands has been projected, mostly ascribed to higher temperatures, but also related to increased inundation (Gedney et al., 2004; Shindell et al., 2004). However, it remains uncertain whether higher methane production rates will necessarily increase methane emission rates due to the complexity of interactions amongst fermentation, methanogenesis and methane oxidation (Conrad et al., 2009).

An important class of effects of warmer temperatures on shallow tropical waters may be more indirect, involving density stratification and mixing patterns. Diurnal stratification and variable degrees of nocturnal mixing are typical of quiescent water columns that are less than 4–5 m deep, and can be marked even when the water is <1 m deep. The propensity for diurnal stratification to develop is greater at higher temperatures due to the temperature–density behaviour of water, which results in greater stability changes for a given heat flux as temperatures increase (Lewis, 1987). A diurnally stratified water column intensifies heat accumulation near the surface, and is more likely to result in thermal stress. Increased occurrence and intensity of diurnal stratification enhances the potential for biogeochemical differentiation between surface and bottom waters, which in turn can increase nutrient limitation of surface waters and oxygen depletion of deeper waters. A sequence of days with incomplete nocturnal mixing can result in anoxic waters and lead to fish kills (Talling & Lemoalle, 1998). In tropical lakes, seasonal changes in wind regimes also can be important for stratification and mixing patterns (Talling & Lemoalle, 1998), although wind may be less important as an agent of mixing in shallow and often vegetated waters of floodplains than it is in larger lakes.

Consumption of dissolved $O_2$ proceeds faster and the solubility of $O_2$ is lower at warmer temperatures, and these factors combine to make $O_2$ depletion more likely (Lewis, 1987), with consequences for aquatic animals, the rates and nature of microbial processes and biogeochemical cycles of redox-sensitive elements including many trace metals (Hamilton et al., 1997). Mercury methylation and consequent contamination of food webs may increase with temperature as well (Mauro et al., 1999). The net effect of these changes may be altered species compositions of plants and animals, accompanied by accelerated microbial processes and elemental cycling, with uncertain implications for nutrient limitation and aquatic ecosystem productivity in shallow waters.

**Changes in hydrological regimes**

The effects of altered hydrological regimes on biogeochemical processes in rivers and floodplains may well prove to be more marked than the effects of warmer water temperatures due to the overriding importance of seasonal inundation patterns in dictating ecological and biogeochemical features of these ecosystems (Junk, 1997; Hamilton, 2008b). Changes in hydrological regimes under a warmer climate begin with the landscape water balance, which responds to changes in precipitation regimes in combination with higher evapotranspiration at warmer temperatures (see above discussion). Thus magnitude and timing of upland runoff, river discharge and floodplain inundation regimes are all subject to change.

Some major implications of altered hydrological regimes are listed in Fig. 3, although the importance of several of these depends on the setting, and they are not unique to tropical river systems. In addition to ecological disequilibria produced by changes in discharge and flood regimes, geomorphological disequilibria invoked by changes in runoff and river discharge can result in enhanced sediment fluxes from uplands to rivers and between rivers and their floodplains and coastal zones (Ericson et al., 2006; Walling, 2006; Syvitsky et al., 2005). Natural sediment fluxes can be important in nutrient supply as well as in maintaining geomorphologically dynamic environments that often support vegetation in early successional stages. However, enhanced sediment and nutrient fluxes due to changing human land use or climate can result in undesirable geomorphic and ecological disequilibria. Day et al. (2008) discuss implications of changes in inputs of freshwater, sediments and nutrients to coastal zones. Changes in the amount of water moving from upland catchments into river systems usually result in changes in dissolved elemental fluxes as well (Neill et al., 2006), particularly for cases where hydrological transport of solutes in soils and groundwater is the limiting factor (e.g. carbonate mineral dissolution; Raymond et al., 2008).

The vegetation and organic carbon stocks of floodplain ecosystems tend to be sensitive to particularly severe floods or dry periods, which can create new vegetation successional processes (e.g. forest dieback or plant colonisation of emergent sediments). Droughts and fires can rapidly eliminate years of organic matter accumulation in normally saturated soils or sediments, as has recently been documented in Indonesian peatlands (Ballhorn et al., 2009), and may also release methane stored as occluded gas bubbles.

## Changes in salinity, tidal influence and marine storms

Salinity is a powerful agent of change in ecological and biogeochemical processes, and in coastal zones, the boundary between saline and fresh waters usually involves regular or episodic seawater penetration (see below). Inland waters also can become seasonally saline by evaporative concentration, typically accompanied by dramatic shifts in biotic composition and production, sediment-water exchanges and nutrient cycling. Salinisation via evaporative concentration is particularly common in arid and semi-arid climates, but can also occur to some degree in more humid climates where broad shallow pools remain isolated after floodplain inundation (Talling & Lemoalle, 1998). Changes in landscape water balances can change the distribution and timing of salinisation of surface waters (Timms, 1999; Hamilton et al., 2005). In addition, some dryland regions are prone to secondary soil salinisation when land cover changes from woodland to grassland or crops, which alters the soil hydrology such that naturally existing soluble salts are displaced from deeper soils towards the surface (Abrol et al., 1988).

Where large rivers flow into oceans, climate change will have profound implications for biogeochemical processes at the freshwater/seawater interface because gradients of salinity are fundamentally important to ecosystem structure and function (Bianchi, 2007). River deltas reflect the balance between fluvial inputs and marine influences including particularly sea level rise, and modern deltas have formed during the current period of relatively stable sea levels, since sea level rise decelerated in the lower Holocene (Ericson et al., 2006). Day et al. (2008) consider three classes of drivers of importance to coastal wetlands, in general, all of which apply to tropical systems and are potentially affected by climate change: sea level rise, changes in storm frequency and duration and changes in freshwater and sediment inputs. These drivers interact to structure the geomorphology and ecology of coastal wetlands, including freshwater portions of river deltas as well as coastal zones in the area of influence of river discharge and sediment loading, which can extend a considerable distance from large river mouths. Tropical river systems include very extensive coastal deltaic wetlands, such as the deltas of the Amazon, Orinoco and La Plata rivers in South America, the Ganges-Brahmaputra and Irawaddy rivers in Asia and the Gulf of Carpenteria region in northern Australia.

Estuarine and coastal wetlands and floodplains usually exist across freshwater–seawater interfaces, reflecting a precarious geomorphological and ecological setting developed over several 1,000 years of relatively stable sea levels. The most certain prediction of how climate change will affect ecosystems across these salinity gradients is via sea level rise, although as noted above the projected timing and magnitude of sea level rise remain a topic of debate. Increased intensity of tropical storms is another relatively confident prediction derived from climate change models. These two drivers interact because storm surges will penetrate further inland with higher sea levels. Effective sea level rise from a combination of climate change and land subsidence results in inundation of coastal wetlands, seawater intrusion into surface and groundwater, and increased vulnerability to storm surges (Ericson et al., 2006). Inland migration of these coastal ecosystems will not always be possible in the face of natural geomorphological barriers as well as anthropogenic land use that is often intense along coasts. The floodplains of the Mary River in northern Australia near Darwin are an example where modest changes in topography allowed salinity intrusion via tidal creeks that dramatically altered coastal freshwater wetlands, killing vast stands of paperbark trees (*Melaleuca* spp.; Knighton et al., 1991).

Changes in river hydrological regimes are a complicating factor that can interact with rising sea levels. The complex and interactive effects of storm surges, discharge variability and river channel modifications are exemplified by the Mekong River delta (Wassmann et al., 2004). Decreased discharge (especially during low-discharge periods) worsens the impacts of salinity intrusions. Increased river discharge as a result of climate change could help reduce salinity intrusion and may be associated with higher sediment loads that foster vegetation growth (including particularly mangroves) and land accretion, which in turn may counteract the effects of rising sea levels and provide more protection from storms (Day et al., 2008).

Decomposition rates are lower in fresh waters than brackish waters of comparable hydrology (Craft, 2007), and any seawater influence can elevate the importance of sulphur biogeochemistry including sulphate reduction. Low-lying coastal land that is most subject to flooding by sea level rise may contain former marine sediments known as acid sulphate soils, and when rewetted such sediments can produce acidic conditions associated with oxidation of sulphide minerals, with negative impacts on freshwater and terrestrial biota (Fitzpatrick & Shand, 2008). Sulphur-linked production of acidity and mobilisation of toxic metals such as Al has been documented in tropical catchments of northern Australia where it is associated with freshwater fish kills (Hart & McKelvie, 1986) as well as negative impacts on coastal marine biota where rivers drain into the Great Barrier Reef ecosystem (Powell & Martens, 2004).

Riverine transport of nutrients to the oceans, which is susceptible to alteration with changes in river hydrological regimes and in the ecosystems they drain, affects marine primary production in coastal waters. Nitrogen delivery is particularly important to coastal marine production (Cotrim da Cunha et al., 2008). Phosphorus, silicon and iron fluxes also affect marine primary production, with effects potentially extending far from river mouths (Subramaniam et al., 2008). Also, deoxygenation of coastal waters is promoted by riverine contributions of nutrients to support phytoplankton growth and of detrital organic carbon, both of which lead to increased respiratory oxygen demands (Cotrim da Cunha et al., 2008).

**Changes in mineral dissolution and precipitation**

Figure 3 includes several specific chemical weathering processes that are directly influenced by temperature and dissolved $CO_2$, and are of interest due to their biogeochemical implications. Dissolution of minerals responds to increased temperature as well as to pH (White & Blum, 1995), and in most natural waters pH tends to be inversely proportional to dissolved $CO_2$ because $CO_2$ reacts in water to produce carbonic acid.

Chemical weathering of silicate minerals is of particular interest as it represents a long-term sink for atmospheric $CO_2$, and because it is the source of base cations, alkalinity and dissolved silicate in river waters. Silicate is of particular interest because its availability affects the growth of diatoms, which are ubiquitous and abundant forms of algae in fresh and marine waters. Tropical rivers account for $\sim 70\%$ of global silicon fluxes from land to oceans (Jennerjahn et al., 2006). Silicate mineral dissolution increases with temperature and carbonic acid in water that contacts soils and rocks, and thus the concentrations of dissolved silicate as well as base cations and alkalinity transported to surface waters may be expected to increase in a warmer and $CO_2$-enriched climate. However, changes in erosion rates and hydrologic flows through soils and rocks are also likely to be important controls on riverine export of solutes from silicate weathering (Kump et al., 2000; Jennerjahn et al., 2006; Subramanian et al., 2006).

Carbonate mineral weathering is less common in many highly weathered tropical landscapes, but is important to riverine chemistry even where only modest amounts of limestone or dolomite exist, strongly affecting the dissolved inorganic carbon and alkalinity concentrations in river waters. In contrast to silicate minerals, carbonate minerals are less soluble at higher temperatures, but more soluble with increased dissolved $CO_2$ (Kump et al., 2000), and their solubility is more likely to be transport-limited (i.e. limited by rate of hydrologic flow through soils and ground water reservoirs). Hence, the net impacts of interacting increases in temperature and $CO_2$ plus accompanying hydrological changes on carbonate mineral weathering are hard to determine.

Subsequent precipitation of calcium carbonate in lakes and marine environments is subject to similar controls, with aquatic metabolism acting as an important regulator of dissolved $CO_2$ concentrations in productive fresh waters. Secondary carbonate precipitation in surface waters has numerous biogeochemical implications, affecting phosphorus and trace metal cycling (Kelts & Hsü, 1978; Hamilton et al., 2009). Many large tropical river systems in humid climates, including the Amazon, Paraná and Orinoco rivers, carry waters that are dilute in dissolved ions and highly supersaturated with dissolved $CO_2$, and therefore are not subject to secondary precipitation of calcium carbonate in their freshwater reaches (Kempe, 1982; Oliveira et al., 2010).

Changes in carbonate dissolution and precipitation impact carbon and other biogeochemical cycles directly, but, in addition, they affect calcification by molluscs in fresh waters of rivers and floodplains, where these organisms can play pivotal roles as grazers of phytoplankton and attached algae (Strayer et al., 1999). Relatively acidic fresh waters in some tropical floodplains including a number of rivers in the Amazon River system support low species diversity of molluscs, possibly reflecting the difficulty of biogenic calcification in such waters (Dillon, 2000; Oliveira et al., 2010). In marine environments, the acidifying effect of increased $CO_2$ overrides the counteracting effect of increasing temperatures, producing potentially severe implications for calcifying organisms including corals and plankton (Raven, 2005).

## Conclusions

The prospect of anthropogenically driven global climate change has many potential implications for biogeochemical processes in tropical rivers and floodplains. The most important changes are likely to involve (Fig. 3):

(1) Aquatic thermal regimes, with implications for thermal optima of plants and animals, rates of microbially mediated biogeochemical transformations, density stratification of water bodies and dissolved oxygen depletion.
(2) Hydrological regimes of discharge and floodplain inundation, which determine the ecological structure and function of rivers and floodplains and the extent and seasonality of aquatic environments.
(3) Freshwater–seawater gradients where rivers meet oceans, affecting the distribution of marine, brackish and freshwater environments and the biogeochemical processing of river water reaching the coastal zone.

These changes will, however, be affected by complex human interactions and feedbacks which are difficult to forecast. In river systems that experience decreasing or more variable hydrological regimes, water withdrawals are likely to increase, further exacerbating the hydrologic changes (Alcamo et al., 2007; Kundzewicz et al., 2008). Tropical regions predicted to suffer the greatest water stress for human populations include southern Mexico, northeastern Brazil, parts of northern and southern Africa, and India, driven particularly by increased human demands, with climate change a secondary factor that alleviates increased demand in some cases (Alcamo et al., 2007). Important stressors that can act synergistically with climate change include population increases and urbanisation. These need to be considered in developing strategies for adaptation to climate change (Palmer et al., 2008).

Future research should strive to improve detection of current trends and projection of future changes in catchment and river hydrology, with greater attention to climate feedbacks and human interactions. Limnological research should include sustained monitoring of thermal and mixing regimes not only in large, deep lakes and rivers but also in the shallow and often ephemeral waters of floodplains and wetlands. Thermal optima and responses of microbes (and the processes they mediate), plants and animals in the upper temperature range (e.g. 30–40°C) is a topic that deserves greater attention as well.

**Acknowledgements** Ralf Conrad, Michael Hobbins, Jeff Shellberg and Tim Jardine provided advice on specific aspects. Jeff Shellberg provided temperature data for the Mitchell River, Australia. The author was supported during the preparation of this review by a Commonwealth Environmental Research Facilities Fellowship from the Australian government, as well as by the Australian Rivers Institute of Griffith University and the Tropical Rivers and Coastal Knowledge (TRaCK) programme. TRaCK receives major funding for its research through the Australian Government's Commonwealth Environment Research Facilities initiative; the Australian Government's Raising National Water Standards Programme; Land and Water Australia; and the Queensland Government's Smart State Innovation Fund.

## References

Abrol, I. P., J. S. P. Yadav & F. I. Massoud, 1988. Salt-affected soils and their management. FAO Soils Bulletin 39. Food and Agriculture Organisation of the United Nations, Rome [available on internet at http://www.fao.org/docrep/x5871e/x5871e00.htm].

Aerts, J. C. J. H., H. Renssen, P. J. Ward, H. de Moel, E. Odada, L. M. Bouwer & H. Goosse, 2006. Sensitivity of global river discharges under Holocene and future climatic conditions. Geophysical Research Letters 33. doi:10.1029/2006GL027493.

Alcamo, J., M. Florke & M. Marker, 2007. Future long-term changes in global water resources driven by socio-economic and climatic changes. Hydrological Sciences Journal – Journal des Sciences Hydrologiques 52: 247–275.

Ballhorn, U., F. Siegert, M. Mason & S. Liwin, 2009. Derivation of burn scar depths and estimation of carbon emissions with LIDAR in Indonesian peatlands. Proceedings of the National Academy of Sciences United States of America, published online before print November 25, 2009. doi:10.1073/pnas.0906457106.

Barlow, J. & C. Peres, 2008. Fire-mediated dieback and compositional cascade in an Amazonian forest. Philosophical Transactions of the Royal Society B – Biological Sciences 363: 1787–1794.

Bates, B. C., Z. W. Kundzewicz, S. Wu & J. P. Palutikof (eds), 2008. Climate Change and Water Technical Paper of the Intergovernmental Panel on Climate Change. IPCC Secretariat, Geneva.

Bergamaschi, P., C. Frankenberg, J. F. Meirink, M. Krol, F. Dentener, T. Wagner, U. Platt, J. O. Kaplan, S. Korner, M. Heimann, E. J. Dlugokencky & A. Goede, 2007. Satellite chartography of atmospheric methane from SCIAMACHY on board ENVISAT: 2. Evaluation based on inverse model simulations. Journal of Geophysical Research 112: D02304. doi:10.1029/2006JD007268.

Berry, J. & O. Björkman, 1980. Photosynthetic response and adaptation to temperature in higher plants. Annual Review of Plant Physiology 31: 491–543.

Betts, R. A., O. Boucher, M. Collins, P. M. Cox, P. D. Falloon, N. Gedney, D. L. Hemming, C. Huntingford, C. D. Jones, D. M. H. Sexton & M. Webb, 2007. Projected increase in continental runoff due to plant responses to increasing carbon dioxide. Nature 448: 1037–1041.

Bianchi, T. S., 2007. Biogeochemistry of Estuaries. Oxford University Press, Oxford.

Bradley, R. S., M. Vuille, H. F. Diaz & W. Vergara, 2006. Threats to water supplies in the tropical Andes. Science 312: 1755–1756.

Cai, W. & T. Cowan, 2008. Evidence of impacts from rising temperature on inflows to the Murray-Darling Basin. Geophysical Research Letters 35: L07701. doi:10.1029/2008GL033390.

CCSP, 2008. Abrupt Climate Change. A report by the U.S. Climate Change Science Program and the Subcommittee on Global Change Research (Clark, P. U., A. J. Weaver (coordinating lead authors), E. Brook, E. R. Cook, T. L. Delworth & K. Steffen (chapter lead authors)). U.S. Geological Survey, Reston, VA: 459.

Chaves, J., C. Neill, H. Elsenbeer, A. Krusche, S. Germer & S. Gouveia Neto, 2008. Land management impacts on runoff sources in small Amazon watersheds. Hydrological Processes 22: 1766–1775.

Clapcott, J. E. & S. E. Bunn, 2003. Can $C_4$ plants contribute to the aquatic food webs of subtropical streams? Freshwater Biology 48: 1105–1116.

Coe, M. T., et al., 2009. The Influence of historical and potential future deforestation on the stream flow of the Amazon River – land surface processes and atmospheric feedbacks. Journal of Hydrology 369: 165–174.

Conrad, R., M. Close & M. Noll, 2009. Functional and structural response of the methanogenic microbial community in rice field soil to temperature change. Environmental Microbiology. doi:10.1111/j.1462-2920.2009.01909.x.

Cook, E. R., P. J. Bartlein, N. Diffenbaugh, R. Seager, B. N. Shuman, R. S. Webb, J. W. Williams & C. Woodhouse, 2008. Hydrological variability and change. In Abrupt Climate Change. A Report by the U.S. Climate Change Science Program and the Subcommittee on Global Change Research. U.S. Geological Survey, Reston, VA: 143–257.

Costa-Cabral, M. C., J. E. Richey, G. Goteti, D. P. Lettenmaier, C. Feldkötter & A. Snidvongs, 2007. Landscape structure and use, climate, and water movement in the Mekong River basin. Hydrological Processes 22: 1731–1746.

Cotrim da Cunha, L., E. T. Buitenhuis, C. Le Quéré, X. Giraud & W. Ludwig, 2008. Potential impact of changes in river nutrient supply on global ocean biogeochemistry. Global Biogeochemical Cycles 21: GB 407. doi:10.1029/2006 GB002718.

Craft, C., 2007. Freshwater input structures soil properties, vertical accretion, and nutrient accumulation of Georgia and U.S tidal marshes. Limnology and Oceanography 52: 1220–1230.

Dai, A., K. E. Trenberth & T. Qian, 2004. A global dataset of Palmer Drought Severity Index for 1870–2002: relationship with soil moisture and effects of surface warming. Journal of Hydrometeorology 5: 1117–1130.

Davidson, E. A. & I. A. Janssens, 2006. Temperature sensitivity of soil carbon decomposition and feedbacks to climate change. Nature 440: 165–173.

Day, J. W., R. R. Christian, D. M. Boesch, A. Yáñez-Arancibia, J. Morris, R. R. Twilley, L. Naylor, L. Schaffner & C. Stevenson, 2008. Consequences of climate change on the ecogeomorphology of coastal wetlands. Estuaries and Coasts 31: 477–491.

de Melo, S. & V. L. M. Huszar, 2000. Phytoplankton in an Amazonian flood-plain lake (Lago Batata, Brasil): diel variation and species strategies. Journal of Plankton Research 22: 63–76.

Denman, K. L., G. Brasseur, A. Chidthaisong, P. Ciais, P. M. Cox, R. E. Dickinson, D. Hauglustaine, C. Heinze, E. Holland, D. Jacob, U. Lohmann, S. Ramachandran, P. L. da Silva Dias, S. C. Wofsy & X. Zhang, 2007. Couplings between changes in the climate system and biogeochemistry. In Solomon, S., D. Qin, M. Manning, Z. Chen, M. Marquis, K. B. Averyt, M. Tignor & H. L. Miller (eds), Climate Change 2007: The Physical Science Basis. Contribution of Working Group I to the Fourth Assessment Report of the Intergovernmental Panel on Climate Change. Cambridge University Press, Cambridge, UK.

Deutsch, C. A., J. J. Tewksbury, R. B. Huey, K. S. Sheldon, C. K. Ghalambor, D. C. Haak & P. R. Martin, 2008. Impacts of climate warming on terrestrial ectotherms across latitude. Proceedings of the National Academy of Sciences 105: 6668–6672.

Dillon, R. T. Jr., 2000. The Ecology of Freshwater Molluscs. Cambridge University Press, Cambridge, UK.

Downing, J. A., M. McClain, R. Twilley, J. M. Melack, J. Elser, N. N. Rabalais, W. M. Lewis Jr., R. E. Turner, J. Corredor, D. Soto, A. Yañez-Arancibia, J. Kopaska & R. W. Howarth, 1999. The impact of accelerating land-use change on the N-cycle of tropical aquatic ecosystems: current conditions and projected changes. Biogeochemistry 46: 109–148.

Ericson, J. P., C. J. Vörösmarty, S. L. Dingman, L. G. Ward & M. Meybeck, 2006. Effective sea-level rise and deltas: causes of change and human dimension implications. Global and Planetary Change 50: 63–82.

Fenchel, T., 2005. Respiration in aquatic protists. In del Giorgio, P. A. & P. J. B. le Williams (eds), Respiration in Aquatic Ecosystems. Oxford University Press, Oxford: 47–56.

Fey, A., K. J. Chin & R. Conrad, 2001. Thermophilic methanogens in rice field soil. Environmental Microbiology 3: 295–303.

Fey, A., P. Claus & R. Conrad, 2004. Temporal change of $^{13}$C-isotope signatures and methanogenic pathways in rice field soil incubated anoxically at different temperatures. Geochimica et Cosmochimica Acta 68: 293–306.

Fitzpatrick, R. & P. Shand, 2008. Inland acid sulfate soils: Overview and conceptual models. In Fitzpatrick, R. & P. Shand (eds), Inland Acid Sulfate Soil Systems Across Australia. CRC LEME Open File Report No. 249 (Thematic Volume). CRC LEME, Perth, Australia: 6–74 [available on internet at http://www.clw.csiro.au/acids ulfatesoils/documents/ass-book/Ch1-Inland-ASS.pdf].

Foley, J. A., A. Botta, M. T. Coe & M. H. Costa, 2002. El Niño-Southern Oscillation and the climate, ecosystems and rivers of Amazonia. Global Biogeochemical Cycles 16(4): Art. No. 1132.

Gedney, N., P. M. Cox & C. Huntingford, 2004. Climate feedback from wetland methane emissions. Geophysical Research Letters 31: L20503. doi:10.1029/2004 GL020919.

Gedney, N., P. M. Cox, R. A. Betts, O. Boucher, C. Huntingford & P. A. Stott, 2006. Detection of a direct carbon dioxide effect in continental river runoff records. Nature 439: 835–838. doi:10.1038/nature04504.

Gerten, D., S. Rost, W. Von Bloh & W. Lucht, 2008. Causes of change in 20th century global river discharge. Geophysical Research Letters 35: L20405. doi:10.1029/2008 GL035258.

Gibson, D. J., 2009. Grasses and Grassland Ecology. Oxford University Press, Oxford.

Goudie, A. S., 2006. Global warming and fluvial geomorphology. Geomorphology 79: 384–394.

Hamilton, S. K., 2008a. Flood plains. In Likens, G. E. (ed.), Encyclopedia of Inland Waters. Elsevier, Oxford.

Hamilton, S. K., 2008b. Floodplain wetlands of large river systems. In Likens, G. E. (ed.), Encyclopedia of Inland Waters. Elsevier, Oxford.

Hamilton, S. K. & W. M. Lewis Jr., 1990. Basin morphology in relation to chemical and ecological characteristics of lakes on the Orinoco River floodplain, Venezuela. Archiv für Hydrobiologie 119: 393–425.

Hamilton, S. K., S. J. Sippel, W. M. Lewis Jr. & J. F. Saunders III, 1990. Zooplankton abundance and evidence for its reduction by macrophyte mats in two Orinoco floodplain lakes. Journal of Plankton Research 12: 345–363.

Hamilton, S. K., W. M. Lewis Jr. & S. J. Sippel, 1992. Energy sources for aquatic animals on the Orinoco River floodplain: Evidence from stable isotopes. Oecologia 89: 324–330.

Hamilton, S. K., S. J. Sippel & J. M. Melack, 1995. Oxygen depletion and carbon dioxide and methane production in waters of the Pantanal wetland of Brazil. Biogeochemistry 30: 115–141.

Hamilton, S. K., S. J. Sippel & J. M. Melack, 1996. Inundation patterns in the Pantanal wetland of South America determined from passive microwave remote sensing. Archiv für Hydrobiologie 137: 1–23.

Hamilton, S. K., S. J. Sippel, D. F. Calheiros & J. M. Melack, 1997. An anoxic event and other biogeochemical effects of the Pantanal wetland on the Paraguay River. Limnology and Oceanography 42: 257–272.

Hamilton, S. K., S. J. Sippel & J. M. Melack, 2002. Comparison of inundation patterns in South American floodplains. Journal of Geophysical Research 107(D20): Art. No. 8038. Available in electronic form; doi 10.1029/2000JD000306.

Hamilton, S. K., S. E. Bunn, M. Thoms & J. C. Marshall, 2005. Persistence of aquatic refugia between flow pulses in a dryland river system (Cooper Creek, Australia). Limnology and Oceanography 50: 743–754.

Hamilton, S. K., D. A. Bruesewitz, G. P. Horst & O. Sarnelle, 2009. Biogenic calcite-phosphorus precipitation as a negative feedback to lake eutrophication. Canadian Journal of Fisheries and Aquatic Sciences 66: 321–342.

Hansen, J. E., 2007. Scientific reticence and sea level rise. Environmental Research Letters 2. doi:10.1088/1748-9326/2/2/024002.

Hart, B. T. & I. D. McKelvie, 1986. Chemical limnology in Australia. In De Decker, P. & W. D. Williams (eds), Limnology in Australia. CSIRO, Melbourne & Dr. W. Junk Publishers, Dordrecht.

Heckman, C. W., 1979. Rice Field Ecology in Northeastern Thailand. In Illies, J. (ed.), Monographiae Biologicae, Vol. 34. Dr. W. Junk Publishers, The Hague.

Hobbins, M. T., J. A. Ramírez & T. C. Brown, 2004. Trends in pan evaporation and actual evapotranspiration across the conterminous U.S.: paradoxical or complementary? Geophysical Research Letters 31: L13503. doi:10.1029/2004GL019846.

Hobbins, M. T., A. Dai, M. L. Roderick & G. D. Farquhar, 2008. Revisiting the parameterization of potential evaporation as a driver of long-term water balance trends. Geophysical Research Letters 35: L12403. doi:10.1029/2008GL033840.

Horton, R., C. Herweijer, C. Rosenzweig, J. Liu, V. Gornitz & A. C. Ruane, 2008. Sea level rise projections for current generation CGCMs based on the semi-empirical method. Geophysical Research Letters 35: L02715. doi:10.1029/2007GL032486.

IPCC, 2007. Climate Change 2007: Synthesis Report. Contribution of Working Groups I, II and III to the Fourth Assessment Report of the Intergovernmental Panel on Climate Change (Core Writing Team, R. K. Pachauri & A. Reisinger (eds)). IPCC, Geneva.

Jennerjahn, T. C., B. A. Knoppers, W. F. L. Souza, G. J. Brunskill, I. L. Silva & S. Adi, 2006. Factors controlling dissolved silica in tropical rivers. In Ittekkot, V., D. Unger, C. Humborg & N. Tac An (eds), The Silicon Cycle: Human Perturbations and Impacts on Aquatic Systems. Island Press, Washington, DC: 29–52.

Junk, W. J. (ed.), 1997. The Central Amazon Floodplain: Ecology of a Pulsing System. Ecological Studies 126. Springer, New York.

Kelts, K. & K. J. Hsü, 1978. Freshwater carbonate sedimentation. In Lerman, A. (ed.), Lakes: Chemistry, Geology, Physics. Springer-Verlag, New York: 292–323.

Kempe, S., 1982. Long-term records of $CO_2$ pressure fluctuations in fresh waters. SCOPE/UNEP Sonderband 52: 91–332.

Khan, S., A. R. Ganguly, S. Bandyopadhyay, S. Saigal, D. J. Erickson III, V. Protopopescu & G. Ostrouchov, 2006. Nonlinear statistics reveals stronger ties between ENSO and the tropical hydrological cycle. Geophysical Research Letters 33: L24402. doi:10.1029/2006GL027941.

Kirk, J. T. O., 1994. Light and Photosynthesis in Aquatic Ecosystems, 2nd ed. Cambridge University Press, Cambridge: 509.

Knighton, A. D., K. Mills & C. Woodroffe, 1991. Tidal-creek extension and saltwater intrusion in northern Australia. Geology 19: 831–834.

Kump, L. R., S. L. Brantley & M. A. Arthur, 2000. Chemical weathering, atmospheric $CO_2$, and climate. Annual Reviews of Earth and Planetary Science 28: 611–667.

Kundzewicz, Z. W., L. J. Mata, N. Arnell, P. Döll, P. Kabat, B. Jiménez, K. Miller, T. Oki, Z. Şen & I. Shiklomanov, 2008. The implications of projected climate change for freshwater resources and their management. Hydrological Sciences Journal – Journal des Sciences Hydrologiques 53: 3–10.

Labat, D., Y. Godderis, J. L. Probst & J. L. Guyot, 2004. Evidence for global runoff increase related to climate warming. Advances in Water Resources 27: 631–642.

Lewis, W. M. Jr., 1987. Tropical limnology. Annual Review of Ecology and Systematics 18: 159–184.

Lewis, W. M. Jr., S. K. Hamilton, M. A. Lasi, M. Rodríguez & J. F. Saunders III, 2000. Ecological determinism on the Orinoco floodplain. Bioscience 50: 681–692.

Li, K. Y., M. T. Coe, N. Ramankutty & R. De Jong, 2007. Modeling the hydrological impact of land-use change in West Africa. Journal of Hydrology 337: 258–268.

Marengo, J. A., C. A. Nobre, J. Tomasella, M. D. Oyama, G. S. de Oliveira, R. de Oliveira, H. Camargo, L. M. Alves & I. F. Brown, 2008. The drought of Amazonia in 2005. Journal of Climatology 21: 495–516. doi:10.1175/2007JCLI1600.1.

Mauro, J. B. N., J. R. D. Guimarães & R. Melamed, 1999. Mercury methylation in a tropical macrophyte: influence of abiotic parameters. Applied Organometallic Chemistry 13: 631–636.

Meehl, G. A., T. F. Stocker, W. D. Collins, P. Friedlingstein, A. T. Gaye, J. M. Gregory, A. Kitoh, R. Knutti, J. M. Murphy, A. Noda, S. C. B. Raper, I. G. Watterson, A. J. Weaver & Z.-C. Zhao, 2007. Global climate projections. In Solomon, S., D. Qin, M. Manning, Z. Chen, M. Marquis, K. B. Averyt, M. Tignor & H. L. Miller (eds), Climate Change 2007: The Physical Science Basis. Contribution of Working Group I to the Fourth Assessment Report of the Intergovernmental Panel on Climate Change. Cambridge University Press, Cambridge, UK.

Melack, J. M., L. L. Hess, M. Gastil, B. R. Forsberg, S. K. Hamilton, I. B. T. Lima & E. M. L. M. Novo, 2004. Regionalization of methane emissions in the Amazon Basin with microwave remote sensing. Global Change Biology 10: 530–544.

Mertes, L. A. K., 2000. Inundation hydrology. In Wohl, E. E. (ed.), Inland Flood Hazards: Human, Riparian, and Aquatic Communities. Cambridge University Press, Cambridge, UK: 145–166.

Milliman, J. D., K. L. Farnsworth, P. D. Jones, K. H. Lu & L. C. Smith, 2008. Climatic and anthropogenic factors affecting river discharge to the global ocean, 1951–2000. Global and Planetary Change 62(2008): 187–194.

Milly, P. C. D., K. A. Dunne & V. Vecchia, 2005. Global pattern of trends in streamflow and water availability in a changing climate. Nature 438: 347–350.

Neill, C., H. Elsenbeer, A. V. Krutsche, J. Lehmann, D. Markewitz & R. O. Figueiredo, 2006. Hydrological and biogeochemical processes in a changing Amazon: results from small watershed studies and the large-scale biosphere-atmosphere experiment. Hydrological Processes 20: 2467–2476.

Nijssen, B., G. M. O'Donnell, A. F. Hamlet & D. P. Lettenmaier, 2001. Hydrologic sensitivity of global rivers to climate change. Climatic Change 50: 143–175.

Nilsson, C., C. A. Reidy, M. Dynesius & C. Revenga, 2005. Fragmentation and flow regulation of the world's large river systems. Science 308: 405–408.

Nohara, D., A. Kitoh, M. Hosaka & T. Oki, 2006. Impact of climate change on river discharge projected by multimodel ensemble. Journal of Hydrometeorology 7: 1076–1089.

Oliveira, M. D., S. K. Hamilton & C. M. Jacobi, 2010. Forecasting the expansion of the invasive golden mussel *Limnoperna fortunei* in Brazilian and North American rivers based on its occurrence in the Paraguay River and Pantanal wetland of Brazil. Aquatic Invasions 5(1). doi:10.3391/ai.2010.5.1.

Palmer, M. A., C. A. Reidy Liermann, C. Nilsson, M. Flörke, J. Alcamo, P. S. Lake & N. Bond, 2008. Climate change and the world's river basins: anticipating management options. Frontiers in Ecology and the Environment 6. doi:10.1890/060148.

Peel, M. C. & T. A. McMahon, 2006. A quality-controlled global runoff data set. Nature 444: E14.

Piao, S., P. Friedlingstein, P. Ciais, N. de Noblet-Ducoudre, D. Labat & S. Zaehle, 2007. Changes in climate and land use have a larger direct impact than rising $CO_2$ on global river runoff trends. Proceedings of the National Academy of Sciences 104: 15242–15247.

Poff, N. L., J. D. Allan, M. B. Bain, J. R. Karr, K. L. Prestegaard, B. D. Richter, R. E. Sparks & J. C. Stromberg, 1997. The natural flow regime. Bioscience 47: 769–784.

Powell, B. & M. Martens, 2004. A review of acid sulfate soil impacts, actions and policies that impact on water quality in Great Barrier Reef catchments, including a case study on remediation at East Trinity. Marine Pollution Bulletin 51: 149–164.

Ramberg, L., P. Wolski & M. Krah, 2006. Water balance and infiltration in a seasonal floodplain in the Okavango Delta, Botswana. Wetlands 26: 677–690.

Raven, J. A., 2005. Ocean Acidification Due to Increasing Atmospheric Carbon Dioxide. Policy Document 12/05. The Royal Society, London, UK.

Raymond, P. A., N.-H. Oh, R. E. Turner & W. Broussard, 2008. Anthropogenically enhanced fluxes of water and carbon from the Mississippi River. Nature 451: 449–452.

Roden, E. E. & R. G. Wetzel, 2002. Competition between Fe(III)-reducing and methanogenic bacteria for acetate in iron-rich freshwater sediments. Microbial Ecology 45: 252–258.

Roehm, C. L., 2005. Respiration in wetland ecosystems. In del Giorgio, P. A. & P. J. B. le Williams (eds), Respiration in Aquatic Ecosystems. Oxford University Press, Oxford: 83–102.

Sheffield, J. & E. F. Wood, 2008. Global trends and variability in soil moisture and drought characteristics, 1950–2000, from observation-driven simulations of the terrestrial hydrologic cycle. Journal of Climate 21: 432–453.

Shindell, D. T., B. P. Walter & G. Faluvegi, 2004. Impacts of climate change on methane emissions from wetlands. Geophysical Research Letters 31: L21202. doi:10.1029/2004GL021009.

Solomon, S., G.-K. Plattner, R. Knutti & P. Friedlingstein, 2009. Irreversible climate change due to carbon dioxide emissions. Proceedings of the National Academy of Sciences 106: 1704–1709.

Steffen, K., P. U. Clark, J. G. Cogley, D. Holland, S. Marshall, E. Rignot & R. Thomas, 2008. Rapid changes in glaciers and ice sheets and their impacts on sea level. In Abrupt Climate Change. A Report by the U.S. Climate Change Science Program and the Subcommittee on Global Change Research. U.S. Geological Survey, Reston, VA: 60–142.

Strayer, D. L., N. F. Caraco, J. J. Cole, S. Findlay & M. L. Pace, 1999. Transformation of freshwater ecosystems by bivalves: a case study of zebra mussels in the Hudson River. Bioscience 49: 19–27.

Subramaniam, A., P. L. Yager, E. J. Carpenter, C. Mahaffey, K. Björkman, S. Cooley, A. B. Kustka, J. P. Montoya, S. A. Sañudo-Wilhelmy, R. Shipe & D. G. Capone, 2008. Amazon River enhances diazotrophy and carbon sequestration in the tropical North Atlantic Ocean. Proceedings of the National Academy of Sciences 105: 10460–10465.

Subramanian, V., V. Ittekot, D. Unger & N. Madhavan, 2006. Silicate weathering in south Asian tropical river basins. In Ittekkot, V., D. Unger, C. Humborg & N. Tac An (eds), The Silicon Cycle: Human Perturbations and Impacts on Aquatic Systems. Island Press, Washington, DC: 3–12.

Syvitsky, J. P. M., C. J. Vörösmarty, A. J. Kettner & P. Green, 2005. Impact of humans on the flux of terrestrial sediment to the global coastal ocean. Science 308: 376–380.

Talling, J. F. & J. Lemoalle, 1998. Ecological dynamics of tropical inland waters. Cambridge University Press, Cambridge, UK.

Timms, B. V., 1999. Local runoff, Paroo floods and water extraction impacts on the wetlands of Currawinya National Park. In Kingsford, R. T. (ed.), A Free-Flowing River: The Ecology of the Paroo River. New South Wales National Parks and Wildlife Service, Huntsville, NSW, Australia: 51–66.

Vörösmarty, C. J., K. Sharma, B. Fekete, A. H. Copeland, J. Holden, J. Marble & J. A. Lough, 1997. The storage and aging of continental runoff in large reservoir systems of the world. Ambio 26: 210–219.

Walling, D. E., 2006. Human impact on land-ocean sediment transfer by the world's rivers. Geomorphology 79: 192–216.

Wassman, R., N. X. Hien, C. T. Hoanh & T. P. Tuong, 2004. Sea level rise affecting the Vietnamese Mekong Delta: Water elevation in the flood season and implications for rice production. Climatic Change 66: 89–107. doi:10.1023/B:CLIM.0000043144.69736.b7.

Weber, K. A., L. A. Achenbach & J. D. Coates, 2006. Microrganisms pumping iron: anaerobic microbial iron oxidation and reduction. Nature Reviews. Microbiology 4: 752–764.

Webster, P. J., G. J. Holland, J. A. Curry & H.-R. Chang, 2005. Changes in tropical cyclone number, duration, and intensity in a warming environment. Science 309: 1844–1846.

Welcomme, R. L., 1985. River Fisheries, FAO Fisheries Technical Paper 262. United Nations Food and Agricultural Organisation, Rome.

White, A. F. & A. E. Blum, 1995. Effects of climate on chemical weathering rates in watersheds. Geochimica et Cosmochimica Acta 59: 1729–1747.

Wu, X. L., M. W. Friedrich & R. Conrad, 2006. Diversity and ubiquity of thermophilic methanogenic archaea in temperate anoxic soils. Environmental Microbiology 8: 394–404.

WWF, 2005. An Overview of Glaciers, Glacier Retreat and Subsequent Impacts in Nepal, India and China. S. C. Rai (coordinator), World Wide Fund for Nature (WWF) Nepal Program [available on internet at http://assets.panda.org/downloads/himalayaglaciersreport2005.pdf].

Yao, H. & R. Conrad, 2000. Effect of temperature on reduction of iron and production of carbon dioxide and methane in anoxic wetland rice soils. Biology and Fertility of Soils 32: 135–141.

Hydrobiologia (2010) 657:37–51
DOI 10.1007/s10750-010-0199-6

GLOBAL CHANGE AND RIVER ECOSYSTEMS

# Dynamics of a benthic microbial community in a riverine environment subject to hydrological fluctuations (Mulargia River, Italy)

Annamaria Zoppini · Stefano Amalfitano · Stefano Fazi · Alberto Puddu

Received: 8 October 2009 / Accepted: 22 February 2010 / Published online: 12 March 2010
© Springer Science+Business Media B.V. 2010

**Abstract** Temporary rivers are characterized by recurrent dry phases, and global warming will stress their hydrology by amplifying extreme events. Microbial degradation and transformation of organic matter (OM) in riverbed sediment are key processes with regard to carbon and nutrient fluxes. In this study, we describe structural and functional changes of benthic microbial communities in a riverine environment subject to hydrological fluctuation. Sampling was carried out in the outlet section of the Mulargia River (Sardinia, Italy) under various water regimes, including one flood event. Overall, sediments were characterized by low bacterial cell abundance (range 0.6–1.8 × $10^9$ cell g$^{-1}$) as a consequence of their low nutrient and OM concentrations. No major differences were found in the community composition. *Alpha-Proteobacteria* dominated during the whole year (range 21–30%) followed by *Beta-Proteobacteria*, *Gamma-Proteobacteria*, and *Cytophaga-Flavobacteria* which always contributed <18%. *Planctomycetes* and *Firmicutes* were found in smaller amounts (<7%). In spring, when the highest total organic carbon content was also detected (0.42% w/w), both bacterial abundance and C production (BCP, 170 nmol C h$^{-1}$ g$^{-1}$) reached relatively high values. During the flood event, an increase in BCP and the highest values of community respiration (CR, 74 nmol C h$^{-1}$ g$^{-1}$) were observed. Moreover, most of the extracellular enzyme activities (EEA) changed significantly during the flood. The variation of the water flow itself can explain part of these changes and other factors also come into play. The presence of different patterns of functional parameters could suggest that the quality of the OM could be the major driving force in nutrient flux.

**Keywords** Temporary rivers · Streambed Bacteria · Bacterial Diversity · Microbial metabolic rates · Water flow · Extracellular enzymes

## Introduction

Recent research and observations of increases in global average air and ocean temperatures have led the Intergovernmental Panel on Climate Change (IPCC, 2007) to conclude that warming of the climate system is unequivocal. Because the saturation vapor pressure of water in air is highly sensitive to temperature, it is expected that global warming will lead to perturbations in the global water cycle (Allen & Ingram, 2002). At the continental, regional, and ocean basin scales, numerous long-term climate

Guest editors: R. J. Stevenson, S. Sabater / Global Change and River Ecosystems – Implications for Structure, Function and Ecosystem Services

A. Zoppini (✉) · S. Amalfitano · S. Fazi · A. Puddu
Water Research Institute (IRSA-CNR), Via Salaria Km 29.300, CP 10, 00015 Monterotondo, Rome, Italy
e-mail: zoppini@irsa.cnr.it

changes were observed from 1900 to 2005, with an increase in the occurrence of extreme events such as increased drying and heavy precipitation (IPCC, 2007). Such changes can have a significant impact on aquatic ecosystems by changing biological processes that obey the physical and chemical principles governing transformations of energy and materials (Gillooly et al., 2001; Brown et al., 2004).

Temporary rivers are dominant in the Mediterranean area and are characterized by a recurrent dry phase of varying duration and spatial extent, usually in summer (Guys & O'Keeffe, 1997). This seasonality significantly changes the dominant processes through the year (Kirkby, 2005) and makes these rivers particularly vulnerable to anthropogenic pressure. Temporary rivers have been exploited by mankind for millennia because they represent an important source of water in semiarid areas. Global warming is going to impose increased stress on temporary rivers by amplifying extreme weather events and by increasing water scarcity associated with intense runoff and flushing (Milly et al., 2005; Gallart et al., 2008). Hence, environmental changes can affect the hydrology of temporary rivers modifying the exploitation of these systems (Guys & O'Keeffe, 1997; Baldwin & Mitchell, 2000; Dahm et al., 2003). Although temporary rivers are widespread in semiarid regions worldwide, few data are available about their hydrological and biogeochemical characteristics. The majority of temporary rivers are found in un-gauged basins, and their geomorphologies differ from the traditional geological norms on which watershed models are based (Tzoraky et al., 2007).

The biogeochemical role of the riverine system in the carbon cycle is typically to export C from terrestrial systems to the ocean although a substantial fraction is respired in freshwater ecosystems (Cole et al., 2007). In different hydrologic conditions river sediments can act as sink or source of nutrients (Bernal et al., 2003; Butturini et al., 2003; Valett et al. 1994). During dry periods, the ecological processes are mainly restricted to sediments where organic matter (OM) and pollutants accumulate and mineralization rates decrease (Fonnesu et al., 2004; Tzoraky et al., 2007; Amalfitano et al., 2008). After a drought, the first rain flushes the sediments, and the resulting remobilization of dissolved and particulate OM affects the chemical characteristics of the receiving water bodies, i.e., estuaries, reservoirs, and lakes (Tzoraky & Nikolaidis, 2007; Tzoraky et al., 2007).

The microbial degradation and transformation of the OM deposited in the river channel bed is a key process with regard to the C-flux in the lotic food web (Fischer & Pusch, 2001; Findlay et al., 2003; Mulholland, 2003). Via the microbial food chain, complex organic substrates are solubilized in a series of steps from particulate OM to high molecular weight dissolved organic carbon and low molecular weight substrates by the enzymatic hydrolysis (Chrost, 1991). The OM thus enters the microbial food chain and is ultimately remineralized to $CO_2$ (Marxsen & Fiebig, 1993; Findlay et al., 2003). In aquatic environments heterotrophic microbial communities therefore represent the link between sedimentary OM and the upper level of the community comprising carnivores (Marxsen, 2006).

In temporary rivers, the high variability of hydrological and physicochemical characteristics can strongly affect microbial community composition and functioning, thus indirectly determining changes in the C-flux (Jones et al., 1995; Butturini et al., 2003; Romani et al., 2006; Zoppini & Marxsen, 2010).

In the European Community, all water resources must be managed to achieve the requirements of the Water Framework Directive (WFD-2000/60/EC) and to date, temporary rivers have posed a significant challenge to the development of sustainable water management plans. This study is part of the EU-funded project TempQsim (EVK1-CT2002-00112) in which seven catchments were investigated to improve the understanding and modeling of water quality in temporary rivers. Recent studies have described the effect of the dry period on biogeochemical processes and on metabolic rates of the benthic microbial community under experimental conditions (Tzoraky et al., 2007; Amalfitano et al., 2008; Fazi et al., 2008; Marxsen et al., 2010). What is still needed is a description of the microbial metabolic rates under natural conditions. To this end, we investigated the response of the benthic microbial communities of the Mulargia River (Italy) in terms of structural and functional parameters, under varying seasonal conditions ranging from baseline water flow to flood. In particular our aim was to analyze the variability of bacterial communities in a variable environment to provide information on how hydrology can drive C-flux mediated by microbial community.

## Materials and methods

Study site and sediment characterization

The Mulargia River is a second-order temporary river representative of the semiarid Mediterranean Region (Fonnesu et al., 2004; Gallart et al., 2008) located in southeastern Sardinia Island (Italy; 39°38′N, 09°11′E). The main reach has a length of 18 km, with a catchment extension of 64.76 km$^2$, spanning an elevation range from 250 to 750 m. The river network has an overall length of around 44 km, and the distance from the origin of the stream to the outlet in the Mulargia reservoir is 18 km. The Mulargia River is one of the main tributaries of the Flumendosa River, which supplies a large majority of the water distribution network of southern Sardinia for civil, agricultural, and industrial purposes. The vegetation is typical for low and high Mediterranean *macchia*. The Mediterranean climate in the area is characterized by an average annual rainfall of 535 mm mostly in autumn and winter. In summer, the temperature can rise to 40°C. Flood events play an important role in the transport of suspended solids and dissolved nutrients (Lo Porto et al., 2006), although the first flushes account for only 10% of total annual flow.

Based on monthly flows measured between 1992 and 2004 by *Ente Acque Sardegna*, three hydrographic periods can be highlighted for Mulargia River: (i) the rainy period, from December to April, characterized by "regular" monthly flows ranging from 1 to $3 \times 10^6$ m$^3$ (0.4–1.2 m$^3$ s$^{-1}$); (ii) a dry period, from July to September, characterized by "very low" monthly flows around $50 \times 10^3$ m$^3$ (0.02 m$^3$ s$^{-1}$); and iii) an intermediate period, in May, June, October, and November, with "low" monthly flows ranging from 0.3 to $1 \times 10^6$ m$^3$ (0.1–0.4 m$^3$ s$^{-1}$).

Water quality characteristics were determined in a parallel study in the period from September 2003 to June 2004, through analyses of inorganic nutrients, particulate P and N and OM concentration in samples collected by an automatic sampling station. Data (provided by Hydrocontrol S.C.R.L., Italy) showed a high variability in water quality, especially for dissolved OM (range 0.06–2.00 g l$^{-1}$, cv 90%), particulate phosphorous (range 0.01–12.60 mg l$^{-1}$, cv 226%), and nitrogen (range 0.04–148.60 mg l$^{-1}$, cv 203%).

In this study we performed bimonthly samplings from January to September 2004 including periods of regular (January and March), low (May), and very low water flow (July and September). An additional sampling was carried out in February 2005 during a flood event.

Sediment samples were collected in the outlet section just upstream of the Mulargia reservoir. The uppermost oxic layers (0.5–2 cm) of three homogeneous patches were sampled with a plastic scoop and immediately sieved by a 2-mm mesh (Amalfitano et al., 2008). Chemical and biological measurements were performed on the fine sediment fraction (<2 mm) where the majority of microbiological activities take place (Goedkoop et al., 1997; Fischer & Pusch, 2001; Hubas et al., 2006). Separation of the fine fraction by sieving can be regarded as a physical normalization to reduce the difference in the granulometric composition of the sediment and to exclude the macrofauna. The sieved fraction was stored in polycarbonate acid washed buckets (3 dm$^3$) and kept refrigerated (4°C) until the beginning of the analyses, within 24 h.

Grain-size distribution was determined in accordance with the soil textural triangle (Gerakis & Baer, 1999). Sediment OM content (ash-free dry weight—AFDW) was determined by subtracting ash weight (500°C, 3 h) from dry weight (105°C, overnight). Total organic carbon (TOC) and total nitrogen (TN) concentrations were determined using a Carlo Erba NA 1500 CHN Analyzer. Subsamples were acidified with 2 N HCl for TOC analysis. Total (TP) and organic phosphorous (TOP) were determined by spectrophotometry (Lambda BIO 20, Perkin Elmer). All analyses were performed in triplicate, and values are expressed as a percentage of dry weight (w/w).

Bacterial abundance and community composition

All sediment samples were treated to detach cells from particles for accurate cell quantification by epifluorescence microscopy. The extraction treatments were performed according to the protocol proposed by Amalfitano & Fazi (2008), which is reported to be particularly efficient for analyses of fine sandy sediments. Briefly, 1 g of sediment (duplicates) was fixed in formalin solution (final concentration 2.0%), and amended with Tween 20 (final concentration 0.5%) and sodium pyrophosphate

($1 \text{ g l}^{-1}$ final concentration), resulting in 10 ml sediment slurry. The slurry was then shaken and sonicated (20 W for 1 min; Microson XL2000 ultrasonic liquid processor with 1.6-mm-diameter microtip probe, Misonix, NY, USA). Thereafter, 1 ml of the resulting slurry was transferred to a 2-ml Eppendorf tube, and 1 ml of the density gradient medium Nycodenz (Nycomed, Oslo, Norway) was placed underneath using a syringe needle. High-speed centrifugation was performed in a swing-out rotor for 90 min at 4°C. Nycodenz-purified subsamples (0.5–1 ml) were filtered on 0.2 µm polycarbonate membranes (47 mm diameter, Nuclepore) by gentle vacuum (<0.2 bar) and washed with 10–20 ml of sterile ultrapure water. One section of each filter was stained for 20 min with DAPI ($1 \text{ µg ml}^{-1}$ final concentration) and the remaining filter was stored at −20°C for further Fluorescence In Situ Hybridization (FISH) analysis. In order to prove the effectiveness of the above-described extraction procedures, as suggested by Amalfitano & Fazi (2008), the remaining sediment slurry (9 ml) was left overnight at 4°C, allowing the coarse particles to settle. Three aliquots (0.1 ml) of the supernatant obtained after the overnight settling were stained for 20 min with DAPI ($1 \text{ µg ml}^{-1}$ final concentration) and collected on black filters (pore size 0.2 µm, 25 mm diameter, Nuclepore Corporation, Pleasanton, CA, USA). Bacterial abundance (BAB) was determined by epifluorescence microscopy in both the Nycodenz purified and unpurified slurry. Because (i) the abundance of cells, expressed per gram of dry sediment (cells $\text{g}^{-1}$), did not differ before (data not shown) versus after Nycodenz purification and (ii) the purification remarkably reduced the occurrence of abiotic particles allowing the straightforward visualization of cells, the bacterial abundance was estimated by counting on FISH-stained filter sections after Nycodenz purification ($n = 12$).

The bacterial biomass (BB) was determined by considering a per-cell C content of 40 fg, as experimentally determined in the same environment (Amalfitano et al., 2008). For community composition, additional filter sections were analyzed by Fluorescence In Situ Hybridization (FISH), in accordance with the protocol of Pernthaler et al. (2001). The following oligonucleotide probes were utilized: EUB338, EUB338-II, and EUB338-III, targeted to most *Bacteria*; ALF968, BET42a, and GAM42a, specific for the *Alpha-*, *Beta-*, and *Gamma- Proteobacteria* subclasses, respectively; CF319a for *Cytophaga-Flavobacterium*; PLA46a for *Planctomycetales*; and LGC354abc for *Firmicutes* (Gram-positive bacteria with low GC content). Further details on the above-mentioned probes are available at probeBase (Loy et al., 2003). All probes, 5′-labeled with Cy3 dye, were commercially synthesized (Biomers.net, Ulm, Germany). Data are expressed as percentage of total DAPI-stained cells.

Bacterial carbon production

Bacterial carbon production (BCP) was estimated with [$^3$H]leucine incorporation measurements following the method proposed by Buesing & Gessner (2003), in combination with the microcentrifugation technique proposed for water samples by Smith & Azam (1992) and applied to soil by Bååth et al. (2001). Wet sediment (0.5 g, four replicates) was transferred into 2-ml screw-cap heat-resistant Sorensen microcentrifuge tubes (Sorenson Bioscience, Salt Lake City, UT, USA). Milli-Q water was added to create a final volume of 1 ml in all tubes. An aqueous solution of radioactive leucine [50 µM: unlabeled, 49.85 µM final concentration (L 8912, Sigma–Aldrich); [$^3$H] labeled, 0.15 µM final concentration (NEN Life Science Products, Boston, Massachusetts, USA)] was added to three replicates of each sediment sample. A preliminary test showed that leucine saturation was reached in all sediment samples at a final concentration of 50 µM. Zero-time controls were run by killing samples with 100% trichloroacetic acid (TCA, 5% final concentration), 15 min before leucine addition. Tubes were homogenized by vortexing and incubating (1 h) at 20°C in the dark. Incubations were stopped by adding 100% TCA (5% final concentration). All tubes were centrifuged at 14,000 rcf for 10 min at room temperature. The supernatant was discarded to separate macromolecules from the non-incorporated label. Four washing steps were then performed adding 1 ml of 5% TCA, unlabeled leucine solution (40 mM), 80% ethanol, and Milli-Q water, respectively. To extract protein, 1 ml of NaOH (1 N) was added to the pellet, and the tubes were heated (1 h at 90°C), cooled down, and recentrifuged. The supernatant (0.1 ml) was transferred into a 2-ml Eppendorf tube, 1 ml of liquid scintillation cocktail (Ultima Gold, Packard Bioscience, Meriden, CT, USA) was added, and radioactivity was detected with the TRICARB 4430 (Packard

Bioscience) scintillation counter. The rates of leucine incorporation were converted into units of C per sediment dry weight ($\mu g\ C\ h^{-1}\ g^{-1}$) by applying the conversion factor of 1.44 kg C produced per mole of incorporated leucine (Buesing & Marxsen, 2005). Considering the exponential growth model (Koch, 1994) and the total biomass (B), bacterial growth rates ($\mu = \ln((BB + BCP)/BB)/h$) and turnover times ($T_2 = (\ln 2)/\mu$) were calculated. The per-cell specific production (BCPs in $fg\ C\ h^{-1}\ cell^{-1}$) was computed by dividing C production by cell abundance.

Respiration of benthic microbial communities

Aerobic respiration of the benthic microbial community (CR) was assessed by measuring the oxygen consumption of sediments enclosed in incubation chambers. This method has been used in the uppermost sediment layer of semiarid streams (Jones et al., 1995; Uehlinger et al., 2002). The choice of the method for respiration rate measurements is tricky because a series of constraints, including interference by algae, the disruption of microbial communities, and excessive manipulation of the sample (Döring, 2007). Consequently this procedure does not reproduce exactly the in situ respiration, but it gives a valuable estimate of the total sediment mineralization rates and permits comparison among samples. In addition, it should be considered that in our sediments, CR most likely represents microbial respiration (Törnblom, 1996; Goedkoop et al., 1997; Bastviken et al., 2003; Hubas et al., 2006), and the sandy nature of our samples (see the granulometric analysis below) reduces the contribution of algae (Romani & Sabater, 2001).

Aliquots ($\sim$400 g w/w) of sieved sediment were utilized to fill the lowest halves of two replicated respiration glass chambers (54 mm × 300 mm). The rest of each chamber was subsequently filled with dissolved oxygen (DO)-saturated river water ($\sim$300 ml) passed through acid washed 0.2 µm Nuclepore filters to remove particulate matter. Each whole chamber was carefully weighed in addition to the empty chamber, the sediment, and the water. The chambers were sealed, inverted several times, and opened to allow any air trapped in the sediments to escape. The stopper was removed cautiously to measure the concentration of DO and temperature using a DO-meter (WTW oxi 330, probe cell oxy 325), and the chamber was resealed again. The chambers were then set down horizontally, gently distributing the sediment within. The incubation, which lasted at least 2 h, was performed in the dark at in situ temperature. The duration of the incubation depended on the metabolic activity of the sediments and was prolonged until a decline of $\geq 1\ mg\ l^{-1}$ was observed. After incubation the final dissolved oxygen concentration was measured by the DO-meter. CR was calculated as dissolved oxygen depletion versus time per unit of dry weight of sediment and then transformed into C units, assuming a respiratory quotient (RQ) of 1.

The bacterial growth efficiency (BGE) is frequently used to describe what proportion of the assimilated carbon is used for biomass production or to meet energy demands by the benthic microbial community (Törnblom, 1996; Goedkoop et al., 1997; Bastviken et al., 2003). BGE was calculated as BCP/(BCP + CR), with (BCP + CR) representing the total biological organic carbon demand in the oxic environment.

Extracellular enzyme activities

Extracellular enzyme activities (EEA) in sediments were determined spectrofluorometrically (Jasco, FP-6200 spectrofluorometer) using fluorescent substrates. Organic substrates (organic phosphorous, proteins, lipids, derived cellulose organics) were selected to model the common constituents of sinking OM. All activities were measured at saturating concentrations of substrate (final concentrations: leucine-4-AMC, 1 mM; 4-MUF-P-phosphate, 0.3 mM; MUF-beta-D-glucopiranoside, 0.5 mM; 4-MUF-oleate, 0.5 mM) following the procedure proposed by Wobus et al. (2003). Incubations were performed in the dark for 1 h (1.5 h for aminopeptidase) at in situ water temperature under continuous shaking. Substrate blanks were also incubated after boiling the sediments at 100°C for 20 min. After incubation, 1 ml of glycine buffer (0.1 M, pH 10) was added to the phosphatase, beta-glucosidase, and lipase assay, and 1 ml Hepes buffer (0.1 M, pH 7.5) to the aminopeptidase assay. Samples were centrifuged at 7,000 × $g$ at 6°C for 10 min. The fluorescence was measured in the supernatant and corrected for the fluorescence of blanks. The hydrolysis rates were calculated after calibration of the spectrofluorometer with a standard alkaline solution of

AMC (7-amino-4-fluoromethyl-coumarin) or MUF (4-methyl-umbelliferone).

Statistical analysis

Bivariate relationships between different parameters were examined using the Pearson correlation (*r*). Principal components analysis (PCA) was performed with the Statistica 7.0 software package (StatSoft Inc., Tulsa, OK). A PCA biplot was calculated using all parameters related to the influence of benthic prokaryotes on C transformation (AFDW, Bacterial Biomass, Bacterial Growth Efficiency, Growth Rate, Doubling Time, Community Respiration) as active variables. Moreover, to better synthesize the matrix of microbiological data, metabolic activities (AMA, APA, lipase, beta-glu) and the percentage abundance of main taxa characterizing the community composition (*Alpha-*, *Beta-*, *Gamma-Proteobacteria*) were used as supplementary variables. All variables were normalized using division by their standard deviations, and the biplot was computed by the data correlation matrix.

## Results

Water and sediment characterization

The variability of the water flow, temperature, and conductivity during the survey are shown in the Fig. 1. The measured flows are coherent with the average historical monthly flows (see above), except for February 2005, when repeated rains gave rise to a peak flow of 6.2 m$^3$ s$^{-1}$. During the entire survey the temperature of the water changed significantly (range 8.5–28.8°C), as did the conductivity (range 128–1,814 µS cm$^{-1}$).

On average sediment samples were classified as coarse sand and composed of sand (80% coarse and 15% fine sands), clay (<2%), and silt (3%). The chemical composition of the sediments is reported in Table 1. Overall nutrients and OM concentrations with the exception of P, which was rather stable, decreased from winter to summer (from regular to very low flow) with the exception of the flood, when almost all parameters increased significantly. Ash-free dry weight varied between 1.80% (January) and 0.41% w/w (September). TOC decreased

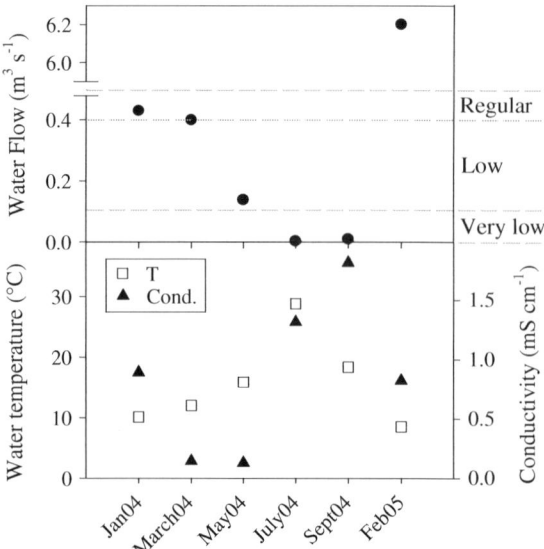

**Fig. 1** Variability of water flow (categorized as regular = 0.4–1.2, low = 0.1–0.4, very low <0.1 m$^3$ s$^{-1}$), temperature, and conductivity at the selected site during the sampling period

significantly, from 0.42% (March) in conditions of regular flow to 0.08% w/w (July) during very low flow. TN decreased from 0.06% (January) to 0.01% w/w (July), but increased about tenfold during the flood (February, 0.24%). TP showed limited changes between sampling periods (0.03 ± 0.004%), whereas TOP showed the smallest contribution in January (0.005%) and a constant value during the other samplings (0.012 ± 0.004%) (Table 1).

Microbial community structure and activity

Bacterial abundance showed a mean value of $0.9 \times 10^9 \pm 0.5 \times 10^9$ cells g$^{-1}$. The highest bacterial abundances were observed in different environmental conditions in March ($1.8 \times 10^9$ cell g$^{-1}$) and September ($1.0 \times 10^9$ cell g$^{-1}$) (Fig. 2a). In winter (January and February), the bacterial abundance assumed similar values in the lowest range (6.0 and $5.7 \times 10^8$ cell g$^{-1}$), although samplings were performed under different water flow conditions.

The percentages of cells hybridized by the generic probe targeting *Bacteria* (EUB I-III) ranged from 61 to 79%. At the division level, *Alpha-Proteobacteria* dominated over the other taxa during the whole year of sampling, with percentages of DAPI-stained cells varying from 30.2 ± 2.1% (January) to a minimum

Table 1 Ash-free dry weight (AFDW), total organic carbon (TOC), total nitrogen (TN), total phosphorous (TP), and total organic phosphorous (TOP) mean contributions (± SD; $n = 3$) to the Mulargia sediments, expressed as percentage of dry weight (w/w)

| Sampling date | AFDW % | TOC % | TN % | TP % | TOP % |
| --- | --- | --- | --- | --- | --- |
| January 19, 2004 | 1.80 ± 0.35 | 0.24 ± 0.04 | 0.06 ± 0.02 | 0.033 ± 0.004 | 0.005 ± 0.002 |
| March 26, 2004 | 0.36 ± 0.13 | 0.42 ± 0.01 | 0.05 ± 0.01 | 0.033 ± 0.002 | 0.012 ± 0.008 |
| May 26, 2004 | 0.80 ± 0.15 | 0.21 ± 0.02 | 0.04 ± 0.00 | 0.033 ± 0.004 | 0.016 ± 0.010 |
| July 20, 2004 | 0.86 ± 0.26 | 0.08 ± 0.01 | 0.01 ± 0.00 | 0.038 ± 0.016 | 0.015 ± 0.015 |
| September 29, 2004 | 0.41 ± 0.28 | 0.14 ± 0.01 | 0.03 ± 0.00 | 0.032 ± 0.005 | 0.007 ± 0.005 |
| February 9, 2005 | 1.56 ± 0.30 | 0.28 ± 0.03 | 0.24 ± 0.08 | 0.041 ± 0.004 | 0.010 ± 0.006 |

of 21.3 ± 3.0% (February) (Fig. 2b). Relatively high percentages of *Beta-* (18.3 ± 1.4%; February) and *Gamma-Proteobacteria* (15.7 ± 0.2%; March) were also detected. *Cytophaga-Flavobacterium* (range 12.0–2.3%), *Planctomycetes* (range 7.7–3.0%), and *Firmicutes* (range 0.45–0.35%) were each found in a smaller amounts. It is interesting to note that cells detected by the probes EUB I-III were almost completely characterized by the use of six generic probes in winter samples. In the other sediment samples, up to 30% of *Bacteria* remained unaffiliated (Fig. 2c).

Bacterial carbon production (Fig. 3a) varied significantly, showing the highest rate in spring, under regular flow conditions (March, 170 ± 9 nmol C g$^{-1}$ h$^{-1}$). The minimum rates were observed in summer under very low flow conditions (July, 69 ± 4 nmol C g$^{-1}$ h$^{-1}$). The per-cell BCP (BCPs) showed the lowest values at regular and very low flow (March and July, 0.09 f mol C h$^{-1}$ cell$^{-1}$), with a significant increase during the flood event (0.27 f mol C h$^{-1}$ cell$^{-1}$). The turnover time ($T_2$) decreased to the lowest value (9.7 h) in February, whereas in July we observed the maximum value (27.9 h).

Mineralization processes were measured by utilizing incubation chambers, which provide a rough estimate of community respiration (CR) rates in response to a complex and dynamic system (Fig. 3a). CR rates showed low variability during the study period (mean 18.2 ± 7.7 nmol C g$^{-1}$ h$^{-1}$), with the exception of the peak value registered during the flood (February, 74.7 nmol C g$^{-1}$ h$^{-1}$). The measurements of BCP and CR allowed us to estimate BGE, assuming that in the fraction of sediments <2 mm in size, microbial contribution to respiration is dominant (Goedkoop et al., 1997; Fischer & Pusch, 2001; Hubas et al., 2006). In this study, BGE always assumed values >0.6. Environmental changes resulted in relatively low efficiencies in July and February (0.6 and 0.7, respectively), when environmental conditions differed significantly in terms of flow, temperature, conductivity, and nutrients.

The potential hydrolysis of phosphorylated organic compounds (APA) and proteinaceous material (AMA) represented the dominant extracellular enzymes in most of the samples (Fig. 3b). APA showed the highest activity in spring, in concomitance with regular flow conditions (March, 140.5 ± 4.2 nmol MUF h$^{-1}$ g$^{-1}$), whereas the highest AMA was found during the flood (February, 160 ± 11.2 nmol MCA h$^{-1}$ g$^{-1}$). Overall the lowest EEA were found in winter under regular flow conditions (January). On average, changes in lipase and beta-glucosidase activities indicated that the hydrolysis of lipids and polysaccharides assumed minor importance for the benthic microbial community, making a more limited contribution. The only exception was observed during the flood (February), when beta-glucosidase activity represented the highest hydrolytic rate (178.7 ± 14.7 nmol MUF h$^{-1}$ g$^{-1}$). In this study, correlation analysis showed that APA accounted for 80% of BAB variability ($P < 0.05$), and AMA determined 91% of BCP variability ($P < 0.05$). Moreover, lipase and beta-glucosidase activities were associated with CR rates ($r = 0.92$ and 0.91, $P < 0.01$).

In synthesis, patterns related to microbial C transformations and the relationships of these patterns to microbial community composition are summarized in a PCA biplot (Fig. 4). PC1 and PC2, respectively, explained 63.5 and 24.1% of the variation of C transformation in sediments collected during the survey. The PC1 axis mainly discriminates the spring conditions combined with regular flow (March), with

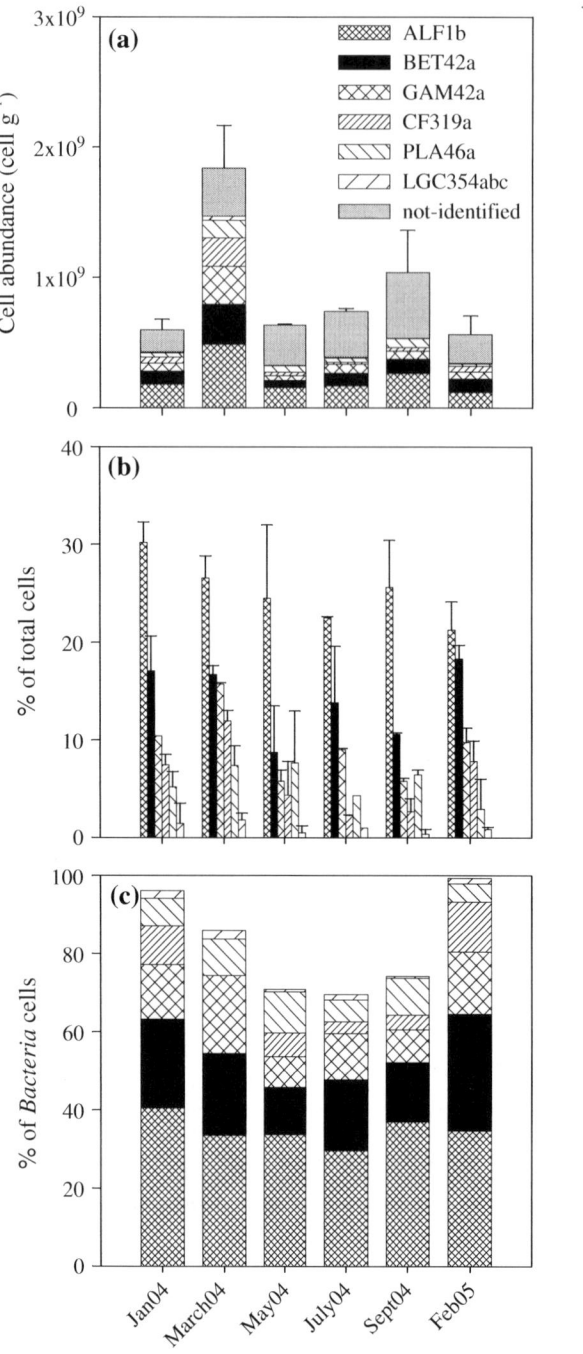

◀ **Fig. 2** Structural parameters of the benthic bacterial community. **a** Bacterial cell abundance per gram of dry sediment and relative contributions of the six analyzed taxa (ALF1b = *Alpha-Proteobacteria*; BET42a = *Beta-Proteobacteria*; GAM42a = *Gamma-Proteobacteria*; CF319a = *Cytophaga-Flavobacterium*; PLA46a = *Planctomycetales*; LGC354abc = *Firmicutes*). Error bars indicate the standard deviation of total DAPI-stained cells, calculated by counting on FISH-stained filter sections ($n = 12$). **b** Taxonomic composition, expressed as percentages of total DAPI-stained cells. Errors bars indicate the range of duplicates. **c** Taxonomic composition expressed as percentages of bacterial cells hybridized by the probe EUB I-III. Values are means of duplicate analyses

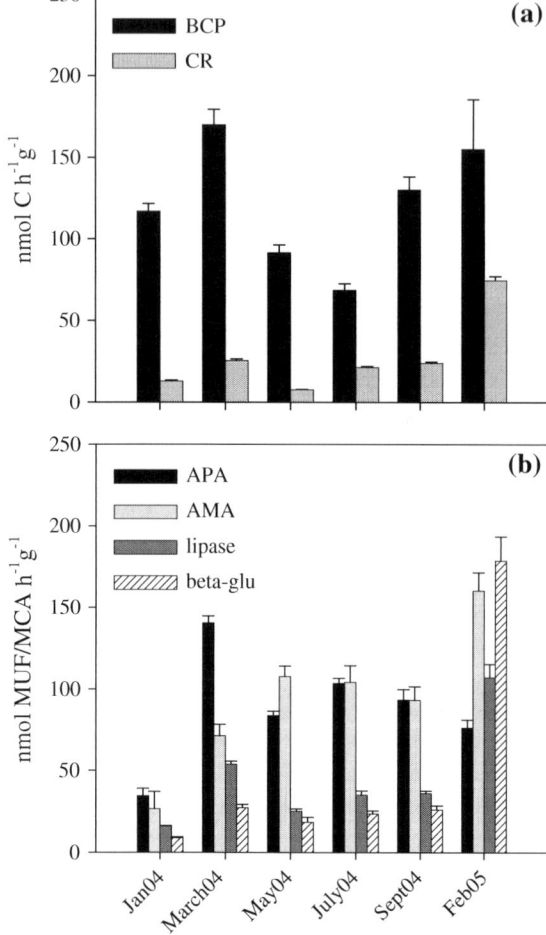

**Fig. 3** Functional parameters of the benthic bacterial community. **a** BCP and CR rates and **b** EEA during the sampling period (APA = alkaline phosphatase; AMA = aminopeptidase; beta-glu = beta-glucosidase). All data are expressed per gram of dry sediment. Error bars indicate standard deviations for BCP and mean deviations for CR and EEA measurements

high values of biomass, from winter conditions combined with flood conditions (February), with high values of CR. Such catabolic functions were associated with specific microbial groups (i.e., *Beta-Proteobacteria*) and enzymatic activities. The PC2 axis distinguishes conditions limited by nutrients

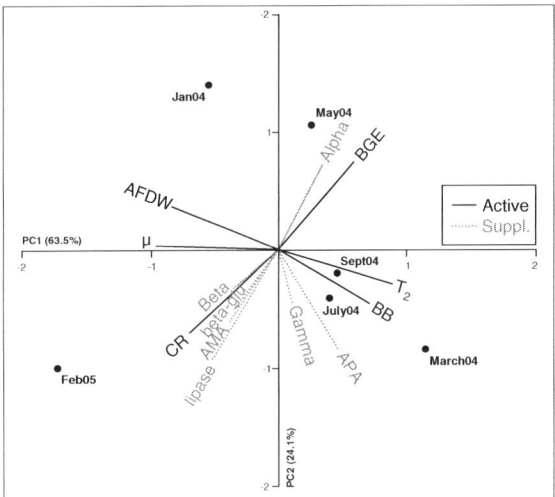

**Fig. 4** Biplot of the PCA carried out via the data correlation matrix. All variables were normalized by their standard deviations. Ash-free dry weight (AFDW), bacterial biomass (BB), bacterial growth efficiency (BGE), growth rate ($\mu$), doubling time ($T_2$), and community respiration (CR) were used as active variables. Enzyme activities—alkaline phosphatase (APA), beta-glucosidase (beta-glu), aminopeptidase (AMA), and Lipase—and the percentages of *Alpha-*, *Beta-*, and *Gamma-Proteobacteria* were projected onto the factor space as supplementary variables

(i.e., OM quality in February or P in March) from those characterized by relatively balanced nutrient conditions (January and May). Interestingly, the percentage of *Alpha-Proteobacteria* was associated with a relatively high value of BGE, whereas high values of biomass were related to a more intense APA and increasing dominance of *Gamma-Proteobacteria*.

## Discussion

In this study we described variabilities among the benthic microbial communities living in a temporary river subject to drastic changes. Our sampling strategy allowed us to analyze communities under different seasonal conditions of water flow. The structural parameters of the microbial community (cell abundance and community diversity) showed reduced variability with respect to functional characteristics. The variation of the water flow itself can explain part of these changes and other factors also come into play.

In this survey, we identified different scenarios that can significantly affect the basic nutrient flux in this ecosystem.

In winter, we observed different microbial responses to flow changes. The most significant variations were observed with EEAs: at regular flow, EEAs were significantly reduced whereas during a flood event, the significant increase of EEAs was associated with the increase of TOC, TN, and TP content in sediments (0.28, 0.24 and 0.041%, respectively). High values of BCPs were also observed during the flood, in the presence of a less abundant bacterial community, along with low values of BGE. These high metabolic rates produced fast bacterial turnover time (less than one half of one day). The differences observed between these two winter samplings can be attributed to the different qualities of the OM transported by the river flow since rapid hydrological changes imply high variability in OM and nutrient availability (Holms et al., 1998; Tzoraky & Nikolaidis, 2007). Unfortunately, we do not have much information on the OM composition in this sediment, but it is known that during floods a large transport of allochthonous OM occurs in this river (Lo Porto et al., 2006). As a matter of fact, the episodic flood events are the major source of the transport of the total suspended solids into the Mulargia reservoir and of 50% of the annual TN and TP inputs.

In spring the highest concentration of TOC (0.42%) sustained the highest values of BCP and bacterial abundance, with consequent low specific BCP rates and high turnover time (one day). Taking into account that in this season photosynthetic biomass and activity are highly stimulated (Wilczek et al., 2005), we can suppose a significant contribution of autotrophic metabolism in fueling the bacterial community with labile organic substrates. The high APA rates could support this hypothesis because this activity is also associated with the autotrophic compartment (Hoppe, 2003; Wilczek et al., 2005).

In summer, relatively low BCP and BCPs rates resulted in the longest turnover time (>1 day). The minimum water flow and the lack of precipitation may also cause a reduction in the input of nutrients and in the renewal of labile OM, consumed as the season progresses, becoming limiting in summer. The low TOC content (0.08%) registered in this period could support this hypothesis. This condition is typical of the semiarid Mediterranean region, in which sediment can become completely dry in the hot season. Unfortunately, we did not have the chance to sample the sediment during a no-flow period as the summer of 2004 was unusually wet. In parallel laboratory experiments where Mulargia River sediments were brought to desiccation, Amalfitano et al. (2008) found that the progressive decrease of water brought to a drastic decrease of bacterial live biomass (only 14% of the initial wet conditions) deprived of its main metabolic function (BCP). Rewetting of Mulargia sediments precipitated a slow recovery of structure and function (BCP), with the exception of EEAs, which promptly reactivated soon after rewetting (Marxsen et al., 2010). Hence summer conditions could represent the bottleneck for the reactivation of ecosystem functioning.

In this context, our results show that bacterial abundance was about one order of magnitude lower in comparison with other freshwater aerobic sediments (Findlay et al., 2002; Fischer et al., 2002b) but comparable with data from sediments with similar low OM content, providing evidence for the oligotrophic nature of this system (Fischer et al., 2002a; Fazi et al., 2005). Bacterial abundances changed moderately throughout this survey, with the exception of the March sample, when a peak value was observed (Fig. 1). Apparently the flood did not affect BAB, as no variations related to flow were observed. Interestingly, no major differences were found in community composition. The community was mainly dominated by *Alpha-Proteobacteria,* which represented about 30% of the bacterial cells, followed by *Beta* and *Gamma-Proteobacteria*. Few studies have assessed the impact of water flow on microbial composition, and most of these have focused on the architecture of benthic biofilms (Battin et al., 2003) or on comparing microbial communities associated with gravels in riffles and with fine sediments in depositional areas (Fazi et al., 2005). Taking into account the results of previous experimental studies on river sediments exposed to drastic changes in water availability (dry-wet) that showed a substantial influence of drying on community composition (Amalfitano et al., 2008; Marxsen et al., 2010), it is possible that the variation in water flow did not represent as strong a selective force as did extreme dryness. However, our results showed an important shift in the functional parameters, and the description of these changes could help elucidate how biota responds to environmental perturbations of different intensities.

Bacterial carbon production rates were in line with the oligotrophic characteristics of the sediments, falling into the lowest range reported in the literature (Kirschner & Velimirov, 1999; Fischer et al., 2002b; Buesing & Gessner, 2006). The BCPs, in accordance with data reported for river sediment (Fischer et al., 2002a; Buesing & Marxsen, 2005), varied between 0.09 f mol C h$^{-1}$ cell$^{-1}$ during spring and summer conditions and 0.27 f mol C h$^{-1}$ cell$^{-1}$ observed in winter during the flood event.

Community respiration is considered a basic measure of OM decomposition and energy flow (Hill & Gardner, 1987; Hall & Meyer, 1998; Hill et al., 2000). Respiration rates result from complex interactions between sources and environmental factors and bacteria have been shown to be key agents of this function, with high seasonality and spatial differences (Jones et al., 1995, Uehlinger & Naegeli, 1998; Mulholland et al., 2001; Mulholland et al., 2005). The data available on the sediment mineralization process are scarce and were obtained by different methods, often rendering difficult any comparison of data (den Heyer & Kalfs, 1998; Hill et al., 2000; Uehlinger et al., 2002; Bastviken et al., 2003; Logue et al. 2004). In aquatic systems, incubation chambers have been used to quantify benthic CR in the uppermost sediment layer of semiarid streams (i.e., Jones et al., 1995; Uehlinger et al., 2002) despite the limitation that they pose (Uehlinger & Brock, 1991; Naegeli et al., 1995; Döring, 2007). In this study, most of the respiration rates in the uppermost oxic layer of the sediment, where most of the mineralization occurs, fell within the range of 0.71–20 nmol C h$^{-1}$ g$^{-1}$ (where rates given in nmol m$^{2}$ day$^{-1}$ were recalculated) reported by den Heyer & Kalfs (1998) for marine and freshwater sediments. The maximum rate observed in the Mulargia River fell out of this range (75 nmol C g$^{-1}$ h$^{-1}$), and it was associated with the flood event, when the other metabolic parameters also showed the highest values. These high rates may have been stimulated by the high inputs of particulate OM of allochthonous origin (see below). BGE values reported in the literature for lakes and rivers vary from 0.03 to 0.8 (Del Giorgio & Cole, 1998), whereas in freshwater sediments a narrower range, from 0.1 to 0.4, has been observed (Bell & Ahlgren, 1987;

Törnblom, 1996; Goedkoop et al., 1997; Bastviken et al., 2003). BGE values >0.8 were observed in pore water supplemented with labile organic substrates under experimental conditions (Cunha et al., 2005). In this study the lowest BGE values were found in July and in February (0.6 and 0.7, respectively). We can hypothesize that the quality of the OM, rather than its quantity, led to these similar results. These two samples are characterized by differences in flow, nutrients (AFDW, TOC, TN), and temperature (Table 1, Fig. 1). In July, very low flow conditions may have determined limited renewal of OM sources (low contributions of TOC and TN) and the accumulation of refractory/exhausted OM. On the contrary, in February, the occurrence of the flood was associated with high OM and nutrient concentrations (see TOC and TN), probably due to the transport of particles of terrestrial origin. High respiration rates are associated with conditions where large debris is abundant (Jones et al., 1995). These hypotheses can be also supported by the changes observed in EEAs during the flood (see below). As a matter of fact, floods in this system are also associated with high transport of readily mobilized particulate OM (Lo Porto et al., 2006). Hence, in this system both the flood event, with its allochthonous inputs of detritus, and the very low flow can shift the metabolism toward mineralization processes, making the system a potential source of $CO_2$. Indeed, the highest BCP rates associated with the relatively low respiration rates resulted in a relatively high BGE (0.8) in March that permitted the observation of the high capacity of the microbial community to transform the dissolved OM in biomass, with the ability to act as a sink for C-flux.

The patterns of EEA may provide specific insights on the available OM sources in natural systems, as well as functional profiles of microbial communities in substrate uptake processes and nutrient cycling (Wobus et al., 2003; Zoppini et al., 2005). In highly dynamic systems, such as the Mulargia River, the OM inputs vary rapidly with respect to the intensity of the flow (Lo Porto et al., 2006), and the assessment of EEA under different flow conditions allowed us to detect changes in the OM utilization. Alkaline phosphatase (APA) and amino peptidase (AMA) activities were the most powerful enzymes in discriminating such differences. The high contributions of APA and AMA implied either a fast phosphorous and nitrogen flux or a nutrient-impoverished environment, because nutrient limitation may induce bacterial cells to synthesize specific enzymes (Taylor et al., 2003). The positive and significant correlation between EEAs (APA and AMA) and bacteria (cell abundance and BCP) could have been indirectly influenced by the presence of algae, the production of labile OM by which may have contributed to the stimulation of the biomass and metabolic rates observed in March. The high contributions of these activities to determining bacterial growth can be related to the oligotrophic conditions of this system. Spring conditions stimulated the highest metabolic rates regarding BCP and APA. Usually labile OM is available during this season, deriving from algae as revealed by its higher chlorophyll/OM ratio (Romani et al., 1998). High APA activity denotes a condition of strong P limitation, where BB and production are strictly dependent on the hydrolyzing capacities of phosphorylated OM. Such conditions of P limitation are associated with high turnover times (>20 h), as shown by the PCA analysis (Fig. 4).

The interpretation of the peptidase activity pattern is complex as it is an amphibolic enzyme potentially important in both carbon degradation and nitrogen acquisition. APA/AMA ratio, utilized as an indicator of inorganic nutrient imbalance in microbial communities (Sala et al., 2001), was found to be rather constant (1.1 ± 0.5). This finding reflects the limited seasonal variability in the quality of OM, at least with respect to the balance between phosphorylated organic compounds and proteinaceous material. The strategic role of nutrients in this environment is apparent in the significant increase of beta-glucosidase and lipase activities coinciding with the increment of contributions of TN and TP to sediments during the flood.

The interdependence observed between CR rates and the activity associated with lipase and beta-glucosidase relies on the specificity of these enzymes to hydrolyze energy-rich compounds. The products of these enzyme activities (e.g., glycerol, fatty acids, and glucose) represent readily usable sources of energy fueling catabolic metabolism (Fig. 4). Flood conditions significantly changed the enzyme profile, stimulating the highest rates of beta-glucosidase and lipase and CR rates. This increase of EEA could be due to the input of allochthonous carbon-rich debris (Jones et al., 1995), as confirmed by the increased contribution of TOC content (Table 1). Flood conditions also greatly stimulated the metabolic rates of

AMA, for which high nutritive values of the hydrolytic products (i.e., amino acids) contribute to short turnover times (<10 h). Hence, the flood event induced a drastic change in the metabolic profile of the microbial community considering that, in the same season but without a flood event (see January, Fig. 2b), we observed EEAs in the lowest range of values. Hence, the ability of AMA and APA to acquire nutrients from OM is determinant of bacterial production of new biomass when the OM source is implemented by either internal (in spring) or external (floods) events, whereas lipase and beta-glucosidase play a major role in fueling catabolic metabolism when external events bring allochthonous resources (Fig. 4).The extracellular enzymes are excreted by different bacterial groups that may differently contribute to the metabolic profile of the whole microbial community (Haynes et al., 2007). Several studies have begun to link microbial community structure to functioning in different aquatic environments using a combination of several methods or focusing on functional genes as markers of functional groups (De Long, 2004; Wilmes & Bond, 2009). However, it is still challenging to relate distinct phylogenetic groups, as identified by the rRNA-based molecular tools, to their respective functional roles because bacterial physiology can vary significantly, and some microbial functions (e.g., nitrogen fixation, denitrification) and the use of certain carbon substrates are widespread throughout the bacterial domain (Logue et al., 2008).

As suggested by the increasing data on cloning of 16S rRNA gene fragments and Fluorescence In Situ Hybridization analysis from a wide range of freshwater ecosystems, the *Alpha*, *Beta*, and *Gamma* subclasses of *Proteobacteria* account for a large portion of the domain *Bacteria* in lotic sediments, whereas members of the other groups, such as *Cytophaga-Flavobacterium*, *Firmicutes*, and *Planctomycetales*, are normally found in lower numbers (Fazi et al., 2005; Gao et al., 2005). Our data on community composition of lotic sediment confirmed these findings. Moreover, when considering patterns related to microbial C transformations (AFDW, BGE, $T_2$, BB, CR) and the relationships of these patterns to microbial community composition (Fig. 4), our data defined an ecological context in which the *Alpha* subclass of *Proteobacteria* clearly dominates the microbial community in terms of biomass and abundance. The relative percentages of this group were associated with relatively higher values of BGE, suggesting a higher efficiency in the nutrient assimilation. The *Beta-Proteobacteria* subclass was consistently lower in abundance than the *Alpha* subclass, but it was associated with CR and specific enzymatic activities (Beta-glu, AMA, Lipase), thus appearing to be involved mainly in substrate mineralization and catabolic processes. These two phylogenetic groups are reported to be particularly active components of freshwater microbial communities associated with sediments and biofilms naturally or experimentally exposed to hydrological fluctuations (Amalfitano et al., 2008; Besemer et al., 2009; Marxsen et al., 2010). The relative abundance of *Gamma-Proteobacteria* was generally low but related to BB and to a more intense APA (Fig. 4). The members of this group, which are well adapted to high nutrient concentration (Glockner et al., 1999), were relatively abundant in March, when extra OM and nutrients may have become available, sustaining the highest BB and bacterial growth.

## Conclusions

Under a global warming scenario, changes in rainfall frequency and distribution could impact the hydrological cycle (Allen & Ingram, 2002; Milly et al. 2005). Moreover, recent results from the analysis of the hydrological regimes in seven catchments of different sizes in Mediterranean Europe, including the Mulargia River, showed a complex response of the catchments to rainfall (Gallart et al., 2008). Hence, in the coming years, changes in the hydrological characteristics of the temporary rivers can be expected, with unpredictable effects of flood and drought. Overall, our results showed that changes in water flow regime can affect the microbial community functioning. Changes in river hydrology seem not to reflect on bacterial biomass or composition, because these communities adapted to live in a dynamic system. However, we observed relevant changes in microbial metabolism. Sediment can act as a potential C "sink" in periods when favorable environmental conditions (i.e., spring) permit the input of organic matter with labile substances. On the other hand, in concomitance with extreme conditions (i.e., floods in winter and minimal flow in summer), the microbial metabolism shifts toward mineralization processes, rendering the river sediments a potential "source" of $CO_2$.

**Acknowledgments** This project was funded by the European Commission, EESD Specific Programme (Contract EVK-CT-2002-00112) "Evaluation and Improvement of Water Quality Models for Application to Temporary Waters in Southern European Catchments—TempQsim." We are grateful to Dr. Anna Barra Caracciolo and Dr. Paola Grenni (IRSA-CNR) for providing C-N-P data for the chemical characterization of sediments and to Hydrocontrol S.C.R.L. (Cagliari, Italy) for providing flow data. We are also grateful to two anonymous reviewers for their precious comments.

# References

Allen, M. & W. I. Ingram, 2002. Constrains on future changes in climate and the hydrologic cycle. Nature 419: 224–232.

Amalfitano, S. & S. Fazi, 2008. Recovery and quantification of bacterial cells associated with streambed sediments. Journal of Microbiological Methods 75: 237–243.

Amalfitano, S., S. Fazi, A. Zoppini, A. Barra Caracciolo, P. Grenni & A. Puddu, 2008. Responses of benthic bacteria to experimental drying in sediments from Mediterranean temporary rivers. Microbial Ecology 55: 270–279.

Bååth, E., M. Pettersson & K. Söderberg, 2001. Adaptation of a rapid an economical microcentrifugation method to measure thymidine and leucine incorporation by soil bacteria. Soil Biology and Biochemistry 33: 1571–1574.

Baldwin, D. S. & A. M. Mitchell, 2000. The effects of drying and reflooding on the sediment and soil nutrient dynamics of lowland river-floodplain systems: a synthesis. Regulated Rivers: Research and Management 16: 457–467.

Bastviken, D., M. Olsson & L. Tranvik, 2003. Simultaneous measurements of organic carbon mineralization and bacterial production in oxic and anoxic lake sediments. Microbial Ecology 46: 73–82.

Battin, T. M., L. A. Kaplan, J. D. Newbold & C. M. E. Hansen, 2003. Contributions of microbial biofilms to ecosystem processes in stream mesocosms. Nature 426: 439–442.

Bell, R. T. & I. Ahlgren, 1987. Thymidine incorporation and microbial respiration in the surface sediment of a hypereutrophic lake. Limnology and Oceanography 32: 476–482.

Bernal, S., A. Butturini, E. Nin, F. Sabater & S. Sabater, 2003. Leaf litter dynamics and nitrous oxide emission in a Mediterranean riparian forest. Journal of Environmental Quality 32: 191–197.

Besemer, K., I. Hödl, G. Singe & T. J. Battin, 2009. Architectural differentiation reflects bacterial community structure in stream biofilms. International Society for Microbial Ecology Journal 3: 1318–1324.

Brown, J. H., J. F. Gillooly, A. P. Allen, V. M. Savage & G. B. West, 2004. Towards a metabolic theory of ecology. Ecology 85: 1771–1789.

Buesing, N. & M. O. Gessner, 2003. Incorporation of radiolabeled leucine into protein to estimate bacterial production in plant litter sediment epiphytic biofilms and water samples. Microbial Ecology 45: 291–301.

Buesing, N. & M. O. Gessner, 2006. Benthic bacterial and fungal productivity and carbon turnover in a freshwater marsh. Applied and Environmental Microbiology 72: 596–605.

Buesing, N. & J. Marxsen, 2005. Theoretical and empirical conversion factors for determining bacterial production in freshwater sediments via leucine incorporation. Limnology and Oceanography 3: 101–107.

Butturini, A., S. Bernal, C. Hellin, E. Nin, L. Rivero, S. Sabater & F. Sabater, 2003. Influences of the stream groundwater hydrology on nitrate concentration in unsaturated riparian area bounded by an intermittent Mediterranean stream. Water Resources Research 39: 1–13.

Chrost, R. J., 1991. Environmental control of the synthesis and activity of aquatic microbial ectoenzymes. In Chrost, R. J. (ed.), Microbial enzymes in aquatic environments. Springer-Verlag, New York: 29–59.

Cole, J. J., Y. T. Prairie, N. F. Caraco, W. H. McDowell, L. J. Tranvik, R. G. Striegl, C. M. Duarte, P. Kortelainen, J. A. Downing, J. J. Middelburg & J. Melack, 2007. Plumbing the global carbon cycle: integrating inland waters into the terrestrial carbon budget. Ecosystems 10: 172–185.

Cunha, M. A., R. Pedro, M. A. Almeida & M. H. Silva, 2005. Activity and growth efficiency of heterotrophic bacteria in a salt marsh (Ria de Aveiro, Portugal). Microbiological Research 160: 279–290.

Dahm, C. N., M. A. Baker, D. I. Moore & J. R. Thibault, 2003. Coupled biogeochemical and hydrological responses of streams and rivers to drought. Freshwater Biology 48: 1219–1231.

Del Giorgio, P. A. & J. J. Cole, 1998. Bacterial growth efficiency in natural aquatic systems. Annual Review of Ecology and Systematics 29: 503–541.

De Long, E. F., 2004. Microbial population genomics and ecology: the road ahead. Environmental Microbiology 6: 875–878.

den Heyer, C. & J. Kalfs, 1998. Organic matter mineralization rates in sediments: A within- and among-lake study. Limnology & Oceanography 43: 695–705.

Döring, M., 2007. Environmental heterogeneity and respiration in a dynamic river corridor: structural properties and functional performance. PhD Dissertation, University of Zurich, ETH No. 17046, pp. 1–137

European Commission & Joint Research Centre, 2005. Climate change and the European Water dimension. In Eisenreich, S. J. (ed.), JRC, Ispra Italy: 253 pp

Fazi, S., S. Amalfitano, J. Pernthaler & A. Puddu, 2005. Bacterial communities associated with benthic organic matter in headwater stream microhabitats. Environmental Microbiology 7: 1633–1640.

Fazi, S., S. Amalfitano, C. Piccini, A. Zoppini, A. Puddu & J. Pernthaler, 2008. Colonization of overlaying water by bacteria from dry river sediments. Environmental Microbiology 10: 2760–2772.

Findlay, S., J. Tank, S. Dye, H. M. Valett, P. J. Mulholland, W. H. McDowell, S. L. Johnson, S. K. Hamilton, J. Edmonds, W. K. Dodds & W. B. Bowden, 2002. A cross-system comparison of bacterial and fungal biomass in detritus pools of headwater streams. Microbial Ecology 43: 55–66.

Findlay, S. E. G., R. L. Sinsabaugh, W. V. Sobczak & M. Hoostal, 2003. Metabolic and structural response of hyporheic microbial communities to variations in supply of dissolved organic matter. Limnology and Oceanography 48: 1608–1617.

Fischer, H. & M. Pusch, 2001. Comparison of bacterial production in sediments, epiphyton and the pelagic zone of a lowland river. Freshwater Biology 46: 1335–1348.

Fischer, H., A. Sachse, C. E. W. Steinberg & M. Pusch, 2002a. Differential retention and utilization of dissolved organic carbon by bacteria in river sediments. Limnology and Oceanography 47: 1702–1711.

Fischer, H., S. C. Wanner & M. Pusch, 2002b. Bacterial abundance and production in river sediments as related to the biochemical composition of particulate organic matter (POM). Biogeochemistry 61: 37–55.

Fonnesu, A., M. Pinna & A. Basset, 2004. Spatial and temporal variations of detritus breakdown rates in the river Flumendosa basin (Sardinia, Italy). International Review of Hydrobiology 89: 443–452.

Gallart, F., Y. Amaxidis, P. Botti, G. Canè, V. Castillo, P. Chapman, J. Froebrich, J. García-Pintado, J. Latron, P. Llorens, A. Lo Porto, M. Morais, R. Neves, P. Ninov, J. L. Perrin, I. Ribarova, N. Skoulikidis & M. G. Tournoud, 2008. Investigating hydrological regimes and processes in a set of catchments with temporary waters in Mediterranean Europe Hydrological Sciences. Journal des Sciences Hydrologiques 53: 618–628.

Gao, X., A. O. Olapade & L. G. Leff, 2005. Comparison of benthic bacterial community composition in nine streams. Aquatic Microbial Ecology 40: 51–60.

Gerakis, A. & B. Baer, 1999. A computer program for soil textural classification. Soil Science Society of America Journal 63: 807–808. http://nowlin.css.msu.edu/software/triangle_form.html/.

Gillooly, J. F., J. H. Brown, G. B. West, V. M. Savage & E. L. Charnov, 2001. Effects of size and temperature on metabolic rate. Science 293: 2248–2251.

Glockner, F. O., B. M. Fuchs & R. Amann, 1999. Bacterioplankton compositions of lakes and oceans: A first comparison based on fluorescence in situ hybridization. Applied Environmental Microbiology 65: 3721–3726.

Goedkoop, W., K. R. Gullberg, R. K. Johnson & I. Ahlgren, 1997. Microbial Response of a Freshwater Benthic Community to a Simulated Diatom Sedimentation Event: Interactive Effects of Benthic Fauna. Microbial Ecology 34: 131–143.

Guys, C. M. & J. O'Keeffe, 1997. Simple words and fuzzy zones: early directions for temporary river research in South Africa. Environmental Management 21: 517–531.

Hall, R. O. Jr. & J. L. Meyer, 1998. The trophic significance of bacteria in a detritus-based stream food web. Ecology 79: 1995–2012.

Haynes, K., T. A. Hofmann, C. J. Smith, A. S. Ball, G. J. C. Underwood & A. Mark, 2007. Diatom-derived carbohydrates as factors affecting bacterial community composition in estuarine sediments. Applied and Environmental Microbiology 73: 6112–6124.

Hill, B. H. & T. J. Gardner, 1987. Benthic metabolism in a perennial and an intermittent Texas prairie stream. Southwestern Naturalist 32: 305–311.

Hill, B. H., R. K. Hall, P. Husby, A. T. Herlihy & M. Dunne, 2000. Interregional comparison of sediment microbial respiration in stream. Freshwater Biology 44: 213–222.

Holms, R. M., S. G. Fisher, N. B. Grimm & B. J. Harper, 1998. The impact of flash floods on microbial distribution and biogeochemistry in the parafluvial zone of a desert stream. Freshwater Biology 40: 641–654.

Hoppe, H. G., 2003. Phosphatase activity in the sea. Hydrobiologia 493: 187–200.

Hubas, C., D. Davoult, T. Cariou & L. F. Artigas, 2006. Factors controlling benthic metabolism during low tide along a granulometric gradient in an intertidal bay (Roscoff Aber Bay, France). Marine Ecology Progress Series 316: 53–68.

IPCC, 2007. Summary for polymakers. In Salomon, S., M. Qin, M. Manning, Z. C. Chen, M. Marquis, K. B. Avergt, M. Tignor & H. L. Miller (eds), Climate Change 2007: The Physical Science Basis. Contribution of Working Group I to the Fourth Assessment Report of the Intergovernmental Panel on Climate Change. Cambridge University Press, Cambridge, UK: 18 pp.

Jones, J. B. Jr., S. G. Fisher & N. B. Grimm, 1995. Vertical hydrologic exchange and ecosystem metabolism in a Sonoran Desert stream. Ecology 76: 942–952.

Kirkby, M. J., 2005. Organisation and process. In Anderson, M. G. (ed.), Encyclopedia of Hydrological Sciences, Vol. 1. John Wiley, Chichester, UK: 41–58.

Kirschner, A. K. T. & B. Velimirov, 1999. Benthic bacterial secondary production measured via simultaneous 3H-thymidine and 14Cleucine incorporation, and its implication for the carbon cycle of a shallow macrophyte dominated backwater system. Limnology and Oceanography 44: 1871–1881.

Koch, A. L., 1994. Growth measurement. In Gerhardt, P., R. G. E. Murray, W. A. Wood & N. R. Krieg (eds), Methods for General and Molecular Bacteriology. American Society for Microbiology, Washington, DC: 248–277.

Logue, J. B., C. T. Robinson, C. Meier & J. R. Van der Meer, 2004. Relationship between sediment organic matter, bacterial composition, and the ecosystem metabolism of alpine streams. Limnology and Oceanography 49: 2001–2010.

Logue, J. B., H. Bürgmann & C. T. Robinson, 2008. Progress in the ecological genetics and biodiversity of freshwater bacteria. Bioscience 58: 103–113.

Lo Porto A., A. Puddu, L. Diliberto, G. Canè, A.M. De Girolamo, A. Zoppini, S. Fazi, S. Amalfitano, A. Barra Caracciolo & P. Grenni, 2006. SW Sardinia, Italy – Flumendosa Mulargia. In J. Froebrich & M. Bauer (eds), Critical Issue in the Water Quality Dynamics of Temporary Waters. Evaluation and Recommendations from the TempQsim Project. University of Hannover, Hannover, Germany: 31–36.

Loy, A., M. Horn & M. Wagner, 2003. ProbeBase: an online resource for rRNA-targeted oligonucleotide probes. Nucleic Acids Research 31: 514–516.

Marxsen, J., 2006. Bacterial production in the carbon flow of a central European stream, the Breitenbach. Freshwater Biology 51: 1838–1861.

Marxsen, J. & D. M. Fiebig, 1993. Use of perfused cores for evaluating extracellular enzyme activity in stream-bed sediments. FEMS Microbial Ecology 11: 1–11.

Marxsen, J., A. Zoppini & S. Wilczek, 2010. Microbial communities in streambed sediments recovering from desiccation. FEMS Microbiology Ecology 71: 374–386.

Milly, P. C. D., K. A. Dunne & A. V. Vecchia, 2005. Global pattern of trends in streamflow and water availability in a changing climate. Nature Letters 438: 17.

Mulholland, P. J., 2003. Large-scale patterns in dissolved organic carbon concentration, flux and sources. In Findlay, S. E. & R. L. Sinsabaugh (eds), Aquatic Ecosystems, Interactivity of Dissolved Organic Matter. Academic Press, Elsevier, San Diego: 139–159.

Mulholland, P. J., C. S. Fellows, J. L. Tank, N. B. Grimm, J. R. Webster, S. K. Hamilton, E. Marti, L. Ashkenas, W. B. Bowden, W. K. Dodds, W. H. McDowell, M. J. Paul & B. J. Peterson, 2001. Inter-biome comparison of factors controlling stream metabolism. Freshwater Biology 46: 1503–1517.

Mulholland, P. J., J. N. Houser & K. O. Maloney, 2005. Stream diurnal dissolved oxygen profiles as indicators of instream metabolism and disturbance effects: Fort Benning as a case study. Ecological Indicators 5: 243–252.

Naegeli, M. W., U. Hartmann, E. I. Meyer & U. Uehlinger, 1995. POM dynamics and community respiration in the sediments of a floodprone prealpine river (Necker, Switzerland). Archiv für Hydrobiologie 133: 339–347.

Pernthaler, J., F. O. Glöckner, W. Schönhuber & R. Amann, 2001. Fluorescence in situ hybridization (FISH) with rRNA-targeted oligonucleotide probes. Methods in Microbiology 30: 207–226.

Romani, A. M. & S. Sabater, 2001. Structure and activity of rock and sand biofilms in Mediterranean stream. Ecology 82: 3232–3245.

Romani, A. M., A. Butturini, F. Sabater & S. Sabater, 1998. Heterotrophic metabolism in a forest stream sediment: surface versus subsurface zones. Aquatic Microbial Ecology 16: 143–151.

Romani, A. M., E. Vasquez & A. Butturini, 2006. Microbial availability and size fractionation of dissolved organic carbon after drought in an intermittent stream: biogeochemical link across the stream-riparian interface. Microbial Ecology 52: 501–512.

Sala, M., M. Karner, L. Arin & C. Marrasé, 2001. Measurement of ectoenzyme activities as an indication of inorganic nutrient imbalance in microbial communities. Aquatic Microbial Ecology 23: 301–311.

Smith, D. C. & F. Azam, 1992. A simple, economical method for measuring bacterial protein synthesis rates in seawater using $^3$H-leucine. Marine Microbial Food Webs 6: 107–114.

Taylor, G. T., J. W. Ying Yu & M. I. Scranton, 2003. Ectohydrolase activity in surface waters of the Hudson River and western Long Island Sound estuaries. Marine Ecology Progress Series 263: 1–15.

Törnblom, E., 1996. Bacterial production and total community metabolism in sediments of a eutrophic lake. Archiv für Hydrobiologie, Special Issues: Advances in Limnology. 48: 207–216.

Tzoraky, O. & N. P. Nikolaidis, 2007. A generalized framework for modelling the hydrologic and biogeochemical response of a Mediterranean temporary river basin. Journal of Hydrology 346: 112–121.

Tzoraky, O., N. P. Nikolaidis, Y. Amaxidis & N. T. H. Skoulikidis, 2007. In-stream biogeochemical processes of a temporary river. Environmental Science and Technology 41: 1225–1231.

Uehlinger, U. & J. T. Brock, 1991. The assessment of river periphyton metabolism: a method and some problems. In Whitton, B. A., E. Rott & G. Friedrich (eds), Use of Algae for Monitoring Rivers. Universität Innsbruck. Institute für Botanik, Innsbruck, Austria: 175–181.

Uehlinger, U. & M. W. Naegeli, 1998. Ecosystem metabolism, disturbance, and stability in a prealpine gravel bed river. Journal of the North American Benthological Society 17: 165–178.

Uehlinger, U., M. W. Naegeli & S. G. Fisher, 2002. A heterotrophic desert stream? The role of sediment stability. Western North American Naturalist 62: 466–473.

Valett, H. M., S. G. Fisher, N. B. Grimm & P. Camill, 1994. Vertical hydrologic exchange and ecological stability of a desert stream ecosystem. Ecology 75: 548–560.

WFD/EUWI, 2006. Mediterranean Joint Process WFD/EUWI, Water Scarcity Drafting Group, Tool Box (Best practices) on water scarcity, Draft Version Number 9, to be modified, 13th February 2006.

Wilczek, S., H. Fischer & M. T. Pusch, 2005. Regulation and seasonal dynamics of extracellular enzyme activities in the sediments of a large lowland river. Microbial Ecology 50: 253–267.

Wilmes, P. & P. L. Bond, 2009. Microbial community proteomics: elucidating the catalysts and metabolic mechanisms that drive the Earth's biogeochemical cycles. Current Opinion in Microbiology 12: 1–8.

Wobus, A., C. Bleul, S. Maassen, C. Scheerer, M. Schuppler, E. Jacobs & I. Röske, 2003. Microbial diversity and functional characterization of sediments from reservoirs of different trophic state. FEMS Microbiology and Ecology 46: 331–347.

Zoppini, A. & J. Marxsen, 2010. Importance of Extracellular Enzymes for Biogeochemical Processes in Temporary River Sediments during Fluctuating Dry-Wet Conditions. In Shukla, G. & A. Varma (eds), Soil Enzymology, Soil Biology Series. Springer, Berlin: in press

Zoppini, A., A. Puddu, S. Fazi, M. Rosati & P. Sist, 2005. Extracellular enzyme activity and dynamics of bacterial community in mucilaginous aggregates of the northern Adriatic Sea. Science of the Total Environment 353: 270–286.

Hydrobiologia (2010) 657:53–70
DOI 10.1007/s10750-010-0319-3

GLOBAL CHANGE AND RIVER ECOSYSTEMS

# Global changes in pampean lowland streams (Argentina): implications for biodiversity and functioning

Alberto Rodrigues Capítulo · Nora Gómez · Adonis Giorgi · Claudia Feijoó

Received: 28 October 2009 / Accepted: 12 June 2010 / Published online: 27 June 2010
© Springer Science+Business Media B.V. 2010

**Abstract** The rivers and streams in the pampean plains are characterized by a low flow rate due to the low slope of the surrounding terrain, high levels of suspended solids, silty sediment in the benthos, and reduced rithron; the riparian forest of this region has been replaced by low-altitude grasslands. Many of these environments contain a wide coverage of aquatic reeds, both submerged and floating, making the pampas limologically extraordinary. These terrains have undergone a gradual transformation in response to the progress of urbanization and agricultural activity in recent years with a resulting loss of biodiversity, leaving only few sites that continue to reflect the original characteristics of the region. Because of human activities in combination with the global climate change, variations have occurred in biological communities that are reflected in the structure and function of populations and assemblages of algae, macrophytes, and invertebrate fauna or in the eutrophication of affected ecosystems. The objective of this article is to describe the principal limnologic characteristics of the streams that traverse the Buenos Aires Province and relate these features with the predicted future global changes for the area under study. Considering the future climate-change scenarios proposed for the pampean region, the projected increment in rainfall will affect the biological communities. Higher rainfall may enhance the erosion and generate floodings; increasing the transport of sediments, nutrients, and contaminants to the ocean and affecting the degree of water mineralization. Changes in discharge and turbidity may affect light penetration in the water column as well as its residence time. The modifications in the use of the soil will probably favor the input of nutrients. This latter effect will favor autotrophy, particularly by those species capable of generating strategies for surviving in more turbid and enriched environments. An accelerated eutrophication will change the composition of the consumers in preference to herbivores and detritivores. The increase in global population projected for the next years will demand more food, and this situation coupled with the new

**Electronic supplementary material** The online version of this article (doi:10.1007/s10750-010-0319-3) contains supplementary material, which is available to authorized users.

Guest editors: R. J. Stevenson, S. Sabater / Global Change and River Ecosystems – Implications for Structure, Function and Ecosystem Services

A. Rodrigues Capítulo (✉) · N. Gómez
Institute of Limnology "Dr. Raul A. Ringuelet"
(UNLP-CONICET La Plata), C.C. 712, 1897 La Plata, Argentina
e-mail: arcapitulo@gmail.com

A. Giorgi
Programa Ecología de Protistas, Universidad Nacional de Luján, C.C. 221, 6700 Lujan, Argentina

C. Feijoó
Programa de investigación en Ecología Acuática, Universidad Nacional de Luján, C.C. 221, 6700 Lujan, Argentina

Reprinted from the journal    53     Springer

scenarios of climate change will lead to profound socioeconomic changes in the pampean area, implying an increase in demand for water resources and land uses.

**Keywords** Pampean streams · Biotic communities · Fisico-chemical · Primary and secondary producers · Landscape

## Introduction

Rising human pressure on water resources and the effects of climate change will probably affect the hydrological and geomorphological state of river systems in many areas of the globe. Hydrological variations will lead to a chain of effects on the structure and functioning of river systems. Hence, these effects are expressed in the loss or malfunction of the ecosystem services that they provide (Sabater, 2008).

As a consequence, freshwater systems have been especially threatened, having suffered a higher proportion of species and habitat losses than terrestrial or marine ecosystems. Moreover, this tendency will probably continue owing to water contamination, flow reduction as a result of irrigation and reservoir construction, and overfishing among other causes (McAllister et al., 1997). Despite the relatively small areas of the earth's surface covered by freshwaters as compared to other ecosystems, these limnic bodies sustain a major biotic diversity. South America is one of the continents with the greatest proportion of freshwater and is accordingly recognized for its great variety of aquatic environments and the extent of its biodiversity (Moyle & Leidy, 1992).

The area occupied by the Argentine pampas is heterogeneous with respect to geology, climate, and extent of land-surface relief (Cabrera & Willink, 1980). The Buenos Aires Province contains the highest demographic and industrial concentration in the country, the greatest agriculture and livestock production as well as the most intense use of agrochemicals, as a result of the great expansion of agriculture within the last 150 years. Because of such intensive human activity, many pampean rivers and streams are impacted by point sources of contamination from sewer effluents and industrial wastes and by diffuse pollution associated with crop cultivation and cattle raising (Salazar et al., 1996; Sala et al., 1998; Gómez & Rodrigues Capítulo, 2001).

Most watercourses that border large cities have suffered significant modifications. For example, the rivers that pass through Buenos Aires city have been either channelized or covered over since 1870. The basins more greatly affected by industrial activity were those of the Reconquista, Matanza-Riachuelo, and Luján rivers (del Giorgio et al., 1991; Rodrigues Capitulo et al., 1997; Gómez, 1998; Rodrigues Capítulo, 1999; Salibián, 2005). With respect to crop cultivation and cattle raising, the previously diversified agriculture (e.g., wheat, corn, sunflowers, and sorghum) has been supplanted by intensive soybean monocultivation; while the earlier free-range grazing of livestock has been partially replaced by the use of feed lots.

The pampas contains ca. 21 million inhabitants, accounts for 90% of the country's soybean production, and has accordingly been the region most affected by the expansion of this crop. Since most crops cultivated in Argentina are transgenic and soybean is the leading one among them that crop also results in a pronounced increase in glyphosate application (Vera et al., 2009). The use of this herbicide in large scale can stimulate the growth of picocyanobacteria (Pérez et al., 2007) or zooplankton (Paggi & José de Paggi, 2001) and thus potentially alter the biotic structure of the bodies of water.

Other common practices that have affected pampean lotic systems have been dredging and channelization of watercourses to avoid flooding. These engineering operations have involved neither an adequate degree of planning nor any recourse to advisory consultation; this oversight has resulted in loss of habitat for a variety of invertebrate fauna, amphibians, and fish. Moreover, these interventions have involved the destruction of natural riverbanks, affecting macrophyes and riparian vegetation, and have often been accompanied by a deterioration of water quality (Bauer, 2009; Licursi & Gómez, 2009). Finally, the advance of urbanization that progressively occupied cultivatable land and caused the displacement of livestock from their traditional areas to marginal lands situated in floodplains has increased the incidence of erosion and the input of particulate material into waterways.

Current models predict temperature and rainfall increases in the pampean plains (Hulme & Sheard,

1999). Under this climatic-change scenario we can expect an increased number of lakes, changes in their runoff patterns, and a greater discharge of water in the streams and rivers. These effects, combined with changes in the land use described above, will lead to a greater input of nutrients and contaminants into streams and rivers. In addition, the increase in sea level will promote the erosion of the current coastal profile, thus adversely affecting the geometry of the mouths of watercourses.

In this article, we discuss how these global changes could affect aquatic communities of the Pampa river systems. We describe the main limnological characteristics of streams in the province of Buenos Aires and relate these characteristics to the future predicted change for the area under study. On the other hand was analyzed a simulation of the increase in nutrients in a plains fluvial system with its headwaters subjected to unhabitual hydrologic irregularities attributable to global climatic changes resulting from human activity. The long-term aim was then to examine the resulting changes in the biological structure of the trophic network and the functioning of the metabolic processes in the resident communities.

**Geology and geomorphology**

The pampean plains are composed of quaternary sediments originating from the erosion of the Andes mountain range and extend over an area of approx. 500,000 km$^2$. This flat landscape has topographic slopes that vary between 0.1 and 1 m/km but is interrupted by occasional high hills that arose during the Tertiary Period, barely exceed altitudes of 1,200 m above sea level, and occupy small areas (e.g., the Tandil and Ventana hills). Different strata of sediments reflect earlier dry and wet periods as well as the incursions of the sea. The subsurface strata are formed by slimes rich in sulfates and chlorides along with sand and clay, which compositions characterize the river and stream beds (Andrade, 1986).

The soils are composed mainly of loess, which component favors the movement of particulates and facilitates the new formation of clays that in turn generate the reducing conditions conducive to the preservation of organic material within the profile of the strata. For this reason, the soils of the area are generally fertile with a high content of nutrients and a marked capacity for cationic interchange, predominantly involving calcium (Papadakis, 1980).

Because of the plains' extensive area—both with respect to latitude and longitude—the temperatures there vary between average annual values of 18°C in the north and 12°C in the south; and the annual rainfall ranges between 600 mm in the west and 1,000 mm in the east, with maximum precipitations occurring toward the end off summer or the beginning of autumn. July, the coldest month, has an average minimum temperature below 10°C and January, the warmest, and an average of 22°C. In contrast, the most westerly region is dry and of a moderate climate, exhibits the greatest overall temperature range, and is characterized by hydrologic deficits during the warmest months.

**Physiognomy of the landscape**

According to Parodi (1942) the intense competition of the graminoids for water in periods of rainfall deficit would have impeded the establishment of hardwood trees within the region. For this reason, grassland is the dominant physiognomy.

In spite of this characteristic feature, large modifications of the pampean landscape have resulted from the introduction of plants and animals by the European settlers (Brailovsky & Foguelman, 1992). The changes occurred mainly in the profiles of the meadowlands as well as through the introduction of thickets of *Baccharis* spp. in elevated areas and of *Solanum glaucophyllum* in the lower regions, among others. There are vestiges of xerophilic woods within limited spaces, such as *tala* (*Celtis tala*) and *espinillo* groves (*Acacia caven*; Burgueño, 2005).

**Hydrology**

The predominant bodies of water in the pampean plains are shallow lakes and ponds. Moreover, in some areas, despite the slight regional topographic slope of 1 m/km, the drainage network exhibits a high density of stream and rivers 0.16 km/km$^2$ (Sala et al., 1998).

The pampean rivers have been grouped into four categories according to their drainage basins and geomorphology (Frenguelli, 1956; Ringuelet, 1962): (1) the Salado-River system (situated in the pampean depression with an overall basin area of 80,000 km$^2$); (2) the closed basins in the south and southwest vertants of the Bravar and Curamamal mountains (the endorheic system of Chasico) and the Vallimanca stream system, it being actually connected to the latter system; (3) the tributaries of the Paraná River and the Río de La Plata (the Paraná-Plata basin); (4) the watercourses of the Atlantic slope and those originating in the Sierra de la Ventana and Tandil hills. The Paraná and Uruguay rivers furthermore drain into the Río de La Plata estuary and influence the pampean hydrographic network changing the regional geomorphology (Fig. 1).

## Physicochemical characteristics

In general, because of the low slope of the plains, the flow rates of the streams are low (<0.4 m s$^{-1}$), a value exceeded only in periods of freshets or in those streams whose sources are localized in the mountainous systems of the area (Rodrigues Capítulo et al., 2002). Giorgi et al. (2004) estimated the transport capacity and the degree of retention of particulate material and nutrients during the spring and summer months of 2003 and 2004 in some first-order streams located in the north east of Buenos Aires Province. At discharges ranging from 0.01 to 10 l s$^{-1}$, the fine

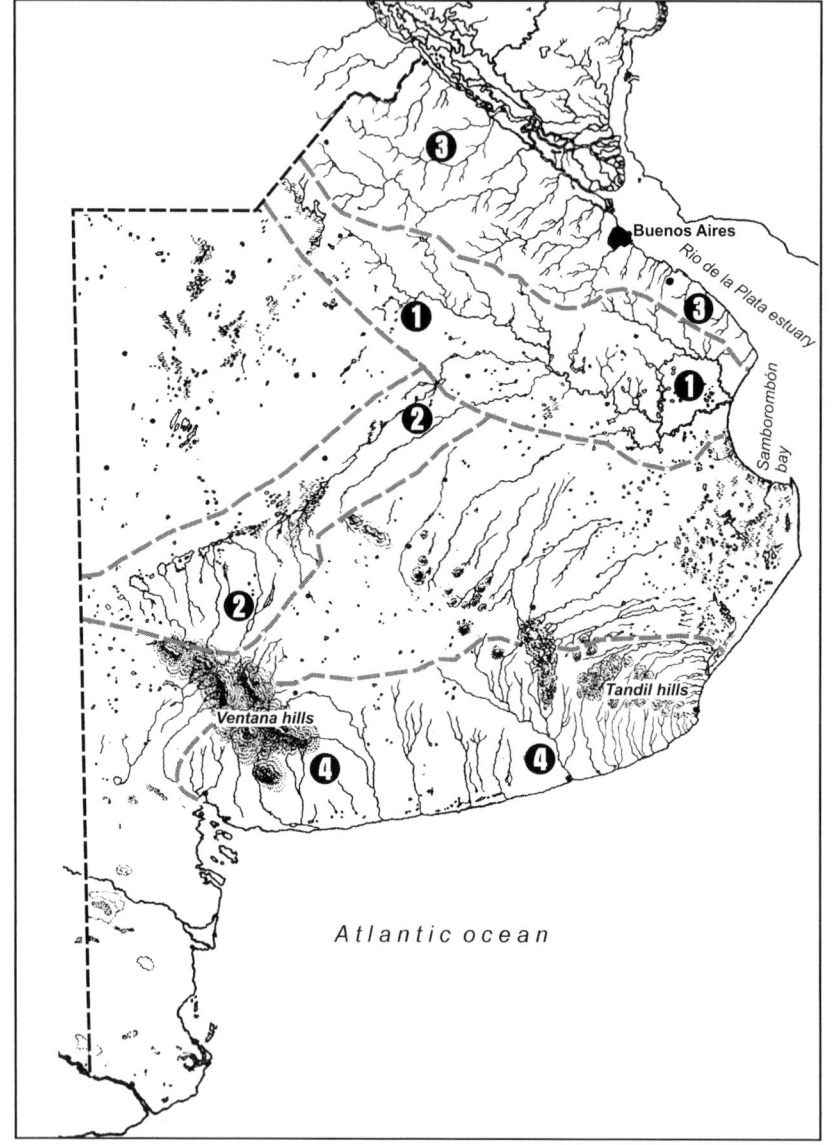

**Fig. 1** Hydrographic basins of the Province of Buenos Aires (modifies of Frenguelli, 1956; Ringuelet, 1962): (*1*) Salado-River system; (*2*) basin without drainage arising from the south and southwest vertants of the Bravar and Curamamal mountains and the Vallimanca stream system; (*3*) tributaries of the Paraná River and the Río de La Plata (the Paraná-Plata basin); and (*4*) the watercourses of the Atlantic slope and those originating in the Sierra de la Ventana and Tandil hills

(between 2 mm and 50 μ) and ultrafine (<50 μ) particulate materials predominate except in the instance of the Pereyra stream where drifting methaphyton become trapped in gross particulate material (>2 mm). The streams generally have a high conductivity as result of the presence of the dissolved substances that are transported, with the total soluble solids fluctuating between 0.5 and 1.0 g l$^{-1}$. The low flow velocities increase the nominal travel time (NTT), the latter being considered an estimation of a stream's hydrologic-retention time (Hauer & Lamberti, 1996) and varying between 45 s$^{-1}$ and 5 h within a reach of 100 m (Table 1).

The sinuosity of the first-order streams is low, ranging between 1 and 1.07; the relative roughness fluctuates between 1.06 and 2.60, while the Reynolds number is variable, ranging from low (12), with laminar flow, to medium-turbulent (>3,000). Moreover, the Froude number, varying from 0 to 0.59—with 1.0 corresponding to turbulence—would also indicate laminar flow.

Given their low flow rates, the streams have limited transport capacity. This parameter, however, becomes modified in accordance with increments in the order of the rivers and after a spate, with the latter usually occurring about twice a year during autumn and the late spring, as a result of higher precipitation. The low infiltration capacity of the soil owing to the high percentage of clay, which component reduces the drainage and elevates the runoff (Fidalgo, 1983), contributes to the autumn and spring spates (Table 2).

## Hydrochemical characteristics

Pampean streams are characterized by alkaline waters with high conductivity and elevated dissolved-oxygen and nutrient concentrations (Feijoó et al., 1999; Bauer et al., 2002; Feijoó & Lombardo, 2007). In a study conducted on 41 streams of Buenos Aires Province, the conductivity correlated with chloride concentrations, which values exceeded by one or two

**Table 1** Mean values and standard deviation (±) of transported materials in pampean streams according to (Giorgi et al., 2004) (for soluble solids the units are g l$^{-1}$, for the others are mg l$^{-1}$

|  | Haras | Gutierrez | Chaña | Pereyra | Nutrias |
| --- | --- | --- | --- | --- | --- |
| Soluble solids | 0.638 ± 0.126 | 0.648 ± 0.090 | 0.583 ± 0.146 | 0.775 ± 0.220 | 0.707 ± 0.214 |
| Ultrafine particles (DW) | 0.096 ± 0.063 | 0.069 ± 0.055 | 0.048 ± 0.042 | 0,079 ± 0.083 | 0.205 ± 0.125 |
| Fine particles (DW) | 0.503 ± 0.307 | 0.123 ± 0.134 | 0.191 ± 0.318 | 0.329 ± 0.437 | 0.004 ± 0.008 |
| Coarse particle (DW) | 0.177 ± 0.251 | 0.103 ± 0.201 | 0.027 ± 0.039 | 3.713 ± 5.343 | 0.002 ± 0.003 |
| Ultrafine particles (AFDW) | 0.009 ± 0.003 | 0.009 ± 0.005 | 0.006 ± 0.004 | 0.010 ± 0.010 | 0.026 ± 0.014 |
| Fine particles (AFDW) | 0.058 ± 0.042 | 0.018 ± 0.021 | 0.031 ± 0.055 | 0.110 ± 0.188 | 0.001 ± 0.003 |
| Coarse particles (AFDW) | 0.054 ± 0.079 | 0.055 ± 0.107 | 0.013 ± 0.025 | 2.737 ± 4.496 | 0.000 ± 0.000 |

**Table 2** Mean values and standard deviation of flow and loads of particulate and dissolved substances at two sampling stations (S1 and S2) in the Las Flores Stream at normal conditions

|  | Normal conditions | | Flooding conditions | |
| --- | --- | --- | --- | --- |
|  | S1 | S2 | S1 | S2 |
| Flow (l/s) | 40.6 ± 27.5 | 79.6 ± 46.8 | 491.9 | 240.2 |
| TDS (mg/s) | 0.7 ± 0.2 | 0.7 ± 0.2 | 0.4 | 0.2 |
| SPM (mg/s) | 637.3 ± 1022.0 | 637.0 ± 570.7 | 27378.0 | 6437.3 |
| POM (mg/s) | 63.7 ± 69.9 | 97.9 ± 79.3 | 1340.3 | 558.4 |
| SRP (mg/s) | 32.3 ± 25.7 | 48.2 ± 35.7 | 228.5 | 105.0 |
| DIN (mg/s) | 173.0 ± 154.7 | 345.2 ± 342.3 | 375.0 | 685.4 |

Values obtained during the flood of April 1993 are indicated separately. *DIN* dissolved inorganic nitrogen ($NO_3^-$ + $NO_2^-$ + $NH_4^+$), *TDS* total dissolved solids, *SPM* suspended particulate materials, *SRP* soluble reactive phosphorous (Giorgi et al., 2005)

orders of magnitude the average concentration worldwide (e.g., 8.3 mg l$^{-1}$) (Berner & Berner, 1987). The chloride levels tended to increase toward the west, where the more arid conditions favor water evaporation and salinization. In general, bicarbonates tended to exceed carbonates (Feijoó & Lombardo, 2007).

The dissolved-phosphorus and nitrogen- concentrations in streamwater are relatively high compared to other lotic systems of the world (Omernik, 1977; Binkley et al., 2004). The mean nutrient values reported for Province of Buenos Aires were 0.17 mg l$^{-1}$ of soluble reactive phosphorus and 1.50 mg l$^{-1}$ of nitrate (Feijoó & Lombardo, 2007). According to the criteria of the Environmental Protection Agency (EPA, 2000), the pampean streams could be classified as eutrophic upon consideration of their phosphorus concentrations, but as mesoeutrophic on the basis of their nitrate levels. The nutrient levels of the pampean streams exceeded those reported for so-called pristine environments, but were not so high as would be expected for intensively cultivated basins (Amuchástegui, 2006)—at least with respect to nitrogen, the nutrient more associated with agricultural activities (Mugni et al., 2005; Feijoó & Lombardo, 2007).

Evidence suggests that eutrophic conditions in the pampean streams are not solely associated with agricultural and cattle-raising activities developed within the region. The analysis of pollen-DNA sequences at some sites within the pampas revealed the existence of bodies of water at an advanced stage of eutrophication with abundant aquatic genera typical of such enriched conditions (Prieto, 1996; Zárate et al., 2000), considerably before the introduction of cattle by the Spaniards during the Colonial period and the rise of agriculture during the nineteenth century.

Nutrient levels in streamwater vary seasonally; especially at high flows (Vilches et al., 2008) and in response to a rainfall, when increments in phosphorus and nitrogen levels are usually observed (Mugni, 2009).

## Decomposers

The information on the identity and abundance of decomposers in the pampean streams is scant and fragmentary. Bacterial densities reported for the sediments and water columns are of the orders $10^6$–$10^8$ and $10^6$–$10^7$ l$^{-1}$, respectively, depending on the variations in the concentration of organic material and the season, with higher values being recorded in the warmer periods (Cochero et al., 2008; Sierra, 2009).

Many of the principal taxonomic groups involved in the mineralization of organic material are the zoosporic organisms; the following genera were among the most frequent encountered in the sediment: *Pythium*, *Catenophlyctis*, *Rhyzophlictis*, *Aphanomyces*, *Achlya*, *Dictyuchus*, *Saprolegnia*, *Rhizophlyctis*, and *Nowakowskiella* (Marano et al., 2008; Sierra, 2009).

## Primary producers

### Phytoplankton

The composition and dynamics of the phytoplankton communities in the rivers that traverse the pampean plains are modified principally by the physiography, the geochemistry, and the various uses of the land (Bauer, 2009). In the streams with current velocities lower than 0.4 m s$^{-1}$—and particularly in those segments with a greater concentration of nutrients—the phytoplankton density can exceed $10^4$ cells ml$^{-1}$. These characteristics are often observed in streams that originate in shallow depressions, with minimal slope and slow, allowing a longer residence time and thus favoring phytoplankton development. In the sources of mountain streams a greater proportion of benthic algae are observed in the water column as a result of the drift of pennate diatoms, particularly *Navicula tripunctata*, *Rhoicosphenia abbreviata*, *Nitzschia fonticola*, and *N. recta*. The phytoplankton composition also varies with the water turbidity: the streams that originate in the plains usually contain a greater proportion of suspended solids, frequently resulting in the development of loricate euglenophytes (e.g., the genera *Strombomonas* and *Trachelomonas*). With an increase in the stream order, the chlorophytes and cyanophytes predominate, and the development of a true potamoplankton becomes observable.

The algal biomass, expressed as the amount of chlorophyll *a* (Chl *a*), is variable and ranges between <10 µg l$^{-1}$ and >400 µg l$^{-1}$ between sites of low and high anthropic impact, respectively.

The natural hydrochemistry of the lotic systems is also recognized as a significant influence on the characteristics of the resident phytoplankton: elevated concentrations of bicarbonate plus carbonate, calcium, and magnesium favor the development of species such as *Fragilaria acus*, *N. tripunctata*, *N. fonticola*, *N. recta*, and *Ulnaria ulna*; whereas low levels of these electrolytes are conducive to the growth of, for example, *Strombomonas treubii*, *Trachelomonas intermedia*, *T. volvocina*, and *Dictyosphaerium subsolitarium* (Bauer, 2009).

The frequent dredging of the streams and rivers of the pampean plains, in order to compensate for the excess to drain water accumulation in the floodlands, influence the characteristics of the phytoplankton. This dredging produces changes in the chemical composition and turbidity—and thus the light penetrability—of the water (Licursi & Gómez, 2009). Among the principal observable alterations are a reduction in the phytoplankton density, an increment in the phaeopigments, and a change in the composition of the phytoplankton community (Bauer, 2009).

Phytobenthos

The phytobenthos of the pampean lotic systems, and particularly those of shallow depth and low velocity, can harbor a diverse biota ranging from the customary benthic organisms to epiphytic species of plankton that survive and develop within the epipelon. The diatoms are one of the groups most widely represented in this community, but whose distribution is influenced by the stream's hydrochemistry. Thus, many of the described species have a preference for high concentrations of $HCO_3^{1-}$, $Ca^{2+}$, and $Mg^{2+}$ (e.g., *Achnanthes minutissima*, *Amphora pediculus*, *Cocconeis placentula*, *Navicula capitatoradiata*, *N. tripunctata*, *N. gregaria*, *N. veneta*, *Nitzschia calida*, *Nitzschia dissipata*, and *Rhoicosphenia abbreviata*); carbonates (*Diadesmis confervacea*, *Navicula trivialis*, *Gomphonema gracile*, *Cymbella silesiaca*, *Pinnularia gibba*, *Gomphonema parvulum*, *Nitzschia brevissima*, *Caloneis bacillum*, and *Luticola mutica*); or chlorides (*Surirella striatula*, *Gyrosigma attenuatum*, *S. acuminatum*, *Nitschia compresa*, and *N. sigma*).

Contamination of the water also can modify the taxonomic profile of the affected stretches of a stream so as frequently to promote the abundant growth of species such as *Navicula subminuscula*, *Navicula pygmaea*, *Diadesmis confervacea*, *Achnanthes hungarica*, *Nitzschia umbonata*, *Sellaphora pupula*, *Gomphonema parvulum*, *Sellaphora seminulum*, and *Nitzschia palea* (Gómez, 1998; Gómez & Licursi, 2001; Licursi & Gómez, 2002). Seasonal variations have also been observed: Chlorophytes (e.g., *Cladophora glomerata* and *Spirogyra* spp.) and the filamentous cyanobacteria (e.g., *Oscillatoria limosa*, *O. tenuis*, and *Lyngbia limnetica*) can reach a major representation during the spring and summer.

The biomass of the phytobenthos (i.e., the Chl *a* levels) can vary as a function of the season (e.g., being greater in spring or fall) and the enrichment of the watercourses with organic material and nutrients. Giorgi (1998) mentioned values over the years ranging from 5 to 14 mg Chl *a* m$^{-2}$ for some pampean watercourses studied (Fig. 2), although Sierra (2009) reported much higher values at more than 700 mg Chl *a* m$^{-2}$ for enriched streams.

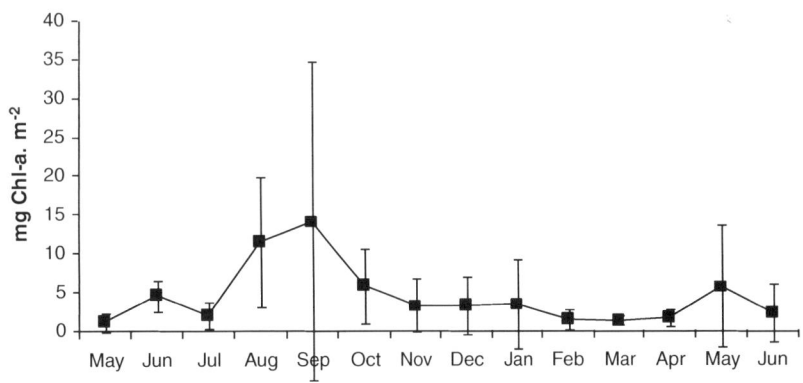

**Fig. 2** Chlorophyll *a* variation of epipelic algae in Las Flores stream reach (1992–1993) (Data from Giorgi, 1998)

## Epiphyton

The taxonomic richness of the epiphyton can vary in accordance with the species of macrophyte that is being colonized. *Ceratophyllum demersum* harbors more species than do *Potamogeton* and *Egeria densa*, with 10–20 species per plant normally being accommodated during the vegetal substrate's growth periods and as many as 100 species per plant when the epiphyton community has become stabilized. The following are among the species most often colonizing the macrophytes: *Cocconeis placentula, Eunotia pectinalis, Gomphonema angustatum, Melosira varians, Roicosphenia curvata, Ulnaria ulna, Spirogyra* sp., and *Oscillatoria* spp. (Giorgi, 1998). The biomass of the epiphyton depends on multiple parameters, with the most influential being: the type and variations in the plant substrate, the presence of floating plants that interfere with the penetration of light, the abundance of herbivorous forms, the competition for nutrients, and the effect of increases in the current flow velocity (Stevenson et al., 1996). Because of these influences the epiphyton biomass is highly variable, ranging from minimal values of 5 mg Chl $a$ g$^{-1}$ dry weight during the growth periods of the vegetal substrate to a maximum of 100 mg Chl $a$ g$^{-1}$ when the community becomes stabilized (Giorgi 1998) (Fig. 3). Finally, the maximal biomass can also vary in accordance with the nature of the substrate colonized: for example, during the same period greater biomasses have been registered for *Egeria densa* than for *Potamogeton* or *Ceratophyllum demersum* (Giorgi et al., 2005) (Fig. 4).

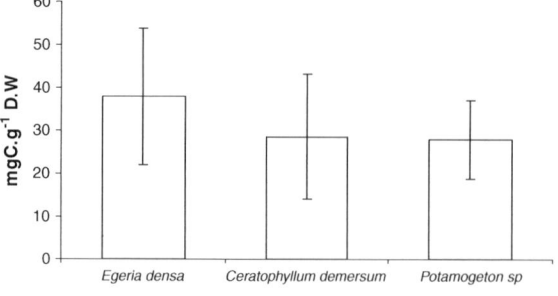

**Fig. 4** Biomass (mean and standard deviations) of the epiphyton on three genera of macrophytes frequently present in pampean streams in natural conditions (colonization for 60 days). Samples of year 2004 at Las Flores stream (Data from Giorgi et al., 2005)

## Metaphyton

The Zygnemateles (*Spirogyra* spp, *Zignema* spp) develop in the headwaters of streams and in certain rivers of greater order. The presence of these algae seems to be favored under conditions of high nutrient concentrations, low current velocities, and a sparse development of macrophytes. The group remains associated with the streambed during the cold season but detaches itself from the bottom at the beginning of spring to form mats of floating filamentous algae (Giorgi, 1998).

## Macrophytes

The low current velocity, the absence of riparian trees, and the high concentrations of nutrients characteristic of pampean streams favor the development of dense macrophyte communities of broad diversity.

**Fig. 3** Chlorophyll *a* variation of epiphytic algae in Las Flores stream reach (1992–1993) (Data from Giorgi, 1998)

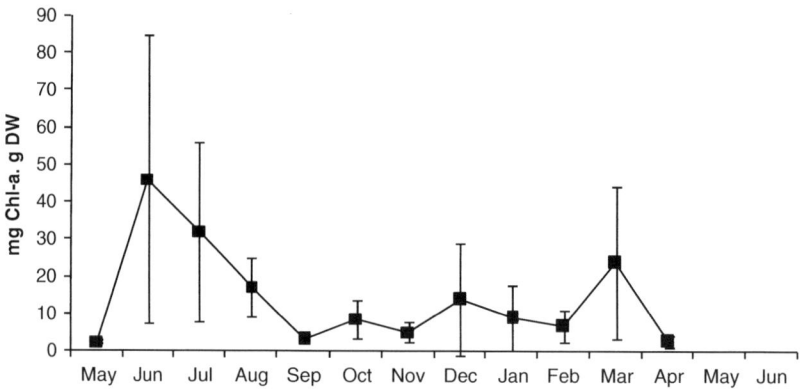

Species richness varies markedly among streams and throughout the year (Feijoó & Lombardo, 2007). For instance, Gantes & Tur (1995) reported for Las Flores stream 16 species of macrophytes (10 rooted, three submerged, and three free-floating). The taxa most frequently observed are *Azolla filiculoides*, *Lemna* spp., *Potamogeton* spp., *Stuckenia striata*, *Ceratophyllum demersum*, *Rorippa nasturtium-aquaticum*, *Schoenoplectus californicus*, *Ludwigia peploides*, and *Hydrocotyle* spp. Almost all these taxa are native or cosmopolitan, and only *R. nasturtium-aquaticum* was introduced from Europe.

Although the pampean streams are typically characterized by low current velocities, this feature is influential in determining the spatial distribution of the aquatic vegetation patches (Gantes & Sánchez Caro, 2001). Species such as *S. californicus*, *L. peploides*, *Alternantera filoxeroides*, and *Typha dominguensis* are often found in middle and lower reaches, while *S. striata* predominates in stream segments with flow rates greater than $0.40 \text{ m s}^{-1}$ (Rodrigues Capítulo et al., 2002). The macrophyte biomass varies throughout the year and is highest during spring and summer when submerged and emergent (rooted) species predominate (Fig. 5).

The composition of aquatic-plant communities within the pampean streams reflects the eutrophic status of their water and has been related to conductivity level and nitrate concentration (Feijoó & Lombardo, 2007).

Although the pampean streams have beds formed with fine sediments and only occasionally with pebbles or rocks; the environmental heterogeneity results not from the different types and sizes of the substrata but rather from the presence of aquatic plants of diverse architecture that shelter a rich and dense invertebrate community, mainly herbivores that feed on epiphytic algae (Rodrigues Capítulo et al., 2002; Tangorra, 2004).

## Primary production

Primary production in the pampean streams can reach high levels because of the good light reception—this feature being favored by the scarcity of trees along the banks—plus adequate concentrations of nutrients. Nevertheless, an increment in humic substances and variations in the stream's discharge can modify the production values markedly. A study comparing the primary production per area of different communities (macrophytes, epiphyton, and phytobenthos) reported that the macrophyte contribution represented some 60% of the total primary production in summer and 40% in winter (Vilches & Giorgi, 2008). This production decreases after a significant increase in the stream's discharge because of the removal of the macrophytes. As a consequence, epiphyton production also decreases so that the epipelon production that remains becomes a significant contribution to the system's overall production. Vilches & Giorgi (2008), investigating the headwaters of a pampean stream, reported production values for macrophytes of above $14 \text{ g C m}^{-2} \text{ day}^{-1}$ during both summer and winter; for epiphyton of around $9 \text{ g C m}^{-2} \text{ day}^{-1}$ in summer, but above $16 \text{ g C m}^{-2} \text{ day}^{-1}$ during winter; and for epipelon below $0.5 \text{ g C m}^{-2} \text{ day}^{-1}$ in summer, but around $0.3 \text{ g C m}^{-2} \text{ day}^{-1}$ in winter. These production data, however, can vary significantly with the

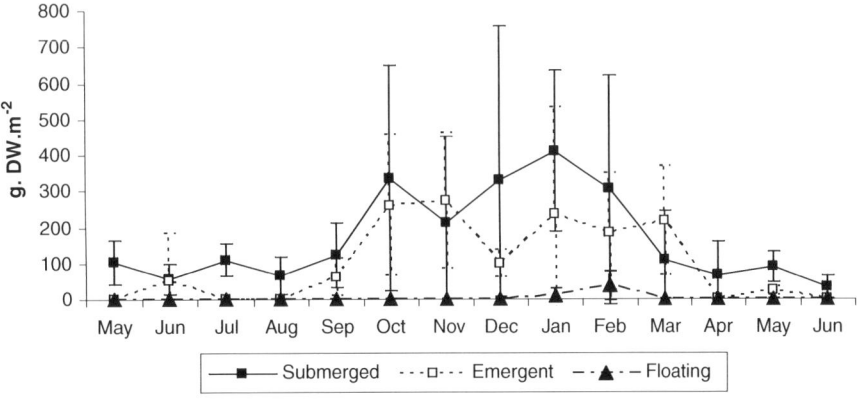

**Fig. 5** Macrophyte biomass variation (mean and standard deviations) in a Las Flores stream reach (1992–1993) (Data from Giorgi, 1998)

nutrient concentrations available in the medium (Sierra, 2009).

A large macrophyte biomass is a significant ecologic feature of pampean streams. Although the plants can be used as a resource by herbivors (e.g., snails and phytophagic insects of terrestrial origin), most of the plant biomass is actually consumed by decomposers and detrivores. The dynamics of this decomposition, however, has still not been characterized.

## Consumers

### Zooplankton

The rotifers constitute one of the more broadly represented groups among the zooplankton and play a key role in the recycling of nutrients and in the secondary production in aquatic environments. Rotifers can use a wide variety of food sources including the algae, protozoa, and detritus and are thus important links for the transfer of carbon within the trophic chain (José de Paggi, 2004). The following species are the most frequently found in pampean streams: *Brachionus angularis*, *B. caudatus*, *B. rotundiformis*, *B. plicatilis*, *Filinia longiseta*, *Keratella cochlearis*, *K tropica*, *Synchaeta* spp., and *Polyarthra vulgaris*. The ciliates are also well represented by *Codonella cratera*, *C. fimbriatus*, and *Tintinidium fluviatile*. The tecamebas of the genera *Arcella*, *Difflugia*, *Centropixis*, and *Euglypha* constitute a tychoplanktonic group habitually present in the zooplankton (Modenutti, 1987). The profile of the planktonic microcrustaceans (copepods and cladocerans) in the rivers is reduced and is, in turn, conditioned by the current velocity and the particular solids in suspension that influence the filtering organisms (José de Paggi, 1984). The most frequently present taxa are: *Notodiaptomus imcompositus*, *Acantocyclops robustus*, *Moina micrura*, *Bosmina huaronensis*, *Ceriodaphnia dubbia*, and *Metacyclops mendocinus*.

### Invertebrates

The low slope of the terrain; the absence of riparian vegetation along the banks; the predominance of clay and slime; and the presence of submerged, emergent, or floating macrophytes are the principal features that condition the development of an invertebrate aquatic fauna—an assemblage that could, in many instances, resemble the one present along the banks of the lentic environments of the region. Only in the mountain headwater streams of Tandil and Ventana and in those of the northeast of the Province of Buenos Aires are possibly found more rheophilic fauna.

The streams that both originate in the plains and are minimally impacted by anthropic activity are characterized by abundant plant detritus; which component attracts the presence of detrivors, such as the tubificid oligochaetes or nematodes, although the nadids of the following genera are also abundant: *Nais*, *Dero*, *Chaetogaster*, *Pristina*, and *Stylaria*. Among the filtering mollusks frequently present within the soft sediment are the nacriferous clam *Diplodon delodontus delodontus* (Pelecypoda) and the gastropod *Heleobia parchappei* (Rodrigues Capítulo et al., 2003).

Associated with the vegetation are the Planorbiidae (*Biomphalaria peregrina* and *Drepanotrema kermatoides*), the Ancylidae (*Uncancylus concentricus*), and the Ampulariidae (*Pomacea canaliculata*. Among the larger-sized crustaceans the decapods *Palaemonetes argentinus* and *Macrobrachium borellii* contribute a significant biomass, while the amphipod *Hyalella curvispina* is well represented particularly in the spring and summer (Fig. 6), (Giorgi, 1998; Rodrigues Capítulo et al., 2002).

Among the abundant microcrustaceans present are the copepods (Cyclopoida, Harpacticoida, and Ostracoda) and the cladocerans Chidoridae, Macrothricidae, and Daphnidae. Likewise, almost always represented are the leeches (Glossiphonidae) and the acarians (Hydrachnidae) (Rodrigues Capítulo et al., 2003).

Among the insects, normally abundant are the larvae of the mayfly families Caenidae (*Caenis nemoralis*), Polymitarcyidae (*Campsurus major*), and Baetidae (*Callibaetis guttatus*) of the order Ephemeroptera along with the preimaginal stages both of the families Ceratopogonidae, Ephydridae, and Chironomidae (*Chironomus*, *Goeldichironomus*, and *Tanypus*) of the order Diptera and of the families Hydrophilidae (*Tropisternus*, *Berosus*), Dytiscidae, and Elmidae of the order Coleoptera. Likewise, frequently encountered are the odonate families Coenagrionidae (*Cyanallagma bonariense*), Aeshnidae (*Aeshna bonariensis*), and Libellulidae (*Mycrathyria dydima*, *Orthemis nodiplaga*, and *Erythordiplax nigricans*) as well as the trichopteron families Polycentropodidae (*Cyrnellus* sp.) and Limnephilidae

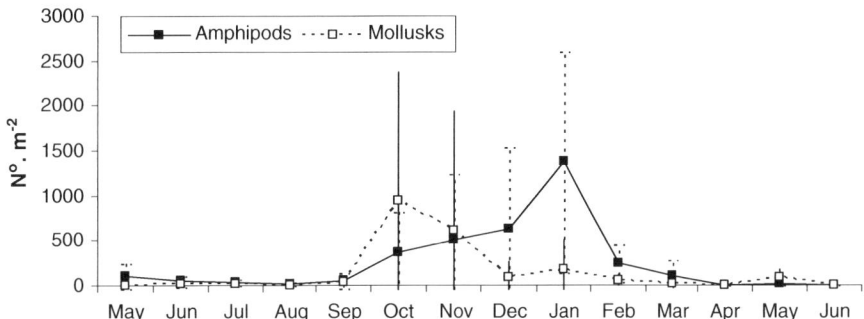

**Fig. 6** Variation of amphipods (*Hyalella* sp.) and mollusks (*Heleobia piscium* and *Uncancylus concentricus bonaeriensis*) (mean and standard deviations) in Las Flores stream reach (1992–1993). (Data from Giorgi, 1998)

(*Verger bruchina*) in addition to a single species of Leptoceridae. Finally, the vegetation is frequented by the hemipteron families Hebridae (*Hebrus*), Belostomatidae (*Belostoma elegans*), and Notonectidae (Rodrigues Capítulo et al., 2003; Ocón & Rodrigues Capítulo, 2004).

The streams with sources in the pampean hills of the Province of Buenos Aires that have a current velocity of >0.5 m s$^{-1}$ develop a more rheophilic aquatic vegetation and have a different fauna. The hydrophyte *Stuckenia striata* provides a high degree of coverage within the segments of greater slope in these systems and is colonized by numerous rheophilic invertebrates, such as the simulids (*Simulium bonariensis*), the mayflies (*Callybaetis*, *Baetis*), the zygopterans (*Oxyagrion hempeli* and *Andinagrion saliceti*), the caddis flies (*Hydroptila sauca*), rheotopic forms of chironomids (*Cricotopus* sp., *Rheotanytarsus* sp., *Hienemanniella* sp., and *Pentaneura* sp.), beetles (Elmidae), and gastropods (*Chilina* sp.) (Rodrigues Capítulo et al., 2002). In the rithronic mesohabitats with greater vegetal coverage, populations of the trichopteran filter *Smicridea pampeana* (Hydropsychidae) are found at high density. Along the banks and within the pools, the emergent plants, such as *Schoenoplectus californicus* and *Thypa dominguensis*, are commonly present; where they support mollusks and odonates. In the lower segments of these hill streams with low current velocity, the macrophytes are abundant, covering between 15 and 60% of the stream beds. There, the macrophytes *Hydrocotyle ranunculoides*, *Alternanthera philoxeroides*, and *Ludwigia peploides* are colonized by the amphipods (*Hyalella curvispina* and *H. pampeana*), the beetles (*Berosus* sp. and *Tropisternus* sp.), the odonates (Libellulidae and Aeshnidae), the gastropods (Planorbidae, Ampullaridae, Hidrobiidae), and the mayflies (Caenidae).

The variation in the composition and relative abundances of the invertebrate assemblages is notable when the summer and winter periods are compared. For example, the Hydrophylidae, Gastropoda, and Chironomidae are the dominant taxa during the coldest months; while the Chironomidae, Simuliidae, and Baetidae are the most common families in summer. For its part, *Hyalella curvispina* is present at high densities within the intermediate stretches of these streams throughout the year (Rodrigues Capítulo et al., 2002) (Fig. 6).

While macrophytes play an important role in the composition of the biota of pampean streams, no obvious difference is apparent among the functional groups that feed directly on them, except for gathering–collector groups that are always dominant. Moreover, no significant changes are evident between the more natural lotic systems and those that have become enriched with organic matter and nutrients (Fig. 7).

Fish

The wider diversity and spatial differences encountered along the watercourses of the Province of Buenos Aires would arise from the direct connection of their lower segments with the Río de La Plata and the Paraná River. The high diversity among the fish that populate the streams of the pampean plains became emphasized in the recent study by Colautti et al. (2010) that identified a total of 24 species throughout four samplings performed at two sites of the La Choza stream in the Buenos Aires province.

The species composition of the ichthyic communities in the pampean plains is strongly linked to the Paraná-Plata basin of South America (Ringuelet, 1975). Of those species, 7% (approximately 24) reach the Salado River (Fig. 1) and its surrounding systems

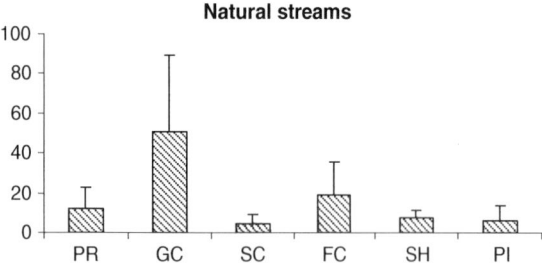

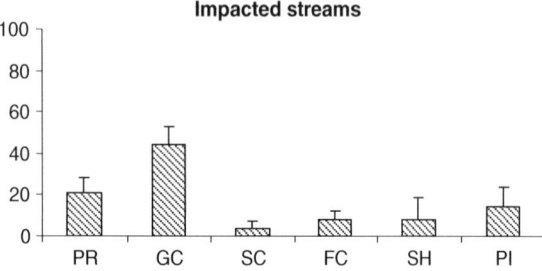

Fig. 7 Principal functional feeding groups of invertebrates (%) in natural and impacted urban streams of pampas. *PR* predators, *GC* gathering collectors, *SC* scrapers, *FC* filtering collectors, *SH* shredders, *PI* piercers (Data from Tangorra, 2004; García et al., 2010)

within the geomorphologic unit known as Depressed Pampa (*Pampa Deprimida*). According to Ringuelet (1975), the distribution of species there appears to coincide with an ecologic area where mainly the low temperatures and excessively salty character of the water become translated into limiting survival conditions.

The most widely represented fish groups are the Characiformes (three families, 10 species) with a predominance of *Cheirodon interruptus*, *Bryconamericus iheringii*, and *Pseudocorynopoma doriae* (Characidae) plus *Cyphocharax voga* (Curimatidae; Almirón et al., 2000; Di Marzio et al., 2003; Remes Lenicov et al., 2005; Fernández et al., 2008; López et al., 2008). Those taxa are predators predominantly of micro- and mesoinvertebrates (Escalante, 1983), and their abundance is favored by the presence of submerged or floating aquatic vegetation.

These characteristics of the habitat in combination with slow current velocities favor the development of the Perciformes (cichlids), likewise predators of micro- and mesofauna. In this regard, the ichthyophage predator *par excellence* within these environments is the species *Hoplias malabaricus* (family Atherinidae) (Ringuelet, 1975), though the ingestion of shrimp and macrophytes is common in its diet (Destefanis & Freyre, 1972; Oliveros, 1980).

The Siluriformes (three families, seven species) in general inhabit slime and clay streambeds and exhibit varied feeding regimens: they are algivorous, consuming micro- and mesoinvertebrates in the benthos, as well as detrivorous (Ringuelet, 1975). Among the Cyprinodontiformes, the species *Jenynsia multidentata* and *Cnesterodon decemmaculatus* have colonized the majority of the basins and would appear to be favored by the presence of hydrophytes, where they forage for food. These two species also seem in general to be relatively tolerant to desiccation, salinization, and lower oxygen concentrations. According to Escalante (1984), the diet of *C. decemmaculatus* is composed basically of the algae Chrysophyta, Chlorophyta, and Cyanophyta; whereas the trophic spectrum of *J. multidentata* is broader, including amphipods (*Hyalella curvispina*), microcrustaceans, and decapods along with the larvae of the chironomids and gastropods. The Symbranchiformes are represented by only *Symbrachus marmoratus* and in general feed on microcrustaceans and small fish in the benthos.

## Climate change

The changes in the climatic variability and the frequency of extreme meteorological events must be considered in determining the probable impacts on flora and fauna and in assessing the adaptive adjustments the latter require to mitigate such perturbations. According to Hulme & Sheard (1999), the mean annual temperature in Argentina increased by 1°C during the last century, with the decade of the 1990s being the warmest. This warming has occurred throughout the entire year, though being more pronounced in winter (June to August). Along with the increase in warming, the frequency of frosts has been diminishing. On the basis of the different alternatives defined in the Special Bulletin Concerning Emission Scenarios (USA) of the Intergovernmental Panel on Climate Change (IPCC), the mean temperature of the Argentine pampean plains is expected to increase by 1°C under conditions of low impact [Scenario B1: a $CO_2$ concentration of 532 parts per million of volume (ppmv) giving a global elevation of 1.2°C by 2080], but will rise by between

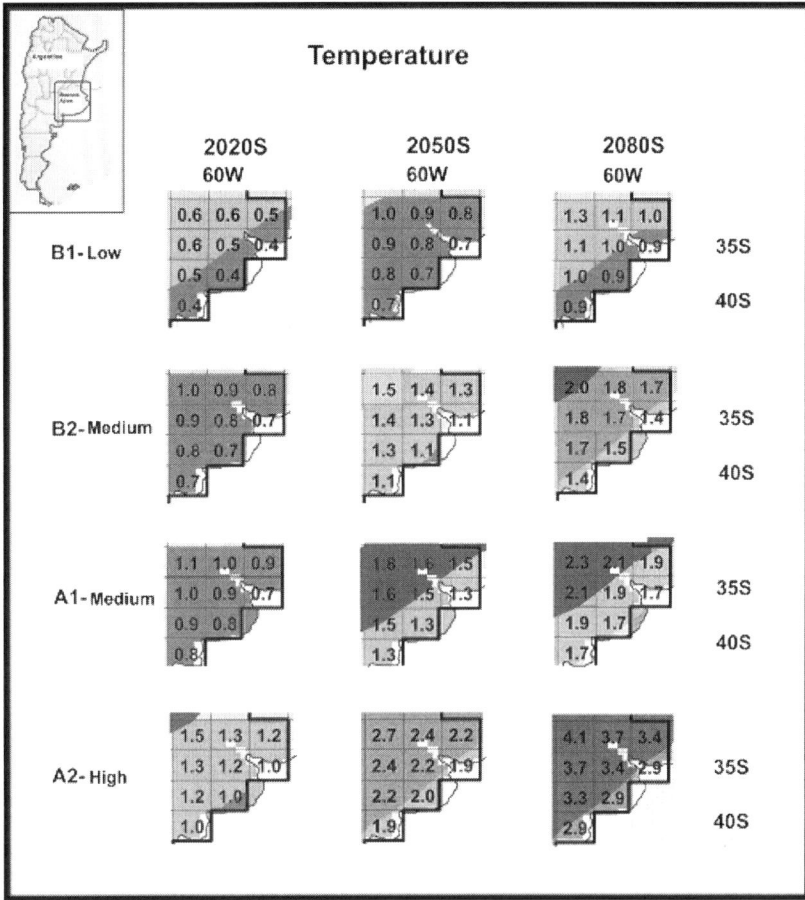

**Fig. 8** The temperatures on different climatic outcomes (in pampean area) defined in the Special Bulletin Concerning Emission Scenarios of the Intergovernmental Panel on Climate Change (IPCC, USA)—pampean region (Modified from Hulme & Sheard, 1999)

2.9 and 3.4°C at high impact (Scenario A2: a $CO_2$ concentration of 721 ppmv yielding an increment of 3.9°C by 2080) (Fig. 8). Under Scenario B1, the precipitation will increase in the pampean region by <5% by 2080, but under Scenario A2 the increment in rainfall will reach nearly 17% by the same year (Fig. 9). These scenarios do not take into consideration "El Niño" events, which are known to affect precipitation in the pampean streams as well (Nuñez, 2009).

## Consequences of global climate changes on pampean fluvial systems

The consequences of alterations in the global climate are especially critical in developing countries from the standpoint of the capacity of their constituent social groups to absorb or mitigate the effects of these changes. This circumstance represents a challenge with respect to the possibility of having the technology, the resources, an appropriate infrastructure, and the basic wherewithal for the adjustments required.

Global climate change will certainly affect the physical, chemical, and hydrologic properties of pampean lotic systems and consequently, the structure and function of the biological communities present.

These future climate-change scenarios proposed for the pampean area project an increment in rainfall that will strongly affect the biological communities. Higher rainfall can increase erosion and generate flooding; thereby increasing the transport of sediments, nutrients, and contaminants to the ocean so as to affect the degree of water mineralization. The changes in the discharge and in the turbidity of the water would affect the residence time and the light penetration of the water column.

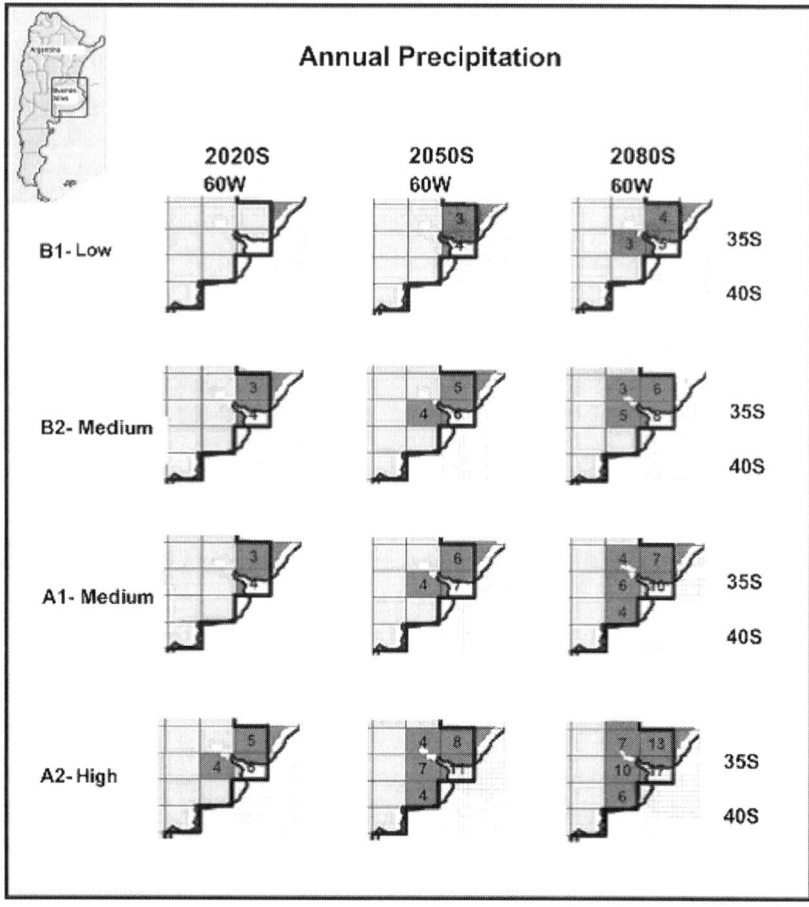

Fig. 9 The precipitations on different climatic outcomes (for pampean area) defined in the Special Bulletin Concerning Emission Scenarios of the Intergovernmental Panel on Climate Change (IPCC, USA)—pampean region (Modified from Hauer & Lamberti, 1996)

The modifications in the use of the soil will probably favor the input of nutrients. This latter effect will benefit autotrophy, particularly by those species capable of generating strategies for surviving in more turbid and enriched environments. An accelerated eutrophication will change the composition of the consumers in favor of herbivores and detritivores. Litter decomposition brought about by bacteria and fungi will become modified by the changes in temperature along with a greater availability of organic matter, both autochthonous and allochthonous. The fungi will make use of a major quantity of vegetal material for decomposition, which possibility would imply an elevation in cellulolytic and ligninolytic enzymes. The alterations in the processes of microbial decomposition will in turn modify the utilization of that material on the part of the invertebrates.

According to Bustingorry (2008), the overflowing of the bodies of water in the pampean plains would evoke a dilution with respect to the salinity without changing the concentration of the nutrients. With the aim of analyzing the effects of increasing nutrient levels within the pampean waterways, a nutrient-enrichment experiment was recently performed in the La Choza Stream, located between 34°39′14″ and 34°47′51″S and 59°10′00″ and 58°3′17″W; the basin's drainage surface is 15,200 ha. Sixty-four per cent of the basin is used for cattle raising and 34% for extensive agriculture, while the remaining 2% is divided between bird raising and urban use. The fertilization experiment was carried out between 2007 and 2008. The immediate objective was to simulate the increase in nutrients in a plains fluvial system with its headwaters subjected to unhabitual hydrologic irregularities attributable to global climatic changes resulting from human activity. The long-term aim was then to examine the resulting changes in the biological structure of the trophic network and the functioning of the metabolic processes in the resident communities. The experiment involved a study of two 100-m stretches of the stream (Control and Treatment) for a period of 9 months prior to the addition of quantities of phosphorus and nitrogen that would triple the average initial concentrations in the treated segment. After the stream fertilization in this manner, both stretches were then monitored for the subsequent 12 months in a follow-up study. This experiment analyzed not only the structural response in the producers (macrophytes, epiphyton, epipelon, and phytoplankton), the consumers (invertebrates and fish), and the decomposers (bacteria), but also both the functional response and the primary production plus the exoenzymatic activities of the epiphyton, phytoplankton, and epipelon communities.

The preliminary results suggest that the bacteria and the algae associated with the sediment responded significantly to the addition of the nutrients (Cochero et al., 2008) by doubling their biomass. An increment in the abundance of the browsers or herbivores within the macroinvertebrate community represented principally by the gastropod mollusks *Pomacea canaliculata* and *Heleobia* sp. also occurred, presumably as a consequence of the increase in the biomass of the primary producers (Rodrigues Capítulo et al., 2008). Changes in the structure of the diatom taxocenosis were also noted with a reduction in the number, diversity, and evenness of species such as *Nitzschia frustulum*, *N. inconspicua*, *N. amphibia* along with an increment in their density relative to the control values (Licursi & Gómez, 2009). In terms of functional responses, fertilization resulted in a duplication in the production of macrophytes and a 68% reduction, on the average, in the concentration of alkaline phosphatase in the treated segment relative to control values (Cochero et al., 2008; Vilches et al., 2008).

Another example of the expected results from a high-temperature scenario is found in the loricariid fish *Hypostomus commersoni*. This species has a low tolerance to cold water, which characteristic limits its southernmost distribution. An increment in the dispersal of this species toward the south and west from 1960s can be explained not only by the implementation of canals to drain off excess water, but also by the accompanying rise in temperature (López & Miquelarena, 1991; Gómez, 2008). Similar predictions can be made with respect to mollusks, insects, and crustaceans, whose distributions were previously relegated to subtropical zones. For example, the displacement of mollusks of the genus *Physa* (Physidae), of Anisoptera such as *Anax amazili*, and of dipteran like *Aedes aegypti* toward the south and west of the pampean area is presently underway (Rodrigues Capítulo, 1981). Another consequences of global warming is an elevation in the sea level. In addition to increasing the erosion of the coastline, this alteration would adversely affect the accessability of the mouths of rivers and streams so as to allow the entrance of estuarial fish into the continental network.

According to Kruse & Mas-Pla (2009), changes in the morphology of the coastal plain of the Pampean region would contribute to the deterioration of groundwater resources within the new climate-change scenarios. Rising sea levels would cause an alteration in groundwater level and a displacement of the border toward the continent, thus causing an incursion of salt water from the ocean into the aquifers. Other negative impacts associated with coastal flooding are storms that would lash the banks of the Rio de la Plata and the Parana River delta and in doing so affect the human activity in that area.

In order to satisfy the need for food resulting from the projected increase in the world population in the years ahead, an increase in agricultural production will be necessary. This situation, coupled with new climate-change scenarios, will lead to profound socioeconomic changes in the pampas region, causing a greater demand for water resources and land uses. These changes in the future are expected on the basis of the increase in pampean agricultural production that has occurred since 1950, partially as the result of an expansion of the agricultural western boundary. During the last decade, there were increases in the cultivated land area of approximately 100,000 km$^2$ and in the population of the province of Buenos Aires of some 8%, with a projected further rise of 4.4% in the latter parameter by the year 2015 (INDEC, 2004).

A final consideration is that even though climate changes can in general be predicted, alterations in the environment introduced by man can also occur at random and for that reason will remain only poorly predictable.

**Acknowledgments** The authors are grateful to Donald Haggerty, PhD, for English language revision of the manuscript and to the anonymous referees for their comments and suggestions that have improved the manuscript. This article is framed in the Project GLOBRIO—(Global Change in fluvial systems) funded by the BBVA.

# References

Almirón, A. E., M. L. García, R. C. Menni, L. C. Protogino & L. C. Solari, 2000. Fish ecology of a seasonal lowland stream in temperate South America. Marine and Freshwater Research 51: 265–274.

Amuchástegui, G., 2006. Relationships among water chemistry, phisiographic features and land use in Pampean streams. Graduate Thesis, National University of Luján.

Andrade, M. I., 1986. Factores de deterioro ambiental en la cuenca del Río Luján. Contribución del Instituto de Geografía, Fac. de Filosofía y Letras (UBA), Buenos Aires.

Bauer, D. E., 2009. Ecología del fitoplancton de arroyos pampeanos y su valor como indicador de la calidad del agua. Tesis doctoral No. 1039, Fac. Cs. Nat. y Museo, Univ. Nac. de La Plata.

Bauer, D. E., J. Donadelli, N. Gómez, M. Licursi, C. Ocón, A. C. Paggi, A. Rodríguez Capítulo & M. Tangorra, 2002. Ecological status of the Pampean plain streams and rivers (Argentina). Verhandlungen Internationale Vereinigung für Theoretische und Angewandte Limnologie 28: 259–262.

Berner, E. K. & R. A. Berner, 1987. The Global Water Cycle. Prentice-Hall, Englewood, NJ.

Binkley, D., G. G. Ice, J. Kaye & C. A. Williams, 2004. Nitrogen and phosphorus concentrations in forest streams of the United States. Journal of the American Water Resources Association 40: 1277–1291.

Brailovsky, A. E. & D. Foguelman, 1992. Memoria Verde. Historia Ecológica de la Argentina. Editorial Sudamericana, Buenos Aires.

Burgueño, G., 2005. Elementos para el plan de manejo del área natural protegida Dique ingeniero Roggiero. Tesis de licenciatura en planificación y diseño del Paisaje, UBA, Buenos Aires.

Bustingorry, S., 2008. Metabolismo de la tierra y cambio climático. Bol. Informativo Fac. Cs. Fcas., Mat y Nat. Buenos Aires.

Cabrera, A. & A. Willink, 1980. Biogeografía de América Latina. OEA, Serie Biología, Monog. No 13, Versión Corregida, Secretaría General de los Estados Americanos, Programa Regional de Desarrollo Científico y Tecnológico, Washington DC.

Cochero, J., M. Licursi, D. E. Bauer, M. E. Sierra, A. Giorgi, S. Sabater & N. Gómez, 2008. Dinámica del microbentos de un arroyo pampeano IV Congreso Argentino de Limnología. Res. CAL 4. Bariloche: 67.

Colautti D. C., M. E. Maroñas, E. D. Sendra, L. C. Protogino, F. Brancolini & D. Campanella, 2010. Ictiofauna del arroyo La Choza, cuenca del río de la Reconquista (Buenos Aires, Argentina). Biología Acuática 26: 55–63.

del Giorgio, P. A., A. L. Vinocur, R. J. Lombardo & H. G. Tell, 1991. Progressive changes in the structure and dynamics of the phytoplankton community along a pollution gradient in a lowland river—a multivariate approach. Hydrobiologia 224: 129–154.

Destefanis, S. & L. R. Freyre, 1972. Relaciones tróficas de peces de la laguna de Chascomús, con un intento de referenciación ecológica y tratamiento bioestadístico del espectro trófico. Acta Zoologica Lilloana 29: 17–33.

Di Marzio, W., M. Tortorelli & L. Freyre, 2003. Diversidad de peces en un arroyo de llanura. Limnetica 22: 71–76.

EPA, 2000. Nutrient Criteria Technical Guidance Manual. Rivers and Streams. EPA-822-B-00-002, Washington DC.

Escalante, A. H., 1983. Contribución al conocimiento de las relaciones tróficas de peces deagua dulce del área platense II: Otros tetragonopteridae. Limnobios 2: 402–479.

Escalante, A. H., 1984. Contribución al conocimiento de las relaciones tróficas de peces de agua dulce del área platense. IV. Dos especies de Cichlidae y Miscelanea. Limnobios 8: 562–578.

Feijoó, C. S. & R. J. Lombardo, 2007. Baseline water quality and macrophyte assemblages in Pampean streams: a regional approach. Water Research 41: 1399–1410.

Feijoó, C., A. Giorgi, M. E. García & F. Momo, 1999. Temporal and spatial variability in streams of a pampean basin. Hydrobiologia 394: 41–52.

Fernández, E., R. Ferriz, C. Bentos & G. López, 2008. Ichthyofauna of two streams in the high basin of the Samborombón River, Buenos Aires province, Argentina. Revista del Museo Argentino de Ciencias Naturales 10: 147–154.

Fidalgo, F., 1983. Algunas características de los sedimentos superficiales en la cuenca del Río Salado y en la Pampa.Ondulada. Actas del Coloquio de Olavarría 2: 1045–1067.

Frenguelli, J., 1956. Rasgos generales de la hidrografía de la Provincia de Buenos Aires. Ministerio de Obras Públicas de la provincia de Buenos Aires 62: 1–19.

Gantes, H. P. & A. Sánchez Caro, 2001. Environmental heterogeneity and spatial distribution of macrophytes in plain streams. Aquatic Botany 70: 225–236.

Gantes, H. P. & N. M. Tur, 1995. Variación temporal de la vegetación acuática en un arroyo de llanura. Revista Brasileira de Biologia 55: 259–266.

García, M. E., A. Rodrigues Capítulo & L. Ferrari, 2010. El ensamble de invertebrados y la calidad del agua: indicadores taxonómicos y funcionales en arroyos pampeanos. Biología Acuática 25: 1–15.

Giorgi, A., 1998. Factores reguladores del fitobentos de arroyos. Tesis doctoral. Tesis No: 0711. Fac. Cs. Nat. y Museo, Univ. Nac. de La Plata.

Giorgi, A., D. Acosta, A. Alasia, C. Martínez, L. Miranda, N. Mufato & J. I. Pamio, 2004. Capacidad de transporte y retención de arroyos pampeanos. Resúmenes IV Congreso Ibérico de Limnología: 29.

Giorgi, A., C. Feijoó & H. G. Tell, 2005. Primary producers in a Pampean stream: temporal variation and structuring role. Biodiversity and Conservation 14: 1699–1718.

Gómez, N., 1998. Use of epipelic diatoms for evaluation of water quality in the Matanza–Riachuelo (Argentina) a pampean plain river. Water Research 32: 2029–2034.

Gómez, S. E., 2008. Notas sobre el cambio ambiental en ictiología. Biología Acuática 24: 1–6.

Gómez, N. & M. Licursi, 2001. The Pampean Diatoms index, (IDP). For assessment of Rivers and streams in Argentina. Aquatic Ecology 35: 173–181.

Gómez, N. & A. Rodrigues Capítulo, 2001. Los bioindicadores y la salud de los ríos. V Seminario Internacional de Ingeniería y Ambiente. Indicadores Ambientales 2000, Fac. Ing. Univ. Nac. de La Plata.: 109–118.

Hauer, F. R. & G. A. Lamberti, 1996. Methods in Stream Ecology. Academic Press, USA.

Hulme, M. & N. Sheard, 1999. Escenarios de Cambio Climático para Argentina. Unidad de Investigación Climática, Norwich, Reino Unido.

INDEC, 2004. Análisis demográficos del Instituto Nacional de Estadística y Censos. Ministerio de Economía y Producción Secretaría de Política Económica. Serie N 30. http://www.indec.gov.ar/default.htm.

José De Paggi, S. B., 1984. Estudios limnológicos en una sección transversal del tramo medio del río Paraná: distribución estacional del zooplancton. Revista de la Asociacion de Ciencias Naturales del Litoral 15: 135–155.

José De Paggi, S. B., 2004. Diversidad de Rotíferos Monogononta del Litoral Fluvial Argentino. Insugeo 12: 185–194.

Kruse, E. & J. Mas-Pla, 2009. Procesos hidrogeológicos y calidad del agua en acuíferos litorales. In Mas-Pla, J. & G. M. Zuppi (eds), En gestión ambiental integrada de áreas costeras. Rubes Editorial, España: 284 pp.

Licursi, M. & N. Gómez, 2002. Benthic diatoms and some environmental conditions in three lowland streams. Annales de Limnologie 38: 1090–1118.

Licursi, M. & N. Gómez, 2009. Effects of dredging on benthic diatom assemblages in a lowland stream. Journal of Environmental Management 90: 973–982.

López, H. L. & A. M. Miquelarena, 1991. Los Hypostominae (Pisces: Loricariidae) de Argentina. In de Castellanos, Z. A. (ed.), Fauna de Agua Dulce la República Argentina, Vol. 40. PROFADU-CONICET, La Plata, Argentina: 1–64.

López, H. L., R. C. Menni, M. Donato & A. M. Miquelarena, 2008. Biogeographical revision of Argentina (Andean and Neotropical Regions): an analysis using freshwater fishes. Journal of Biogeography 35(9): 1564–1579.

Marano, A. V., M. Barrera, M. M. Steciow, J. Donadelli & M. C. N. Saparrat, 2008. Frequency, abundance and distribution of zoosporic organisms from Las Cañas stream (Buenos Aires, Argentina). Mycologia 100: 691–700.

Mcallister, D. E., A. L. Hamilton & B. Harvey, 1997. Global freshwater biodiversity: striving for the integrity of freshwater ecosystems. Sea Wind 11: 1–145.

Modenutti, B. E., 1987. Caracterización y variación espacial del zooplancton del arroyo Rodriguez (Provincia de buenos Aires, Argentina). Anales del Instituto Ciencias del Mar y Limnologia, Universidad Nacional Autonoma de Mexico 14: 21–28.

Moyle, P. B. & R. A. Leidy, 1992. Loss of biodiversity in aquatic ecosystems: evidence from fish faunas. In Fiedler, P. L. & S. K. Jain (eds), Conservation Biology: The Theory and Practice of Nature Conservation, Preservation, and Management. Chapman and Hall, New York.

Mugni, H. D., 2009. Concentración de nutrientes y toxicidad de pesticidas en aguas superficiales de cuencas rurales. Tesis No. 1010. Fac. Cs. Nat. y Museo, Univ. Nac. La Plata.

Mugni, H., S. Jergentz, R. Schulz, A. Maine & C. Bonetto, 2005. Phosphate and nitrogen compounds in streams of Pampean Plain areas under intensive cultivation (Buenos Aires, Argentina). In Serrano, L. & H. L. Golterman (eds), Phosphates in Sediments. Backhuys Publishers, The Netherlands: 163–170.

Nuñez, S., 2009. El cambio climático global. Una problemática compleja págs 77–110. In: Calidad ambiental, una responsabilidad compartida. Informe sobre desarrollo Humano en la Provincia de Buenos Aires 2008-2009-EUDEBA-Banco Provincia.

Ocón, C. & A. Rodrigues Capítulo, 2004. Presence and abundance of Ephemeroptera in relation with habitat conditions in pampean streams (Buenos Aires, Argentina). Archiv für Hydrobiologie 159: 473–487.

Oliveros, O., 1980. Campaña limnologica Keratella I en el Rio Paraná Medio: aspectos tróficos de los peces de ambientes leníticos. Ecología 4: 115–126.

Omernik, J. M., 1977. Nonpoint Source-Stream Nutrient Level Relationships: A Nationwide Study. EPA-600/3-77-105, US Environmental Protection Agency, Environmental Research Laboratory, Corvallis, Oregon.

Paggi, J. C. & S. José de Paggi, 2001. Efecto del Herbicida Glifosato sobre el Zooplancton de Agua Dulce: Un Experimento a nivel de mesocosmo. Resúmenes Actas del V Congreso Latinoamericano de Ecología 10–15: 19.

Papadakis, J., 1980. El suelo. Albatros, Buenos Aires.

Parodi, L. R., 1942. ¿Por qué no existen bosques naturales en la llanura bonaerense? Revista del Centro de Estudiantes de la Facultad de Agronomía de la Universidad Nacional de Buenos Aires 30: 387–390.

Pérez, G. L., A. Torremorel, H. Mugni, M. Rodrigues, S. Vera, M. Do Nacimento, L. Allende, R. Bustingorry, M. Escaray, M. Ferraro, I. Izaguirre, H. Pizarro, C. A. Bonetto, D. P. Morris & H. Zagarese, 2007. Effects of the herbicide roundup of freshwater microbial communities: a mesocosm study. Ecological Applications 17: 2310–2322.

Prieto, A. R., 1996. Late quaternary vegetational and climatic changes in the Pampa grassland of Argentina. Quaternary Research 45: 73–88.

Remes Lenicov, M., D. C. Colautti & H. L. López, 2005. Ictiofauna de un ambiente lótico suburbano: el arroyo Rodríguez (Buenos Aires, Argentina). Biología Acuática 22: 223–230.

Ringuelet, R. A., 1962. Ecología Acuática Continental. EUDEBA, Buenos Aires.

Ringuelet, R. A., 1975. Zoogeografía y ecología de los peces de aguas continentales de Argentina y consideraciones sobre las áreas ictiológicas de América del Sur. Ecosur 2: 1–122.

Rodrigues Capítulo, A., 1981. Presencia de Anax amazili (Odonata Anactinae) en la República Argentina, Algunos datos acerca de su comportamiento y cálculo respirométrico a diferentes temperaturas. Limnobios 2: 207–214.

Rodrigues Capítulo, A., 1999. Los macroinvertebrados como indicadores de La calidad de ambientes lóticos en el área pampeana. Revista de la Sociedad de Entomología Argentina 58: 2008–2217.

Rodrigues Capitulo, A., A. C. Paggi, I. I. César & M. Tassara, 1997. Monitoreo de la calidad ecológica de la cuenca Matanza Riachuelo a partir de los meso y macroinvertebrados. Resúmenes II Congreso Argentino de Limnología. Buenos Aires: 138.

Rodrigues Capítulo, A., A. C. Paggi & C. S. Ocón, 2002. Zoobenthic communities in relation with slope, substrate heterogeneity and urban disturbances on pampean hills streams (Argentina). Verhandlungen Internationale Vereinigung für Theoretische und Angewandte Limnologie 28: 1267–1273.

Rodrigues Capítulo, A., C. S. Ocón & M. Tangorra, 2003. Una visión bentónica de ríos y arroyos pampeanos. Biología Acuática 21: 1–17.

Rodrigues Capítulo, A., C. Ocón, A. Cortelezzi, I. Muñoz, V. López, X. Benito Granell & S. Torres, 2008. Estequiometría ecológica e isótopos estables en macroinvertebrados de un arroyo de llanura. Resumenes Cal 4. Bariloche: 47.

Sabater, S., 2008. Alterations of the global water cycles and their effects on river structure, function and services. Freshwater Reviews 1: 75–88.

Sala, J. M., E. E. Kruse, A. Rojo, P. Laurencena & L. Varela, 1998. Condiciones hidrológicas en la Provincia de Buenos Aires y su problemática. Cátedra de Hidrología General, Facultad de Ciencias Naturales y Museo, UNLP, Publicación Especial.

Salazar, R. H., D. Luzzi & C. Lacoste, 1996. Cuencas Hidricas Contaminación, evaluación de riesgo y saneamiento. Inst. Prov. Medio Ambiente Prov. Buenos Aires: 184 pp.

Salibián, A., 2005. Ecotoxicological assessment of the highly polluted Reconquista River of Argentina. Reviews of Environmental Contamination and Toxicology 185: 35–65.

Sierra, M. V., 2009. Microbentos de sistemas lóticos pampeanos y su relación con la calidad del agua: respuestas estructurales y funcionales. Tesis Doctoral No. 1014. Fac. Cs. Nat. y Museo, Univ. Nac. de La Plata.

Stevenson, R. J., M. I. Bothwell & R. L. Lowe. 1996. Algal ecology. Academic Press.

Tangorra, M., 2004. Colonización y descomposición de especies vegetales por invertebrados en sistemas lóticos pampásicos. Tesis doctoral No. 0868 Fac. Cs. Nat. Univ. Nac. La Plata.

Vera, M. S., L. Lagomarsino, M. Sylvester, G. L. Pérez, P. Rodríguez, H. Mugni, R. Sinistro, M. Ferraro, C. Bonetto, H. Zagarese & H. Pizarro, 2009. New evidences of Roundup® (glyphosate formulation) impact on the periphyton community and the water quality of freshwater ecosystems. Ecotoxicology. doi:10.1007/s10646-009-0046-7.

Vilches, C. & A. Giorgi, 2008. Metabolismo de productores de un arroyo pampeano. Biología Acuática 24: 87–93.

Vilches, C., A. Giorgi, L. Leggieri, N. Ferreiro, E. Troitiño, V. Sierra & S. Sabater, 2008. Efecto del enriquecimiento de nutrientes en el metabolismo de macrófitos y fitobentos. Res. V Cong. de Ecología y Manejo de Ecosistemas Acuáticos Pampeanos: 22.

Zárate, M., R. A. Kemo, M. Espinosa & L. Ferrero, 2000. Pedosedimentary and palaeo-environmental significance of a Holocne alluvial sequence in the southern Pampas, Argentina. The Holocene 10: 481–488.

Hydrobiologia (2010) 657:71–88
DOI 10.1007/s10750-010-0296-6

GLOBAL CHANGE AND RIVER ECOSYSTEMS

# Factors regulating epilithic biofilm carbon cycling and release with nutrient enrichment in headwater streams

Stream biofilm carbon cycling

Susan E. Ziegler · David R. Lyon

Received: 17 July 2009 / Accepted: 4 May 2010 / Published online: 20 May 2010
© Springer Science+Business Media B.V. 2010

**Abstract** This study uses the results from in situ $^{13}$C-labeling experiments conducted in six streams representing a gradient in nutrient enrichment to explore how nutrient availability, stoichiometry, and the composition of active biofilm phototrophs may regulate C cycling in epilithic biofilms. Carbon cycling was tracked through epilithic biofilm communities by assessing net primary production (NPP) and $^{13}$C-labeling of biofilm phospholipids fatty acids (PLFA), and stream water dissolved organic carbon (DOC) within light and dark enclosure incubations. We used generalized linear models coupled with an information-theoretic approach for model selection to assess which factors most influenced C exchange within and DOC release from these biofilms. The ratio of new C incorporated into heterotrophic bacterial PLFA ia15:0 to total polyunsaturated fatty acids (PUFA) indicated that greatest algal–bacterial exchange occurred in the two most nutrient-poor streams. Further, this ratio was best predicted by newC$_{18:3\omega3}$/PUFA suggesting increased relative activity of some algae relates to reduced algal–bacterial C exchange within these biofilms. Net release of DOC represented 2–45% of NPP with greatest release of DOC having occurred in the two most nutrient-rich streams. Further, the model selection indicated that newC$_{18:3\omega3}$/PLFA was the only highly plausible explanatory factor for net DOC release, while a combination of NPP and newC$_{18:3\omega3}$/PLFA was a strong predictor of the quantity of new C released as DOC. The results presented here indicate factors regulating or correlating with the activity of green algae in these biofilms regulated the exchange of C within and DOC release from these biofilms. This suggests increased algal exudation and greater biofilm development with nutrient enrichment may increase DOC release but reduce bacterial use of autochthonous C within these biofilms.

**Keywords** Dissolved organic carbon ·
Bacterial–algal exchange · Exudation ·
$^{13}$C-PLFA · Carbon cycling · Epilithic biofilm

Guest editors: R. J. Stevenson, S. Sabater / Global Change and River Ecosystems – Implications for Structure, Function and Ecosystem Services

S. E. Ziegler (✉)
Department of Earth Sciences, Memorial University of Newfoundland, 4063 Alexander Murray Building, St. John's, NL A1B 3X5, Canada
e-mail: sziegler@mun.ca

D. R. Lyon
Environmental Dynamics Program, University of Arkansas, 113 Ozark Hall, Fayetteville, AR 72701, USA

## Introduction

Global environmental change and its impacts on the terrestrial landscape include direct and indirect increases in nutrient input to watersheds where impacts are typically greatest in the most biogeochemically active regions—headwater streams (Vitousek et al.,

1997). Direct anthropogenic nutrient sources as well as altered nutrient cycling in the landscape due to human activities such as agriculture, urbanization and forestry practices, insect infestation, and climate change have all resulted in the globally ubiquitous nutrient enrichment of aquatic ecosystems. The consequences of this nutrient enrichment include both localized phenomena, such as eutrophication, species extinction, water quality degradation, toxic algal blooms and fish kills, as well as global scale consequences, such as the frequent development and persistence of dead zones (Rabalais, 2002) and altered N-cycling and resulting increased $N_2O$ fluxes (Seitzinger & Kroeze, 1998).

Headwater streams are the hot spots for nutrient and organic matter processing within watersheds, this key role in the landscape is primarily due to the large substrate surface area relative to water volume which results in a strong effect on the downstream export of nutrients and organic matter (Alexander et al., 2000; Peterson et al., 2001). Microbial biofilms, attached to these surfaces in headwater streams, are largely responsible for the nutrient and organic matter uptake and transfer to higher trophic levels (Grimm, 1987; Dodds et al., 2000). Changes in the biological and physical structure of biofilms can alter the exchange of stream water nutrients and cause variation in substrate availability for microbial activity across the biofilm interior and perimeter (Arnon et al., 2007). Further, the organic carbon supplies from algae and cyanobacteria can be a significant source of energy for heterotrophic bacteria in biofilms making such associations critical to biofilm function (Murray et al., 1986; Neely & Wetzel, 1995). Though structurally and functionally variable, such biofilm communities do respond readily to external factors including nutrient availability in stream water (Bothwell, 1985; Lohman et al., 1991; Guasch et al., 1995; Stelzer & Lamberti, 2001).

Nutrient enrichment of the aquatic environment can significantly alter the structure of phototrophic communities (Pringle, 1990), with the identification and community structure of phototrophs having been used as indicators of nutrient pollution and impact (Patrick, 1972; Patrick & Palavage, 1994). Further, studies have indicated changes in algal species richness with increased nutrient availability (Mccormick & Stevenson, 1991; Biggs & Smith, 2002; Passy, 2008) with emphasis on benthic community nutrient content as the key indicator of nutrient availability given the role of N-fixation in N-deplete streams (Peterson & Grimm, 1992; Biggs & Smith, 2002). Changes in stream biofilm phototrophic composition can be important to both trophodynamics and biogeochemistry of streams due to the varied palatability of different algae and cyanobacteria to grazers (Gregory, 1983; Mccormick & Stevenson, 1991; Steinman, 1996) and potential differences in how they support heterotrophic bacteria or generate stream water DOC. At this point, however, there is limited understanding of the linkage among nutrient enrichment, biofilm phototrophic communities, and carbon cycling in streams.

The extent of nutrient cycling in headwater streams typically depends on the availability of dissolved organic carbon (DOC) for heterotrophic use (Bernhardt & Likens, 2002). Autotrophic-derived DOC can be a key energy source for heterotrophic activity in streams (Kaplan & Bott, 1982, 1989). Furthermore, positive relationships between algal and bacterial biomass within stream biofilms have been observed in streams (Romani & Sabater, 1998; Rier et al., 2007) and attributed to habitat structure such as substrate surface area for algae (Pusch et al., 1998) as well as the use of algal-released DOC by heterotrophic bacteria (Haack & Mcfeters, 1982; Carr et al., 2005). The source of autochthonous DOC in streams may be regulated by nutrient availability. For example, nutrient enrichment can regulate the release of DOC by algae typically decreasing the fraction of primary production released as DOC (Fogg, 1983). In pelagic systems, elevated nutrients have been shown to stimulate algal DOC exudation and subsequent bacterial activity during the late stages of blooms when algae become nutrient limited (Norrman et al., 1995; Van Den Meersche et al., 2004). In stream biofilms, however, stimulation of algal production occurs with nutrient enrichment but this increased production can become decoupled from bacterial activity (Scott et al., 2008). This suggests that potential DOC release by algae can become decoupled from heterotrophic bacterial production in stream biofilms.

Recent studies conducted in pairs of contrasting streams indicate that the nutrient status of streams may regulate both the exchange of C within epilithic biofilms and between the biofilm and stream water DOC pool (Lyon & Ziegler, 2009; Ziegler et al., 2009). In each of these investigations, a greater proportion of autotrophic C was incorporated into heterotrophic bacterial biomass within the epilithic biofilms in the nutrient-poor relative to the nutrient-rich streams. In addition to regulating the exchange of C between heterotrophs and autotrophs, general nutrient status appears to influence the fraction of NPP released as DOC. Factors, such as nutrient availability and limitation or microbial community composition, that regulate this observed variation in C flow remain unknown but are important to our understanding of how environmental change may impact the biogeochemistry of stream ecosystems.

Here, we investigate the relative importance of nutrient status and the composition of active phototrophs in driving the previously observed differences in C cycling and exchange, by analyzing results of $^{13}$C-labeling in stream enclosure experiments conducted within the same six headwater streams, ranging in nutrient status, over three seasons. This study differs from these previous two studies in that it uses a new set of data collected in fall 2005 in combination with the existing data sets across all six study streams to directly assess how nutrient availability and community composition of active biofilm phototrophs varies among these sites and how these attributes may regulate the variation in C flow and release in these stream ecosystems. The limited number of times these in situ $^{13}$C-labeling experiments were conducted across these six streams precludes us from statistically testing many potential factors regulating C cycling and flow in these biofilms. Rather the goal of this study was to consider multiple working hypotheses (sensu Chamberlain, 1897) in order to assess the relative importance of nutrient availability and the composition of active phototrophic groups in regulating C flow and exchange in stream biofilms. Specifically, we exploit these data to assess the relative importance of nutrient availability and the relative phototrophic activity of a single group (green algae) in regulating; (1) DOC source and flux, and (2) algal–bacterial C exchange in these biofilms. The significance of this work, determining the relative importance of tracking autotrophic C in stream biofilms, lies in its use to refine future approaches to better understand the effects of global change on stream biogeochemistry.

## Materials and methods

### Study sites

Biofilm-colonized rock substrate was collected from and incubated at mid-stream sites in six headwater streams (Huey Hollow, Cecil Creek, Mill Creek, Spavinaw, Moore Creek, and Columbia Hollow) located in northwest Arkansas region of the Ozark Plateau, USA. Details regarding the stream and watershed land cover for Cecil, Mill, Spavinaw, and Columbia can be found in Lyon & Ziegler (2009), and for Huey and Moore they can be found in Ziegler et al. (2009). Briefly, the watersheds ranged from 11 to 57 km$^2$, and all streams were alkaline with high dissolved inorganic carbon (DIC) concentrations and covered over a 100-fold range of total dissolved nitrogen and soluble reactive phosphorus (SRP) concentrations representative of the wide range of nutrient loading existing in the region (Table 1). All research sites were located in deciduous forested regions with wooded riparian zones except Moore and Columbia, which were both located in pasture with primarily open to sparsely treed riparian zones.

### Field enclosure experiments for investigating biofilm C cycling

Data analyzed in this study were collected from in situ field enclosure experiments that took place in the winter, summer, and spring of 2003 in Huey and Moore as part of a study focused on one pair of streams contrasting in nutrient enrichment history (Ziegler et al., 2009) and in spring 2006 in Cecil, Mill, Spavinaw, and Columbia as part of a separate study of two pairs of streams contrasting, again, in nutrient enrichment history (Lyon & Ziegler, 2009). Additionally, the data used in this study included results from four new experiments conducted in Cecil, Mill, Spavinaw, and Columbia in fall 2005. Details of the methods for the enclosure experiments can be found in Ziegler et al. (2009) and Lyon & Ziegler (2009). Briefly, rocks from the stream benthos representative of the typical size, surface area, and biofilm colonization of the stream were randomly distributed into three transparent and three opaque 12 L polycarbonate enclosures following removal of large invertebrates. Some grazers such as small snails and chironomids, however, remained in the enclosures. An additional set of rocks were immediately sampled for initial biofilm properties [dry weight, %C, %N, chlorophyll content, $\delta^{13}$C, $\delta^{15}$N, phospholipid fatty acids (PLFA) concentration and C isotopic composition] by scraping, brushing, and rinsing the rocks. Eight liters of ambient stream

**Table 1** Average water pH, dissolved inorganic nitrogen (sum of nitrate, nitrite, and ammonium), dissolved soluble reactive phosphorus (DIP), dissolved organic carbon (DOC), biofilm biomass given as total biofilm C in the light enclosures (mmol C), molar carbon to nitrogen ratio of epilithic biofilms, chlorophyll a content given as µg Chl a per mg biofilm C, stable nitrogen isotopic values of biofilms in ‰, and the polyunsaturated phospholipid fatty acid (PUFA) content given as µg PUFA per g biofilm C at each of the six study streams in each season the experiments were conducted. All values are given as mean ± 1 standard deviation. All measurements where concentrations or rates were below detection are designated BD. The photosynthetically active radiation (PAR) measured at the water surface within 1 h of solar noon during the incubation is provided as a means to compare light availability among the sites and seasons

| Site | Season | pH | DIN (µM) | DIP (µM) | DOC (µM) | Biofilm biomass | Biofilm C:N | Chl a:C | Biofilm $\delta^{15}$N | PUFA | PAR (mW cm$^{-2}$) |
|---|---|---|---|---|---|---|---|---|---|---|---|
| Huey | W | 7.7 | 3.5 ± 0.3 | 0.3 ± 0.01 | 19 ± 1 | 30 ± 10 | 11.7 ± 0.4 | 8 ± 0.2 | −1.0 ± 0.6 | 0.8 ± 0.2 | 2.2 |
|  | Sp | 7.9 | 1.5 ± 0.3 | 0.3 ± 0.04 | 61 ± 4 | 55 ± 12 | 11.4 ± 0.1 | 4 ± 0.5 | −0.5 ± 0.9 | 1.3 ± 0.3 | 4.4 |
|  | Su | 7.9 | 3.6 ± 0.7 | 0.3 ± 0.03 | 80 ± 3 | 21 ± 8 | 11.6 ± 0.4 | 5 ± 0.7 | −0.9 ± 0.4 | 1.4 ± 0.1 | 3.4 |
| Cecil | Sp | 8.3 | 4.3 ± 0.3 | 0.2 ± 0.02 | 165 ± 3 | 20 ± 3 | 15.2 ± 1.1 | 4 ± 1.4 | 1.4 ± 0.4 | 0.9 ± 0.2 | 3.4 |
|  | Fall | 9.0 | 2.7 ± 0.4 | 0.02 ± 0.001 | 67 ± 6 | 66 ± 9 | 17.3 ± 0.5 | 5 ± 0.8 | 0.1 ± 0.2 | NA | NA |
| Mill | Sp | 8.0 | 27 ± 0.7 | BD | 198 ± 9 | 157 ± 29 | 19.9 ± 0.7 | 7 ± 1.0 | 6.3 ± 0.1 | 1.1 ± 0.2 | 3.2 |
|  | Fall | 9.0 | 63 ± 0.9 | BD | 82 ± 6 | 70 ± 14 | 17.3 ± 1.4 | 6 ± 0.5 | 6.6 ± 0.2 | NA | NA |
| Spavinaw | Sp | 7.6 | 176 ± 2 | 0.7 ± 0.03 | 41 ± 2 | 47 ± 14 | 13.8 ± 1.2 | 6 ± 1.1 | 5.1 ± 0.6 | 1.0 ± 0.1 | 3.3 |
|  | Fall | 8.3 | 239 ± 2 | 0.6 ± 0.01 | 40 ± 1 | 53 ± 5 | 11.1 ± 0.3 | 12 ± 1.1 | 4.3 ± 0.2 | NA | NA |
| Moore | W | 7.4 | 24 ± 2 | 8.7 ± 1.1 | 284 ± 7 | 45 ± 12 | 8.2 ± 0.5 | 22 ± 3.2 | 5.9 ± 0.4 | 2.0 ± 0.1 | 4.6 |
|  | Sp | 7.3 | 3.0 ± 0.5 | 11.5 ± 0.3 | 644 ± 4 | 86 ± 23 | 8.2 ± 0.7 | 11 ± 1.0 | 8.2 ± 0.3 | 2.5 ± 0.5 | 7.1 |
|  | Su | 7.5 | 1.3 ± 0.2 | 1.2 ± 0.1 | 613 ± 5 | 54 ± 19 | 11.7 ± 0.8 | 7 ± 1.9 | 7.5 ± 0.6 | 2.0 ± 0.4 | 7.8 |
| Columbia | Sp | 7.5 | 566 ± 8 | 14.8 ± 0.1 | 390 ± 7 | 25 ± 9 | 7.4 ± 0.1 | 18 ± 2.1 | 23.1 ± 2.0 | 2.9 ± 0.4 | 5.2 |
|  | Fall | 7.8 | 1800 ± 0 | 18.2 ± 0.001 | 381 ± 0.1 | 42 ± 15 | 6.8 ± 0.1 | 25 ± 3.7 | 4.5 ± 1.0 | NA | 4.4 |

water, previously spiked with a solution of NaH$^{13}$CO$_3$ (final $\delta^{13}$C-DIC approximately 750‰) and equilibrated, was dispensed into each enclosure, and circulation of water (0.26 l min$^{-1}$) within the individual enclosures was started using silicone tubing and individual peristaltic pumps (APT Industries Model: SP301C.0045SP). At 15 min and 8–8.5 h after commencing circulation, measurements of DO, temperature, and pH were made and water samples collected for DIC, DOC, ammonium (NH$_4^+$), nitrate + nitrite (NO$_{2/3}^-$), soluble reactive phosphorus (SRP), $\delta^{13}$C-DOC, and $\delta^{13}$C-DIC capturing the initial and final samples for each incubation. The entire biofilm community from each enclosure was collected for final biofilm properties as described for the initial biofilm sample. The chlorophyll a content of the biofilms were determined by extraction of lyophilized biofilm in 90% acetone followed by centrifugation and absorbance measurements made on a Shimadzu UV-1201 spectrophotometer (Lorenzen, 1967; Jeffrey & Humphrey, 1975).

In order to determine the importance of stream water respiration and net primary production (NPP), twelve 300 ml biochemical oxygen demand (BOD) bottles were filled with stream water at the beginning of the incubation period. Four of the bottles were measured for initial DO by Winkler titration on an automated titrator with potentiometric endpoint detection, and the remaining bottles, half of which were wrapped with foil to block light exposure, were incubated during the experiment at approximately the same depth as the enclosures and measured for final DO after 8 h by Winkler titration.

Primary production and carbon fluxes

Total enclosure NPP was determined from the decrease in DIC concentration in each individual light enclosure. Stream water primary production within each enclosure was estimated by converting the net DO uptake measured in the BOD bottles (µmol O$_2$ l$^{-1}$ h$^{-1}$) to the same units as for the total enclosure NPP (mmol C mol biofilm C$^{-1}$ h$^{-1}$) by assuming a photosynthetic quotient of 1.0 and normalizing to biofilm C using the stream water volume and biofilm C within individual enclosures.

Net DOC release was estimated for each enclosure from the difference between initial and final DOC concentrations during the course of the incubation within both light and dark enclosures. The dark enclosures were used as controls to determine if C fluxes were the result of light-mediated processes. The net DOC flux derived from C fixed during the incubation (new C) was estimated for all positive fluxes (release of DOC into stream water) using the equation:

$$\text{new C released as DOC} = \frac{\delta^{13}C_{DOC_f} - \delta^{13}C_{DOC_i}}{\delta^{13}C_{DIC_i} - \delta^{13}C_{DOC_i}} \times [DOC]_f, \qquad (2)$$

where $\delta^{13}C_{DOC_f}$ is the $\delta^{13}C$ of DOC in the enclosure at the end of the incubation, $\delta^{13}C_{DOC_i}$ is the $\delta^{13}C$ of DOC in the enclosure at the start of the incubation period, $\delta^{13}C_{DIC_i}$ is the $\delta^{13}C$ of DIC in the enclosure at the start of the incubation period, and $[DOC]_f$ is the concentration of DOC at the end of the incubation period. This estimate of the new C released as DOC may be conservative because the isotopic fractionation associated with the uptake and fixation of the DIC during the incubations was not included in our calculations. This factor can be governed by stream flow (France, 1995) which would be one factor consistent among the enclosure incubations. However, this fractionation likely represents a small part of the change in $\delta^{13}C$ of the newly fixed C, given that $^{13}C$ enrichment of the labeled DIC pool was approximately 50 times the level of fractionation expected.

Because of the high spatial variability of stream benthic substrate surface area and type (e.g., sediment, gravel, and rock surface), flux measurements per unit stream area or stream reach could not be adequately determined from the enclosure experiments. All flux measurements were, therefore, normalized to biofilm carbon (e.g., NPP is reported in mmol C mol biofilm $C^{-1}$ $h^{-1}$) to enable comparison of flux rates associated with the biofilm communities among study sites and across seasons.

Phospholipid fatty acid analysis

Details on the sample handling and analytical methods used to determine the concentration and $\delta^{13}C$ of individual phospholipid fatty acids (PLFA) in the epilithic biofilms in these enclosure experiments are provided in Lyon & Ziegler (2009) and Ziegler et al. (2009). Briefly here, lyophilized biofilms were extracted, lipid class separated, saponified, and derivatized to fatty acid methyl esters (FAMEs) according to (Dobbs & Findlay, 1993; White & Ringelberg, 1998). FAMEs were quantified by gas chromatography (GC) using a flame ionization detector (Agilent 6890) and identified with a GC-mass spectrometer (Agilent 6890 and Agilent $5970_{inert}$ mass selective detector). The $\delta^{13}C$ of individual PLFA was determined by analyzing the FAMEs on a GC (Agilent 6890) interfaced (via a GC/CIII, ThermoFinnigan, Bremen, Germany) with an isotope ratio mass spectrometer (Delta$_{Plus}$, Thermo-Finnigan, Bremen, Germany). When used, a correction for the addition of the methyl carbon from the $BF_3$-methanol derivatization was calculated for each fatty acid by mass balance using standards. Stable isotopic ratios were measured relative to high purity, calibrated, and reference gas standards expressed relative to international standard PDB. All GCs used the same parameters and capillary column (50 m SGE BPX-70).

Polyunsaturated PLFA (16:2ω4, 16:2ω6, 18:2ω6, 18:3ω3, and 20:5ω3) were attributed to algae with 18:3ω3 and 20:5ω3 primarily attributed to green algae and diatoms, respectively (Wood, 1988; Findlay & Dobbs, 1993). Terminally branched fatty acids (i15:0 and a15:0) were attributed to heterotrophic bacteria as they are not commonly found in cyanobacteria (Parker et al., 1967; Merritt et al., 1991; Ahlgren et al., 1992). Due to its high concentration in many freshwater cyanobacteria (Sallal et al., 1990; Merritt et al., 1991; Ahlgren et al., 1992), PLFA 16:1ω7 was used to track changes in the contribution of cyanobacteria. The relative abundance of these diagnostic PLFA can vary among microbial grouping given that relative phospholipid content varies among microbial groups (e.g., White, 1983; Cooksey et al., 1987; Findlay et al., 1989). Therefore, it is important to note that the quantities or percentages of these PLFA, or newly fixed C incorporated into these PLFA, can only be used to compare among samples and not to assess differences in abundance or activity between microbial groups. Individual PLFA concentrations were converted to nmols or μmol PLFA C and used to calculate the flow of autotrophic or newly fixed C into the total pool of PLFA-C (newC$_{PLFA}$) and individual PLFA extracted from the enclosure biofilms. The quantity of newC$_{PLFA}$ was determined by the change in $\delta^{13}$C of the total PLFA pool between the initial and final biofilm samples according to the following equation:

$$\text{newC}_{PLFA} = \frac{\delta^{13}C_{PLFA_f} - \delta^{13}C_{PLFA_i}}{\delta^{13}C_{DIC_i} - \delta^{13}C_{PLFA_i}} \times C_{PLFA_f}, \quad (3)$$

where $\delta^{13}C_{PLFA_f}$ and $\delta^{13}C_{PLFA_i}$ are the $\delta^{13}$C of the total PLFA C pool at the start and end of the incubation, respectively, and $C_{PLFA_f}$ is the total quantity of PLFA-C (μmol C) in the biofilm within the enclosure at the end of the incubation. The same equation was applied to individual PLFA to obtain the quantity of new C incorporated into individual PLFA (16:1ω7, 18:3ω3, and 20:5ω3), bacterial PLFA as sum of i15:0 and a15:0 (ia15:0) and the sum of all polyunsaturated fatty acids (PUFA) to obtain the flow of autotrophic C into different microbial groups during the enclosure incubations. Further, the proportion of newly fixed C incorporated into these individual or groups of PLFA relative to the newly fixed C incorporated into the total PLFA pool was calculated to obtain a relative activity level of microbial groups in these biofilms using the following equation provided for PLFA 18:3ω3 as an example:

$$\text{newC}_{18:3\omega3/PLFA}(\%) = \frac{\text{newC}_{18:3\omega3}}{\text{newC}_{PLFA}} \times 100, \quad (4)$$

where newC$_{18:3\omega3}$ is determined for the PLFA 18:3ω3 from Eq. 3. Additionally, the quantity of newly fixed C incorporated into ia15:0 relative to PUFA (newC$_{ia15:0/PUFA}$) was used as a relative measure of newly fixed C incorporated by bacteria relative to C fixed by eukaryotic algae during the light enclosure incubations:

$$\text{newC}_{ia15:0/PUFA} = \frac{\text{newC}_{ia15:0}}{\text{newC}_{PUFA}} \quad (5)$$

where newC$_{PUFA}$ is the new C incorporated into the sum of all polyunsaturated fatty acids.

Statistical analyses and modeling

Paired $t$ tests were used to determine if initial and final measurements of DIC and DOC concentration and quantities of newly fixed C as DOC from the enclosures were significantly different before using them in the NPP and net release calculations. Since NPP was found to be the major factor regulating autotrophic C flow rates in previous analyses (Lyon & Ziegler, 2009; Ziegler et al., 2009) all modeling was conducted using NPP as a predictor variable and using total C released or incorporated per unit biofilm biomass per unit time. Here, we use site and season averages and include NPP as a predictor variable to assess the relative importance of nutrient availability and stoichiometry as well as active phototroph composition in predicting the variation in C cycling in these biofilms.

The dependent variables used to assess C cycling in these biofilm experiments were modeled using chi-square ($\chi^2$) statistics from analysis of deviance within the framework of the generalized linear model (GLM). The assumptions of homogeneity and normal errors for each GLM (Lindsey, 1997) were checked by plotting residuals versus fits and by plotting residuals as probability plots. In cases where the violations were strong, a different error structure was applied to the GLM. The exponential error distribution was found to produce acceptable residuals in most cases where normal error assumptions were violated. The exponential distribution is a special case of the gamma distribution and like the gamma distribution its effect is to reduce the weight given to observations exhibiting greater error (Lindsey, 1997). In the GLMs using the exponential error distribution, the deviance residual by predicted plot was assessed to determine if the assumptions of the model were met before further statistical analysis (Hoffmann, 2004). Those GLMs where an exponential error distribution was employed are listed in model results tables provided. All GLMs were calculated using JMP 8.1 (SAS; Cary, NC). All analyses employed an alpha level of 0.05 with a Bonferroni correction applied to reduce alpha to avoid inflating Type I error resulting from the multiple tests conducted (Zar, 1999).

Predictor variables were chosen among those variables captured in these experiments which best represented nutrient availability (biofilm C:N), stoichiometry (stream water DIN:DIP), and a measure of the relative composition of active phototrophs in these biofilms (new C incorporated into green algal PLFA 18:3ω3 as percent of new C incorporated into total PLFA; newC$_{18:3\omega3/PLFA}$). C:N or C:P ratios are considered to be more integrative measures of stream nutrient status relative to DIP or SRP concentrations

(Dodds, 2003). In this study, streamwater DIN and DIP were both correlated to NPP ($P = 0.0438$ and $P = 0.0024$, respectively), while biofilm C:N and stream water DIN:DIP were not, nor were biofilm C:N and DIN:DIP correlated. Therefore, the role of nutrient status was assessed by investigating biofilm C:N and stream water DIN:DIP to capture the most integrated measures of nutrient status and avoid confounding correlations. Further, we used the green algae PLFA 18:3ω3 because newC$_{18:3\omega3/PLFA}$, unlike newC$_{16:1\omega7/PLFA}$ which was correlated to NPP ($P = 0.0021$), was not correlated to NPP ($P = 0.6949$), biofilm C:N ($P = 0.6564$), or DIN:DIP ($P = 0.8695$). Since 18:3ω3 is a polyunsaturated PLFA, it is possible that newC$_{18:3\omega3/PLFA}$ may covary with the quantity of new C incorporated into total polyunsaturated PLFA (newC$_{PUFA}$). Since 18:3ω3 as a percent of total polyunsaturated PLFA (newC$_{18:3\omega3/PUFA}$) would not be expected to vary with the quantity of newC$_{PUFA}$, we used newC$_{18:3\omega3/PUFA}$ in place of newC$_{18:3\omega3/PLFA}$ in the models investigating variation in the ratio of new C incorporated into heterotrophic bacterial PLFA ia15:0 to newC$_{PUFA}$. This was likely an over cautious approach since both newC$_{18:3\omega3/PLFA}$ and newC$_{18:3\omega3/PUFA}$ were not found to be correlated to newC$_{PUFA}$ ($P = 0.3283; P = 0.3670$, respectively), however, we wanted to avoid testing factors that may be related to denominator or numerator of the ratio tested.

Because of the exploratory nature of this study, a multiple-hypothesis approach, similar to that used by Hay et al. (2008), was used to fit all possible GLMs for each dependent variable. The uncertainty of model selection was expected to be high because of the large number of predictor variables (3 or 4) relative to the number of observations (10 or 14). We, therefore, chose to use a model-based inference procedure using an information theoretic approach to rank the likelihood of the models given the data to assess the importance of different sets of predictors when compared against each other (Burnham & Anderson, 2002). The multi-model inference approach used here ranks predictors by their relative importance in predicting the data given, thereby allowing us to determine the strength of evidence for the importance of each predictor. Further, averaging over all subsets of possible models avoids the use of marginal statistics in selecting predictor variables and therefore the problems associated with multicollinearity (Graham, 2003). Models were evaluated using Akaike's Information Criterion (AIC; Burnham & Anderson, 1998). Because of the low ratio of sample size to number of predictor variables, we used the Akaike Information Criterion modified for small sample size (AICc; Wedderburn, 1974; Burnham & Anderson, 1998). Where global models exhibited overdispersion, the quasi-AIC$_c$ (QAIC$_c$) was employed (Burnham & Anderson, 2002). Akaike weights ($w_i$) were also calculated (Burnham & Anderson, 2002) and provide a measure of the weight of evidence in favor of one model relative to the others (White, 2001).

## Results

### Net primary production and biofilm carbon fixation

Total enclosure NPP, measured as dissolved inorganic carbon (DIC) uptake in the light enclosure incubations ranged from 1.2 to 25.6 mmol C mol biofilm C$^{-1}$ h$^{-1}$ (Fig. 1). Results of least squares analysis of NPP versus new C incorporated into the biofilms yielded a slope = $0.4700 \pm 0.034$ and $r^2 = 0.9424$ ($P < 0.0001$, $n = 14$) indicating that incorporation of new C was roughly 50% of NPP measured as DIC uptake across all six study streams

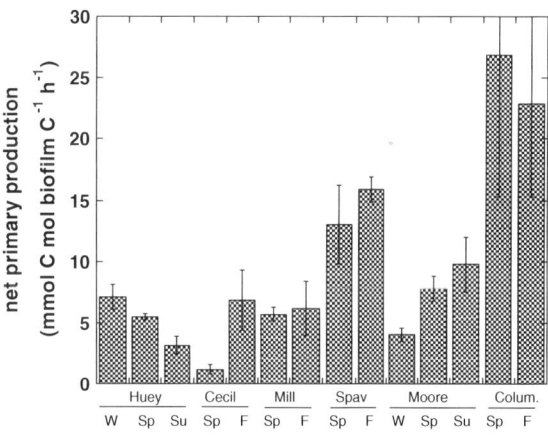

**Fig. 1** Net primary production measured as dissolved inorganic carbon uptake in the replicate light enclosures and normalized to total enclosure biofilm biomass C for each of the six study streams. Season is indicated below each site and corresponds to winter (W), spring, (Sp), summer (Su), and fall (F). Values are provided as mean ± 1 standard deviation ($n = 3$)

and seasons and that variation in this incorporation of new C was largely explained by NPP. The reduced incorporation of new C into biofilm relative to NPP was likely due to a combination of loss upon sampling and loss to DOC. Total NPP measured as DIC uptake was, therefore, used here since it represented our best estimate of total C fixation during these incubation experiments.

Stream water NPP, estimated from dissolved oxygen production in light BOD bottle incubations conducted simultaneously with enclosure incubations, was typically below detection at all stream sites with the exception of Spavinaw in spring ($0.19 \pm 0.01$ mmol C mol biofilm $C^{-1}$ $h^{-1}$), Moore in summer ($0.01 \pm 0.0002$ mmol C mol biofilm $C^{-1}$ $h^{-1}$), and Columbia in spring ($7.6 \pm 0.04$ mmol C mol biofilm $C^{-1}$ $h^{-1}$) and fall ($7.0 \pm 0.67$ mmol C mol biofilm $C^{-1}$ $h^{-1}$). When measured and calculated for the entire volume of stream water in the enclosures, stream water NPP represented <2% of the total enclosure NPP in Spavinaw and Moore and between 30 and 32% of total enclosure NPP in Columbia.

Carbon incorporation into epilithic biofilm phototroph PLFA

The absolute quantities of polyunsaturated phospholipid fatty acids was highest in the two most nutrient-rich sites and followed the trend in biofilm Chl $a$:C ratios among the study streams (Table 1). However, the incorporation of newly fixed C, as determined from the $^{13}$C-label, into the polyunsaturated fatty acids relative to total PLFA (newC$_{PUFA/PLFA}$) as a relative measure of eukaryotic algal photosynthetic activity in these epilithic biofilms did not exhibit any trend and ranged from 1 to 22% among the study streams (Fig. 2a). The quantity of new C incorporated into PUFA was generally higher in the two more nutrient-rich streams; however, Huey Hollow exhibited elevated newC$_{PUFA}$ in stream coinciding with higher light availability at that time. Further, the second-order AIC$_c$ values for the models investigated for newC$_{PUFA/PLFA}$ indicated that all three single factor models were highly plausible given the data; however, the $P$ values for each of these models indicate that they are not significant models for predicting newC$_{PUFA/PLFA}$ (Table 2). Here and in

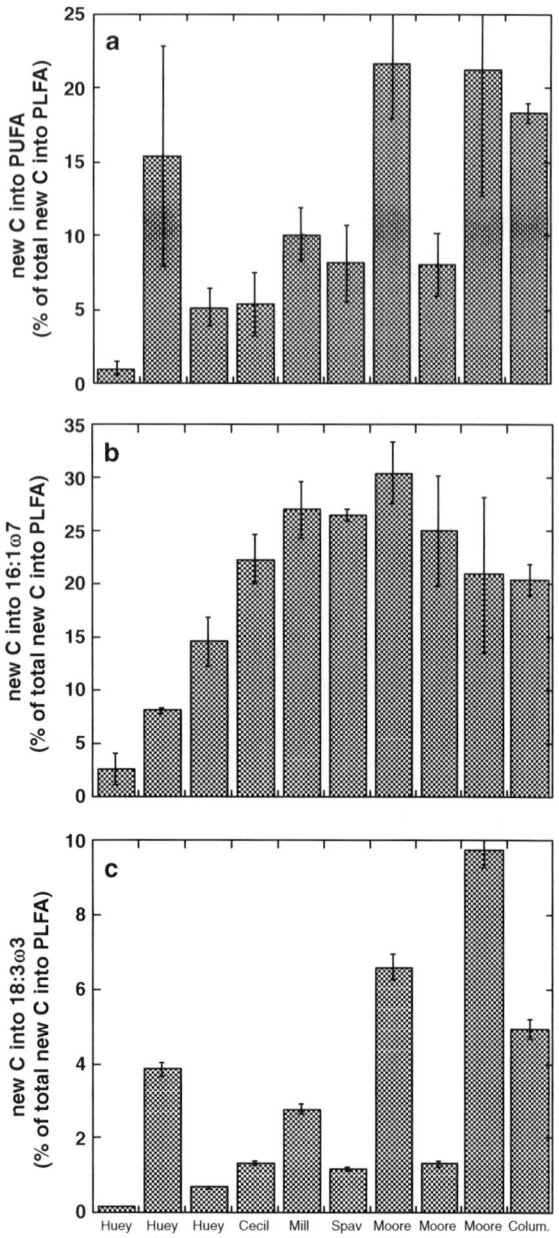

Fig. 2 The incorporation of newly fixed C into total polyunsaturated fatty acids (PUFA; **a**), cyanobacterial PLFA 16:1ω7 (**b**), and green algal PLFA 18:3ω3 (**c**) as a percent of the new C incorporated into the total PLFA during the incubation of the light enclosures in each of the six study streams. Season is indicated below each site and corresponds to winter (W), spring, (Sp), summer (Su), and fall (F). Values are provided as means $\pm$ 1 standard deviation ($n = 3$)

subsequent reporting of the modeling results, we follow the protocols of Anderson et al. (2001) on the presentation of analyses that involve model selection.

**Table 2** Log-likelihood and the second-order Akaike Information Criterion, corrected for small sample size (AIC$_c$; Wedderburn, 1974; Burnham & Anderson, 1998) for the general linearized models (GLMs) for new C incorporated into polyunsaturated phospholipid fatty acids (PUFA) as a % of new C incorporated into total PLFA (newC$_{PUFA/PLFA}$), new C incorporated into cyanobacterial phospholipid fatty acid (PLFA) 16:1$\omega$7 as a % of new C incorporated into total PLFA (newC$_{16:1\omega7/PLFA}$), and new C incorporated into green algal phospholipid fatty acid (PLFA) 18:3$\omega$3 as a % of new C incorporated into total PLFA (newC$_{18:3\omega3/PLFA}$) measured in the 10 experiments conducted across the 6 study streams. $K$ equals the number of parameters in the model including the error term ($E$). Delta AIC$_c$ ($\Delta_i$) values <4 are considered plausible, those <2 are considered highly plausible (Burnham & Anderson, 2002). Thus, in all three cases here models one, two, and three are all highly plausible. The measure of the weight of evidence in favor of the given model over all others is given as $w_i$. The parameters used in the models investigated were net primary production (NPP), the ratio of carbon to nitrogen in the biofilm (C:N), and the ratio of stream water dissolved inorganic nitrogen to dissolved inorganic phosphorus (DIN:DIP). All models exhibited overdispersion; however, the overdispersion parameter for the global model was <1 in each case therefore AIC$_c$ was used to assess these models. The Bonferroni corrected $\alpha = 0.0083$

| | Log-likelihood | K | AIC$_c$ | $\Delta_i$ | $w_i$ | P |
|---|---|---|---|---|---|---|
| Model (newC$_{PUFA/PLFA}$) | | | | | | |
| NPP + E | −45.76 | 2 | 97.2 | 0.0 | 0.34 | 0.6947 |
| DIN:DIP + E | −45.78 | 2 | 97.3 | 0.0 | 0.33 | 0.7367 |
| C:N + E | −45.83 | 2 | 97.4 | 0.1 | 0.32 | 0.8835 |
| NPP + C:N + NPP * C:N + E | −45.04 | 4 | 106.1 | 8.8 | <0.01 | 0.6602 |
| NPP + DIN:DIP + NPP * DIN:DIP + E | −45.33 | 4 | 106.7 | 9.4 | <0.01 | 0.7952 |
| NPP + C:N + DIN:DIP + NPP * C:N + NPP * DIN:DIP + E | −45.01 | 6 | 130.0 | 32.8 | <0.01 | 0.8952 |
| Model (newC$_{16:1\omega7/PLFA}$) | | | | | | |
| DIN:DIP + E | −34.79 | 2 | 75.30 | 0.00 | 0.35 | 0.3001 |
| C:N + E | −34.86 | 2 | 75.43 | 0.14 | 0.33 | 0.3333 |
| NPP + E | −34.94 | 2 | 75.60 | 0.30 | 0.30 | 0.3800 |
| NPP + C:N + NPP * C:N + E | −33.75 | 4 | 83.51 | 8.21 | 0.01 | 0.3697 |
| NPP + DIN:DIP + NPP * DIN:DIP + E | −34.02 | 4 | 84.04 | 8.75 | <0.01 | 0.4554 |
| NPP + C:N + DIN:DIP + NPP * C:N + NPP * DIN:DIP + E | −32.42 | 6 | 104.84 | 29.6 | <0.01 | 0.3247 |
| Model (newC$_{18:3\omega3/PLFA}$) | | | | | | |
| NPP + E | −25.20 | 2 | 56.12 | 0.00 | 0.36 | 0.5319 |
| C:N + E | −25.27 | 2 | 56.25 | 0.13 | 0.34 | 0.6079 |
| DIN:DIP + E | −25.38 | 2 | 56.47 | 0.35 | 0.30 | 0.8497 |
| NPP + DIN:DIP + NPP * DIN:DIP + E | −24.62 | 4 | 65.24 | 9.12 | <0.01 | 0.6696 |
| NPP + C:N + NPP * C:N + E | −25.08 | 4 | 66.16 | 10.0 | <0.01 | 0.9257 |
| NPP + C:N + DIN:DIP + NPP * C:N + NPP * DIN:DIP + E | −22.64 | 6 | 85.28 | 29.2 | <0.01 | 0.3563 |

Investigation of relative photosynthetic activity of individual phototrophic components of the epilithic biofilms studied using $^{13}$C-labeled PLFA was limited to the use of the cyanobacterial PLFA 16:1$\omega$7 and green algal PLFA 18:3$\omega$3 since the diatom PLFA 20:5$\omega$3 was not found in sufficient quantities for stable isotopic analysis in 9 of the 42 replicate incubation experiments which reduced the degrees of freedom well below our ability to use them in the tests conducted here. The PLFA 20:5$\omega$3 although an often reliable biomarker for diatoms is often not found in large quantities relative to other PLFA (Napolitano, 1999) and may be found in lower concentrations under nutrient or light limiting conditions and (Ahlgren et al., 1992). The proportion of new C incorporated into cyanobacterial PLFA 16:1$\omega$7 relative to new C incorporated into total PLFA (newC$_{16:1\omega7/PLFA}$) ranged from 3 to 31% with proportions being similar among most sites except Huey Hollow where they were lowest overall (Fig. 2b). The second-order AIC$_c$ values for the models investigated for newC$_{16:1\omega7/PLFA}$ indicated that all three single factor models were highly plausible given these data. However, the $P$ values

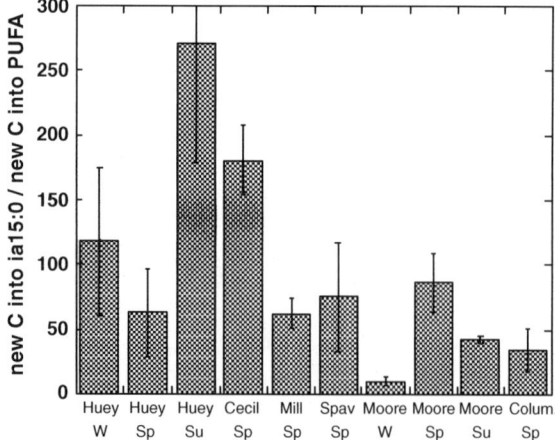

Fig. 3 The ratio of newly fixed C into heterotrophic bacterial PLFA ia15:0 to new C incorporated into total polyunsaturated fatty acids (PUFA) measured during the incubation of light enclosures in each of the six study streams. Season is indicated below each site and corresponds to winter (W), spring, (Sp), summer (Su), and fall (F). Values are provided as means ± 1 standard deviation ($n = 3$)

for each of these models indicate they are not significant models for predicting $newC_{16:1\omega7/PLFA}$ in these experiments (Table 2). The relative incorporation of new carbon into 18:3ω3 ($newC_{18:3\omega3/PLFA}$) ranged from 0.2 to 10% of new C incorporated into total biofilm PLFA with highest proportions found in Moore and Columbia (Fig. 2c). Similar to the results for $newC_{16:1\omega7/PLFA}$, the second-order $AIC_c$ values for the models investigated for $newC_{18:3\omega3/PLFA}$ indicated that all three single factor models were highly plausible given the data. However, the $P$ values for each of these models indicate they are not significant models for predicting $newC_{18:3\omega3/PLFA}$ (Table 2).

Autotrophic carbon incorporation into heterotrophic bacterial PLFA

The quantity of new C incorporated into ia15:0 ($newC_{ia15:0}$; mmol C mol biofilm $C^{-1}$) was used to assess the variation in the uptake of autotrophic C by heterotrophic bacteria in these biofilms and ranged from 0.3 to 8.2 mmol C mol biofilm $C^{-1}$. The most elevated $newC_{ia15:0}$ having been observed in Columbia in spring where highest NPP was measured and lowest having been observed in Moore in winter. The $QAIC_c$ values for the models investigated for $newC_{ia15:0}$ indicate that NPP was the only highly plausible explanatory factor given these data with an Akaiki weight (weight of evidence in favor of) of 0.94 (Table 3).

The ratio of new C incorporated into ia15:0 relative to new C incorporated into PUFA ($newC_{ia15:0/PUFA}$) was used to assess the exchange of C between algae and heterotrophic bacteria in these biofilms. This ratio exhibited a range of 10 to 272 mmol C mol $C^{-1}$ with highest ratios observed in the experiments conducted in the two lowest nutrient streams (Fig. 3). The second-order $AIC_c$ values for the models investigated for $newC_{ia15:0/PUFA}$ indicate that $newC_{18:3\omega3/PUFA}$ was a highly plausible explanatory factor, but that NPP and biofilm C:N were also plausible models. However, the model with $newC_{18:3\omega3/PUFA}$ as a single predictor was roughly four times as important as a predictor of $newC_{ia15:0/PUFA}$ for these data relative to the other two plausible models (Table 4). The $P$ values for these models were all high with the exception of the single factor model with $newC_{18:3\omega3/PUFA}$ ($P = 0.0336$), however, this value still exceeded the Bonferroni adjusted $\alpha = 0.0083$ suggesting this model is not a significant fit for this data either.

Release of dissolved organic carbon from epilithic biofilms

Net release of DOC into the stream water occurred in all the light enclosure incubations (all $P < 0.05$). The net changes in DOC concentrations within all dark enclosure incubations, however, were not significant (all $P > 0.1$); therefore, these data were not used or reported further. The spring experiments in Spavinaw, however, represented the one exception to the lack of significant DOC fluxes in the dark enclosures. At this time, a significant release of DOC (146 ± 10 μmol C; $P = 0.0010$) was measured in the dark enclosure incubations and represented 65% of the light enclosure release of DOC (225 ± 34 μmol C). Net release of DOC in the light ranged from 0.07 to 5.6 mmol C mol biofilm $C^{-1}$ $h^{-1}$ with highest releases found in Moore Creek and Columbia, the two streams with the most elevated dissolved inorganic phosphorus (DIP) levels (Fig. 4a). Net release of DOC ranged between 2 and 45% of total enclosure NPP with highest proportion of NPP released as DOC having occurred in both the two most nutrient-rich and two most nutrient-poor streams. The $QAIC_c$ values for the models investigated for net

**Table 3** Statistical results (see Table 2 for details) for GLMs of new C incorporated into heterotrophic bacterial (PLFA) ia15:0 (newC$_{ia15:0}$; mmol C mol biofilm C$^{-1}$), net DOC release (mmol C mol biofilm C$^{-1}$), and new C released as dissolved organic carbon (newC$_{DOC}$; mmol C mol biofilm C$^{-1}$). The global model is the sum of all parameters given in the previous seven models in each case. $K$ equals the number of parameters in the model including the error term ($E$) plus one to account for the estimation of the overdispersion parameter ($c$; for global model in each case) which was used to calculate QAIC$_c$ (Burnham & Anderson, 2002). Delta QAIC$_c$ values ($\Delta_i$) values <4 are considered plausible, those <2 are considered highly plausible (Burnham & Anderson, 2002). Thus, for newC$_{ia15:0}$ and net DOC release there is only one model of the set that is a likely predictor of the data while there are two for newC$_{DOC}$. The parameters used in the models investigated were the same as in Table 2 with the addition of green algal phospholipid fatty acid 18:3ω3 as a % of total new C incorporated into all biofilm PLFA (newC$_{18:3\omega3/PLFA}$). All models for newC$_{ia15:0}$ exhibited overdispersion with the exception of models one, three, and six which all used an exponential error distribution. For net DOC release, models two and eight (global model) exhibited overdispersion and all other models were based upon an exponential error distribution. For newC$_{DOC}$ models seven and eight (global model) exhibited overdispersion. The Bonferroni corrected $\alpha = 0.0083$ for newC$_{ia15:0}$ and $\alpha = 0.0063$ for both net DOC release and newC$_{DOC}$

| | Log-likelihood | $K$ | QAIC$_c$ | $\Delta_i$ | $w_i$ | $P$ |
|---|---|---|---|---|---|---|
| Model (newC$_{ia15:0}$) | | | | | | |
| NPP + E | −14.62 | 3 | 23.4 | 0.0 | 0.94 | 0.0070 |
| C:N + E | −21.96 | 3 | 30.1 | 6.7 | 0.03 | 0.4299 |
| newC$_{18:3\omega3/PLFA}$ + E | −22.26 | 3 | 30.4 | 7.0 | 0.03 | 0.8230 |
| NPP + C:N + NPP * C:N + E | −11.84 | 5 | 35.8 | 12.5 | 0.00 | 0.0001 |
| NPP + newC$_{18:3\omega3/PLFA}$ + NPP * newC$_{18:3\omega3/PLFA}$ + E | −14.57 | 5 | 38.3 | 15.0 | 0.00 | 0.0859 |
| Global ($c = 2.185$) | −11.35 | 7 | 80.4 | 57.0 | 0.00 | 0.0006 |
| Model (net DOC release) | | | | | | |
| newC$_{18:3\omega3/PLFA}$ + E | −33.78 | 3 | 26.6 | 0.0 | 0.78 | 0.0430 |
| NPP + newC$_{18:3\omega3/PLFA}$ + NPP * newC$_{18:3\omega3/PLFA}$ + E | −25.06 | 5 | 31.0 | 4.4 | 0.09 | <0.0001 |
| NPP + E | −43.96 | 3 | 32.1 | 5.5 | 0.05 | 0.0130 |
| C:N + E | −44.01 | 3 | 32.2 | 5.5 | 0.05 | 0.0139 |
| DIN:DIP + E | −45.70 | 3 | 33.1 | 6.4 | 0.03 | 0.0525 |
| NPP + C:N + NPP * C:N + E | −42.51 | 5 | 40.5 | 13.8 | <0.01 | 0.0492 |
| NPP + DIN:DIP + NPP * DIN:DIP | −42.55 | 5 | 40.5 | 13.8 | <0.01 | 0.0501 |
| Global ($c = 3.703$) | −13.90 | 9 | 70.5 | 43.9 | <0.01 | <0.0001 |
| Model (newC$_{DOC}$) | | | | | | |
| NPP + E | −5.22 | 3 | 10.1 | 0.0 | 0.41 | <0.0001 |
| newC$_{18:3\omega3/PLFA}$ + E | −6.10 | 3 | 10.4 | 0.3 | 0.35 | <0.0001 |
| C:N + E | −11.60 | 3 | 12.2 | 2.1 | 0.14 | <0.0001 |
| DIN:DIP + E | −14.54 | 3 | 13.2 | 3.1 | 0.09 | 0.0350 |
| NPP + C:N + NPP * C:N + E | −2.37 | 5 | 18.3 | 8.2 | 0.01 | <0.0001 |
| NPP + DIN:DIP + NPP * DIN:DIP + E | −3.86 | 5 | 18.8 | 8.6 | 0.01 | <0.0001 |
| NPP + newC$_{18:3\omega3/PLFA}$ + NPP * newC$_{18:3\omega3/PLFA}$ + E | −11.12 | 5 | 21.2 | 11.1 | <0.01 | <0.0001 |
| Global ($c = 6.030$) | −17.66 | 9 | 68.9 | 58.7 | <0.01 | <0.0001 |

DOC release indicate that newC$_{18:3\omega3/PLFA}$ was the only highly plausible explanatory factor for net DOC release given these data (Table 3).

New C released as DOC, defined as the amount derived from C fixed during the incubation and determined from the $^{13}$C-labeling of the DOC released, ranged from 0.03 to 9.4 mmol C mol biofilm C$^{-1}$ and represent between 0.6 and 21% of net DOC released in the light incubations (Fig. 4b). The release of newly fixed C as DOC was highest in Columbia throughout all three seasons and in Moore Creek in summer, the two streams having exhibited the highest DIP concentrations. The QAIC$_c$ values for the models investigated indicate that the single-factor

**Table 4** Statistical results (see Table 2 for details) for the GLMs for the ratio of new C incorporated into heterotrophic bacterial (PLFA) ia15:0 to new C incorporated into total polyunsaturated PLFA (new$C_{ia15:0:PUFA}$) and for new C released as dissolved organic carbon as a percent of net DOC release (new $C_{DOC\%}$). The global model is the sum of all parameters given in the previous seven models. For models for new $C_{ia15:0:PUFA}$ one is highly plausible and models two, three, and four are plausible, while for new$C_{DOC\%}$ there is only one model of the set that is a likely predictor of the data. The parameters used in the models investigated were the same as Table 2 in addition to new C incorporated into green algal phospholipid fatty acid 18:3ω3 as a % of total new C incorporated into all biofilm PLFA (new$C_{18:3\omega3/PLFA}$). All models for (new$C_{ia15:0:PUFA}$) exhibited overdispersion with the exception of model 6. For new$C_{DOC\%}$ models seven and eight (global model) exhibited overdispersion and all other models used an exponential error distribution. The overdispersion parameter for the global model was <1 in both cases therefore $AIC_c$ was used to assess these models. The Bonferroni corrected $\alpha = 0.0083$

| | Log-likelihood | K | $AIC_c$ | $\Delta_i$ | $w_i$ | P |
|---|---|---|---|---|---|---|
| Model (new$C_{ia15:0:PUFA}$) | | | | | | |
| new$C_{18:3\omega3/PLFA}$ + E | −55.62 | 2 | 117.0 | 0.0 | 0.59 | 0.0336 |
| NPP + E | −57.08 | 2 | 119.9 | 2.9 | 0.14 | 0.4140 |
| DIN:DIP + E | −57.18 | 2 | 120.1 | 3.1 | 0.13 | 0.6258 |
| C:N + E | −57.23 | 2 | 120.2 | 3.2 | 0.12 | 0.6329 |
| NPP + new$C_{18:3\omega3/PLFA}$ + NPP * new$C_{18:3\omega3/PLFA}$ + E | −54.18 | 4 | 124.4 | 7.4 | 0.01 | 0.1005 |
| NPP + DIN:DIP + NPP * DIN:DIP + E | −55.00 | 4 | 126.0 | 9.0 | 0.01 | 0.8148 |
| NPP + C:N + NPP * C:N + E | −57.05 | 4 | 130.1 | 13.1 | 0.00 | 0.8735 |
| Global | −51.63 | 8 | 263.3 | 146.3 | 0.00 | 0.1247 |
| Model (new$C_{DOC\%}$) | | | | | | |
| new$C_{18:3\omega3/PLFA}$ + E | −32.32 | 2 | 69.7 | 0.0 | 0.97 | 0.7046 |
| NPP + new$C_{18:3\omega3/PLFA}$ + NPP * new$C_{18:3\omega3/PLFA}$ + E | −32.10 | 4 | 76.7 | 6.9 | 0.03 | 0.9011 |
| C:N + E | −40.78 | 2 | 86.7 | 16.9 | 0.00 | 0.2176 |
| NPP + E | −40.81 | 2 | 86.7 | 17.0 | 0.00 | 0.2257 |
| DIN:DIP + E | −41.15 | 2 | 87.4 | 17.7 | 0.00 | 0.3798 |
| NPP + C:N + NPP * C:N + E | −39.74 | 4 | 91.9 | 22.2 | 0.00 | 0.3082 |
| NPP + DIN:DIP + NPP * DIN:DIP + E | −43.58 | 4 | 99.6 | 29.9 | 0.00 | 0.1928 |
| Global | −29.89 | 8 | 104.6 | 34.8 | 0.00 | 0.6581 |

models for both NPP and new$C_{18:3\omega3/PLFA}$ are highly plausible. The model with the highest Akaiki weight ($w_i$) is between three and four times more important than biofilm C:N and streamwater DIN:DIP, both of which are plausible single factor models given their $\Delta QAIC_c$ values, for the quantity of new C released as DOC (mmol C mol biofilm $C^{-1}$; Table 3). In order to assess the potential role of the factors, considered in these experiments, controlling the source of DOC released from these biofilms, we additionally analyzed the same set of models for new C released as DOC as a percent of the net DOC release in the light enclosure incubations. The second-order $AIC_c$ values for the models investigated for new C released as DOC as a percent of net DOC release indicate that only the single factor model with new$C_{18:3\omega3/PLFA}$ was plausible and exhibited a weight of evidence of 0.97 (Table 4). The high P values for the individual models tested, however, indicate that these models were non-significant for the data investigated.

## Discussion

Previous work focusing on a comparison of two contrasting streams during three seasons (Ziegler et al., 2009) and two stream pairs within a single season (Lyon & Ziegler, 2009) suggest that nutrient status of streams may regulate both the exchange of C within the biofilm, and between the biofilm and stream water DOC. Specifically, results indicated that newly fixed C released as DOC appeared highest in the most nutrient-rich streams, while a larger proportion of NPP was incorporated into biofilm

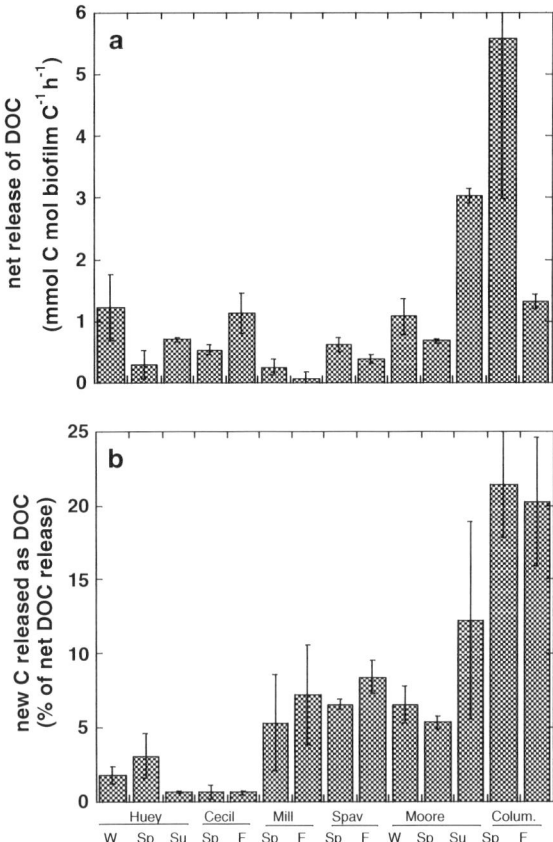

Fig. 4 Net dissolved organic carbon (DOC) release normalized to total enclosure biofilm biomass C (a) and release of newly fixed C as a percent of net DOC released (b) measured in the replicate light enclosures at each of the six study streams. Season is indicated below each site and corresponds to winter (W), spring, (Sp), summer (Su), and fall (F). Values are provided as means ± 1 standard deviation ($n = 3$)

predictor of C exchange within and net DOC release from these biofilms. These results suggest increased algal exudation, greater biofilm development (e.g., thickness, density, proportion of extrapolymeric substances), and perhaps development of specific algal genera with nutrient enrichment may have regulated C cycling in the biofilms across this nutrient gradient of sites.

The increasing PUFA content in these biofilms with nutrient enrichment was not observed in the newC$_{PUFA}$ data suggesting that activity and retention of newly fixed C by eukaryotic algae may not be simply controlled by nutrient availability in these streams. The relative composition of the active phototrophs in these epilithic biofilms, however, did vary with nutrient enrichment. Low statistical power caused by the small sample size and our inability to test other factors, such as light availability, substrate, and grazing activities, and their possible combined effects with nutrient availability likely prevented the detection of any strong explanatory factors for the specific phototrophic group activities measured. For example, elevated nutrients have been noted to stimulate algal growth (Stevenson et al., 2006) and interact with other factors, such as light, causing changes in grazing rates, and algal community composition (Steinman et al., 1991; Walton et al., 1995; Liess et al., 2009).

The differences observed in the PLFA representing two phototrophic groups strongly suggest the need to investigate the role of nutrient availability and stoichiometry in controlling the relative activity of specific groups of phototrophs in stream biofilms (Frost et al., 2002). The distinct differences in how newC$_{16:1\omega7/PLFA}$ and newC$_{18:3\omega3/PLFA}$ varied among the study sites suggest differences in the response of the relative activity of cyanobacteria versus green algae to nutrient enrichment. Relative activity level of cyanobacteria was fairly similar among the study streams with the exception of lower levels in the most nutrient-poor stream. Whereas, relative levels of green algal activity exhibited a general increase with nutrient enrichment congruent with previous work where increases in Chl *b* along this nutrient gradient suggested increases in biomass of green algae (Lyon & Ziegler, 2009). Differences in the composition of active stream biofilm phototrophs may be important to both trophodynamics and biogeochemistry of streams due to the varied palatability of different

heterotrophic bacterial biomass in the lowest nutrient streams. Further, the release of both total DOC and newly fixed C as DOC varied with DIN and DIP but differently for two contrasting streams. This current study analyzed these data sets in combination with new data from the same nutrient gradient of four stream sites to directly assess how nutrient availability and stoichiometry, and community composition of active biofilm phototrophs may regulate the observed variation in C flow in these streams. Our analyses here do not enable us to directly link nutrient enrichment or stoichiometry to the variation in the activity of specific autotrophic groups within these biofilms. However, we found the structure of active biofilm autotrophs can be a relatively important algae and cyanobacteria to consumers (Gregory, 1983; Brown et al., 2003) as well as the potential differences in how they support heterotrophic bacteria (Romani & Sabater, 1998) or generate stream water DOC.

Not surprisingly, variation in the incorporation of newly fixed C into heterotrophic bacteria, as indicated by the quantity of $newC_{ia15:0}$, was best explained by NPP. NPP is after all a measure of the total quantity of C incorporated by phototrophs in these experimental enclosures and was found to be significantly correlated to new C incorporated into the biofilm. Algal–heterotrophic bacteria C exchange, however, decreased with nutrient enrichment and reconfirms previous analyses at some of these sites and other studies indicating a reduction in algal–bacterial interaction with increased nutrient enrichment (Scott et al., 2008; Ziegler et al., 2009). Potential mechanisms for this reduced algal–bacterial C exchange may include: (1) structural attributes within the biofilm linked to its development and composition that reduce algal–bacterial exchange (Scott et al., 2008) and (2) increased use of stream water DOC by biofilm bacteria in these streams exhibiting higher DOC concentrations and agricultural sources.

Results of this study suggest the composition of the active phototrophs can be an important factor regulating the proportion of autotrophic C incorporated by heterotrophic bacterial in these epilithic biofilms. Although these results are limited by the number of factors we could test, the fact that $newC_{18:3\omega3/PUFA}$ was both highly plausible and found to be the single best model to explain variation in $newC_{ia15:0/PUFA}$ indicates that the relative activity of green algae can be important in regulating direct algal–bacterial C exchange in these biofilms. Given the disturbance regime of streams in this region (Lohman et al., 1991) and the interactive effects of nutrients and biofilm development observed in other streams (Bertrand et al., 2009), it is likely that the negative relationship between green algal activity and algal–bacterial C exchange observed here is attributed to variation in biofilm development. Specifically, the composition of active phototrophs, such as relative activity of green algae, may be regulated by the extent of biofilm development in these streams (Power et al., 2008). Less developed stream biofilms exhibit higher rates of labile C uptake (Augspurger et al., 2008) suggesting that less developed biofilms in the lowest nutrient streams may have lead to greater incorporation of labile algal C sources as observed here. Such development could in turn be regulated by a combination of physiochemical parameters that may include nutrient enrichment and represent one reason we have observed the trend in increasing biofilm $newC_{18:3\omega3/PLFA}$ with nutrient enrichment among these streams.

Results of our investigation of biofilms from these six streams suggest that the composition of active phototrophs have the potential not only to regulate C cycling within but also DOC release from these biofilms. Net release of DOC from the biofilms with increasing nutrient enrichment suggests some role of nutrient availability in regulating DOC release. The variation in net DOC release, however, appears best explained by the relative activity of green algae. The role of the activity of this one phototrophic group tested here suggests that the composition of active phototrophs in stream biofilms can influence DOC release and highlight the need for further analysis of the algal community to complement the PLFA approach used. Further, our findings for DOC release were congruent with those for algal–heterotrophic bacterial C exchange. The observed trends of increasing $newC_{18:3\omega3/PLFA}$, net DOC release, and new C released as DOC as a percent of net DOC release and decreasing $newC_{ia15:0/PUFA}$ with nutrient enrichment are supported by the models investigated. There are three plausible and nonexclusive mechanisms that may explain the possible role of green algal activity in regulating biofilm C cycling as revealed in the data and models investigated: (1) mechanisms responsible for releasing autotrophic C sources from biofilms, which may include algal exudation and grazing activity, may be enhanced by greater green algal activity; (2) green algal activity may be indicative of development of specific biofilm genera as well as structure (e.g., thickness, density, proportion of extrapolymeric substances) that may enhance release of DOC; and (3) increased net DOC release to stream water via such mechanisms would decrease direct access and use of autotrophic C by biofilm heterotrophic bacteria.

Enhanced algal exudation in combination with reduction in heterotrophic bacterial uptake of labile C within more developed biofilms (Augspurger et al., 2008) is likely responsible for the observed relationship between DOC release and green algal activity observed here. Grazing activity has been found to be an important mechanism controlling the release of allochthonous DOC in streams (Meyer & O'hop, 1983) and the same may be the case for autochthonous C sources. Although increased activity of green algae may enhance grazing activities by increasing palatability of the biofilm and/or providing larger algal food sources (Walton et al., 1995), these biofilms are more likely regulated by bottom up controls due to disturbance regimes in these streams (Lohman et al., 1991; Fritz & Dodds, 2002). Further, green algae may exhibit high rates of exudation relative to other algae and cyanobacteria particularly with high light availability in streams with unforested riparian zones. The importance of the active phototrophic community composition in regulating C cycling in stream biofilms revealed by the data investigated here suggests the need for studying C cycling in conjunction with studies of algal–nutrient (Stevenson et al., 2006) and possibility nutrient–algal–grazer interactions (Steinman et al., 1991; Rosemond et al., 2000) in streams.

In addition to active phototroph composition and grazer activity, light availability could have played a significant role in regulating C cycling in these experiments. Results from the comparison of Moore and Huey Hollow suggest light levels may also regulate DOC release in these streams (Ziegler et al., 2009). Adequately assessing the role of light level and quality as an additional factor, however, would have required light level manipulations. As another important result of land use change in watersheds, the quantity and quality of light and it effect on C cycling in streams needs future consideration as it has also been found to regulate photochemical transformations of stream DOC (Brisco & Ziegler, 2004; Brooks et al., 2007; Cory et al., 2007) as well as trophic and nutrient interactions (Dodds et al., 1999; Hessen et al., 2002).

## Conclusions

Algal–heterotrophic bacterial C exchange and DOC release in stream biofilms are both important to stream ecosystem health and biogeochemical dynamics. The findings of this study suggest nutrient enrichment can significantly alter C exchange and cycling in stream biofilms. Further, the modeling approach taken here suggests that composition and structure of stream biofilms are likely sensitive to nutrient enrichment and represent key factors regulating stream C cycling. The potential effects of nutrient enrichment on the composition of active phototrophs in streams in combination with the results of this study signify the biogeochemical implications of the current trends in P enrichment of streams in the central US (Sprague & Lorenz, 2009). The results here indicate a strong need for integrative research coupling studies of nutrient stoichiometry and availability with microbial composition and activity, and carbon cycling dynamics. This future research is particularly important given the global phenomena of watershed nutrient enrichment (Howarth et al., 2006), the vulnerability of headwater streams (Vitousek et al., 1997), and the key role streams play in watershed scale nutrient processing (Peterson et al., 2001).

**Acknowledgments** We gratefully acknowledge Andrea Kopecky for her work in the field and laboratory. We also thank Sherri Townsend, Lindsey Conaway, Erik Pollock, Glenn Piercey, and Tom Millican for their assistance. We are also thankful for Brain Haggard's assistance with field site selection. Much thanks goes to Y. Wiersma and D. Schneider for all of their assistance and advise on the statistical analyses. The review of a previous draft by two anonymous reviewers was helpful in improving this manuscript. The National Park Service granted permission to conduct research at the Buffalo National River. Funding was provided by National Science Foundation (DEB 0445357) and Natural Sciences and Engineering Research Council (NSERC).

## References

Ahlgren, G., I. B. Gustafsson & M. Boberg, 1992. Fatty-acid content and chemical-composition of fresh-water microalgae. Journal of Phycology 28: 37–50.

Alexander, R. B., R. A. Smith & G. E. Schwarz, 2000. Effect of stream channel size on the delivery of nitrogen to the Gulf of Mexico. Nature 403: 758–761.

Anderson, D. R., W. A. Link, D. H. Johnson & K. P. Burnham, 2001. Suggestions for presenting the results of data analyses. Journal of Wildlife Management 65: 373–378.

Arnon, S., K. A. Gray & A. I. Packman, 2007. Biophysicochemical process coupling controls nitrate use by benthic biofilms. Limnology and Oceanography 52: 1665–1671.

Augspurger, C., G. Gleixner, C. Kramer & K. Kusel, 2008. Tracking carbon flow in a 2-week-old and 6-week-old stream biofilm food web. Limnology and Oceanography 53: 642–650.

Bernhardt, E. S. & G. E. Likens, 2002. Dissolved organic carbon enrichment alters nitrogen dynamics in a forest stream. Ecology 83: 1689–1700.

Bertrand, K. N., K. B. Gido, W. K. Dodds, J. N. Murdock & M. R. Whiles, 2009. Disturbance frequency and functional identity mediate ecosystem processes in prairie streams. Oikos 118: 917–933.

Biggs, B. J. F. & R. A. Smith, 2002. Taxonomic richness of stream benthic algae: effects of flood disturbance and nutrients. Limnology and Oceanography 47: 1175–1186.

Bothwell, M. L., 1985. Phosphorus limitation of lotic periphyton growth rates: an intersite comparison using continuous-flow troughs (Thompson River system, British Columbia). Limnology and Oceanography 30: 527–542.

Brisco, S. & S. Ziegler, 2004. Effects of solar radiation on the utilization of dissolved organic matter (DOM) from two headwater streams. Aquatic Microbial Ecology 37: 197–208.

Brooks, M. L., D. M. Mcknight & W. H. Clements, 2007. Photochemical control of copper complexation by dissolved organic matter in Rocky Mountain streams, Colorado. Limnology and Oceanography 52: 766–779.

Brown, R. J., S. D. Rundle, T. H. Hutchinson, T. D. Williams & M. B. Jones, 2003. Small-scale detritus-invertebrate interactions: influence of detrital biofilm composition on development and reproduction in a meiofaunal copepod. Archiv fur Hydrobiologie 157: 117–129.

Burnham, K. P. & D. R. Anderson, 1998. Model selection and interference: a practical information theoretical approach. Springer, New York.

Burnham, K. P. & D. R. Anderson, 2002. Model selection and multimodel inference: a practical-theoretic approach. Springer-Verlag.

Carr, G. M., A. Morin & P. A. Chambers, 2005. Bacteria and algae in stream periphyton along a nutrient gradient. Freshwater Biology 50: 1337–1350.

Chamberlain, T. C., 1897. The method of multiple working hypotheses. Journal of Geology 6(5): 837–848.

Cooksey, K. E., J. B. Guckert, S. A. Williams & P. R. Callis, 1987. Fluorometric-determination of the neutral lipid-content of microalgal cells using Nile Red. Journal of Microbiological Methods 6: 333–345.

Cory, R. M., D. M. Mcknight, Y. P. Chin, P. Miller & C. L. Jaros, 2007. Chemical characteristics of fulvic acids from Arctic surface waters: microbial contributions and photochemical transformations. Journal of Geophysical Research – Biogeosciences 112: 8142–8149.

Dobbs, F. C. & R. H. Findlay, 1993. Analysis of microbial lipids to determine biomass and detect the response of sedimentary microorganisms to disturbance. In Kemp, P. F., B. F. Sherr, E. B. Sherr & J. J. Cole (eds), Current Methods in Aquatic Microbial Ecology. Lewis Publishers, Baton Rouge: 347–358.

Dodds, W. K., 2003. Misuse of inorganic N and soluble reactive P concentrations to indicate nutrient status of surface waters. Journal of the North American Benthological Society 22: 171–181.

Dodds, W. K., B. J. F. Biggs & R. L. Lowe, 1999. Photosynthesis-irradiance patterns in benthic microalgae: variations as a function of assemblage thickness and community structure. Journal of Phycology 35: 42–53.

Dodds, W. K., M. A. Evans-White, N. M. Gerlanc, L. Gray, D. A. Gudder, M. J. Kemp, A. L. López, D. Stagliano, E. A. Strauss & J. L. Tank, 2000. Quantification of the nitrogen cycle in a prairie stream. Ecosystems 3: 574–589.

Findlay, R. H. & F. C. Dobbs, 1993. Quantitative description of microbial communities using lipid analysis. In Kemp, P. F., B. F. Sherr, E. B. Sherr & J. J. Cole (eds), Current Methods in Aquatic Microbial Ecology. Lewis Publishers, Boca Raton: 271–284.

Findlay, R. H., G. M. King & L. Watling, 1989. Efficacy of phospholipid analysis in determining microbial biomass in sediments. Applied and Environmental Microbiology 55: 2888–2893.

Fogg, G. E., 1983. The ecological significance of extracellular products of phytoplankton photosynthesis. Botanica Marina 26: 3–14.

France, R., 1995. Critical examination of stable isotope analysis as a means for tracing carbon pathways in stream ecosystems. Canadian Fisheries and Aquatic Sciences 52: 651–656.

Fritz, K. M. & W. K. Dodds, 2002. Macroinvertebrate assemblage structure across a tallgrass prairie stream landscape. Archiv fur Hydrobiologie 154: 79–102.

Frost, P. C., R. S. Stelzer, G. A. Lamberti & J. J. Elser, 2002. Ecological stoichiometry of trophic interactions in the benthos: understanding the role of C:N:P ratios in lentic and lotic habitats. Journal of the North American Benthological Society 21: 515–528.

Graham, M. H., 2003. Confronting multicollinearity in ecological multiple linear regression. Ecology 84: 2809–2815.

Gregory, S. V., 1983. Plant-herbvoir interactions in stream systems. In Barnes, J. R. & G. W. Minshall (eds), Stream Ecology: Applications and Testing of General Ecology. Plenum Press, New York: 399.

Grimm, N. B., 1987. Nitrogen dynamics during succession in a desert stream. Ecology 68: 1157–1170.

Guasch, H., E. Marti & S. Sabater, 1995. Nutrient enrichment effects on biofilm metabolism in a Mediterranean stream. Freshwater Biology 33: 373–383.

Haack, T. K. & G. A. Mcfeters, 1982. Nutritional relationships among microorganisms in an epilithic biofilm community. Microbial Ecology 8: 115–126.

Hay, C. H., T. G. Franti, D. B. Marx, E. J. Peters & L. W. Hesse, 2008. Macroinvertebrate drift density in relation to abiotic factors in the Missouri River. Hydrobiologia 598: 175–189.

Hessen, D. O., P. J. Faerovig & T. Andersen, 2002. Light, nutrients, and P:C ratios in algae: grazer performance related to food quality and quantity. Ecology 83: 1886–1898.

Hoffmann, J. P., 2004. Generalized Linear Models an Applied Approach, 1st edn. Pearson, Boston, MA: 204 pp.

Howarth, R. W., D. P. Swaney, E. W. Boyer, R. Marino, N. Jaworski & C. Goodale, 2006. The influence of climate on average nitrogen export from large watersheds in the Northeastern United States. Biogeochemistry 79: 163–186.

Jeffrey, S. W. & G. F. Humphrey, 1975. New spectrophotometric equations for determining chlorophylls a, b, c and c2 in higher plants, algae, and natural phytoplankton. Biochemistry and Physiology of Plants 167: 191–194.

Kaplan, L. A. & T. L. Bott, 1982. Diel fluctuations of DOC generated by algae in a piedmont stream. Limnology and Oceanography 27: 1091–1100.

Kaplan, L. A. & T. L. Bott, 1989. Diel fluctuations in bacterial activity on streambed substrata during vernal algal blooms: effects of temperature, water chemistry, and habitat. Limnology and Oceanography 34: 718–733.

Liess, A., K. Lange, F. Schulz, J. J. Piggott, C. D. Matthaei & C. R. Townsend, 2009. Light, nutrients and grazing interact to determine diatom species richness via changes to productivity, nutrient state and grazer activity. Journal of Ecology 97: 326–336.

Lindsey, J. K., 1997. Applying Generalized Linear Models. Springer, Berlin: 271 pp.

Lohman, K., J. R. Jones & C. Baysingerdaniel, 1991. Experimental-evidence for nitrogen limitation in a Northern Ozark Stream. Journal of the North American Benthological Society 10: 14–23.

Lorenzen, C. J., 1967. Determinations of chlorophyll and phaeo-pigments: spectrophotometric equations. Limnology and Oceanography 12: 343–346.

Lyon, D. R. & S. E. Ziegler, 2009. Carbon cycling within epilithic biofilm communities across a nutrient gradient of headwater streams. Limnology and Oceanography 54(2): 439–449.

Mccormick, P. V. & R. J. Stevenson, 1991. Mechanisms of benthic algal succession in lotic environments. Ecology 72: 1835–1848.

Merritt, M. V., S. P. Rosenstein, C. Loh, R. H. S. Chou & M. M. Allen, 1991. A comparison of the major lipid classes and fatty-acid composition of marine unicellular cyanobacteria with fresh-water species. Archives of Microbiology 155: 107–113.

Meyer, J. L. & J. O'hop, 1983. Leaf-shredding insects as a source of dissolved organic carbon in headwater streams. American Midland Naturalist 109: 175–183.

Murray, R. E., K. E. Cooksey & J. C. Priscu, 1986. Stimulation of bacterial DNA synthesis by algal exudates in attached algal-bacterial consortia. Applied and Environmental Microbiology 52: 1177–1182.

Napolitano, G. E., 1999. Fatty acids as trophic and chemical markers in freshwater ecosystems. In Arts, M. T. & B. C. Wainman (eds), Lipids in Freshwater Ecosystems. Springer, Berlin: 21–44.

Neely, R. K. & R. G. Wetzel, 1995. Simultaneous use of 14C and 3H to determine autotrophic production and bacteria protein production in periphyton. Microbial Ecology 30: 227–237.

Norrman, B., U. L. Zweifel, C. S. Hopkinson Jr & B. Fry, 1995. Production and utilization of dissolved organic carbon during an experimental diatom bloom. Limnology and Oceanography 40: 898–907.

Parker, P. L., C. Van Baalen & L. Maurer, 1967. Fatty acids in eleven species of blue-green algae: geochemical significance. Science 155: 707–708.

Passy, S. I., 2008. Continental diatom biodiversity in stream benthos declines as more nutrients become limiting. Proceedings of the National Academy of Sciences 105: 9663–9667.

Patrick, R., 1972. Aquatic communities as indices of pollution. In Thomas, W. A. (ed.), Indicators of Environmental Quality. Plenum Publishing Corp., New York: 93–100.

Patrick, R. & D. M. Palavage, 1994. The value of species as indicators of water quality. Proceedings of the Academy of Natural Sciences of Philadelphia 145: 55–92.

Peterson, C. G. & N. B. Grimm, 1992. Temporal variation in enrichment effects during periphyton succession in a nitrogen-limited desert stream ecosystem. Journal of the North American Benthological Society 11: 20–36.

Peterson, B. J., W. M. Wollheim, P. J. Mulholland, J. R. Webster, J. T. Meyer, J. L. Tank, E. Marti, W. B. Boweden, H. M. Valett, A. E. Hershey, W. H. McDowell, W. K. Dodds, S. K. Hamilton & S. D. Gregory, 2001. Control of nitrogen export from watersheds by headwater streams. Science 292: 86–90.

Power, M. E., M. S. Parker & W. E. Dietrich, 2008. Seasonal reassembly of a river food web: floods, droughts, and impacts of fish. Ecological Monographs 78: 263–282.

Pringle, C. M., 1990. Nutrient spatial heterogeneity – effects on community structure, physiognomy, and diversity of stream algae. Ecology 71: 905–920.

Pusch, M., D. Fiebig, I. Brettar, H. Eisenmann, B. K. Ellis, L. A. Kaplan, M. A. Lock, M. W. Naegeli & W. Traunspurger, 1998. The role of microorganisms in the ecological connectivity of running waters. Freshwater Biology 40: 453–495.

Rabalais, N. N., 2002. Nitrogen in aquatic ecosystems. Ambio 31: 102–112.

Rier, S. T., K. A. Kuehn & S. N. Francoeur, 2007. Algal regulation of extracellular enzyme activity in stream microbial communities associated with inert substrata and detritus. Journal of the North American Benthological Society 26: 439–449.

Romani, A. M. & S. Sabater, 1998. A stromatolitic cyanobacterial crust in a Mediterranean stream optimizes organic matter use. Aquatic Microbial Ecology 16: 131–141.

Rosemond, A. D., P. J. Mulholland & S. H. Brawley, 2000. Seasonally shifting limitation of stream periphyton: response of algal populations and assemblage biomass and productivity to variation in light, nutrients, and herbivores. Canadian Journal of Fisheries and Aquatic Sciences 57: 66–75.

Sallal, A. K., N. A. Nimer & S. S. Radwan, 1990. Lipid and fatty-acid composition of fresh-water cyanobacteria. Journal of General Microbiology 136: 2043–2048.

Scott, J. T., J. A. Back, J. M. Taylor & R. S. King, 2008. Does nutrient enrichment decouple algal bacterial production in periphyton? Journal of the North American Benthological Society 27: 332–344.

Seitzinger, S. P. & C. Kroeze, 1998. Global distribution of nitrous oxide production and N inputs in freshwater and coastal marine ecosystems. Global Biogeochemical Cycles 12: 93–113.

Sprague, L. A. & D. L. Lorenz, 2009. Regional nutrient trends in streams and rivers of the United States, 1993–2003. Environmental Science and Technology 43: 3430–3435.

Steinman, A. D., 1996. Effects of grazers on benthic freshwater algae. In Bothwell, R. J. & R. L. Lowe (eds), Algal Ecology – Freshwater Benthic Ecosystems. Academic Press, London: 341–373.

Steinman, A. D., P. J. Mulholland & D. B. Kirschtel, 1991. Interactive effects of nutrient reduction and herbivory on biomass, taxonomic structure, and P-uptake in lotic periphyton communities. Canadian Journal of Fisheries and Aquatic Sciences 48: 1951–1959.

Stelzer, R. S. & G. A. Lamberti, 2001. Effects of N:P ratio and total nutrient concentration on stream periphyton community structure, biomass, and elemental composition. Limnology and Oceanography 46: 356–367.

Stevenson, R. J., S. T. Rier, C. M. Riseng, R. E. Schultz & M. J. Wiley, 2006. Comparing effects of nutrients on algal biomass in streams in two regions with different disturbance regimes and with applications for developing nutrient criteria. Hydrobiologia 561: 149–165.

Van Den Meersche, K., J. J. Middelburg, K. Soetaert, P. Van Rijswijk, H. T. S. Boschker & C. H. R. Heip, 2004. Carbon–nitrogen coupling and algal–bacterial interactions during an experimental bloom: modeling a $^{13}$C tracer experiment. Limnology and Oceanography 49: 862–878.

Vitousek, P. M., H. A. Mooney, J. Lubchenco & J. M. Melillo, 1997. Human domination of Earth's ecosystems. Science 277: 494–499.

Walton, S. P., E. B. Welch & R. R. Horner, 1995. Stream periphyton response to grazing and changes in phosphorus concentration. Hydrobiologia 302: 31–46.

Wedderburn, R. W. M., 1974. Quasi-likelihood functions, generalized linear models, and the Gauss-Newton method. Biometrika 61: 439–447.

White, D. C., 1983. Analysis of microorganisms in terms of quantity and activity in natural environments. Symposium of the Society for General Microbiology 34: 37–48.

White, D. C. & D. B. Ringelberg, 1998. Signature lipid biomarker analysis. In Burlage, R. S., R. Atlas, D. Stahl, G. Geesey & G. Sayler (eds), Techniques in Microbial Ecology. Oxford University Press, New York: 255–272.

White, G. C., 2001. Statistical models: keys to understanding the natural world. In Shenk, T. M. & A. B. Franklin (eds), Modeling in Natural Resource Management: Development, Interpretation, and Application. Island Press, Washington, DC: 35–56.

Wood, B. J. B., 1988. Lipids of algae and protozoa. In Ratledge, C. & S. G. Wilkinson (eds), Microbial Lipids. Harcourt Brace Jovanovich, New York: 807.

Zar, J. H., 1999. Biostatistical analysis. Upper Saddle River, NJ, USA, Prentice Hall.

Ziegler, S. E., D. R. Lyon & S. Townsend, 2009. Carbon release and cycling within epilithic biofilms in two contrasting headwater streams. Aquatic Microbial Ecology 55: 285–300.

Hydrobiologia (2010) 657:89–105
DOI 10.1007/s10750-009-9984-5

GLOBAL CHANGE AND RIVER ECOSYSTEMS

# Periphyton biomass and ecological stoichiometry in streams within an urban to rural land-use gradient

Patrick J. O'Brien · John D. Wehr

Received: 20 January 2009 / Accepted: 2 November 2009 / Published online: 20 November 2009
© Springer Science+Business Media B.V. 2009

**Abstract** This study examined the effects land use on biomass and ecological stoichiometry of periphyton in 36 streams in southeastern New York State (USA). We quantified in-stream and land-use variables along a N–S land-use gradient at varying distances from New York City (NYC). Streams draining different landscapes had fundamentally different physical, chemical, and biological properties. Human population density significantly decreased ($r = -0.739$; $P < 0.00001$), while % agricultural land significantly increased ($r = 0.347$; $P = 0.0379$) with northing. Turbidity, temperature, conductivity, and dissolved Mg, Ca, SRP, pH, DOC, and Si significantly increased in more urban locations, but $NO_3^-$ and $NH_4^+$ did vary not significantly along the gradient. Periphyton biomass (as AFDM and Chl-$a$) in rural streams averaged one-third to one-fifth that measured in urban locations. Periphyton biomass in urban streams averaged $18.8 \pm 6.0$ g/m$^2$ AFDM and $75.6 \pm 28.5$ mg/m$^2$ Chl-$a$. Urban Chl-$a$ levels ranging between 100 and 200 mg/m$^2$, are comparable to quantities measured in polluted agricultural streams in other regions, but in our study area was not correlated with % agricultural land. Periphyton nutrient content also varied widely; algal C varied >20-fold (0.06–1.7 µmol/mm$^2$) while N and P content varied >6-fold among sites. Algal C, N, and P correlated negatively with distance from NYC, suggesting that periphyton in urban streams may provide greater nutrition for benthic consumers. C:N ratios averaged 7.6 among streams, with 91% very close to 7.5, a value suggested as the optimum for algal growth. In contrast, periphyton C:P ratios ranged from 122 to >700 (mean = 248, twice Redfield). Algal-P concentrations were significantly greater in urban streams, but data suggest algal growth was P-limited in most streams regardless of degree of urbanization. GIS models indicate that land-use effects did not easily fit into strict categories, but varied continuously from rural to urban conditions. We propose that the gradient approach is the most effective method to characterize the influence of land use and urbanization on periphyton and stream function.

**Keywords** Benthic algae · Periphyton · Rivers · Land use · Ecological stoichiometry · Urban–rural gradient · Nutrients · New York

Guest editors: R. J. Stevenson, S. Sabater / Global Change and River Ecosystems—Implications for Structure, Function and Ecosystem Services

P. J. O'Brien · J. D. Wehr (✉)
Louis Calder Center—Biological Station and Department of Biological Sciences, Fordham University, Armonk, NY 10504, USA
e-mail: wehr@fordham.edu

## Introduction

The array of ecological variables that drive or limit algal production in streams, such as nutrient supply,

Reprinted from the journal     89      Springer

light availability, physical disturbance, and grazing, have been studied extensively, both through correlative and experimental approaches (Bothwell, 1988; Chessman et al., 1992; Stevenson et al., 1996; Wehr & Sheath, 2003). The ongoing challenge in studying these systems lies in the inherent complexity of stream habitats and their communities, but also in understanding the multiple scales of factors that regulate algal production and composition (Biggs, 1995; Snyder et al., 2002). Interest in algal production has increased in light of recent studies (e.g., Finlay et al., 2002; Torres-Ruiz et al., 2007) that have demonstrated a greater importance for autochthonous matter in lotic food webs than was suggested in earlier models (e.g., Vannote et al., 1980). These studies indicate that benthic algae consist of higher quality organic matter than that terrestrial matter, which is essential for consumers in stream food webs.

The factors that affect the quantity and quality of this production are of importance to stream ecosystem theory. Biggs (1996) proposed a two-tiered conceptual model to characterize the multiple factors that regulate benthic algal production and structure in streams. Proximate variables directly regulate biomass accrual and loss, and include physical and chemical factors, water quality (dissolved nutrients) temperature, optics (light availability, turbidity), and hydrography. The main factors predicted to lead more directly to biomass accrual are nutrients and light availability, while the main factors leading to loss of production are disturbance, especially floods and droughts. Larger-scale environmental or landscape features, ultimate variables, include climate, topography, land use, geology, and human impacts. The connection between the effects of proximate and ultimate variables as they affect stream periphyton has not received extensive study. However, Snyder et al. (2002) demonstrated that periphyton biomass and diatom community structure in broad (>5th order) streams in Idaho (USA) were most affected by N and P supply, and that these were in turn affected by location, presumably reflecting land-use differences. Blinn & Bailey (2001) demonstrated that diatom community structure was strongly correlated with land-use practices, especially irrigation practices and dryland farming, in streams in Victoria, Australia. Carr et al. (2005) tested the ability to use land-use variables to replace local water quality variables in predictive models of periphyton chlorophyll-$a$, using a 21-year database of rivers in Alberta, Canada. Land use, especially human population density, explained roughly 25–28% of the variability in periphyton Chl-$a$, but the best models included both land use and local nutrient data. They suggested that within ecoregions, land use can be a good surrogate for nutrient data in predicting lotic periphyton Chl-$a$ concentration.

Stream periphyton assemblages also vary in their nutritional quality. There is evidence suggesting that the importance of periphyton in stream food webs may be more a function of quality than quantity (Cross et al., 2003). Laboratory data suggest that the optimal stoichiometry of C:N:P content in freshwater periphyton should be around 119:17:1 to avoid nutrient limitation of growth (Hillebrand & Sommer, 1999), a ratio very close to that suggested for oceanic plankton: 106:16:1 (Redfield, 1958). However, unlike plankton in comparatively stable open oceans, autotrophs in more variable and spatially structured systems, like streams, show marked deviations from this C:N:P ratio (Wetzel, 2001; Hillebrand et al., 2004). Both the biomass produced and concentrations of essential nutrients contained in algal assemblages subsequently affect consumer growth rates (Frost & Elser, 2002; Stelzer & Lamberti, 2002), as well as nutrient cycling properties within the ecosystem (Dodds et al., 2004). A better understanding of the influence of different land-use conditions on periphyton stoichiometry in stream ecosystems is needed.

Urbanization exerts profound effects on the landscape and associated aquatic systems, such as re-direction of rainfall by impervious surfaces (Hirsch et al., 1990), increased surface runoff (McMahon & Cuffney, 2000), increased sediment load, and decreases in sediment particle size (Paul & Meyer, 2001). Temperature changes have been attributed to removal of riparian vegetation, decreased recharge of groundwater, and the urban "heat-island" effect (Pluhowski, 1970; Pickett et al., 2001). Oxygen demand, conductivity, turbidity, and dissolved metals also tend to increase with urbanization (Paul & Meyer, 2001). Inputs of sewage, wastewater, and fertilizers, which typify many urban streams, result in greater dissolved N and P concentrations (Meybeck, 1998; Winter & Duthie, 2000). Elevated levels of base cations (Ca, Mg, Na, and K) may also cause an increase in specific conductance (Paul & Meyer, 2001).

These physical and chemical changes can have important effects on stream periphyton. Walker &

Pan (2006) demonstrated that diatom species composition in streams in the Portland region (Oregon, USA) significantly correlated with differences in water chemistry and land use along an urban–rural gradient. Periphyton chlorophyll-$a$ accrual rates in one urban stream in Catalonia, Spain did not differ significantly among experimental nutrient (N, P) treatments, apparently due to shaded, canopy conditions (Schiller et al., 2007). Interestingly, substrata in the urban site also experienced accumulations of fine organic matter (detritus) on their surfaces, which was suggested to have inhibited periphyton growth. However, $NH_4^+$ treatments did result in significantly reduced $NO_3^-$ uptake rates in this stream. Periphyton growth in one urban stream near College Station, Texas was strongly affected by the high frequency of floods, but still reached 30 times the designated nuisance level (>100 mg Chl-$a$/m$^2$), and was composed of edible, early-stage algal species rather than late-stage, and less edible taxa (Murdock et al., 2004). How such changes may be mirrored by periphyton nutrient stoichiometry remains to be seen.

Here, we examine a suite of land-use factors that may influence the biomass and ecological stoichiometry of stream periphyton. We aim to link in-stream or proximate (e.g., temperature and nutrients) and ultimate (e.g., land use) variables to understand the key factors affecting streams along a land-use gradient at varying distances from a large urban center, NYC. We predict that (1) there will be significant changes in physical and water chemistry (proximate) variables in concert with identifiable land-use (ultimate) variables, and that (2) periphyton biomass and nutrient stoichiometry will be significantly affected by physical and chemical changes in these streams. Our goal is to identify which variables show greatest sensitivity to land-use changes and which may be most important to stream periphyton. We also aim to identify if periphyton nutrient stoichiometry data may be used to assess nutrient limitation among different land-use conditions.

## Materials and methods

Design and site selection

The study area is a 6,000 km$^2$ region in southeastern New York State east of the Hudson River and north from Yonkers to Troy NY. Land use ranges from dense urban districts in the south to sparsely populated rural areas. Rural regions are mosaics of forested (mixed hardwood) and light- to moderate-intensity agriculture. Following a pilot study of 20 streams, a power analysis determined that a sample size of at least 32 streams was required to detect local and landscape effects. We identified 70 streams as potential sites which met the following criteria: (1) first to third order; (2) stream width ≥3 m; (3) cobble-boulder substratum with at least three riffles; (4) current velocity in riffles ≥25 cm/s; (5) streambed neither completely shaded nor fully open to sunlight. From this set, a stratified-random method (stratified by watershed) was used to select 36 streams of sample in 2001. All streams were sampled within a 6-week period during average summer base flow (no major rainfall events). These criteria ensured that streams had predominantly rocky substrata and differed mainly by land use along a gradient from urban to rural conditions (verified using GIS, below).

Field sampling

Most field methods follow Stevenson & Bahls (1999). Geographic locations were determined using a Garmin 12-XL GPS unit. Stream width was measured with a measuring tape and depth with a meter stick. Current velocity was measured with a General Oceanics 2030 current meter from five riffles in the center of each reach. Water temperature and conductivity were measured with a YSI 30 S/C/T meter and canopy cover was measured with a Forest Model-C spherical densiometer. For periphyton samples, five rocks between 10 and 35 cm diameter were arbitrarily chosen from separate riffles, placed in 4 l plastic bags and stored on ice. Unfiltered water was collected for pH (25 ml polypropylene bottles) and turbidity (20 ml borosilicate vials) measurements and stored on ice. Two water chemistry samples were syringe-filtered in situ (Nalgene nylon membrane 0.2 μm poresize) and preserved by acidification (U. S. Environmental Protection Agency, 1987) to pH <2.0 with HCl (for dissolved organic carbon, DOC) or $H_2SO_4$ (for $NO_3^-$, $NH_4^+$, soluble-reactive-P [SRP], Si [as $SiO_3$]) and refrigerated (4°C) until analysis.

Periphyton processing and analyses

Periphyton was scraped from rocks with razor blades and brushes to a final volume of 250–500 ml; processing was completed within 1 day of collection. Periphyton suspensions were homogenized and mixed using a small hand blender, and divided into aliquots for (1) dry mass (DM) and ash-free dry mass (AFDM), (2) Chlorophyll-a (Chl-a), (3) C and N analyses, (4) P analysis, and (5) taxonomic identification (preserved with Lugol's iodine). DM and AFDM were determined after filtration onto pre-ashed and pre-weighed (to 0.0001 g) 47 mm glass fiber filters (Whatman GF/F), dried ($\geq$18 h) at 80°C, weighed (=DM), then ashed at 450°C for 2–3 h and re-weighed (AFDM) to the nearest 0.0001 g (AFDM = difference between DM and mass of ash after combustion; American Public Health Association, 1985). Pheophytin-corrected Chl-a was measured after extraction with 90% buffered acetone and absorbances measured at 750, 665, and 664 nm (Lorenzen, 1967). The Autotrophic Index (AI) of each periphyton assemblage was calculated as the ratio of AFDM/Chl-a, after Biggs & Close (1989). This method classifies samples with ratios <200 ("low") to consist mainly of autotrophic algae, while "moderate" ratios (200–400) are regarded to be a mixture of autotrophic and heterotrophic periphyton, and values >400 to be dominated by heterotrophic organisms and/or detritus. In our study, we use this index to estimate algal versus non-algal periphyton in streams draining different landscapes. Periphyton C and N concentrations were measured from homogenized periphyton dried at 80°C in 9 × 10 mm tin cups and measured using a Perkin Elmer 2400 Series II CHNS/O analyzer. Periphyton-P was measured after digestion following Solórzano & Sharp (1980) and the digest analyzed as reactive-P as described below. After all scrapings were completed, rocks were measured for determination of surface area (SA), based on linear dimensions and formulas for appropriate geometric shapes. We defined "colonizable area" as the upper 50% of each rock area as all were embedded in streambed at time of collection.

Water chemistry

Streamwater pH was measured with an Orion Model 720 pH Meter and turbidity with a Turner TD-40 nephelometer; both were analyzed in the lab immediately after returning from the field (<6 h). DOC was measured using a Shimadzu TOC-5050A TOC analyzer (American Public Health Association, 1985). Dissolved ammonium ($NH_4^+$–N) concentrations were measured as using the phenol–hypochlorite method (American Public Health Association, 1985; Bran+Luebbe Analyzing Technologies, 1986b); nitrite (measured as $NO_2^-$–N) using the sulfanilamide-NNED (American Public Health Association, 1985; Bran+Luebbe Analyzing Technologies, 1987a); nitrate ($NO_3^-$–N) by reduction to nitrite using a Cd–Cu column and analyzed as nitrite; reactive silica using the molybdate–ascorbate method (American Public Health Association, 1985; Bran+Luebbe Analyzing Technologies, 1987b); and soluble-reactive phosphate (SRP) using the antimony–ascorbate–molybdate method (American Public Health Association, 1985; Bran+Luebbe Analyzing Technologies, 1986a). Total dissolved phosphorus (TDP) was predigested using acid persulfate (Eisenreich et al., 1975) and analyzed as SRP. Nutrients were measured using a Bran & Luebbe TRAACS 800 autoanalyzer. Dissolved Ca and Mg were measured using a Perkin-Elmer 1100B atomic absorption spectrophotometer (American Public Health Association, 1985).

Data variables

Nutrient stoichiometry of periphyton assemblages was calculated on a molar basis from nutrient content data and used as a broad measure of nutrient limitation (Redfield, 1958; Hillebrand & Sommer, 1999). The proximate stream variables stream width, maximum depth, current velocity, turbidity, temperature, and percent canopy cover were defined as physical properties. Conductivity, pH, DOC, $NH_4^+$, $NO_3^-$, Si, SRP, TDP, Ca, and Mg were defined as (proximate) water chemistry variables. The following were defined as periphyton variables (on DM and SA basis): % organic matter, AFDM, Chl-a, C, N, P, and C:N, C:P, and N:P ratios. Landscape-level variables were land use, bedrock geology, and population density. Data were obtained from the New York State GIS Clearinghouse and analyzed using Arc Info. Population density data (persons per square km) were derived from a layer of the U.S. Department of Commerce, Bureau of the Census 1990 coverage. Land use data were derived from a layer of the U.S. Geological Survey (USGS) 1990 1:250,000 Scale

Land use and Land Cover for New York, Hartford and Albany Quadrangles. Data were presented in the Anderson level 2 classification system (general land use and subdivisions), but for the purpose of this stream study only the level 1 classification was used, as suggested by Wang & Yin (1997). Bedrock geology data were derived from a layer of the New York Geologic Survey 1999 Bedrock Geology Lower Hudson and Hudson Mohawk coverages. Additional hydrography and boundary coverages for Westchester, Putnam, Dutchess, Columbia, and Rennssalaer counties were obtained from the U.S. Department of Commerce, Bureau of the Census 1998 TIGER Files.

Data analysis

Analyses addressed correlations between northing (distance from NYC) and physical properties, water chemistry and periphyton variables, to establish if distance from a large urban area had a general influence on these variables. Multiple stepwise linear regression was used to determine which variables most affected periphyton biomass and nutrient content ($\alpha \leq 0.05$ for inclusion, $\alpha \geq 0.10$ for exclusion into the model). Multicollinearity was assessed by comparing individual $t$-values and $r^2$ with overall $r^2$ and $P$ values in each model (Graham, 2003); no issues were detected. Predictor (independent) variables included physical (temperature, canopy cover, turbidity, depth, width, and current velocity), and water chemistry (DOC, $NO_3^-$–N, $NH_4^+$–N, SRP, TDP, Ca, and Mg) variables. GIS analysis, correlation, and ANOVA were used to determine the effect of land use and urbanization on stream and periphyton variables. Data from physical, water chemistry, and periphyton variables for each stream were entered into an Arc Info data table, visualized as points by using location data, northing and easting, and then combined with population density, land use and geology coverages. For GIS, a 150-m buffer zone (radius) was created around each sample point in Arc Info (Tufford et al., 1998). Arc Info output files for population density, land use, and geology attributes were generated for the buffer zones of each stream. The weighted mean population density, bedrock geological data, and percent land use categories were determined for each 150-m buffer zone. Correlations were used to determine links between population density and the physical, water chemistry, and periphyton variables. ANOVA followed by Tukey post-hoc HSD tests were used to determine the effects of land use (Sokal & Rohlf, 1995). Each stream-reach site was classified into a predominant land-use category, determined by the landscape that occupied the largest percentage within each buffer zone (typically >60% of total area). Statistical analyses were performed using Systat 10.0 with $\alpha \leq 0.05$ set for a type I error. All quantitative data were tested for normality; non-normal data were transformed using standard transformation methods (Sokal & Rohlf, 1995). Most physical, chemical, and periphyton variables required $\log_{10}$ transformation with the exception of Si (square root transformation) and DOC and pH (normal). Of the landscape variables, population density was $\log_{10}$ transformed.

**Results**

Landscape patterns

Streams situated in urban landscapes were concentrated in the southern portion of the study area. Percent urban area (as determined by GIS) within respective 150-m buffer zones (radii) at each site, ranged from <0.01 to 100%, as did percent agriculture and forest area. These landscape variables were compared with geographic distance from NYC (northing). Human population density significantly decreased ($r = -0.739$; $P < 0.00001$; Fig. 1A), while % agricultural land use significantly increased ($r = 0.347$; $P = 0.0379$) with northing (Table 1). Percent urban and % forested land showed similar trends but were NS ($P > 0.10$; Table 1). Therefore, northing (distance from NYC) was used as a proxy variable to further test the effects of urbanization.

Among physical variables, stream water turbidity and temperature significantly decreased with northing (i.e., greater values in urban streams; Table 1; Fig. 1B). Turbidity ranged from 0.83 to 5.63 NTU, while summer temperatures ranged from 8.5 to 25.9°C. Percent canopy cover, stream depth, stream width, and current velocity were not significantly correlated with northing. Most of the water chemistry variables were negatively correlated with northing (i.e., greater in urban streams), including conductivity (Fig. 1C) and dissolved Mg, Ca, SRP, pH, DOC, and

**Fig. 1** Plots of the significant correlations between geographic distance from NYC (northing) and key land use (**A** human population density; $r = -0.739$, $P < 0.001$), physical (**B** stream turbidity; $r = -0.587$, $P < 0.001$), and water chemistry (**C** conductivity; $r = -0.722$, $P < 0.001$) variables measured at each of 36 stream sites along a putative urban to rural land-use gradient (see Table 1 for complete statistics)

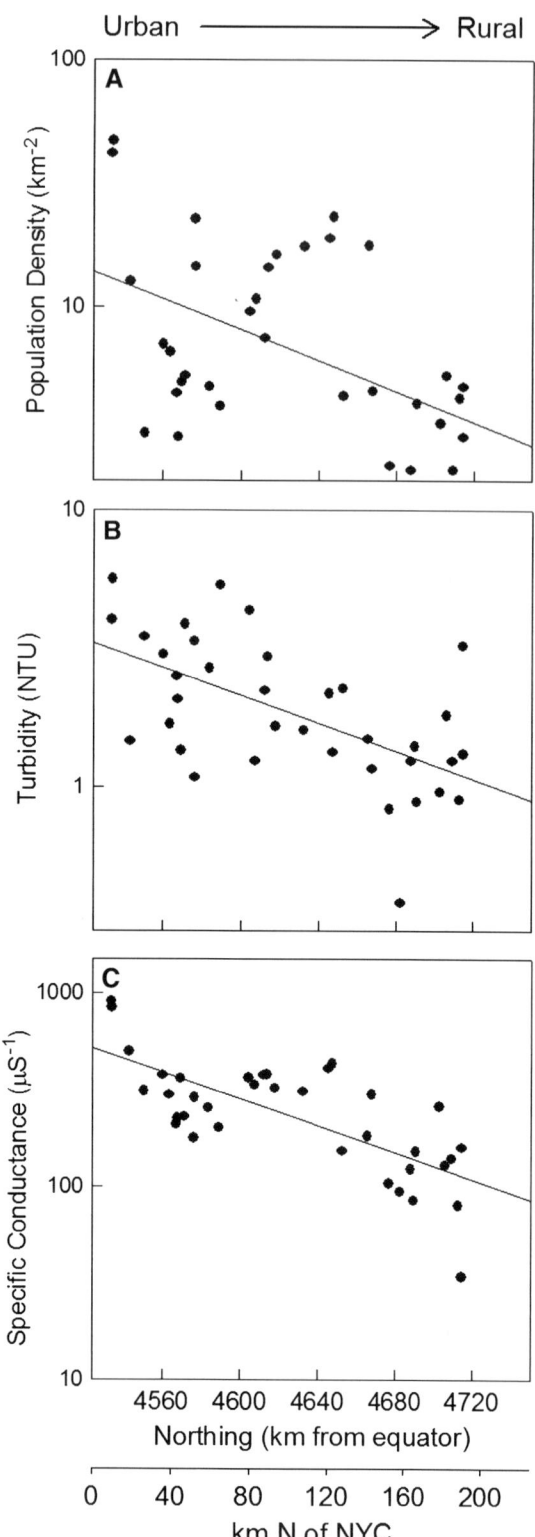

Si, but were non-significant for dissolved $NO_3^-$ and $NH_4^+$. Water chemistry variables that differed most among streams were pH (6.1–8.9), Mg (0.6–22 mg/l), conductivity (34–908 μS/cm), and SRP (6.4–66 μg P/l). Each of these was significantly greater in urban streams.

Periphyton biomass and stoichiometry

Periphyton biomass, measured as Chl-*a* (per unit area), varied by two orders of magnitude among the 36 streams and by approximately one order when measured as AFDM (Table 2). Similarly, carbon content of algal periphyton ranged nearly 30-fold, from 0.06 to 1.7 μmol/mm² (unit area basis), but only about 6.6-fold on a DM basis (14.2–93.6 μmol/mg). Periphyton AFDM and Chl-*a* concentrations were very highly correlated ($r = 0.8625$; $P < 0.00001$), and the AI of these measures varied strongly among streams (mean $425 \pm 44$ [SE]), with 58% of the streams (21 of 36) with low (<200) or moderate (200–400) ratios.

Periphyton C, N, and P content also varied among streams (Table 2) and created a range of periphyton C:N ratios from 4.4 to 12.5 (mean = $7.6 \pm 1.5$; Redfield = 6.6) and C:P ratios from 122 to 706 (mean = $248 \pm 109$; Redfield = 106). The periphyton C:N ratios in 29 of 36 streams were greater than the Redfield ratio of 6.6, and all C:P ratios were greater than the predicted ratio of 106. The average C:N:P ratio of 191C:24N:1P was more C-rich than the idealized Redfield ratio of 106:16:1.

Periphyton biomass and nutrient content correlated either positively or negatively with northing (Table 3). Periphyton biomass, measured as AFDM ($r = -0.485$, $P = 0.003$; Fig. 2A) was significantly greater in streams at the urban end of the gradient (=negative correlation with northing), although Chl-*a* concentration per SA, while showing a similar trend, was non-significant. Correlations between calculated AI and

**Table 1** Correlations between northing (distance from NYC) and land-use properties, physical variables, and water chemistry from 36 streams from an urban–rural land-use gradient north of New York City ($r$, correlation coefficient; values with $P \leq 0.05$ are printed in bold)

| Northing versus | $r$ | $P$ |
| --- | --- | --- |
| Land use | | |
| % Agricultural land | **+0.347** | **0.0379** |
| % Forested land | −0.245 | 0.1497 |
| % Urban land | −0.128 | 0.4564 |
| Population density | **−0.739** | **<0.0001** |
| Physical variables | | |
| Canopy cover | −0.179 | 0.2972 |
| Current velocity | −0.035 | 0.8397 |
| Depth | −0.114 | 0.5067 |
| Temperature | **−0.555** | **0.0004** |
| Turbidity | **−0.587** | **0.0002** |
| Width | −0.096 | 0.5792 |
| Water chemistry | | |
| Ca | **−0.575** | **0.0002** |
| Conductivity | **−0.722** | **<0.0001** |
| DOC | **−0.418** | **0.0111** |
| Mg | **−0.653** | **0.0002** |
| $NH_4^+$ | +0.171 | 0.3195 |
| $NO_3^-$ | −0.183 | 0.2843 |
| pH | **−0.525** | **0.0010** |
| Si | **−0.341** | **0.0416** |
| SRP | **−0.558** | **0.0004** |
| TDP | −0.217 | 0.2038 |

Negative correlations indicate greater values in urban streams

land-use measures were weak or inconsistent (correlations with northing, human population density, and % urban land NS; $P > 0.250$), but AI did correlate positively with % forested land ($r = 0.537$; $P = 0.0007$).

Periphyton in urban streams had significantly greater C, N, and P content (=negative correlation with northing; Fig. 2B; Table 3). Periphyton also had greater algal C:N in urban streams (negative correlation with northing) and lower N:P ratios (positive correlation with northing; Fig. 2C). Further, periphyton N:P ratios in a majority of streams were greater than Redfield predictions, although those in urban locations were closer to the predicted value of 16. The trend for periphyton C:P was positive with northing (relatively greater P in urban streams), but the trend was non-significant ($r = 0.277$; $P = 0.1068$).

Influence of in-stream variables on periphyton biomass and nutrient stoichiometry

Multiple linear regression (MLR) models were next used to identify the most important proximate predictors (local physical and water chemistry variables) of variation in periphyton biomass and nutrient content among the study streams (Table 4). Variation in AFDM ($r^2 = 0.646$) and Chl-$a$ ($r^2 = 0.561$) per unit area were more effectively predicted than was percent organic matter ($r^2 = 0.119$) for these streams. Collectively, the models suggest that benthic algal biomass was negatively influenced by percent canopy cover, stream depth, and in one measure (% organic matter), turbidity. These variables were all significantly greater in urban streams (decreased with northing). None of the major nutrients associated with urbanization (SRP, TDP, $NH_4^+$, $NO_3^-$) were included in any of the models, despite several positive bivariate correlations (e.g., SRP versus AFDM; $r = +0.366$, $P = 0.0304$; $NO_3^-$ vs. Chl-$a$; $r = +0.434$, $P = 0.0092$).

The influence of proximate, in-stream variables on periphyton nutrient content and stoichiometry were also examined using MLR (Table 5). Nutrient content was also best predicted by aqueous cation (Mg or Ca) concentration and canopy cover. However, stoichiometric ratios, particularly those with phosphorus, were predicted to be a function of aqueous P (as TDP).

Influence of land-use variables

Differences in human population density and landscape type were clearly identified within the study area (Table 1). From this, seven of nine major land use classes were identified from the NY State GIS database: urban, agricultural, rangeland, forested, water (lakes, ponds, and streams), wetlands, and barren land (non-vegetated but not impervious surfaces). Of these, only urban, agricultural and forested land occurred in large enough frequencies for analysis. Their potential influence on periphyton was first examined using correlations between landscape variables (three land-use types and human population density) and the biomass and nutrient stoichiometry of stream periphyton (Table 6). Biomass (as AFDM) was positively correlated with increasing population density (Fig. 3A), with percent of urban land, based on GIS estimates (Fig. 3B), and negatively with

Table 2 Summary data for biomass, nutrient content, and stoichiometry of algal periphyton sampled from 36 streams from an urban to rural land-use gradient north of New York City (SD, standard deviation; AFDM, ash-free dry mass; DM, dry mass; SA, surface area; atomic stoichiometric ratios calculated from nutrient per unit area)

|  | Min | Max | Mean | SD |
|---|---|---|---|---|
| Biomass | | | | |
| AFDM ($g/m^2$) | 1.6 | 47.0 | 9.8 | 10.4 |
| Chl-$a$/DM (µg/g DM) | 2.0 | 63.4 | 16.8 | 16.6 |
| Chl-$a$/SA ($mg/m^2$) | 1.7 | 226.4 | 37.5 | 49.9 |
| Ratio Chl-$a$/AFDM (%) | 0.2 | 6.3 | 1.7 | 1.7 |
| % Organic matter | 23.2 | 60.0 | 41.5 | 4.7 |
| Nutrient content | | | | |
| C/DM (µmol/mg) | 14.2 | 93.60 | 38.1 | 24.1 |
| C/SA ($µmol/mm^2$) | 0.06 | 1.70 | 0.39 | 0.41 |
| N/DM (µmol/mg) | 0.008 | 0.209 | 0.050 | 0.047 |
| N/SA ($µmol/mm^2$) | 1.8 | 12.2 | 0.05 | 0.05 |
| P/DM (µmol/mg) | 0.0016 | 0.0138 | 0.0020 | 0.0030 |
| P/SA ($µmol/mm^2$) | 0.0170 | 0.1360 | 0.0777 | 0.0274 |
| Nutrient stoichiometry | | | | |
| C:N | 4.4 | 12.5 | 7.6 | 1.5 |
| C:P | 122 | 706 | 248 | 109 |
| N:P | 14.3 | 93.8 | 33.7 | 15.6 |

Table 3 Correlations between northing (distance from NYC) and biomass, nutrient content and stoichiometry of algal periphyton sampled from 36 streams from an urban–rural land-use gradient north of New York City (see Table 2 for abbreviations, values with $P \leq 0.05$ are printed in bold)

| Northing versus | r | P |
|---|---|---|
| Biomass | | |
| AFDM | **−0.485** | **0.0031** |
| Chl-$a$/DM | 0.023 | 0.8976 |
| Chl-$a$/SA | −0.301 | 0.0792 |
| Ratio Chl-$a$/AFDM (%) | −0.272 | 0.1080 |
| % Organic matter | 0.254 | 0.1408 |
| Nutrient content | | |
| Algal C/DM | −0.223 | 0.1982 |
| Algal C/SA | **−0.459** | **0.0056** |
| Algal N/DM | −0.141 | 0.4197 |
| Algal N/SA | **−0.425** | **0.0110** |
| Algal P/DM | −0.088 | 0.6171 |
| Algal P/SA | **−0.501** | **0.0022** |
| Nutrient stoichiometry | | |
| C:N / SA | **−0.334** | **0.0500** |
| C:P / SA | 0.277 | 0.1068 |
| N:P / SA | **0.402** | **0.0167** |

Negative correlations indicate greater values in urban streams

greater forest land cover (Fig. 3C). Biomass measured as Chl-$a$ concentration correlated negatively with % forested land (Table 6). Periphyton C, N, and P concentrations each correlated positively with human population density (Fig. 4) and with % urban land cover (Table 6). Despite a wide range of % agricultural land across the streams-sites (from 0 to 100%; mean: 44.4 ± 43.5%), periphyton biomass was unrelated to this land-use attribute (Table 6). Periphyton N:P ratio correlated negatively with human population density, but none of the stoichiometric ratios correlated with any of the three main land-use categories.

The influence of land-use type on periphyton biomass and nutrient stoichiometry was contrasted among the three major GIS-based land-use categories, designated as either forested, agriculture or urban, based on the greatest land-use percentage within a 150-m radius at each geographic location. Not all periphyton variables exhibited clear patterns, but algal biomass (as Chl-$a$) was significantly different among the three land-use areas, with greatest amounts measured in urban streams (Table 7; Fig. 5). Similarly, periphyton in urban streams also had significantly greater C and N content. In each of

**Table 4** Multiple linear regression analysis identifying stream-variable (proximate) predictors of benthic algal biomass, measured as ash-free dry mass (AFDM), Chl-$a$ per unit dry mass (Chl-$a$/DM), Chl-$a$ per unit surface area (Chl-$a$/SA), and percent organic matter

|  | Slope ± SE | $t$-Score | $P$ |
|---|---|---|---|
| **AFDM** | | | |
| Mg | 0.697 ± 0.100 | 6.968 | <0.00001 |
| Percent canopy cover | −0.006 ± 0.002 | −2.729 | 0.01024 |
| Depth | −0.001 ± 0.004 | −2.488 | 0.01826 |
| ANOVA: $P < 0.00001$ | | Adjusted $r^2 = 0.646$ | |
| **Chl-$a$/DM** | | | |
| DOC | −0.029 ± 0.016 | −1.841 | 0.07521 |
| pH | 0.143 ± 0.042 | 3.390 | 0.00192 |
| Percent canopy cover | −0.003 ± 0.001 | −2.221 | 0.03376 |
| Depth | −0.007 ± 0.002 | −2.746 | 0.00994 |
| ANOVA: $P = 0.00003$ | | Adjusted $r^2 = 0.497$ | |
| **Chl-$a$/SA** | | | |
| pH | 0.592 ± 0.116 | 5.122 | <0.00001 |
| Percent canopy cover | −0.009 ± 0.003 | −2.968 | 0.00564 |
| Depth | −0.020 ± 0.006 | −3.280 | 0.00251 |
| ANOVA: $P < 0.00001$ | | Adjusted $r^2 = 0.561$ | |
| **% Organic matter** | | | |
| Turbidity | −11.341 ± 4.739 | −2.393 | 0.02236 |
| ANOVA: $P = 0.02236$ | | Adjusted $r^2 = 0.119$ | |

Independent variables were physical and water chemistry variables measured at each stream (independent variables listed in order of relative importance; $r^2$ = coefficient of determination for complete model)

**Table 5** Multiple linear regression analysis identifying stream-variable (proximate) predictors of periphyton nutrient content and stoichiometric ratios (per unit SA)

|  | Slope ± SE | $t$-Score | $P$ |
|---|---|---|---|
| **C/SA** | | | |
| Mg | 0.665 ± 0.111 | 5.984 | <0.00001 |
| Percent canopy cover | −0.006 ± 0.002 | −2.677 | 0.01162 |
| ANOVA: $P < 0.00001$ | | Adjusted $r^2 = 0.571$ | |
| **N/SA** | | | |
| Ca | 0.830 ± 0.140 | 5.923 | <0.00001 |
| Percent canopy cover | −0.005 ± 0.002 | −2.423 | 0.02121 |
| ANOVA: $P < 0.00001$ | | Adjusted $r^2 = 0.562$ | |
| **P/SA** | | | |
| Ca | 1.138 ± 0.182 | 6.256 | <0.00001 |
| ANOVA: $P < 0.00001$ | | Adjusted $r^2 = 0.529$ | |
| **C:N** | | | |
| Mg | 4.431 ± 1.484 | 2.986 | 0.00539 |
| Conductivity | −4.408 ± 1.962 | −2.247 | 0.03169 |
| ANOVA: $P = 0.00587$ | | Adjusted $r^2 = 0.230$ | |
| **C:P** | | | |
| TDP | −0.258 ± 0.098 | −2.620 | 0.01334 |
| Si | −0.297 ± 0.095 | −3.130 | 0.00372 |
| ANOVA: $P = 0.00087$ | | Adjusted $r^2 = 0.316$ | |
| **N:P** | | | |
| TDP | −0.317 ± 0.098 | −3.238 | 0.00281 |
| Si | −0.348 ± 0.094 | −3.688 | 0.00083 |
| ANOVA: $P = 0.00008$ | | Adjusted $r^2 = 0.410$ | |

Independent variables were physical and water chemistry variables measured at each stream (independent variables listed in order of relative importance; $r^2$ = coefficient of determination for complete model)

**Fig. 2** Plots of the significant correlations between geographic distance from NYC (northing) and periphyton biomass (**A** ash-free dry mass per unit area; $r = -0.485$, $P = 0.003$), periphyton nutrient content (**B** phosphorus concentration per unit area; $r = -0.501$, $P = 0.002$), and periphyton stoichiometry (**C** N:P ratio; $r = 0.402$, $P = 0.17$) measured at each of 36 stream sites along a putative urban to rural land-use gradient (see Table 3 for complete statistics)

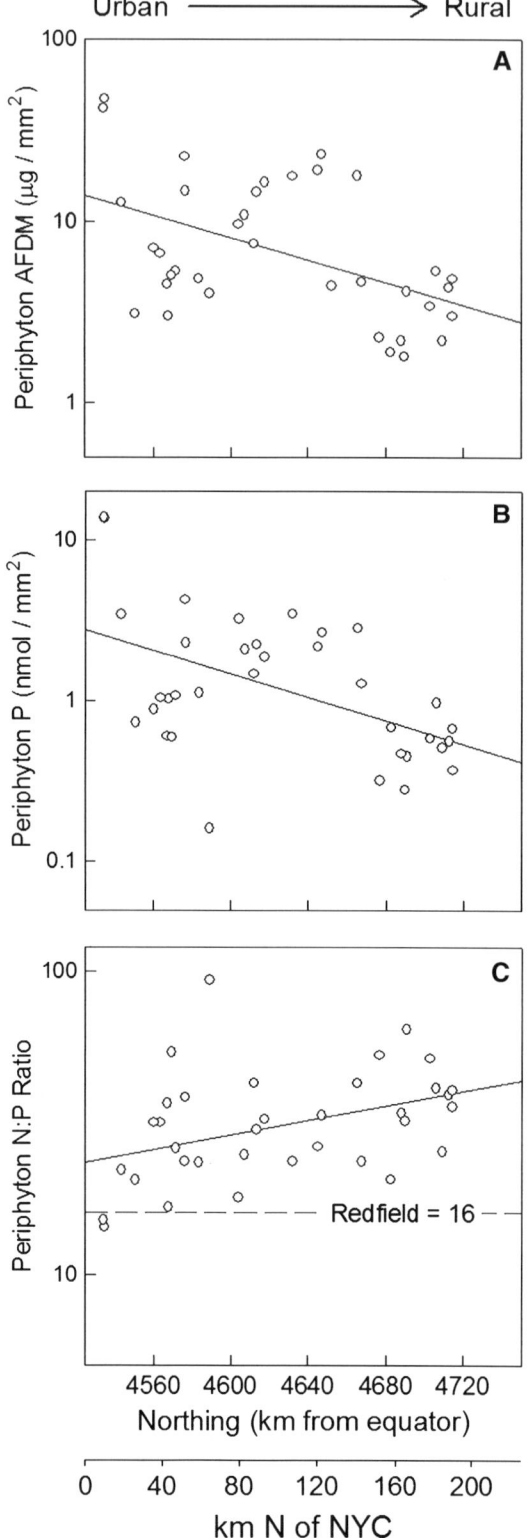

these comparisons, those streams draining rural forested land were classified with the least biomass and nutrient content. Although the nutrient stoichiometric ratios varied along the urban–rural gradient (based on northing; Table 3), no significant differences were revealed when compared by land-use categories (Table 7).

## Discussion

Streams in contrasting landscapes

Our data demonstrate that streams draining different landscapes in southern New York State have correspondingly different physical, chemical, and biological properties. Watersheds along this urban–rural gradient differ little with regard to soil type, geology, or forest type, but exhibit profound ecological differences, making this region a useful setting in which to test human impacts (McDonnell et al., 1997). In our study, urban streams were more nutrient-rich, had higher pH, greater concentrations of dissolved cations, and greater conductivity. The range of conductivities and nutrient concentrations equal or exceed that observed along other urban–rural land-use gradients in the U.S. (Walker & Pan, 2006; Sprague et al., 2007; Ponader et al., 2008).

All of the significant trends in water chemistry along the urban–rural gradient increased from rural to urban areas (Table 1; Figs. 1, 3). The elevated SRP concentrations likely resulted from sewage inputs, as well as fewer wetlands and altered soils (Walsh et al., 2005). Sonoda & Yeakley (2007) demonstrated that soils adjacent to urban streams have lower capacities for retaining P than those in non-urban areas. We observed a highly significant correlation between SRP in stream water and distance from an urban center, as well significant trends for dissolved Mg, Ca, Si, DOC, and pH, but not dissolved $NH_4^+$ or $NO_3^-$ (Table 1). Our data partly support earlier contentions (Pluhowski,

1970; Pickett et al., 2001) that streams along an urbanization gradient are warmer (Table 1: $r = -0.555$ versus northing). However, when compared among specific GIS land-use categories, stream temperatures were not significantly different (forested: $18.7 \pm 1.0$°C versus urban: $18.8 \pm 3.9$°C). While evidence of a "heat-island" effect was not seen, our data do agree with prior studies showing many other physical and chemical effects of urbanization (Paul & Meyer, 2001; Strayer et al., 2003), effects that likely have important effects on stream periphyton.

Periphyton biomass and land use

Periphyton biomass was significantly greater in streams draining urbanized landscapes and in locations with greater human population density. Periphyton in rural, forested streams averaged roughly one-third to one-fifth the amount in urban locations (Figs. 3, 5; Tables 3, 6). Average biomass in urban streams of $18.8 \pm 6.0$ g/m$^2$ AFDM and $75.6 \pm 28.5$ mg/m$^2$ Chl-$a$, and maximal Chl-$a$ levels exceeding 100 mg/m$^2$, are comparable to quantities measured in agricultural streams elsewhere (Biggs & Close, 1989; Chételat et al., 1999; Godwin & Carrick, 2008) and in experimentally nutrient-enriched streams (e.g., Greenwood & Rosemond, 2005). However, variation in stream algal biomass in our study area was not correlated with % agricultural land, despite dominating the watersheds of 13 streams. Instead, biomass was most closely correlated with human population density and % urban land (Table 6). Landscapes classified as agricultural in

**Table 6** Correlations between landscape variables and periphyton biomass, nutrient content and stoichiometry ($r$, correlation coefficient, with probabilities; * $P \leq 0.05$, ** $P < 0.01$, values with $P \leq 0.05$ are printed in bold)

|  | Population density $r$ | Percent urban land $r$ | Percent agricultural land $r$ | Percent forested land $r$ |
|---|---|---|---|---|
| Biomass |  |  |  |  |
| AFDM/SA | **0.429*** | **0.414*** | −0.091 | −0.256 |
| Chl-$a$/SA | **0.356*** | **0.362*** | 0.159 | **−0.465**** |
| % Organic matter | −0.185 | 0.132 | 0.093 | −0.205 |
| Nutrient content |  |  |  |  |
| Algal C/SA | **0.443**** | **0.401*** | −0.052 | −0.284 |
| Algal N/SA | **0.445**** | **0.404*** | 0.005 | **−0.344*** |
| Algal P/SA | **0.491**** | **0.389*** | −0.037 | −0.289 |
| Nutrient stoichiometry |  |  |  |  |
| C:N | 0.168 | 0.164 | −0.263 | 0.128 |
| C:P | −0.286 | −0.116 | −0.020 | 0.117 |
| N:P | **−0.338*** | −0.170 | 0.101 | 0.041 |

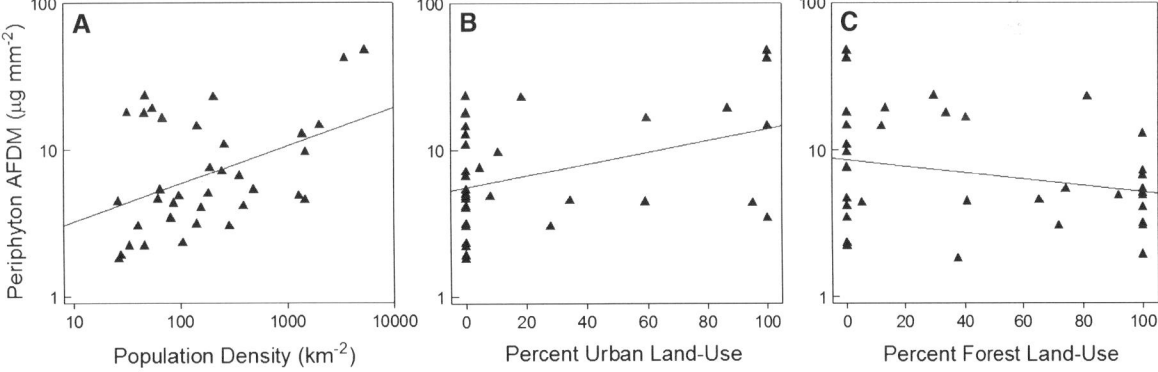

Fig. 3 Relationships between periphyton biomass (ash-free dry mass per unit area) versus human population density, percent urban land use and percent forest land use across 36 stream sites along a putative urban to rural land-use gradient

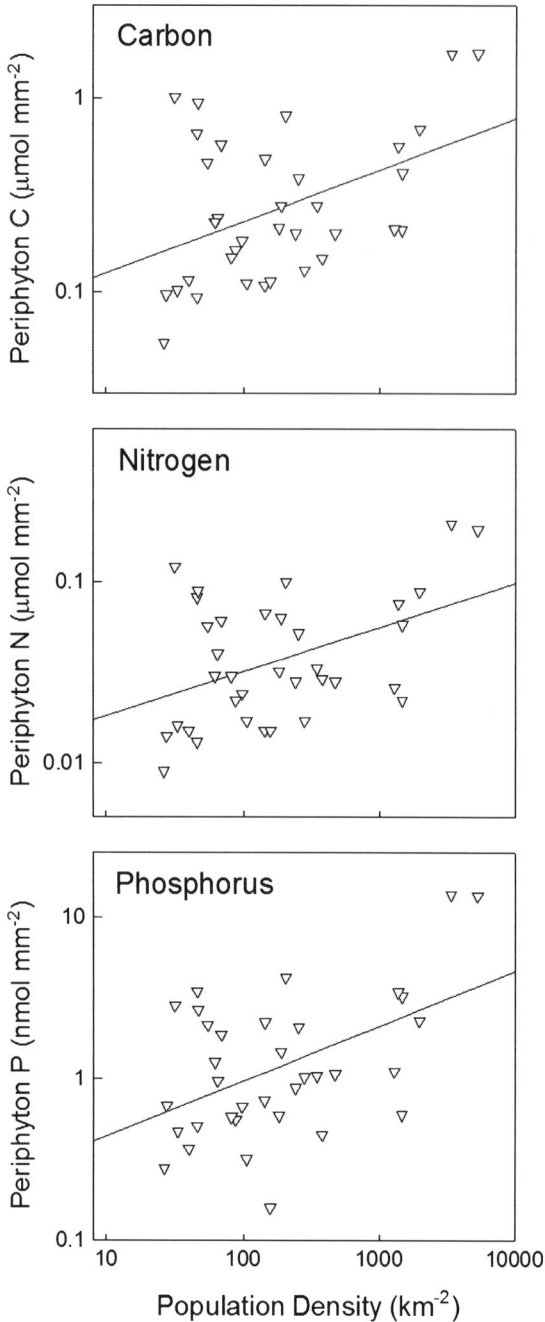

**Fig. 4** Relationships between algal carbon, nitrogen, and phosphorus (per unit area) versus human population density across 36 stream sites along a putative urban to rural land-use gradient

biomass (>20 g AFDM/m$^2$, 100–300 mg Chl-$a$/m$^2$; Biggs & Close, 1989). We attribute this difference to a lower intensity form of agriculture in our region, which is a mixture of grain crops and fallow fields.

Few studies have characterized periphyton biomass and stoichiometry in urban streams. However, a meta-analysis of data from more than 300 streams and rivers from the USGS National Stream Water-Quality Monitoring Network revealed significant positive correlations between benthic Chl-$a$ and % urban land area (Dodds et al., 2002), which agrees with our data for the NYC urban–rural gradient. Murdock et al. (2004) reported that one urban stream in Texas accumulated periphyton biomass at very rapid rates, despite frequent floods that were capable of removing most of this production. Average total biomass in their system occasionally exceeded 500 mg Chl-$a$/m$^2$, which was five or more times the recognized nuisance level (Welch et al., 1988; Murdock et al., 2004). Just three streams in the present study (all urban) exceeded that criterion.

It has been suggested that algal biomass may exhibit inconsistent responses to increasing urbanization of streams (Walsh et al., 2005), but the present data and those cited earlier suggest that urban land use usually results in greater periphyton biomass. Some of this increased biomass may be facilitated by reduced canopy cover in the riparian zone, providing greater light for primary production (Table 4). Urban streams in our study area also had significantly greater dissolved SRP, Si, Ca, and Mg (negative correlations with northing; Table 1), suggesting that greater SRP supply may also have exerted a positive influence. Hill & Fanta (2008) have demonstrated that stream periphyton growth can be co-limited by P and irradiance, although experiments in other streams suggest that light may be a more important factor (Schiller et al., 2007). Canopy cover values estimated for our streams by spherical densitometer measurements indirectly measure light availability, but have the advantage of reflecting longer-term landscape properties of each site (e.g., Fitzpatrick et al., 1998; Pan et al., 1999).

One may also ask whether urban streams are more autotrophic, based on the presence of elevated periphyton biomass. In the present study, no direct measures of productivity were made. However, periphyton AFDM and Chl-$a$ concentrations in our streams were

our study area had intermediate levels of periphyton biomass (mean AFDM: 9.1 ± 2.0 mg/m$^2$). Agricultural streams in other regions have much greater algal

**Table 7** Results of analysis of variance comparing mean periphyton biomass and nutrient stoichiometry (per unit surface area) among three major land-use categories (Tukey HSD post-hoc analyses run following significant general ANOVAs; individual land use listed in order of their means; classes sharing underlines were judged not statistically different [$P > 0.05$])

|  | $F$ | $P$ | Post hoc tests | | |
|---|---|---|---|---|---|
| Biomass | | | | | |
| AFDM/SA | 3.057 | 0.06053 | | | |
| Chl-$a$/SA | 5.615 | 0.00796 | Forest | Agricultural | Urban |
| % Organic matter | 1.613 | 0.21455 | | | |
| Nutrient content | | | | | |
| Algal C/SA | 3.484 | 0.04276 | Forest | Agricultural | Urban |
| Algal N/SA | 4.016 | 0.02780 | Forest | Agricultural | Urban |
| Algal P/SA | 2.874 | 0.07116 | | | |
| Nutrient stoichiometry | | | | | |
| C:N | 0.893 | 0.41928 | | | |
| C:P | 0.086 | 0.91784 | | | |
| N:P | 0.236 | 0.79117 | | | |

highly correlated ($r = 0.862$; $P < 0.00001$), suggesting that these streams have an algal-based metabolism. The geographic pattern of periphyton AFDM : Chl-$a$ (AI), a relative measure (although approximate) of the degree of heterotrophy or autotrophy (Biggs & Close, 1989), was inconsistent among stream sites. The AI was significantly greater (=less autotrophic) in more rural streams in forested landscapes, but did not correlate with any measures of urbanization. However, the C:Chl-$a$ ratio of epilithic periphyton among all of our streams averaged $\approx 200$, which is broadly indicative of a relatively high algal content in the mixed periphyton, and is substantially less than a global average of 405 across many freshwater systems (Frost et al., 2005). Among our streams, this ratio was least (greater algal content) in urban streams (139 ± 25) and greatest in rural, forested streams (299 ± 42). It is possible that the degree of autotrophy may not be directly responsive to urbanization, but our AFDM, Chl-$a$ and carbon data suggested that the majority of this carbon is algal based. For this reason, nutrient stoichiometry of stream periphyton may be important in studies of urban-to-rural gradients.

Nutrient effects and algal stoichiometry

Nutrient content of periphyton varied widely among our study streams, with algal carbon ranging more than from 20-fold, from 0.06 to 1.7 μmol/mm$^2$, while N and P content each varied by more than 6-fold among streams (Table 2). Concentrations of all three elements correlated significantly along the land-use gradient (northing, distance from NYC, % urban land), with greater concentrations in urban streams (Tables 3, 6; Figs. 4, 5). This trend of greater N and P content suggests that algal matter in urban streams may provide greater nutrition for consumers, which require specific quantities of N and P for metabolism and growth. Mass gain in the larval caddisfly *Glossosoma nigrior* was positively correlated with greater periphyton N content on which they grazed (Hart, 1987). Similarly, growth of the stream-dwelling snail *Elimia flavescens* was significantly greater when provided with P-enriched periphyton (40% greater than controls), but only when food supply was low (Stelzer & Lamberti, 2002). The nutrient content of periphyton was also shown to affect the rates of excretion and retention of N and P by heptageniid mayfly larvae, although the animals retained or accumulated P in excess of immediate needs (Rothlisberger et al., 2008). A shift from nutrient-limited to nutrient-sufficient conditions in a stream, as occurs during urbanization, could alter interactions between grazers and algal food sources, and have important consequences for nutrient cycling.

Greater accumulation of P and N by periphyton in urban streams cannot be fully attributed to elevated aqueous sources of these nutrients, as dissolved $NH_4^+$ and $NO_3^-$ concentrations did not correlate with northing (Table 1) or appear in any of the multiple regression models for periphyton biomass (Tables 4, 5). The absence of aqueous nutrients in some of the regression models may be due to the fact that some of the carbon in periphyton was detrital rather than algal. However, we did observe significantly greater SRP concentrations in urban streams, which could have enhanced periphyton growth, as well as P and N content. An alternative explanation is that algal growth effectively diluted the detrital content (high C:P) of the periphyton, and thereby increased the net P concentration of the assemblage. Nonetheless, experiments using nutrient-diffusing substrata have shown directly that increased P can have a direct stimulating effect (Schiller et al., 2007). Such effects may also result when either N or P is limiting algal growth, and one nutrient is preferentially assimilated and stored (Liess & Hillebrand, 2006).

The C:N:P stoichiometry of periphyton from our study streams varied both above and below predicted Redfield ratios. However, C:N ratios averaged approximately 7.6 across our streams (Table 2), which is very similar to a median of 7.5 from a series of laboratory growth experiments with optimum growth rates (Hillebrand & Sommer, 1999). These authors also suggested that a range of C:N ratios between 5 and 10 are optimum for algal growth, while ratios >10 are indicative of N limitation. More than 91% of our measured periphyton C:N ratios (33 of 36) fell between 5 and 10, suggesting that most were within their "optimal" range. In contrast, periphyton C:P ratios in our streams were more variable, ranging from 122 to more than 700 (mean = 248; Table 2). While Redfield (1958) suggested an optimum or average value of around 106, Hillebrand & Sommer's (1999) results suggest that algal C:P ratios around 130 would achieve maximum growth rates. Only 5 of 36 streams had C:P ratios ≤130 (although all were >106), and were located within all three land-use types. By this evidence, it would appear that algal growth may have been P-limited in streams in all land-use types. In situ experiments indicate that P supply may often limit stream periphyton growth, although N + P or light + P co-limitation is more common, and such responses can be season-dependent (Bothwell, 1985; Francoeur et al., 1999; Francoeur, 2001; Hill & Fanta, 2008). However, other experiments have observed ≥90% maximal algal growth rates at SRP concentrations as low as 16 μg/l (Rier & Stevenson, 2006), a concentration that was exceeded in 44% (16 of 36) of our streams. While our data are suggestive of P limitation, they should be viewed with some caution, especially as periphyton assemblages likely have varying amounts of algal, bacterial, and detrital material.

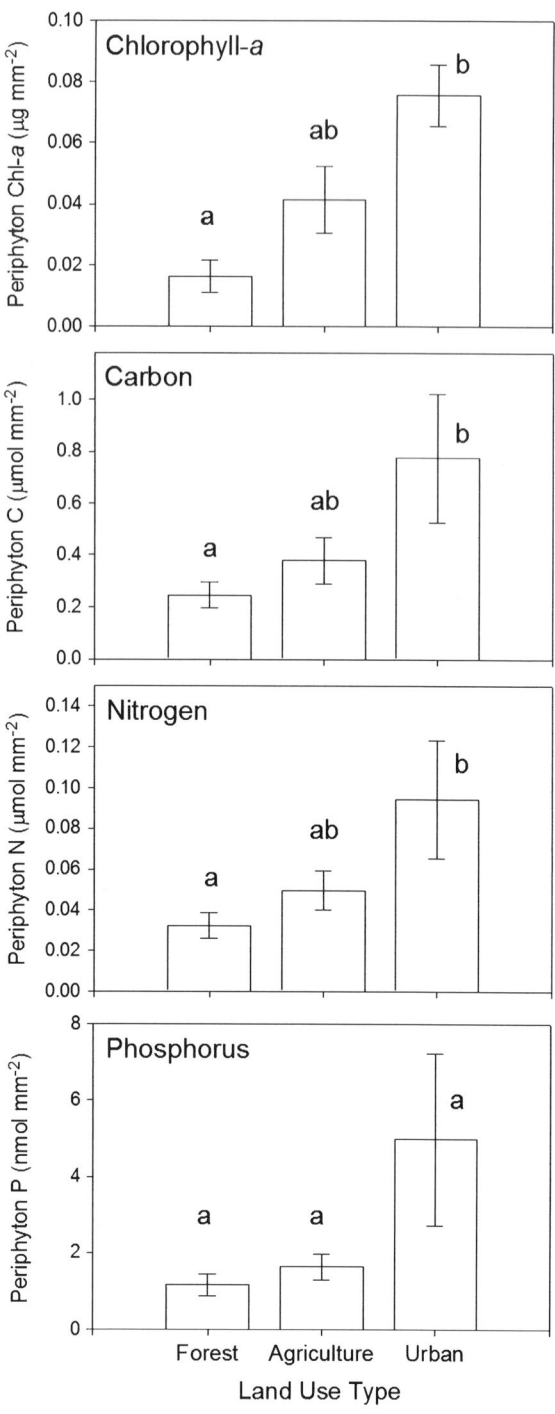

Fig. 5 Mean periphyton biomass (μg chlorophyll-$a$/mm$^2$), and periphyton nutrient content (nmol or pmol C, N, P/mm$^2$) measured in streams located in one of three major land-use types (*bars* represent means ± 1 SE; $n_{forest}$ = 15; $n_{agricultural}$ = 13; $n_{urban}$ = 8; *bars* sharing the *same letter* were judged not significantly different, based on Tukey's HSD test [$P > 0.05$])

Schiller et al. (2007) examined nutrient limitation and algal stoichiometry in three streams in NE Spain and observed that periphyton from a stream located in a forested site had greatest C:N ratios, with intermediate ratios at an urban site, and lowest ratios (greater relative N) in the agricultural stream. We observed a different pattern among our 36 streams in New York, with greatest periphyton C:N in urban streams (Tables 3, 6), as well as greater absolute concentrations of N and P (Fig. 5). However, the range of C:N ratios in stream periphyton was not broad, ranging only from 4.4 to 12.5, with most near Redfield predictions. These ratios were not significantly different among the three major land-use categories (Table 7). Perhaps unlike other parts of the world, urban streams in SE New York State received greater inputs of dissolved N and P than either rural forested or agricultural streams. This difference may be attributed to differences in the intensity of agriculture in southeastern New York state, but perhaps also from differences in perspective. Changes in the land use may not easily fit into strict categories, but more likely vary continuously from rural to urban conditions. As such we suggest that the gradient approach may reveal more about the influence of land use on stream function.

**Acknowledgments** We thank Alissa Perrone for technical assistance in the laboratory, John Tirpak for advice with GIS analyses, and Drs. William Giuliano and Amy Tuininga for comments on an earlier draft of this article. POB received financial support from Fordham University.

## References

American Public Health Association, 1985. Standard Methods for the Analysis of Water and Wastewater, 16th ed. American Public Health Association, Washington, DC.

Biggs, B. J. F., 1995. The contribution of flood disturbance, catchment geology and land use to the habitat template of periphyton in stream ecosystems. Freshwater Biology 33: 419–438.

Biggs, B. J. F., 1996. Patterns in benthic algae of streams. In Stevenson, R. J., M. L. Bothwell & R. L. Lowe (eds), Algal Ecology: Freshwater Benthic Ecosystems. Academic Press, San Diego, CA: 31–56.

Biggs, B. J. F. & M. E. Close, 1989. Periphyton biomass dynamics in gravel bed rivers: the relative effects of flows and nutrients. Freshwater Biology 22: 209–231.

Blinn, D. W. & P. C. Bailey, 2001. Land-use influence on stream water quality and diatom communities in Victoria, Australia: a response to secondary salinization. Hydrobiologia 466: 231–244.

Bothwell, M. L., 1985. Phosphorus limitation of lotic periphyton growth rate: an intersite comparison using continuous-flow troughs (Thompson River system, British Columbia). Limnology and Oceanography 30: 527–542.

Bothwell, M. L., 1988. Growth rate responses of lotic periphytic diatoms to experimental phosphorus enrichment: the influence of temperature and light. Canadian Journal of Fisheries and Aquatic Sciences 45: 261–270.

Bran+Luebbe Analyzing Technologies, 1986a. Ortho-phosphate in Water and Seawater. Industrial Method No. 812-86T. Bran+Luebbe Analyzing Technologies, Buffalo Grove, IL.

Bran+Luebbe Analyzing Technologies, 1986b. Ammonium in Water and Seawater. Industrial Method No. 804-86T. Bran+Luebbe Analyzing Technologies, Buffalo Grove, IL.

Bran+Luebbe Analyzing Technologies, 1987a. Nitrate/Nitrite in Water and Seawater. Industrial Method No. 818-87T. Bran+Luebbe Analyzing Technologies, Buffalo Grove, IL.

Bran+Luebbe Analyzing Technologies, 1987b. Silicates in Water and Seawater. Industrial Method No. 785-86T. Bran+Luebbe Analyzing Technologies, Buffalo Grove, IL.

Carr, G. M., P. A. Chambers & A. Morin, 2005. Periphyton, water quality, and land use at multiple spatial scales in Alberta rivers. Canadian Journal of Fisheries and Aquatic Sciences 62: 1309–1319.

Chessman, B. C., E. Primrose, H. Burch & J. M. Burch, 1992. Limiting nutrients for periphyton growth in sub-alpine, forest, agricultural and urban streams. Freshwater Biology 28: 349–361.

Chételat, J., F. R. Pick, A. Morin & P. B. Hamilton, 1999. Periphyton biomass and community composition in rivers of different nutrient status. Canadian Journal of Fisheries and Aquatic Sciences 56: 560–569.

Cross, W. F., J. P. Benstead, A. D. Rosemond & J. B. Wallace, 2003. Consumer-resource stoichiometry in detritus-based streams. Ecology Letters 6: 721–732.

Dodds, W. K., V. H. Smith & K. Lohman, 2002. Nitrogen and phosphorus relationships to benthic algal biomass in temperate streams. Canadian Journal of Fisheries and Aquatic Sciences 59: 865–874.

Dodds, W. K., E. Mart, J. L. Tank, J. Pontius, S. K. Hamlton, N. B. Grimm, W. B. Bowden, W. H. McDowell, B. J. Peterson, H. M. Valett, J. R. Webster & S. Gregory, 2004. Carbon and nitrogen stoichiometry and nitrogen cycling rates in streams. Oecologia 140: 458–467.

Eisenreich, S. J., R. T. Bannerman & D. E. Armstrong, 1975. A simplified phosphorus analysis technique. Environmental Letters 9: 43–53.

Finlay, J. C., S. Khandwala & M. E. Power, 2002. Spatial scales of carbon flow in a river food web. Ecology 83: 1845–1859.

Fitzpatrick, F. A., I. Waite, P. J. D'Arconte, M. R. Meador, M. A. Maupin & M. E. Gurtz, 1998. Revised Methods for Characterizing Stream Habitat in the National Water-Quality Assessment Program. U.S. Department of the Interior, U.S. Geological Survey, Water Resources Division, Raleigh, NC: 67 pp.

Francoeur, S. N., 2001. Meta-analysis of lotic nutrient amendment experiments: detecting and quantifying subtle responses. Journal of the North American Benthological Society 20: 358–368.

Francoeur, S. N., B. J. F. Biggs, R. A. Smith & R. L. Lowe, 1999. Nutrient limitation of algal biomass accrual in streams: seasonal patterns and a comparison of methods. Journal of the North American Benthological Society 18: 242–260.

Frost, P. C. & J. J. Elser, 2002. Growth responses of littoral mayflies to the phosphorus content of their food. Ecology Letters 5: 232–240.

Frost, P. C., H. Hillebrand & M. Kahlert, 2005. Low algal carbon content and its effect on the C:P stoichiometry of periphyton. Freshwater Biology 50: 1800–1807.

Godwin, C. M. & H. J. Carrick, 2008. Spatio-temporal variation of periphyton biomass and accumulation in a temperate spring-fed stream. Aquatic Ecology 42: 583–595.

Graham, M. H., 2003. Confronting multicollinearity in ecological multiple regression. Ecology 84: 2809–2815.

Greenwood, J. L. & A. D. Rosemond, 2005. Periphyton response to long-term nutrient enrichment in a shaded headwater stream. Canadian Journal of Fisheries and Aquatic Sciences 62: 2033–2045.

Hart, D. D., 1987. Experimental studies of exploitative competition in a grazing stream insect. Oecologia 73: 41–47.

Hill, W. R. & S. E. Fanta, 2008. Phosphorus and light colimit periphyton growth at subsaturating irradiances. Freshwater Biology 53: 215–225.

Hillebrand, H. & U. Sommer, 1999. The nutrient stoichiometry of benthic microalgal growth: Redfield proportions are optimal. Limnology and Oceanography 44: 440–446.

Hillebrand, H., G. de Montpellier & A. Liess, 2004. Effects of macrograzers and light on periphyton stoichiometry. Oikos 106: 93–104.

Hirsch, R. M., J. F. Walker, J. C. Day & R. Kallio, 1990. The influence of man on hydrologic systems. In Woleman, M. G. & H. C. Riggs (eds), Surface Water Hydrology. Geological Society of America, Boulder, CO: 329–359.

Liess, A. & H. Hillebrand, 2006. Role of nutrient supply in grazer-periphyton interactions: reciprocal influences of periphyton and grazer nutrient stoichiometry. Journal of the North American Benthological Society 25: 632–642.

Lorenzen, C. J., 1967. Determination of chlorophyll and pheopigments: spectrophotometric equations. Limnology & Oceanography 12: 343–346.

McDonnell, M. J., S. T. A. Pickett, P. Groffman, P. Bohlen, R. V. Pouyat, W. C. Zipperer, R. W. Parmelee, M. M. Carreiro & K. Medley, 1997. Ecosystem processes along an urban-to-rural gradient. Urban Ecosystems 1: 21–36.

McMahon, G. & T. F. Cuffney, 2000. Quantifying urban intensity in drainage basins for assessing stream ecological conditions. Journal of the American Water Resources Association 36: 1247–1262.

Meybeck, M., 1998. Man and river interface: multiple impacts on water and particulates chemistry illustrated in the Seine River Basin. Hydrobiologia 373: 1–20.

Murdock, J., D. Roelke & F. Gelwick, 2004. Interactions between flow, periphyton, and nutrients in a heavily impacted urban stream: implications for stream restoration effectiveness. Ecological Engineering 22: 197–207.

Pan, Y., R. J. Stevenson, B. H. Hill, P. R. Kaufman & A. T. Herlihy, 1999. Spatial patterns and ecological determinants of benthic algal assemblages in mid-Atlantic highland streams, USA. Journal of Phycology 35: 460–468.

Paul, M. J. & J. L. Meyer, 2001. Streams in the urban landscape. Annual Review of Ecology and Systematics 32: 333–365.

Pickett, S. T. A., M. L. Cadenasso, J. M. Grove, C. H. Nilon, R. V. Pouyat, W. C. Zipperer & R. Costanza, 2001. Urban ecological systems: linking terrestrial ecological, physical and socio-economic components of metropolitan areas. Annual Review of Ecology and Systematics 32: 127–157.

Pluhowski, E. J., 1970. Urbanization and its effect on the temperature of streams in Long Island, New York. USGS Professional Paper 627-D.

Ponader, K. C., D. F. Charles, T. J. Belton & D. M. Winter, 2008. Total phosphorus inference models and indices for coastal plains streams based on benthic diatom assemblages from artificial substrates. Hydrobiologia 610: 139–152.

Redfield, A. C., 1958. The biological control of chemical factors in the environment. American Scientist 46: 205–221.

Rier, S. T. & R. J. Stevenson, 2006. Response of periphytic algae to gradients in nitrogen and phosphorus in streamside mesocosms. Hydrobiologia 561: 131–147.

Rothlisberger, J. D., M. A. Baker & P. C. Frost, 2008. Effects of periphyton stoichiometry on mayfly excretion rates and nutrient ratios. Journal of the North American Benthological Society 27: 497–508.

Schiller, D. V., E. Martí, J. L. Riera & F. Sabater, 2007. Effects of nutrients and light on periphyton biomass and nitrogen uptake in Mediterranean streams with contrasting land uses. Freshwater Biology 52: 891–906.

Snyder, E. B., C. T. Robinson, G. W. Minshall & S. R. Rushforth, 2002. Regional patterns in periphyton accrual and diatom assemblage structure in a heterogeneous nutrient landscape. Canadian Journal of Fisheries and Aquatic Sciences 59: 567–577.

Sokal, R. R. & F. J. Rohlf, 1995. Biometry, 3rd ed. W.H. Freeman and Company, NY.

Solórzano, L. & J. H. Sharp, 1980. Determination of total dissolved phosphorus and particulate phosphorus in natural waters. Limnology and Oceanography 25: 754–758.

Sonoda, K. & J. A. Yeakley, 2007. Relative effects of land use and near-stream chemistry on phosphorus in an urban stream. Journal of Environmental Quality 36: 144–154.

Sprague, L. A., D. A. Harned, D. W. Hall, L. H. Nowell, N. J. Bauch, & K. D. Richards, 2007. Response of stream chemistry during base flow to gradients of urbanization in selected locations across the conterminous United States, 2002–04. USGS Scientific Investigations Report # 2007–5083.

Stelzer, R. S. & G. A. Lamberti, 2002. Ecological stoichiometry in running waters: periphyton chemical composition and snail growth. Ecology 83: 1039–1051.

Stevenson, R. J. & L. L. Bahls, 1999. Periphyton protocols. In Barbour, M. T., J. Gerritsen, B. D. Snyder & J. B. Stribling (eds), Rapid Bioassessment Protocols for Use in Streams and Wadeable Rivers: Periphyton, Benthic Macroinvertebrates and Fish, 2nd ed. EPA 841-B-99-002, USEPA, Washington, DC: 6-1–6-23.

Stevenson, R. J., M. L. Bothwell & R. L. Lowe, 1996. Algal Ecology: Freshwater Benthic Ecosystems. Academic Press, San Diego, CA.

Strayer, D. L., R. E. Beighley, L. C. Thompson, S. Brooks, C. Nilsson, G. Pinay & R. J. Naiman, 2003. Effects of land cover on stream ecosystems: roles of empirical models and scaling issues. Ecosystems 6: 407–423.

Torres-Ruiz, M., J. D. Wehr & A. A. Perrone, 2007. Trophic relations in a stream food web: importance of fatty acids for macroinvertebrate consumers. Journal of the North American Benthological Society 26: 509–522.

Tufford, D. L., H. N. McKellar Jr. & J. R. Hussey, 1998. In-stream nonpoint source nutrient prediction with land-use proximity and seasonality. Journal of Environmental Quality 27: 100–111.

U.S. Environmental Protection Agency. 1987. Handbook of Methods for Acid Deposition Studies. Laboratory Analysis for Surface Water Chemistry. EPA600/4-87/026. Washington, DC.

Vannote, R. L., G. W. Minshall, K. W. Cummins, J. R. Sedell & C. E. Cushing, 1980. The river continuum concept. Canadian Journal of Fisheries and Aquatic Sciences 37: 130–137.

Walker, C. E. & Y. Pan, 2006. Using diatom assemblages to assess urban stream conditions. Hydrobiologia 561: 179–189.

Walsh, C. J., A. H. Roy, J. W. Feminella, P. D. Cottingham, P. M. Groffman & R. P. Morgan II, 2005. The urban stream syndrome: current knowledge and the search for a cure. Journal of the North American Benthological Society 24: 706–723.

Wang, X. & Z.-Y. Yin, 1997. Using GIS to assess the relationship between land use and water quality at a watershed level. Environment International 23: 103–114.

Wehr, J. D. & R. G. Sheath, 2003. Freshwater habitats of algae. In Wehr, J. D. & R. G. Sheath (eds), Freshwater Algae of North America. Academic Press, San Diego, CA: 11–57.

Welch, E. B., J. M. Jacoby, R. R. Horner & M. R. Seeley, 1988. Nuisance biomass levels of periphytic algae in streams. Hydrobiologia 157: 161–168.

Wetzel, R. G., 2001. Limnology: Lake and River Ecosystems, 4th ed. Academic Press, San Diego, CA.

Winter, J. G. & H. C. Duthie, 2000. Epilithic diatoms as indicators of stream total N and total P concentration. Journal of the North American Benthological Society 19: 32–49.

GLOBAL CHANGE AND RIVER ECOSYSTEMS

# The physico-chemical habitat template for periphyton in alpine glacial streams under a changing climate

U. Uehlinger · C. T. Robinson · M. Hieber · R. Zah

Received: 10 May 2009 / Accepted: 5 October 2009 / Published online: 20 November 2009
© Springer Science+Business Media B.V. 2009

**Abstract** The physico-chemical habitat template of glacial streams in the Alps is characterized by distinct and predictable changes between harsh and relatively benign periods. Spring and autumn were thought to be windows of favorable environmental conditions conducive for periphyton development. Periphyton biomass (measured as chlorophyll *a* and ash-free dry mass) was quantified in five glacial and three non-glacial streams over an annual cycle. One glacial stream was an outlet stream of a proglacial lake. In all glacial streams, seasonal patterns in periphyton were characterized by low biomass during summer high flow when high turbidity and transport of coarse sediment prevailed. With the end of icemelt in autumn, environmental conditions became more favorable and periphyton biomass increased. Biomass peaked between late September and January. In spring, low flow, low turbidity, and a lack of coarse sediment transport were not paralleled by an increase in periphyton biomass. In the non-glacial streams, seasonal periphyton patterns were similar to those of glacial streams, but biomass was significantly higher. Glacier recession from climate change may shift water sources in glacier streams and attenuate the glacial flow pulse. These changes could alter predicted periods of optimal periphyton development. The window of opportunity for periphyton accrual will shift earlier and extend into autumn in channels that retain surface flows.

**Keywords** Alpine · Algae · Primary production · Stream · Flow regime · Glacier · Ecological windows

Guest editors: R. J. Stevenson, S. Sabater / Global Change and River Ecosystems—Implications for Structure, Function and Ecosystem Services

U. Uehlinger (✉) · C. T. Robinson
Department of Aquatic Ecology, Swiss Federal Institute of Aquatic Science and Technology (Eawag), 8600 Dübendorf, Switzerland
e-mail: uehlinger@eawag.ch

U. Uehlinger · C. T. Robinson
Institute of Integrative Biology, ETH Zürich, Zürich, Switzerland

M. Hieber
INTERTEAM, Untergeissenstein 10/12, 6005 Luzern, Switzerland

R. Zah
Department Sustainable Information Technology, Swiss Federal Institute for Material Testing (EMPA), Lerchenfeldstrasse 5, 9014 St. Gallen, Switzerland

## Introduction

Glacial streams are common landscape features of high latitudes and mountain ranges ascending above the permanent snow line. Glacial streams are year-round cold habitats with a characteristic fauna (Ward, 1994; Füreder, 1999), and in temperate latitudes, the only remnants for cold-adapted lotic organisms since the end of the last ice age. However, these habitats are under severe threat as ongoing global warming has dramatically accelerated glacier recession. For instance, glaciers in Switzerland lost ~20% of their area from 1985 to 1998 (Paul, 2003) and 80 to >90% of the Alpine glacier mass may be lost by 2100 (Watson et al., 1997; Zemp et al., 2006). The consequences of such a scenario would be the widespread loss of Alpine glacial streams, with most streams then being fed by snowmelt or groundwater (Milner et al., 2009).

The ecology of alpine streams, glacial streams in particular, has attracted much interest driven by scientific curiosity and concerns about the impact of climate change (e.g., Milner & Petts, 1994; Ward & Uehlinger, 2003; Milner et al., 2009). However, most studies have focused on benthic invertebrates and largely ignored the energy base of these systems (e.g., Burgherr, 2000; Castella et al., 2001; Robinson et al., 2001; Brown et al., 2006). Primary production in wetted channels and allochthonous inputs of organic matter from adjacent riparian zones provide the basal energy supporting heterotrophic communities in lotic ecosystems. High latitude streams and alpine headwaters drain catchments with sparse or even absent vegetation, and consequently, inputs of particulate organic matter from the riparian zone into stream channels are low or lacking (McKnight & Tate, 1997; Zah & Uehlinger, 2001). Some studies suggest that benthic algae are the dominant energy source in these streams (Lavandier & Décamps, 1984, McKnight & Tate, 1997; Uehlinger & Zah, 2003), and only a few investigations exist on the structure of benthic algal communities of high altitude alpine streams (see review by Rott et al., 2006). For instance, benthic algae in glacial streams of Antarctica have been examined by various investigators (Howard-Williams et al., 1986; Howard-Williams & Vincent, 1989; Vincent & Howard-Williams, 1989; Hawes et al., 1992), but seasonal patterns are presumably unlike those of glacial streams in temperate latitudes because of differences in flow regime, including periods of scouring flow, and photoperiod.

In glacial streams of the Alps, the radiation and temperature controlled release of water between early summer and autumn occurs in the form of a distinct and predictable flow pulse (Tockner et al., 2000; Uehlinger et al., 2003) during which bed load transport, high concentration of suspended solids (high turbidity), and relatively low temperatures prevail (Milner & Petts, 1994). In winter, stream channels can be covered by snow, freeze or fall dry (Malard et al., 2006). In contrast to the harsh environment in summer and winter, more benign conditions prevail during spring and autumn when flow is relatively low (no bed load transport), temperature moderate, and light conditions little affected by suspended solids. Therefore, it has been hypothesized that spring and autumn are ecological windows of opportunity for benthic algae, i.e., periods favoring the accrual of algal biomass otherwise constrained by factors such as moving bed sediments and limited light availability (Uehlinger et al., 2002; Milner et al., 2009). In a year-long study of a glacial river, periphyton dynamics reflected to some extent this seasonal change between harsh and benign habitat conditions (Uehlinger et al., 1998).

The objectives of this study were to (1) examine seasonal periphyton development in glacial streams over an annual cycle with regard to the environmental features that characterize the physico-chemical habitat template of Alpine glacial streams; i.e., to what extent annual patterns in periphyton biomass correspond to the above mentioned concept of ecological windows (Uehlinger et al., 2002) and (2) discuss how ongoing climate change may affect the physico-chemical habitat template and, as a consequence, periphyton, which is assumed to be the autotrophic energy base of these streams. This study was part of a comprehensive investigation of high altitude Alpine streams that focused on benthic invertebrates and benthic algae (Hieber et al., 2001, 2002, 2005; Robinson et al., 2001; Zbinden et al., 2008). In order to better evaluate the glacial influence, we used data from three non-glacial streams that were at similar elevations and with year-round data available.

**Methods**

Study sites

The glacial streams studied were located downstream of the termini of five rapidly receding valley glaciers in the Swiss Alps (inset Fig. 1). Site elevations varied from 1,210 to 2,159 m a.s.l., and catchment areas from 14 to 35 km$^2$ (Table 1). Between 40 and 70% of individual catchments were covered by glaciers.

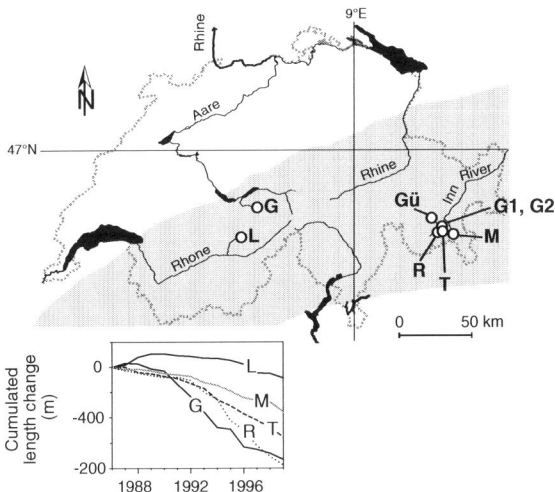

**Fig. 1** Map of the study sites (*open circles*) within the Swiss Alps (*shaded area*). Major lakes are *black*. *L* Lang Glacier, *G* Oberer Grindelwald Glacier, *R* Lake outlet of the proglacial Roseg Lake, *T* Tschierva Glacier, *M* Morteratsch Glacier. *G1* and *G2* Groundwater-fed streams in the Val Roseg. *Gü* Güglia. The *inset* shows the cumulated length change of the five glaciers since 1987 (data provided by the World Glacier Monitoring Service, Zurich)

Morphological data and catchment characteristics are given in Table 1. One stream (Roseg) is the outlet of a proglacial lake; i.e., the Roseg Glacier ends in a 1.2-km long proglacial lake (area 0.22 km$^2$) confined by the lateral moraine of the adjacent Tschierva Glacier. Melt waters of both glaciers merge in the foreland of the Tschierva Glacier to form the Roseg River. In four of the five catchments, the underlying bedrock was granite or granitoids. Sedimentary rock (limestone) only occurred in the lowest part of the upper Grindelwald Glacier drainage. In glacial streams, sampling took place from July 1998 to September/October 1999. Study sites were visited at about monthly intervals from spring to autumn. During winter, sampling was less frequent because snow restricted site access. The sites below the Lang Glacier could not be sampled from December to April. The non-glacial streams comprised two perennial groundwater-fed streams (G1 and G2) located in the Val Roseg flood plain, at about 2,030 m a.s.l., and the Güglia, an alpine stream fed by snowmelt and groundwater, at 2,205 m a.s.l. The groundwater-fed streams were sampled from July 1996 to January 1999, and the Güglia was sampled from July 1999 to August 2000.

## Physics and chemistry

Gauging stations of the Federal Office for Water and Geology recorded discharge 8–15 km downstream of each glacier stream. At these stations, the recorded discharge also included water from glacial tributaries joining the study river between the lowermost sampling sites and the gauging station. In order to estimate discharge for the study reaches, we allocated the discharge to study rivers and tributaries proportional to the glacier surface. For the Tschierva and Roseg glacier streams, discharge recorded at the gauging station in Pontresina was allocated to each stream using temperature as a tracer (Uehlinger et al., 2003). Continuous discharge records of non-glacial streams were not available. Discharge of these streams was measured a few times using the tracer dilution method (Gordon et al., 1992) or the velocity area method (Davis et al., 2001).

As an indicator of bed sediment stability, we estimated the critical discharge for initiation of bed sediment transport based on channel geometry, average slope, and grain size distribution according to Günter (1971) and Gessler (1965). In each stream, one temperature logger (StowAway, Onset Computer Corp., Pocasset MS, USA or Minilog, Vemco Ltd., Shad Bay N.S., Canada) was installed between the upper and lower sites. Loggers were enclosed in stainless steel housings and fixed with a chain to a metal rod on the bank.

Snow cover at the study sites was estimated each time when the sites were visited, but more comprehensive snow data are lacking. Information on light availability at the channel bottom at the study sites was not available.

Specific conductance (μS cm$^{-1}$ at 20°C) was measured with a conductivity meter (LF323, WTW, Weilheim, Germany). Turbidity (nephelometric turbidity units: NTU) was determined using a portable turbidity meter (Cosmos, Züllig, Rheineck, Switzerland). Surface water (1 l) was collected from each stream on each sample date, returned to the laboratory in a cooler, and then filtered through pre-ashed glass fiber filters (GF/F, Whatman). The filtrate was analyzed for $NH_4$–N, $NO_3$–N, soluble reactive phosphorus (SRP), and particulate phosphorus (PP). Analytical methods are described in detail by Tockner et al. (1997).

**Table 1** Morphological and hydrological characteristics of the investigated glacial and non-glacial streams

| | Sampling station with elevation (m a.s.l.) | | Maximum elevation (m a.s.l.) | Catchment | | Channel width (m) | Channel slope (%) | Average summer discharge (m³/s) | Critical discharge (initiation of bed load transport) (m³/s)[e] |
|---|---|---|---|---|---|---|---|---|---|
| | | | | Area (km²) | Glacierized (%) | | | | |
| *Glacial streams* | | | | | | | | | |
| Morteratsch | Upstream | 2,000 | 4,050 | 34 | 54 | 15 | 1 | 12[a] | 31–47 |
| | Downstream | 1,970 | | 35 | 52 | 6 | 5 | | 5–8 |
| Upper Grindelwald | Upstream | 1,240 | 3,741 | 17 | 64 | 3–7 | 8 | 10[a] | 2–4 |
| | Downstream | 1,210 | | 17 | 62 | 6–8 | 6 | | 3 |
| Tschierva | | 2,100 | 4,050 | 14 | 53 | 4–10 | 4 | 3[b] | 3–4 |
| Roseg (lake outlet) | | 2,159 | 3,937 | 20 | 52 | 10–12 | 2 | 4[b] | 27–41 |
| Lang | Upstream | 1,990 | 3,892 | 19 | 59 | 6–10 | 8 | 10[a] | 2–3 |
| | Downstream | 1,910 | | 29 | 50 | 8–12 | 6 | | 3–5 |
| *Non-glacial streams* | | | | | | | | | |
| G1 | | 2,035 | – | – | 0 | 0.5–1 | 2 | <0.01[c] | |
| G2 | | 2,012 | – | – | 0 | 1–7 | 2 | 0.02–0.05[c] | |
| Güglia | | 2,205 | 3,280 | 7 | 0 | 4–6 | 4 | 0.3[d] | |

[a] Estimates, see text
[b] Uehlinger et al. (2003)
[c] U. Uehlinger, unpublished data
[d] M. Hieber, unpublished data
[e] Estimated according to Günter (1971) and Gessler (1965)

Periphyton

For determination of periphyton biomass, 10 rocks were randomly collected at each site along a 20-m long section of channel. Collected rocks were transported to the laboratory in a cooler. Algae were removed from each stone with a brass wire brush and rinsed into a bucket. Aliquots of the suspension were filtered through glass fiber filters (Whatman GF/F) for analysis of chlorophyll $a$ and ash-free dry mass (AFDM). The remaining suspensions were composited, and an aliquot of 20 ml preserved in 2% formalin for later identification of algae (a detailed description has been given by Hieber et al., 2001). The filters for chlorophyll $a$ analysis were transferred to individual screw-cap vials filled with 6 ml ethanol (90% v/v), boiled at 76°C for 10 min, and stored in the dark at 4°C until analyzed. Chlorophyll $a$ was determined by HPLC (Meyns et al., 1994). For determination of AFDM, filters were dried at 60°C, weighed, ashed for 3 h at 500°C, and reweighed. Area values of chlorophyll $a$ (mg/m$^2$) and AFDM were calculated as described in Uehlinger (1991).

Data analysis

In order to evaluate the effect of stream type (glacial and non-glacial) and season, we used analysis of variance (ANOVA) followed by Tukey's multiple comparison test. Prior to analysis, data were transformed $\log(x + 1)$ to improve normality (Zar, 1996). Effects were considered significant at $P < 0.05$. We defined the length of each season as follows: (1) spring: April and May; (2) summer: June 1–September 15; (3) autumn: September 16–November 15; and (4) winter: November 16–March 31. These season lengths approximately correspond to the glacial hydrograph: snowmelt/flow increase from April to end of May, summer high flow/ice melt until the first half of September, subsequent autumnal flow decline, and the extended winter low flow (Fig. 2).

Results

The physico-chemical environment

The overall discharge pattern (vernal flow increase, summer high flow, autumnal flow recession, winter low flow) was similar in all glacial streams including the Roseg Lake outlet. Cold weather periods (reduced icemelt) and spates induced by rain storms resulted in substantial (stochastic) flow variation in summer and autumn (Fig. 2). The estimated critical flow for initiation of bed sediment transport was frequently exceeded during summer, except for the proglacial Roseg Lake outlet and the upper Morteratsch site. Both of these sites are characterized by wide channels and relatively small slopes (Table 1). Discharge in the groundwater-fed streams of the Val Roseg varied from 1996 to 1999 between 0.005 in autumn/winter and 0.06 m$^3$ s$^{-1}$ in late spring (snow melt) and between 0.24 and 0.37 m$^3$ s$^{-1}$ in the Güglia from 1999 to 2000. Movement of bed sediments can be excluded in the groundwater-fed streams, but bed load transport cannot be excluded during snow melt in May/June and rainstorm induced spates in the Güglia.

Figure 3 illustrates seasonal patterns of daily mean temperatures in three glacial and two non-glacial streams. Daily mean water temperatures during summer varied between 0.5 and 2.1°C in the glacial streams below Morteratsch, upper Grindelwald, Lang, and

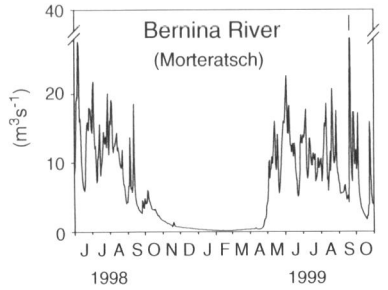
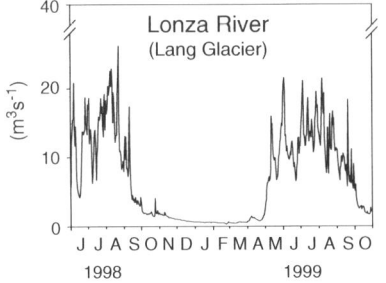
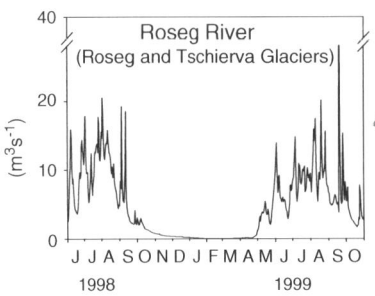

**Fig. 2** Daily discharge downstream of four study sites from June 1998 to October 1999. Gauging stations were 5 km (Bernina River), 6 km (Lonza), and 10.5 km (Roseg) downstream of the study sites. At the Roseg River gauging station discharge included the meltwater of both Tschierva and Roseg Glaciers

**Fig. 3** Daily temperatures based on logger records in three glacial (**A**) and two non-glacial streams (**B**)

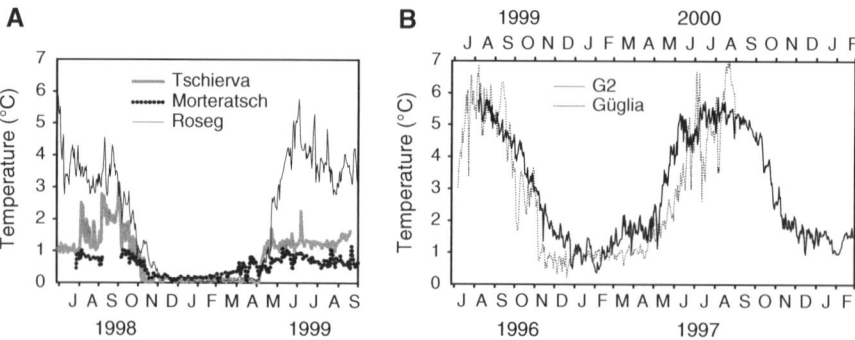

Tschierva Glacier, and between 2.6 and 7.0°C in the Roseg Lake outlet. Winter temperatures ranged from 0.0 to 1.6°C (glacial and non-glacial streams). Daily mean temperatures in the non-glacial streams ranged from 3.7 to 5.9°C in the Val Roseg floodplain and from 3.9 to 7.1°C in the Güglia during summer.

In glacial streams, turbidity was significantly higher (annual average 125 ± 128 NTU) than in non-glacial streams (annual average 1.9 ± 3.4 NTU) and significantly varied with season (Fig. 4A). Turbidity was low from December to April (<5 NTU) except from the Roseg Lake outlet, where it ranged from 30 to 55 NTU, began to increase by the end of April, peaked between June and mid-September, and was still relatively high in autumn. The investigated non-glacial streams exhibited no significant seasonal variation.

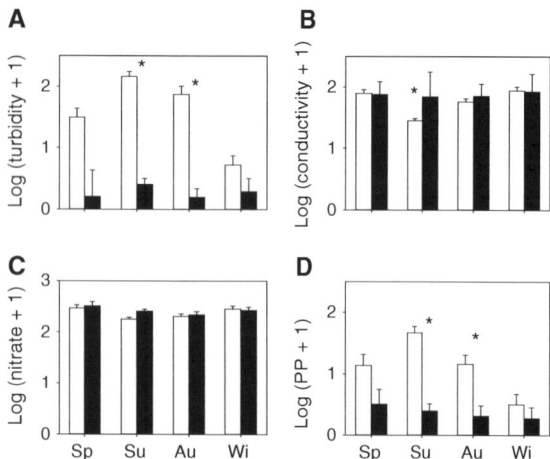

**Fig. 4** Seasonal variability of turbidity (**A**), conductivity (**B**), nitrate (**C**) and PP (**D**) in glacial and non-glacial streams. *Bar plots* with standard errors. Open bars = glacial streams. Filled bars = non-glacial streams. *Asterisks* indicate significant differences between the two stream types (post-hoc Tukey's test). *Sp* Spring, *Su* summer, *Au* Autumn, *Wi* winter

Snow covered the channel below the Tschierva Glacier from January to April 1999. The upper and lower sites below the Grindelwald Glacier were snow covered by 90 and 60%, respectively, in January 1999, and by 100 and 90% until April after a heavy snowfall in February 1999. The Tschierva site was snow covered in January and February (presumably to mid of April). The Roseg Lake outlet site was open during winter as well as the upstream Morteratsch site. At the lower Morteratsch site, snow free areas were restricted to a few areas (about 1–2 m$^2$) until a large snowfall in February deeply covered the stream until mid April 1999. The two groundwater-fed channels in the Roseg flood plain as well as the Güglia site were not snow covered during winter.

In glacial streams, conductivity reflected the seasonality in glacial influence (Fig. 4B). In non-glacial streams, conductivity was higher and lacked seasonal variation (Table 2). In both stream types, nitrate concentrations were relatively high with maximum concentrations during spring (Table 2; Fig. 4C). In glacial streams, ammonia concentrations were 2–8 fold higher than in non-glacial streams (Table 2); the difference mainly resulted from the significantly elevated concentrations during the summer ice melt. Concentrations of SRP were generally low (Table 2), often below the detection limit (<1 μg P l$^{-1}$), and showed no seasonal pattern in both stream types. PP was high in glacial streams and varied seasonally like turbidity (Table 2; Fig. 4D). The low PP concentrations in non-glacial streams lacked any seasonality.

Periphyton

Figure 5 depicts seasonal patterns of chlorophyll *a* in the investigated streams (ash-free dry mass showed

**Table 2** Conductivity, inorganic nitrogen and phosphorus compounds (mean and standard deviation) of the five streams during the study period

|  | Conductivity (μS cm$^{-1}$) | NH$_4$–N (μg l$^{-1}$) | NO$_3$–N (μg l$^{-1}$) | SRP (μg l$^{-1}$) | PP (μg l$^{-1}$) |
|---|---|---|---|---|---|
| *Glacial streams* | | | | | |
| Grindelwald | 88 ± 52 | 14 ± 12 | 234 ± 107 | 1 ± 1 | 49 ± 58 |
| Lang | 53 ± 28 | 20 ± 12 | 138 ± 59 | 1 ± 1 | 112 ± 66 |
| Tschierva | 52 ± 32 | 8 ± 9 | 329 ± 120 | 3 ± 1 | 130 ± 224 |
| Morteratsch | 47 ± 33 | 25 ± 30 | 263 ± 65 | 2 ± 2 | 69 ± 127 |
| Roseg | 45 ± 19 | 21 ± 21 | 174 ± 42 | 2 ± 1 | 31 ± 19 |
| *Non-glacial streams* | | | | | |
| G1 | 98 ± 26 | 4 ± 8 | 365 ± 114 | 2 ± 1 | 3 ± 3 |
| G2 | 64 ± 20 | 4 ± 3 | 266 ± 95 | 1 ± 1 | 2 ± 2 |
| Güglia | 88 ± 12 | 3 ± 4 | 176 ± 52 | <1 ± <1 | 1 ± 1 |

*G1* and *G2* are groundwater-fed streams in the Roseg floodplain

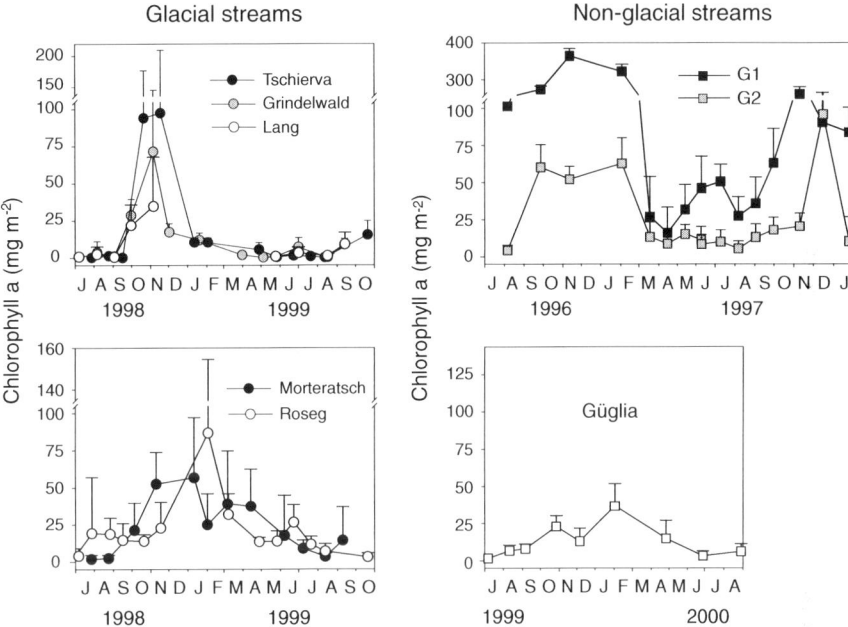

**Fig. 5** Chlorophyll *a* time series in five glacial and three non-glacial streams located in the Alps

similar patterns). In the three glacial streams, Tschierva, Grindelwald and Lang, chlorophyll was low during summer (average values 0.6–4.9 mg Chl. *a* m$^{-2}$), increased in September, peaked in October/November (35–98 mg Chl. *a* m$^{-2}$), rapidly decreased through December/January (no data from Lang), and subsequently declined to summer low values. In two glacial streams, Morteratsch and Roseg Lake outlet, seasonal patterns were similar but relatively high biomass persisted until March/April (no snow cover at the Roseg and upper Morteratsch sites); average summer biomass equaled 8.1 ± 6.7 mg Chl. *a* m$^{-2}$ at the Morteratsch sites and 14.5 ± 8.5 mg Chl. *a* m$^{-2}$ at the Roseg site (stable bed sediments at the Roseg and upper Morteratsch site, Table 1). High biomass in autumn/winter and relatively low biomass in late spring and summer characterized the non-glacial streams; in G1 and G2, Chl. *a* peaked in November at 329 ± 0.39 and 120 ± 41 mg m$^{-2}$, respectively. Average annual biomass in these two groundwater-fed streams exceeded that in the snowmelt-fed Güglia by two- to sevenfold.

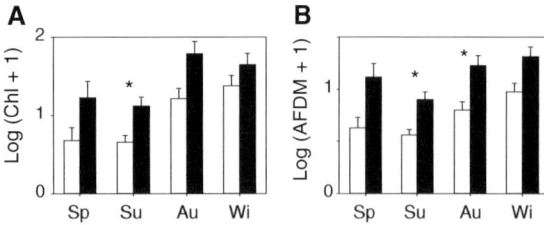

**Fig. 6** Effect of season and stream type on chlorophyll *a* (**A**) and ash-free dry mass (**B**). *Bar plots* with standard errors. Open bars = glacial streams. Filled bars = non-glacial streams. Asterisks indicate significant differences between the two stream types (post-hoc Tukey's test). *Sp* spring, *Su* summer, *Au* Autumn, *Wi* winter

In glacial and non-glacial streams, seasonal differences in chlorophyll *a* and ash-free dry mass were significant, and the overall seasonal patterns were similar in both stream types; i.e., low biomass during summer and high biomass in autumn and winter (Fig. 6). In non-glacial streams, biomass was significantly higher than in glacial streams (ANOVA, $P < 0.05$).

Parallel to our study, Hieber et al. (2001) analyzed the structure of the algal communities of the five investigated glacial streams. Algal abundances showed strong seasonal variation, but seasonality in community composition (genera richness) was minor. Algal communities of the investigated glacial streams were dominated by *Hydrurus foetidus* during autumn and winter. The genera *Achnanthes*, *Cymbella*, and *Fragilaria* prevailed among diatoms, *Lyngbya*, *Chamaesiphon*, and *Oscillatoria* among cyanobacteria. Genera richness observed during the study varied between three (Lang, Tschierva) and 22 (Roseg).

## Discussion

Periphyton biomass in the investigated Alpine glacial streams was low in summer and relatively high in autumn and, if sites were not covered by snow, also in winter. Apparently, suitable conditions in respect to flow, turbidity, and temperature did not coincide with enhanced vernal periphyton accrual. In non-glacial streams, vernal periphyton accrual was lacking and biomass continued to be relatively low during summer. This pattern does not support the hypothesis of two windows of opportunity for periphyton. Environmental conditions are apparently less suitable for periphyton in spring than in autumn. As seasonal biomass patterns were similar in non-glacial alpine streams to that of glacial streams, the low biomass in summer and high biomass in autumn/winter may be a general feature of periphyton dynamics in high alpine streams. This similarity in biomass patterns also suggests that besides turbidity, temperature, flow, and related factors such as shear stress and bed movement, other parameters should also be examined as potential constraints on periphyton.

The physico-chemical habitat template

The discharge regime of glacial streams is characterized by a distinct and predictable flow pulse (Uehlinger et al., 2003). In spring, rising air temperatures and solar radiation increase the release of melt water. Discharge in glacial streams peaks in July but remains high until mid-September. The glacial flow pulse usually coincides with substantial sediment transport that includes fine inorganic particles as well as bed load. Sediment stability plays a crucial role in periphyton accrual (Peterson, 1996) as moving bed sediments are a major constraint of periphyton accrual (Uehlinger et al., 1996). However, data on sediment transport or information on flow thresholds for sediment transport in glacial streams are usually lacking (e.g., Milner et al., 2001). Instead, channel stability indices based on hydraulic parameters, size and shape of grains, and vegetation have been used as a surrogate of substratum stability (Pfankuch, 1975; Hieber et al., 2002). In this study, evidence for the transport of coarse sediment includes lateral channel shifts (below the Tschierva and Lang glaciers) and burying of logging instruments in all glacial streams except Roseg.

Our calculations of critical discharge (initiation of bed load transport) suggest that bed load transport can occur at most glacial sites during summer high flow, but since discharge estimates for the study reaches are relatively rough, uncertainties about frequency and extent of periods with bed load transport are substantial. The forelands of receding glaciers are a rich source of sediment susceptible to fluvial transport during summer high flow. A proglacial lake such as Lake Roseg interrupts the downstream transport of coarse sediments from recently deglacierized forelands. As a consequence, corresponding outlet streams are considered to be relatively benign environments, at least in respect to the abrasive impact of moving sediments, unless coarse sediment is supplied from other sources (Hieber et al., 2002).

Solar radiation includes photosynthetically active radiation (PAR) as well as UV radiation with PAR as the ultimate source of energy for algae. High turbidity from high loads of suspended solids characterizes glacial streams during summer high flow. Concentrations of suspended solids positively correlate with discharge at a seasonal scale, but sediment concentrations can suddenly change without any noticeable change in discharge at the scale of hours (Gurnell, 1987). Depending on depth and turbidity, light availability to the stream bed can be strongly reduced. The relationship between turbidity and PAR attenuation determined in the Roseg catchment indicates that a turbidity of 250 NTU will attenuate about 95% of the incident light in a water column of 0.5 m depth (U. Uehlinger, unpublished data). In shallow streams (depth <0.5 m), an increase of 25 NTU is expected to decrease primary production by 13–50% (Lloyd et al., 1987). A reduction in light intensity by >90% largely reduces benthic primary production but may have a minor influence on periphyton standing crops (Hill et al., 2001). In the absence of scouring flow, biomass can still be relatively high although light attenuation is high. For example, in the turbid Roseg Lake outlet (estimated average light attenuation about 80% and lack of bed load transport), ash-free dry mass was several times higher than in the adjacent glacial stream below the Tschierva glacier (estimated average light attenuation about 90%) during summer, i.e., $8.0 \pm 3.4$ vs. $1.0 \pm 04$ g m$^{-2}$.

Snow cover during winter is presumably more efficient in intercepting light than turbidity. Measurements in the Roseg catchment showed that a snow-pack of 60 cm may reduce incident PAR light by 99% (U. Uehlinger, unpublished data); snow depth in winter often exceed 1 m at elevations >2,000 m a.s.l. We hypothesize that light exclusion by snow may be an important factor responsible for low biomass during winter. The extent to which glacial streams will be snow covered depends on local factors such as a lake (e.g., Lake Roseg outlet stream) or the upwelling of relatively warm groundwater (e.g., upper site at Morteratsch) (Schütz et al., 2001). Elucidating the availability of light as constraint of periphyton accrual in high alpine streams requires continuous monitoring of incident light, assessment of vertical light attenuation in the water column and snow pack, and quantitative monitoring of snow and ice cover in combination with periphyton colonization experiments using natural or artificial substrata.

In the Alps, global UV (direct and diffuse radiation) increases by about 11% 1,000 m$^{-1}$ (Schmucki & Philipona, 2002). Moreover, high reflectivity of the ground and low aerosol concentrations further elevates UV-radiation levels in alpine environments. The intensity of UV radiation is subject to distinct seasonal variation: UV-A varies about 6-fold and U-B around 20-fold (Blumthaler et al., 1992). In the Alps, at an elevation of ~2,000 m a.s.l., the average intensity of incident UV radiation in spring (April, May) is about 2.5 times higher than in autumn (September, October) (Blumthaler et al., 1992). In spring, snow cover additionally increases the UV radiation load (Caldwell et al., 1980). Relatively high UV radiation and PAR intensities, shallow water, and low concentrations of dissolved organic matter may affect periphyton accrual in spring. Algae are more sensitive to UV stress at low temperature and relatively high PAR (Roos & Vincent, 1998); i.e., under environmental conditions prevailing at high altitude. UV exclusion experiments have demonstrated that UV can suppress periphyton accrual (Bothwell, 1985, 1989; Bothwell et al., 1993; Vinebrooke & Leavitt, 1996; Francoeur & Lowe, 1998; Vinebrooke & Leavitt, 1999).

Some authors have also found no significant effect of UV on periphyton (DeNicola & Hoagland, 1996; Hill et al., 1997) and, over a long term, UV exposure even increased periphyton biomass presumably by reducing grazing impacts or bacterial competition for nutrients (Bothwell et al., 1993; Francoeur & Lowe, 1998). Based on studies in alpine lakes, Vinebrooke and Leavitt (1996, 1999) hypothesized that in cold unproductive systems, the indirect impact of UV via the food web (e.g., reduced grazing) is low compared to the overall dominance of abiotic community regulation. A significant negative relationship between PAR and periphyton accrual was found in a stream-side channel experiment performed in the Alps at an elevation of 2,200 m a.s.l. (Wellnitz & Ward, 2000). The investigation of Wellnitz & Ward (2000) took place in September when light intensities were similar to those in spring. Field studies are needed for a decisive assessment of the potential UV/PAR-impact that comprises experimental manipulation of UV and PAR.

Glacial streams are cold systems; temperatures are near 0°C even during summer at the snout of a glacier (Gíslason et al., 2001). Glacial streams reach maximum temperatures early in the year (May/early June). Rising solar radiation and air temperatures increase stream temperatures in spring, but with the release of cold meltwater, temperatures are persistently reduced compared to non-glacial streams at the same elevation (Uehlinger et al., 2003). In the Alps, water temperatures rapidly increase downstream (mean annual temperature by about $0.005°C\ m^{-1}$), which limits the longitudinal extent of cold habitats (metakryal and hypokryal) and the phenomenon of the summer temperature depression (Uehlinger et al., 2003). Proglacial lakes increase outlet stream temperatures during summer compared to non-outlet glacial streams by several degrees (Hieber et al., 2002; Uehlinger et al., 2003). Low temperatures slow down algal growth rates, but periphyton biomass apparently reaches high values if discharge and turbidity are low and snow cover is lacking. For example, biomass increased to $30 \pm 14\ g\ AFDM\ m^{-2}$ in March at the open upper Morteratsch site. Low temperature is apparently not a primary constraint for the formation of autotrophic biofilms in glacial streams.

The Alps receive relatively high amounts of nitrogen compounds by atmospheric deposition (Rhim, 1996). As a consequence, headwater streams in the Alps, including glacial streams, are characterized by nitrate concentrations typically >200 µg $NO_3$–N $l^{-1}$ (this study, Hieber et al., 2002; Robinson et al., 2002, this study, Tockner et al., 2002). These concentrations are high considering the fact that the transition from nitrogen limitation to nitrogen saturation is in the range of 50–60 µg $l^{-1}$ for inorganic nitrogen compounds (Grimm & Fisher, 1986, Newbold, 1992). Experiments using nutrient diffusing substrata performed in alpine streams, including the proglacial reach below the Tschierva Glacier, indicated that periphyton accrual is not nitrogen limited (Robinson et al., 2002).

In contrast, phosphorus may become temporarily limiting in glacial and non-glacial streams. Concentrations of SRP varied from 0 (below detection limits) to a maximum of 6 µg P $l^{-1}$, which is in the range reported to limit algal growth (Bothwell, 1985; Newbold, 1992). Experiments with nutrient diffusing substrata performed in the Roseg River and other glacial streams indicate potential phosphorus limitation in spring but not in autumn; the results of summer experiments were inconsistent presumably due to the interference of abiotic factors such as current and glacial flour (Rinke et al., 2001; Robinson et al., 2002). Glacial flour can be a source or a sink of bio-available phosphorus (Bretschko, 1966; Hodson et al., 2004). Hodson et al. (2004) found that between 0.2 and 7% of the total P of glacial flour of different provenience may be potentially available for algae. However, we lack information about environmental conditions favoring desorption of P from glacial flour in the rivers investigated.

Biotic processes

Invertebrate grazing can negatively affect periphyton accrual in alpine streams (Wellnitz & Ward, 2000). The few year-round investigations of glacial streams showed that benthic invertebrates can reach high densities at various times during the annual cycle (Burgherr & Ward, 2000, Robinson et al., 2001; Schütz et al., 2001). Therefore, the hypothesis that grazing influences annual patterns of periphyton biomass in glacial streams, apart from abiotic factors, should be considered. Invertebrate densities in glacial streams are likely to be low during summer high flow and high between autumn and spring (Kowanacki, 1991; Gíslason et al., 2001, Robinson et al., 2001; Schütz et al., 2001; Burgherr et al., 2002), although deviations from this overall pattern can be substantial. For example, annual patterns of benthic grazers were quite different in some of the streams investigated in this study (Fig. 6 in Robinson et al., 2001). Overall, there exists no unequivocal coincidence of low invertebrate densities with high periphyton biomass and vice versa. A conclusive evaluation of a grazing impact in glacial streams would require field experiments similar to the stream-side channel study of Wellnitz & Ward (2000) and was beyond the scope of this study.

Ecological windows of opportunity for periphyton?

The conceptual diagram in Fig. 7 is an attempt to summarize the potential relationships between regional climate and abiotic habitat conditions in Alpine glacial streams and the response of periphyton. Regional climate, topography, and geology are the ultimate factors controlling proximate factors such as

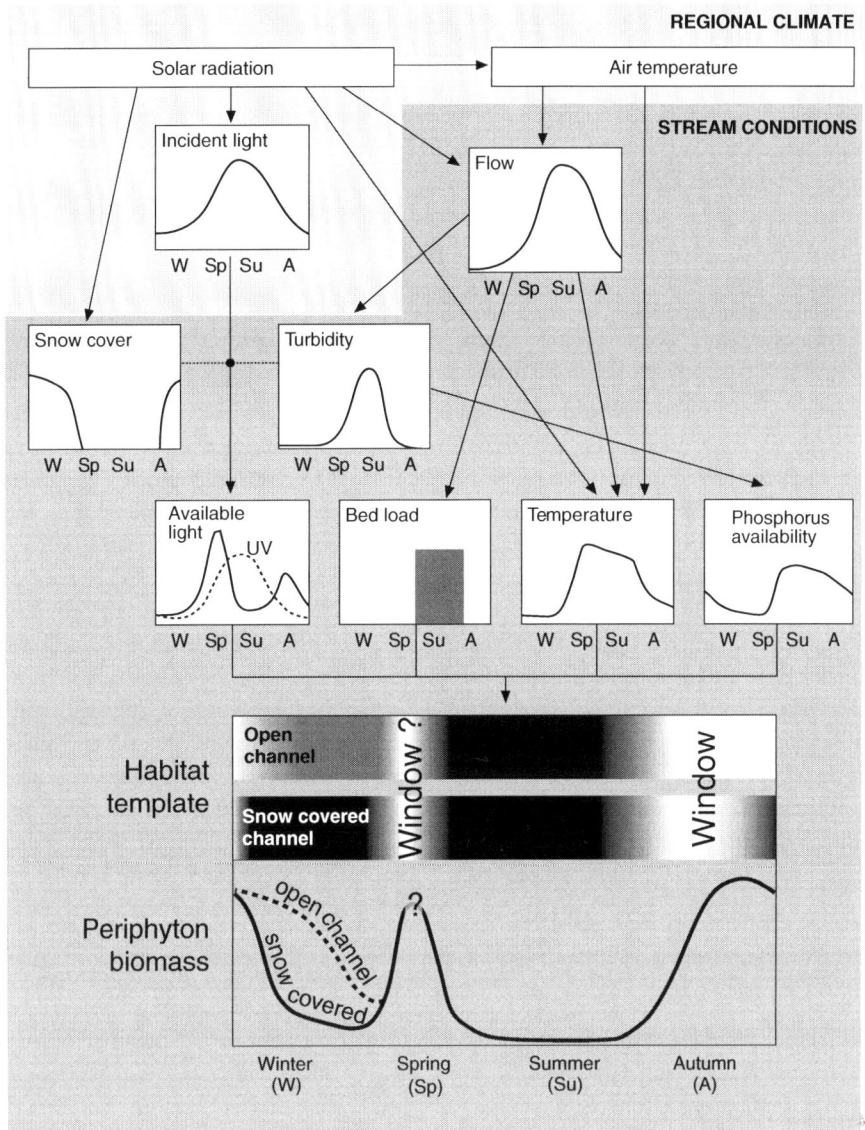

**Fig. 7** Windows of opportunity for periphyton accrual in the physico-chemical habitat template of glacial streams in the Alps (conceptual diagram). Solar radiation and air temperature control discharge (melt of snow/ice, transient precipitation storage) and stream temperature. The cold meltwater halts the vernal temperature increase and keeps summer stream temperatures low. Glacial melt is paralleled by high turbidity (glacial flour) reducing PAR at the stream bottom. PAR availability also depends on the seasonally changing sun angle and the interception by snow if channels become snow covered; PAR exclusion by snow imposes a major constraint on algal growth. High levels of UV radiation may suppress periphyton in spring and summer. Transport of coarse sediments (bed load) occurs when flow is high, e.g., when warm summer weather enhances ice melt. The abrasive impact of moving sediment severely impedes periphyton accrual. Superimposing annual patterns of the proximate factors of influence delimits a relatively favorable period for periphyton growth in autumn and eventually in spring and, if the stream channel remains open, also in winter

incident PAR and UV radiation, discharge (flow regime), shear stress, transport of fine (turbidity) and coarse sediments, and snow cover. Super-imposing annual patterns of the proximate factors result in periods during which environmental conditions are expected to favor or constrain periphyton accrual.

The period in autumn (mid-September to mid-November) is characterized by relatively low discharge (no transport of coarse sediments), low turbidity, moderate temperatures (considering constraints of the regional climate), and moderate solar radiation with respect to the impact, UV in particular. Low temperatures and low PAR intensities during winter (low solar angle, shading by aspect) apparently impose minor constraints on periphyton unless channels become covered by snow (>99% PAR exclusion). UV radiation, in combination with low temperature and relatively high PAR intensities, and/or phosphorus limitation may suppress periphyton accrual in spring despite otherwise suitable flow conditions. However, Rott et al. (2006) reported spring periphyton peaks in a glacial stream of the Austrian Alps, in contrast to the findings of our study, suggesting the existence of a vernal opportunity window.

The conceptual model linking habitat conditions and periphyton patterns should be considered as a hypothesis for glacial streams in the Alps and presumably other high mountain ranges at temperate latitudes. We are aware that the database for generalizations about periphyton in these systems is still relatively small, and that further studies are needed to corroborate or refute assumptions, on which the model is based, such as, for example, the role of UV for periphyton accrual in spring or the lack of substantial biotic interactions.

Climate change perspectives

Predictions for catchments in the Swiss Alps suggest a decline in annual precipitation that is paralleled by increasing winter precipitation, decreasing summer precipitation, increasing evapotranspiration and major loss of glacierized areas (Horton et al., 2006; Zemp et al., 2006). In the catchment of the Roseg River, the glacier covered area is expected to decrease by >90% and discharge by about 20% (median of regional climate model experiments, A2 and B2 scenarios defined by IPCC) to the end of this century (Horton et al., 2006). The shift from a glacier-driven to a snowfall/rainfall-driven flow regime will increase inter-annual flow variability (Horton et al., 2006) and, thus, reduce the predictability of the annual flow pulse. The flow pulse will also be shifted toward spring because of the earlier onset of snow melt and hydrographs may become more influenced by unpredictable rainstorms. Precipitation typically has a negative influence on glacial runoff, whereas in largely deglacierized catchments, precipitation rapidly turns into runoff (but see Röthlisberger & Lang, 1987). In largely glacierized basins, the enhanced water yield due to increased air temperatures is expected to intensify the glacial flow pulse but only during an initial phase (Braun et al., 2000).

The loss of glaciers will result in a shift of water sources from snow/ice melt dominated to snow melt, rain and groundwater dominated during summer, and the flow regime will be similar to the nivo-pluvial regimes of lower Alpine regions (Braun et al., 2000). The reduced water yield in combination with minor subsurface water storage, shallow or lacking aquifers that typically characterize high Alpine basins, is expected to increase surface flow intermittency with declining glacierization. Today, flow intermittency of high elevation low-order Alpine streams can already be substantial. For example, Robinson and Matthaei (2007) documented that the wet channel network of a non-glacierized Swiss alpine catchment contracted by more than 60% in late autumn, and Tonolla (2005) showed that about 90% of the surface channels fell dry by autumn in the Roseg catchment. The receding glaciers expose large amounts of unconsolidated sediments susceptible to fluvial transport during the meltwater peak and rainfall induced spates presumably increase the devastating impact of such events for sessile organisms. The shift in water source also will cause water temperatures to become warmer as well. All these changes will likely influence the distributions and abundances of macroinvertebrates with some lower elevation species already colonizing high elevation streams in the Alps (D. Finn, unpublished data) as well as benthic primary producers. The amount and kinds of riparian vegetation are also expected to change with potential effects on carbon and nutrient relationships in adjacent streams.

Based on these expectations, we predict in the long term that the windows of opportunity for periphyton growth to shift in running waters of the Alps with decreasing glacial influence and raising temperatures. In spring, periphyton development should be more limited by higher flows that mobilize sediments, reduce transparency, and thus presumably confound the impact of high solar UV. The high flows should also limit the effects of grazing by invertebrates. On the other hand, if the channel snow cover disappears earlier than today, the UV impact during the spring low flow period will be smaller and periphyton may accumulate before increasing discharge impose constraints. Already, early summer flows will be influenced by periodic extreme events that may scour periphyton and reduce biomass, although recovery should be relatively rapid. Water clarity should be improved with the reduction in glacial meltwater and inputs of glacial flour. Lower flows should occur earlier in late summer when light and temperature conditions become more optimal for periphyton growth, but this effect may be offset by high UV radiation. The autumnal window of opportunity may shift earlier in alpine catchments and extend into late autumn in channels that retain surface flows. Flowing water channels also may stay open longer and not become snow covered until late winter, further extending the window for periphyton growth. The counter effects by grazing invertebrates may limit somewhat periphyton development. In perennial systems with no or minor glacial influence, seasonal biomass pattern may still be similar to those observed today in non-glacial streams apart from an eventually early vernal and an extended autumnal/winter peak. However, major changes in seasonal patterns are expected when streams fall dry, e.g., if surface flow already ends in late summer and only resumes in spring. When glaciers disappear, this becomes a realistic scenario for rivers, the surface flow of which strongly depends on the recharge of the valley aquifer by the glacial flow pulse (Malard, 2003).

**Acknowledgements** We thank R. Illi and B. Ribi for analyses of water chemistry and chlorophyll *a*, and the Swiss Federal Office for Water and Geology for discharge data of several streams. We appreciate the helpful comments of two anonymous reviewers.

# References

Blumthaler, M., W. Ambach & W. Rehwald, 1992. Solar UV-A and UV-B radiation fluxes at two Alpine stations at different altitudes. Theoretical and Applied Climatology 46: 39–44.

Bothwell, M. L., 1985. Phosphorus limitation of lotic periphyton growth rates: an intersite comparison using continuous-flow troughs (Thompson River system, British Columbia). Limnology and Oceanography 30: 527–542.

Bothwell, M. L., 1989. Phosphorus-limited growth dynamics of lotic periphytic diatom communities: areal biomass and cellular growth rate responses. Canadian Journal of Fisheries and Aquatic Sciences 46: 1293–1301.

Bothwell, M. L., D. Sherbot, A. C. Roberge & R. J. Daley, 1993. Influence of natural ultraviolet radiation on lotic periphytic diatom community growth, biomass accrual, and species composition: short-term versus long-term effects. Journal of Phycology 29: 24–35.

Braun, L. N., M. Weber & M. Schulz, 2000. Consequences of climate change for runoff from Alpine regions. Annals of glaciology 31: 19–25.

Bretschko, G., 1966. Untersuchung zur Phosphatführung zentralalpiner Gletscherabflüsse. Archiv Fur Hydrobiologie 62: 327–338.

Brown, L. E., A. M. Milner & D. M. Hannah, 2006. Stability and persistence of alpine stream macroinvertebrate communities and the role of physical habitat variables. Hydrobiologia 560: 159–173.

Burgherr, P., 2000. Spatio-temporal community patterns of lotic zoobenthos across habitat gradients in an alpine glacial stream ecosystem. PhD thesis. Swiss Federal Institute of Technology, Zurich, Zürich.

Burgherr, P. & J. V. Ward, 2000. Zoobenthos of kryal and lake outlet biotopes in a glacial floodplain. Verhandlungen der Internationalen Vereinigung für Theoretische und Angewandte Limnologie 27: 1587–1590.

Burgherr, P., J. V. Ward & C. T. Robinson, 2002. Seasonal variation in zoobenthos across habitat gradients in an alpine glacial floodplain (Val Roseg, Swiss Alps). Journal of the North American Benthological Society 21: 561–4575.

Caldwell, M. M., R. Robberecht & W. D. Billings, 1980. A steep latitudinal gradient of solar ultraviolet-B radiation in the Arctic-Alpine life zone. Ecology 61: 600–611.

Castella, E., H. Adalsteinsson, J. E. Brittain, G. M. Gíslason, A. Lehmann, V. Lencioni, B. Lods-Crozet, B. Maiolini, A. M. Milner, J. S. Ólafsson, S. J. Saltveit & D. L. Snook, 2001. Macrobenthic invertebrate richness and composition along a latitudinal gradient of European glacier-fed streams. Freshwater Biology 46: 1811–1831.

Davis, J. C., G. W. Minshall, C. T. Robinson & P. Landres, 2001. Monitoring wilderness stream ecosystems. General Technical Report RMRS-GTR-70. United States Department of Agriculture, Forest service. Rocky Mountain Research Station, Ogden.

DeNicola, D. M. & K. D. Hoagland, 1996. Effects of solar spectral irradiance (visible to UV) on a prairie stream epilithic community. Journal of the North American Benthological Society 15: 155–169.

Francoeur, S. N. & R. L. Lowe, 1998. Effects of ambient ultraviolet radiation on littoral periphyton: biomass accrual and taxon-specific responses. Journal of Freshwater Ecology 13: 29–37.

Füreder, L., 1999. High alpine streams: cold habitats for insect larvae. In Margesin, R. & F. Schinner (eds), Cold-Adapted Organisms. Ecology, Physiology, Enzymology and Molecular Biology. Springer, Berlin: 181–195.

Gessler, J., 1965. Der Geschiebetriebbeginn bei Mischungen untersucht an natürlichen Abpflästerungserscheinungen in Kanälen. Mitteilungen der Versuchsanstalt für Wasser- und Erdbau der ETH Zürich 69: 1–67.

Gíslason, G. M., H. Adalsteinsson, I. Hanson & K. Svavarsdóttir, 2001. Longitudinal changes in macroinvertebrate assemblages along a glacial river system in central Iceland. Freshwater Biology 46: 1737–1751.

Gordon, N. D., T. A. McMahon & B. L. Finlayson, 1992. Stream Hydrology. An Introduction for Ecologists. Wiley, Chichester.

Grimm, N. B. & S. G. Fisher, 1986. Nitrogen limitation in a Sonoran Desert (Arizona, USA) stream. Journal of the North American Benthological Society 5: 2–15.

Günter, A., 1971. Die kritische mittlere Sohlenschubspannung bei Geschiebemischungen unter Berücksichtigung der Deckschichtbildung und der turbulenzbedingten Sohlenschubspannungsschwankunen. Mitteilungen VAW 3. Laboratory of Hydraulics, Hydrology and Glaciology, Swiss Federal Institute of Technology, Zürich.

Gurnell, A. M., 1987. Suspended sediment. In Gurnell, A. M. & M. J. Clark (eds), Glacio-Fluvial Sediment Transfer. Wiley, Chichester: 305–354.

Hawes, I., C. Howard-Williams & W. F. Vincent, 1992. Desiccation and recovery of Antarctic Cyanobacterial Mats. Polar Biology 12: 587–594.

Hieber, M., C. T. Robinson, S. R. Rushforth & U. Uehlinger, 2001. Algal communities associated with different alpine stream types. Arctic, Antarctic, and Alpine Research 33: 447–456.

Hieber, M., C. T. Robinson, U. Uehlinger & J. V. Ward, 2002. Are alpine lake outlets less harsh than other alpine streams? Archiv Fur Hydrobiologie 154: 199–223.

Hieber, M., C. T. Robinson, U. Uehlinger & J. V. Ward, 2005. A comparison of benthic macroinvertebrate assemblages among different types of alpine streams. Freshwater Biology 50: 2087–2100.

Hill, W. R., S. M. Dimick, A. E. McNamara & C. A. Branson, 1997. No effect of ambient UV radiation detected in periphyton and grazers. Limnology and Oceanography 42: 769–774.

Hill, W. R., P. J. Mulholland & E. R. Marzolf, 2001. Stream ecosystem response to forest leaf emergence in spring. Ecology 82: 2306–2319.

Hodson, A., P. Mumford & D. Lister, 2004. Suspended sediment and phosphorus in proglacial rivers: bioavailability and potential impacts upon the P status of ice-marginal receiving waters. Hydrological Processes 18: 2409–2422.

Horton, P., B. Schaefli, A. Mezghani, B. Hingray & A. Musy, 2006. Assessment of climate-change impacts on alpine discharge regimes with climate model uncertainty. Hydrological Processes 20: 2091–2109.

Howard-Williams, C. & W. F. Vincent, 1989. Microbial communities in southern Victoria Land streams (Antarctica). I. Photosynthesis. Hydrobiologia 172: 27–38.

Howard-Williams, C., C. L. Vincent, P. A. Broady & W. F. Vincent, 1986. Antarctic stream ecosystems: variability in environmental properties and algal community structure. Internationale Revue der Gesamten Hydrobiologie 71: 511–544.

Kowanacki, A., 1991. Zonal distribution and classification of the invertebrate community in high mountain streams in South Tyrol (Italy). Verhandlungen der Internationalen Vereinigung für Theoretische und Angewandte Limnologie 24: 2010–2014.

Lavandier, P. & H. Décamps, 1984. Estaragne. In Whitton, B. A. (ed.), Ecology of European Rivers. Blackwell, Oxford: 237–263.

Lloyd, D. S., J. P. Koenings & J. D. LaPerriere, 1987. Effects of turbidity in fresh waters of Alaska. North American Journal of Fisheries Management 7: 18–33.

Malard, F., 2003. Groundwater–surface water interactions. In Ward, J. V. & U. Uehlinger (eds), Ecology of a Glacial Flood Plain. Kluwer Academic Publishers, Dordrecht: 37–56.

Malard, F., U. Uehlinger, R. Zah & K. Tockner, 2006. Flood-pulse and riverscape dynamics in a braided glacial river. Ecology 87: 704–716.

McKnight, D. M. & C. M. Tate, 1997. Canada stream: a glacial meltwater stream in Taylor Valley, South Victoria Land, Antarctica. Journal of the North American Benthological Society 16: 14–17.

Meyns, S., R. Illi & B. Ribi, 1994. Comparison of chlorophyll-a analysis by HPLC and spectrophotometry: where do the differences come from? Archiv Fur Hydrobiologie 132: 129–139.

Milner, A. M. & G. E. Petts, 1994. Glacial rivers: physical habitat and ecology. Freshwater Biology 32: 295–307.

Milner, A. M., J. E. Brittain, E. Castella & G. E. Petts, 2001. Trends of macroinvertebrate community structure in glacier-fed rivers in relation to environmental conditions: a synthesis. Freshwater Biology 46: 1833–1847.

Milner, A. M., L. E. Brown & D. M. Hannah, 2009. Hydroecological response of river systems to shrinking glaciers. Hydrological Processes 23: 62–77.

Newbold, D. J., 1992. Cycles and spirals of nutrients. In Calow, P. & G. E. Petts (eds), The Rivers Handbook. Hydrological and Ecological Principles. Blackwell, London: 379–408.

Paul, F., 2003. The new Swiss glacier inventory 2000. PhD. University of Zurich, Zurich.

Peterson, C. G., 1996. Response of benthic algal communities to natural physical disturbance. In Stevenson, R. J., M. L. Bothwell & R. L. Lowe (eds), Algal Ecology. Freshwater Benthic Ecosystems. Academic Press, San Diego: 375–402.

Pfankuch, D. J., 1975. Stream Reach Inventory and Channel Stability Evaluation. U.S.D.A. Forest Service, Missoula, Montana.

Rhim, R., 1996. Critical Loads of Nitrogen and their Exceedances. Environmental Series 275, Swiss Agency for the Environment, Forests and Landscape. Berne, Switzerland.

Rinke, K., C. T. Robinson & U. Uehlinger, 2001. A note on abiotic factors that constrain periphyton growth in Alpine glacier streams. International Review of Hydrobiology 86: 361–366.

Robinson, C. T. & S. Matthaei, 2007. Hydrological heterogeneity of an Alpine streamlake network in Switzerland. Hydrological Processes 21: 3146–3154.

Robinson, C. T., U. Uehlinger & M. Hieber, 2001. Spatio-temporal variation in macroinvertebrate assemblages of glacial streams in the Swiss Alps. Freshwater Biology 46: 1663–1672.

Robinson, C. T., U. Uehlinger, F. Guidon, P. Schenkel & R. Skvarc, 2002. Limitation and retention of nutrients in alpine streams of Switzerland. Verhandlungen der Internationalen Vereinigung für Theoretische und Angewandte Limnologie 28: 263–272.

Roos, J. C. & W. F. Vincent, 1998. Temperature dependence of UV radiation effects on Antarctic cyanobacteria. Journal of Phycology 43: 118–125.

Röthlisberger, H. & H. Lang, 1987. Glacial hydrology. In Gurnell, A. M. & M. J. Clark (eds), Glacio-Fluvial Sediment Transfer. Wiley, Chichester: 207–284.

Rott, E., M. Cantonati, L. Füreder & P. Pfister, 2006. Benthic algae in high altitude streams of the Alps—a neglected component of the aquatic biota. Hydrobiologia 562: 195–216.

Schmucki, D. & R. Philipona, 2002. Ultraviolet radiation in the Alps: the altitude effect. Optical Engineering 41: 390–3095.

Schütz, C., M. Wallinger, R. Burger & L. Füreder, 2001. Effects of snow cover on the benthic fauna in a glacier-fed stream. Freshwater Biology 46: 1691–1704.

Tockner, K., F. Malard, P. Burgherr, C. T. Robinson, U. Uehlinger, R. Zah & J. V. Ward, 1997. Physico-chemical characterization of channel types in a glacial floodplain ecosystem (Val Roseg, Switzerland). Archiv Fur Hydrobiologie 140: 433–463.

Tockner, K., F. Malard & J. V. Ward, 2000. An extension of the flood pulse concept. Hydrological Processes 14: 2861–2883.

Tockner, K., F. Malard, U. Uehlinger & J. V. Ward, 2002. Nutrients and organic matter in a glacial river floodplain system (Val Roseg, Switzerland). Limnology and Oceanography 47: 521–535.

Tonolla, D., 2005. Characterization of the temporary hydrosystem of a high alpine catchment (Val Roseg, Switzerland) by GIS mapping and assessment of morphological features. Diploma Thesis, Swiss Federal Institute of Technology, Zurich.

Uehlinger, U., 1991. Spatial and temporal variability of the periphyton biomass in a prealpine river (Necker, Switzerland). Archiv Fur Hydrobiologie 123: 219–237.

Uehlinger, U. & R. Zah, 2003. Organic matter dynamics. In Ward, J. V. & U. Uehlinger (eds), Ecology of a Glacial Floodplain. Kluwer Academic Publishers, The Hague: 199–215.

Uehlinger, U., H. Bührer & P. Reichert, 1996. Periphyton dynamics in a floodprone prealpine river: evaluation of significant processes by modelling. Freshwater Biology 36: 249–263.

Uehlinger, U., R. Zah & J. V. Ward, 1998. The Val Roseg Project: temporal and spatial patterns of benthic algae in an Alpine stream ecosystem influenced by glacier runoff. In Kovar, K., U. Tappeiner, N. E. Peters & R. G. Craig (eds), Hydrology, Water Resources and Ecology in Headwaters. IAHS Press, Wallingford, U.K: 419–424.

Uehlinger, U., K. Tockner & F. Malard, 2002. Ecological windows in glacial stream ecosystems. EAWAG News 54e: 16–17, 20–21.

Uehlinger, U., F. Malard & J. V. Ward, 2003. Thermal patterns in the surface waters of a glacial river corridor (Val Roseg, Switzerland). Freshwater Biology 48: 284–300.

Vincent, W. F. & C. Howard-Williams, 1989. Microbial communities in southern Victoria Land streams (Antarctica) II. The effects of low temperature. Hydrobiologia 172: 39–49.

Vinebrooke, R. D. & P. R. Leavitt, 1996. Effects of ultraviolet radiation on periphyton in an alpine lake. Limnology and Oceanography 41: 1035–1040.

Vinebrooke, R. D. & P. R. Leavitt, 1999. Differential responses of littoral communities to ultraviolet radiation in an alpine lake. Ecology 80: 223–237.

Ward, J. V., 1994. Ecology of alpine streams. Freshwater Biology 32: 277–294.

Ward, J. V. & U. Uehlinger (eds), 2003. Ecology of a Glacial Flood Plain. Kluwer Academic Publishers, Dordrecht.

Watson, R. T., M. C. Zinyowera & R. H. Moss, 1997. The Regional Impacts of Climate Change: An Assessment of Vulnerability. Cambridge University Press, Cambridge.

Wellnitz, T. A. & J. V. Ward, 2000. Herbivory and irradiance shape periphytic architecture in a Swiss alpine stream. Limnology and Oceanography 45: 64–75.

Zah, R. & U. Uehlinger, 2001. Particulate organic matter inputs to a glacial stream ecosystem in the Swiss Alps. Freshwater Biology 46: 1597–1608.

Zar, J. H., 1996. Biostatistical Analysis. Prentice Hall, Upper Saddle River, N.J.

Zbinden, M., M. Hieber, C. T. Robinson & U. Uehlinger, 2008. Short-term colonization patterns of macroinvertebrates in alpine streams. Fundamental and Applied Limnology 171: 75–86.

Zemp, M., W. Haeberli, M. Hoelzle & F. Paul, 2006. Alpine glaciers to disappear within decades. Geophysical Research Letters 33: L13501–L13504.

GLOBAL CHANGE AND RIVER ECOSYSTEMS

# The periphyton as a multimetric bioindicator for assessing the impact of land use on rivers: an overview of the Ardières-Morcille experimental watershed (France)

B. Montuelle · U. Dorigo · A. Bérard ·
B. Volat · A. Bouchez · A. Tlili · V. Gouy ·
S. Pesce

Received: 28 July 2009 / Accepted: 18 January 2010 / Published online: 4 February 2010
© Springer Science+Business Media B.V. 2010

**Abstract** Developing new biological indicators for monitoring toxic substances is a major environmental challenge. Intensive agricultural areas are generally pesticide-dependent and generate water pollution due to transfer of pesticide residues through spray-drift, run-off and leaching. The ecological effects of these pollutants in aquatic ecosystems are broad-ranging owing to the variety of substances present (herbicides, fungicides, insecticides, etc.). Biofilms (or periphyton) are considered to be early warning systems for contamination detection and their ability to reveal effects of pollutants led researchers to propose a variety of methods to detect and assess the impact of pesticides. The present article sought to provide new insights into the ecological significance of biofilm microbial communities and to discuss their bioindication potential for water quality and land use by reporting on 4 years of research performed on the French Ardières-Morcille experimental watershed (AMEW). Various biological indicators have been applied during several surveys on AMEW, allowing the characterisation of (i) the structure and diversity of biofilm communities [community level finger printing (CLFP) such as PCR–DGGE and pigment classes], (ii) functions associated with biofilm [community level physiological profiles (CLPP) such as extracellular enzymes, pesticides biodegradation or carbon sources biodegradation] and (iii) biofilm tolerance assessment (pollution-induced community tolerance, PICT) of the main contaminant in the AMEW (copper and diuron). Approaches based on CLFPs and PICT were consistent with each other and indicated the upstream–downstream impact due to the increasing land use by vineyards and the adaptation of algal and bacterial communities to the pollution gradient. CLPPs gave a contrasted bioindication because some parameters (most of the tested extracellular enzymes activities) did not detect a pollution gradient. Such CLPPs, CLFPs and PICT methods applied to biofilm could constitute the basis for a relevant in situ assessment both for chemical effects and aquatic ecosystem resilience.

**Keywords** Biofilms · River · Biological indication · Pollution · Community level physiological profile · Finger prints · Pollution-induced community tolerance

Guest editors: R. J. Stevenson, S. Sabater / Global Change and River Ecosystems – Implications for Structure, Function and Ecosystem Services

B. Montuelle (✉) · B. Volat · A. Tlili ·
V. Gouy · S. Pesce
Cemagref, UR MALY, 3 quai Chauveau CP220,
Lyon 69336 Cedex 09, France
e-mail: bernard.montuelle@cemagref.fr

U. Dorigo · A. Bouchez
INRA UMR CARRTEL, Laboratoire RITOXE, BP 511,
74203 Thonon Cedex, France

A. Bérard
INRA, UMR INRA/UAPV Climat Sol Environnement,
Site Agroparc, 84914 Avignon Cedex, France

## Introduction

In setting objectives for achieving a 'good ecological status' for aquatic systems, the European Water Framework Directive (WFD 2000/60/EC) has raised numerous scientific questions: What is a 'good ecological status'? How can we improve the chemical and biological quality of aquatic environments? Some authors doubt that these objectives can be reached for priority substances or persistent organic pollutants (Fuerhacker, 2009). This aim requires in particular being able to identify causal relationships between contaminants and biological effects: a difficult task in situ, because of multiple co-occurring pollutants and confounding factors, etc. Bioindicators supporting French and European legislation (invertebrates, fishes macrophytes and diatoms) have been defined to characterise trophic pressure: the macrophytes biological index for rivers (Haury et al., 2006), the biological diatom index (BDI) (Lenoir and Coste, 1996) and the standardised global biological index (AFNOR NF T 90 350, 2004). These three standardised indicators are used to evaluate water quality in France. However, much research work is in progress to design or propose new bioindicators appropriate for toxic substances or newly emerging contaminants like drugs and hormones, to allow early warning of such pollution.

In low-order streams (Strahler order below 3 or 4) draining rural watersheds, domestic and agricultural discharges often result in eutrophication (N and P inputs) and toxic contamination through point or diffuse pollution (e.g. pesticides). Because of the generally low dilution in small rivers, such discharges may temporarily or permanently impair biodiversity and dynamics in aquatic ecosystems. Intensive agricultural areas that are generally pesticide-dependent generate water pollution due to transfer of organic or mineral pesticide residues through spray-drift, run-off and leaching (Landry et al., 2004; Vu et al., 2006). Their ecological effects in aquatic ecosystems are broad-ranging owing to the variety of substances present (herbicides, fungicides, insecticides, etc.) (DeLorenzo et al., 2001).

In small aquatic systems, trophic webs are generally reduced and most microorganisms live on submerged substrates such as biofilms (periphyton). These biofilms are complex assemblies of microbial communities embedded in a polysaccharide and protein matrix. In such environments, living organisms, both prokaryotic (bacteria) and eukaryotic (mostly microalgae, but also fungi), interact strongly (Rier & Stevenson, 2002; Barranguet et al., 2003) and are responsible for most of the energy input through primary production and nutrient cycling (Battin et al., 2003a, b). They play a marked ecological role in biochemical processes such as organic matter degradation (Romani et al., 2004) or nitrogen biotransformation (Teissier & Torre, 2002).

The ability of biofilms to reveal the effect of pollutants has been the subject of research for some years (Admiraal et al., 1999; Sabater, 2000). As reviewed by Sabater et al. (2007), there is a variety of current methods to detect and assess the impact of pesticides on functional and structural targets, and periphyton is considered as an early warning system for contamination detection. Benthic algae have been the subject of a great deal of work in this area, based on taxonomy (e.g. diatoms: Stevenson & Pan, 1999; Gold et al., 2003; Morin et al., 2007), algal pigments or their photosynthetic capacity (Dorigo et al., 2007; Schmitt-Jansen & Altenburger, 2008; Villeneuve, 2008). Conversely, little work has been published on periphytic bacterial communities (Lyautey et al., 2003; Pesce et al., 2006, 2008; Dorigo et al., 2007, 2009). The ability of microbenthic communities to adapt to pollutants (short generation time, high taxonomic and functional diversity) has produced the concept of pollution-induced community tolerance (PICT; Blanck et al., 1988). PICT is based on the principle that an ecosystem in contact with a toxicant results in changes at the community level due to various toxicant-induced phenomena. These latter include individual acclimation, physiological or genetic adaptation, and loss of sensitive species. Overall, this adaptation results in a lowered sensitivity to this contaminant and higher effective concentration ($EC_x$) values. This approach has already been successfully applied to evaluating the tolerance of algal communities to pesticides (Dorigo et al., 2004; Bérard et al., 2003) and of bacterial communities to metals (Boivin et al., 2005), demonstrating upstream–downstream gradients in a wine-growing drainage basin (Dorigo et al., 2007, Pesce et al., 2010) and differentiating between chronic and acute effects (Tlili et al., 2008). Such methods, associated with molecular tools for biodiversity studies, have lent new perspectives to bioindication based on microbial communities. These methods allow a better understanding of ecological responses of rivers, as global change effects are often confused because of the complexity of interactions among anthropic drivers acting on a mosaic of natural hydrogeomorphological and climatic settings. With this aim, the work reported here sought to provide new insight into the ecological significance of benthic microbial communities (biofilms), and discuss their bioindication potential for water quality and land use. We report here on 4 years of research performed on the French Ardières-Morcille experimental watershed (AMEW) (part of the Rhône Basin long-term ecological research watershed http://www.graie.org/zabr/index.tm). Several chemical and biological surveys were carried out to give a comprehensive overview of the microbial dynamics in play along this impacted river and to provide tools for assessing the effects of intense agriculture (as a particular element of global change) on freshwater ecological services.

## Materials and methods

### Field site and sampling

The Morcille River is located in the Beaujolais wine-growing area of eastern France (46.150°N, 4.600°E). Its watershed consists of three nested sub-watersheds (Fig. 1), defined by three sampling stations: St Joseph, Versauds and St Ennemond (upstream to downstream). In this region, which is characterised by an intensive wine-producing activity (vineyards occupy 79% of the catchment area), the routine application of pesticides to the vineyards generates a pesticide contamination gradient from the upstream section of the river down to its mouth (Collectif, 2008).

Most of the samples analysed and discussed here were taken in 2007 except for CLLP measurements (2009). Other surveys with some of these parameters and with specific objectives were performed between 2006 and 2009 (cf. Dorigo et al., 2007, 2009; Tlili et al., 2008; Rabiet et al., 2008; Pesce et al., 2009a, b, 2010; Villeneuve et al., 2010).

### Physical characteristics and land use

General characteristics of the soils, slopes, crops and climate were acquired through various sources such as the 1: 25,000 topographical map of IGN (French

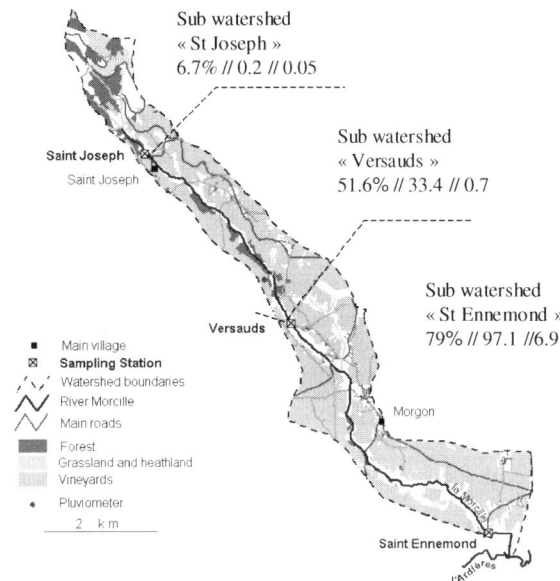

**Fig. 1** GIS of Morcille watershed and its three sub-basins with the main land uses. Data indicate, respectively: vineyard area as % of sub-basin surface//length of ditches connected to the river (km)//surface area of housing (ha)

Institute of Geography), a digitized map of the agricultural field boundaries, soil maps, local measurements of soil characteristics, rainfall and water flow and field identification of the main surface water pathways from the plots to the stream, especially the numerous ditches originally dug to limit erosion. Complementary inquiries among the farmers with the support of agricultural advisers (Chambre d'Agriculture du Rhône) were compiled to make an inventory of the commonly used pesticides and periods of application.

### Chemistry

Chemical variables were regularly analysed in the course of annual surveys. The data presented here are those from 2007 and are typical of the water chemical quality of the river Morcille ($n = 33$ for pesticides, $n = 33$ for metals and $n = 45$ for nutrients).

#### Nutrients

DOC, $NO_2^-$, $NO_3^-$, $NH_4^+$, $PO_4^{3-}$, conductivity and pH were analysed using French standard operating procedures and protocols (Association Française de Normalisation, AFNOR).

## Pesticides

Using standardised protocols, the eight most frequently found pesticides and some of their degradation products were analysed in the water samples by ESI–LC–MS/MS (API 4000, Applied Biosystems) at the Water Chemistry Laboratory in Cemagref (Lyon). The herbicide diuron together with two of its breakdown products (DCPMU and 3,4-DCA) and fungicides azoxystrobin, carbendazim, tebuconazole, procymidone and dimetomorph were analysed and quantified. The quantification thresholds of the method ranged between 0.02 and 0.08 µg/l according to the compound analysed.

## Metals

Cu and As were analysed by filtration on PVDF membrane (0.2 µm), and then acidification with $HNO_3$ SUPRAPUR 0.5%. The analyses were performed by ICP-MS (THERMO ELECTRON X7 Series 2) to meet the standard NF EN ISO 17294-2. The quantification threshold was 0.05 µg/l for both compounds.

## Biological variables

Biofilms were generally sampled on non-embedded stones (diameter 1–4 cm) in the river bottom, but artificial substrata in the form of glass discs immersed in the river Morcille for 8 weeks were also used for PCR–DGGE and pigment analysis. Sample triplicates were made and several microbial variables were measured to describe the biological and functional quality of the periphyton and its responses to changes in the chemical quality of the river water. All the measurements were made on periphyton suspensions after scraping and dilution in 0.2 µm filtered river water.

## Biomass

Periphytic biomass was evaluated by calculating ash-free dry weight (AFDW). Suspensions (2 ml) of biofilm replicates were filtered through individual, previously dried, 25 mm CF/C Whatman glass fibre filters (pore size 1.2 µm). Each filter was dried for 24 h at 105°C and weighed to calculate dry matter. The filters were then burned to ash at 480°C (Nabertherm P320) for 1 h and weighed again. The AFDW was calculated by subtracting the mineral matter from the total dry matter. Results were expressed in $g\ m^{-2}$.

## Community level finger printing: pigment analysis

For each biofilm sample, one glass disc was placed in a centrifuge tube (Corning) containing 4 ml of methanol/0.5 M ammonium acetate (98/2 v/v) solution and sonicated by means of a 4 mm probe for 1 min at 180 W and at 50% activity (Vibracell, Bioblock Scientific 375W). The tubes were then centrifuged for 6 min at $6 \times g$ and 0°C. The supernatant was collected and filtered through a 0.2 µm filter syringe (Cameo 3N-syringe nylon filter; Micron Separation Inc.); 100 µl of this extract was injected to determine the lipophilic pigment composition by high pressure liquid chromatography. Pigments were separated on a $4.6 \times 250$ mm column (Waters Spherisorb ODS5 25 µm) and identified from retention time and absorption spectrum using DAD according to SCOR.

Chlorophyll *a* can be considered as a proxy of the total periphyton biomass beside the AFDW. In this case, biomasses are given as µg of chl *a* per $cm^2$. A quantitative method was used, obtained from a calculation model based on published ratios for monocultures (Dorigo et al., 2004). A table was constructed, taking into account the relative abundance of each pigment in a given sample (expressed as the percentage of the sum of the area of all pigments in a sample) (see Dorigo et al., 2007 for detailed method and data). Furthermore, the total number of pigments and the total number of degraded pigments per sample were counted.

## Community level finger printing: PCR–DGGE

Each biofilm sample was collected by scraping six glass discs and was suspended in 2 ml of 0.2 µm filtered river water. Biofilm suspensions were centrifuged at $14,000 \times g$ for 30 min and nucleic acid extraction was performed on the biofilm pellets (Dorigo et al., 2009). The integrity of the total DNA was checked by agarose gel electrophoresis and the nucleic acid concentration determined by 260 nm absorbance. PCR amplification of eukaryotic 18S rRNA gene fragments, bacterial 16S rRNA gene fragments and their DGGE analysis was performed according to Tlili et al. (2008). After migration, separated PCR products were stained for 45 min in the dark with SYBRGold (molecular probes), visualised on a UV transilluminator (Claravision), then photographed and digitalized using Microsoft Photo Editor software. Each band at a given height in each lane was scored 1 or 0 (presence or absence). This data set was used to perform correspondence analysis (COA) using ADE-4 software. A similarity index (Jaccard index) was also built using DGGE fingerprints based on presence/absence data. A similarity value of 1 indicates that two DGGE banding patterns are identical, whereas a value of 0 indicates that there are no common bands.

*Community level physiological profile*

Two variables were used to describe the community level physiological profiles (CLPPs) based on organic matter degradation: extracellular enzyme activities and carbon mineralisation (substrate-induced respiration method).

The three extracellular enzymes $\beta$-D-glucosidase ($\beta$Glu), $\beta$-xylosidase ($\beta$Xyl) involved in cellulose degradation, and leucine aminopeptidase (Lap) involved in amino acid degradation, were analysed as in Romani et al. (2004). Activities were analysed by fluorimetry, using substrate analogues (4-methyl-umbelliferyl-$\beta$-D-glucopyranoside (750 µM), 4-methyl-umbelliferyl-xylopyranoside (1,000 µM) and L-leucine-4-methylcoumarinyl-7-amide HCl (1,000 µM), respectively) to predetermine saturation curves and for experimental measurements. For all enzyme assays 6 ml of substrate solution was added to triplicate biofilm samples (non-disrupted biofilm in place on stone) and formaldehyde-killed control samples incubated for 30 min with formol 40% before assay. Incubation was performed at 20°C for 20 min under continuous shaking in the dark. Substrate blanks were also prepared with filter-sterilised stream water. The reaction was stopped in boiling water and each tube was then centrifuged for 10 min at 5,000×$g$. The fluorescent products released by enzyme activities were measured after adding 0.05 M glycine buffer pH 10.4, using a microplate reader (SAFIRE, TECAN Group Ltd, Switzerland) with excitation and emission wavelengths of 363 and 441 nm, respectively, for MethylUmbelliFeryl (MUF) and with 343 nm (excitation) and 436 nm (emission) for MethylCoumarineAcid (MCA). Quantification was achieved using a standard solution of MUF and MCA. The intensity of fluorescence of the blanks was subtracted from all samples to correct for non-enzymatic hydrolysis. Activity in formaldehyde-killed controls was subtracted to correct for abiotic activity. Results were expressed in nmol of hydrolysed compound per hour and per cm$^2$ surface area.

Basal (BR) and substrate-induced respiration (SIR) were assessed using the MicroResp$^{TM}$ system of Campbell et al. (2003), consisting of a 96-deep-wells microplate (Nunc 278012) housing periphyton suspension and aqueous carbon sources, sealed individually to a colorimetric $CO_2$-trap microplate. Mineralisation of 11 carbon sources was tested (glucose, fructose, sucrose, ribose, galactose, maltose, arginine, glycine, lysine, glutamic acid and citric acid). The carbon stock solutions were prepared at 120 mg/ml and adjusted to river pH (7) to avoid any substrate–pH effects on microbial communities and minimise chemical artefacts due to carbonate-derived $CO_2$. Colorimetric $CO_2$ traps were prepared in 96 microplate wells. The indicator dye with the gel detector plate consisted of cresol red dye (12.5 ppm), potassium chloride (150 mM) and sodium bicarbonate (2.5 mM) set in a 1% gel of noble agar (150 ml per well). Periphyton suspension (500 µl) was added to the 96-deep-well plate after 30 µl of each C source had been dispensed (four wells per substrate = SIR plus four water per plate = BR). Each deep-well microplate was sealed to the $CO_2$-trap microplate with a silicone seal and incubated in the dark at 25°C. $CO_2$-trap absorbance was measured at 570 nm (Biotek Synergy HT spectrophotometer) immediately prior to sealing to the soil deep-well plate, and after 15 h incubation. A calibration curve of absorbance against headspace equilibrium $CO_2$ concentration (measured on a gas chromatograph) was fitted to a regression model.

*Organic pesticide biodegradation*

The ability of aquatic microbial communities to mineralise diuron was determined by radiorespirometry as described by Pesce et al. (2009a). Samples were treated with 1.45 kBq of $^{14}$C uniformly (ring)-labelled diuron [specific activity 567 MBq/mmol;

99% radiochemical purity (Sigma-Aldrich)]. An epilithon suspension was prepared with the collected stones using filtered river water. Final concentration was adjusted to approximately 5 cm$^2$ of stone biofilm per millilitre of suspension and 50 ml was used for each sample. Epilithon samples were supplemented with 0.5 μg of diuron to reach a final concentration of 10 μg per litre of river water. They were processed in triplicate and incubated under artificial light (13 h photoperiod) at 20°C for 16 weeks; $^{14}CO_2$ resulting from the mineralisation of $^{14}C$-diuron was trapped in 5 ml of 0.2 M NaOH solution and analysed by liquid scintillation counting using ACS II (Amersham) scintillation fluid.

*Tolerance assessment*

Short-term laboratory experiments were performed to assess the natural tolerance of the photoautotrophic communities to the main toxicants found in the river, namely Cu and diuron. The PICT concept makes the assumption that communities exposed to contaminants become tolerant to these contaminants by adaptation or species changes (Blanck et al., 1988; Bérard et al., 2002). During laboratory toxicity assays, exposed communities will be characterised by higher $EC_{50}$ values than reference communities with respect to the toxicant tested. The effects of contaminants on periphyton were assessed using $^{14}C$ photosynthetic assimilation as the endpoint (Guasch & Sabater, 1998). A stock solution containing 100 μM diuron (MW 233 g/mole) (Sigma high grade standard 99.5%) was prepared in water and stored at $-20°C$ prior to use. A semi-logarithmic series of concentrations was freshly prepared by serial dilution of the stock solution in 0.2 μm filtered river water. Final test concentrations ranged from 0 to 10 μM of diuron (one blank and nine increasing concentrations, i.e. from 0 to 2.3 mg/l). For copper bioassays, final concentrations in the test vessels ranged from 0 to 100 μM (five blanks and nine increasing concentrations, i.e. from 0 to 6.3 mg Cu), starting from a stock solution of 1,000 μM Cu ($CuSO_4$, Merck high purity grade). The measurements of photosynthesis activity by $^{14}C$ incorporation were made as described in Dorigo et al. (2007). Data were fitted to a logistic equation using the least squares method, and used to plot a dose–response curve and determine photosynthetic $EC_{50}$ values for each biofilm and period.

## Results

Description of land uses

Description of land uses is summarised in a geographic information system (GIS) (Fig. 1, Gouy, unpublished data). The three sampling points are associated with a specific sub-watershed, with vineyards occupying an increasing percentage of the catchment area along the upstream to downstream gradient (from ca. 7% to nearly 80%). Different land use indicators allowed a better characterisation of the landscape features involved in the formation of pollution flows (Collectif, 2008). In these, allowance was made for drainage channels (sum total of drainage ditches collecting run-off water, helping the rapid transfer of pesticides from treated plots to waterways), and the building surface areas measuring the potential upstream/downstream increase in domestic pressure. Hence, these two indicators could help to understand the changes in chemical quality of the river Morcille, as they indicate a level of anthropogenic pressure.

Water chemistry

A fairly strong upstream/downstream increase (Fig. 2) was significant for most of the variables (DOC, $PO_4$, conductivity), very likely indicating domestic wastewater inputs. Median values of DOC, $PO_4$ and conductivity increased by about 125, 300 and 150%, respectively, from upstream to downstream. All the chemical parameters were significantly different between sites (ANOVA 1; $P > 0.001$), except for pH and nitrates (ANOVA 1; $P = 0.569$ and 0.106, respectively). However, the chemical quality of the upstream water (St Joseph) was more stable, and the variability more marked at St Ennemond. Overall the water quality was in line with French standards of good chemical status for the nutrients (DCE, 2005).

Copper and arsenic concentrations increased gradually from upstream to downstream, irrespective of season (Fig. 3, yearly data), with maxima also strongly increasing (more than 16 μg/l for Cu and more than 40 μg/l for As). The annual median values (out of 33 analyses) rose, respectively, between St Joseph and St Ennemond, six-fold for Cu and five-fold for As. The French water quality standards specify 'probable no-effect concentrations' (PNEC)

**Fig. 2** Boxplot of water chemistry from upstream to downstream in the river Morcille (date 2007; $n = 44$ for each parameters). Boxplots indicate mean value, first and third quartile, mini and maxi. J: St Joseph; V: Versauds; E: St Ennemond

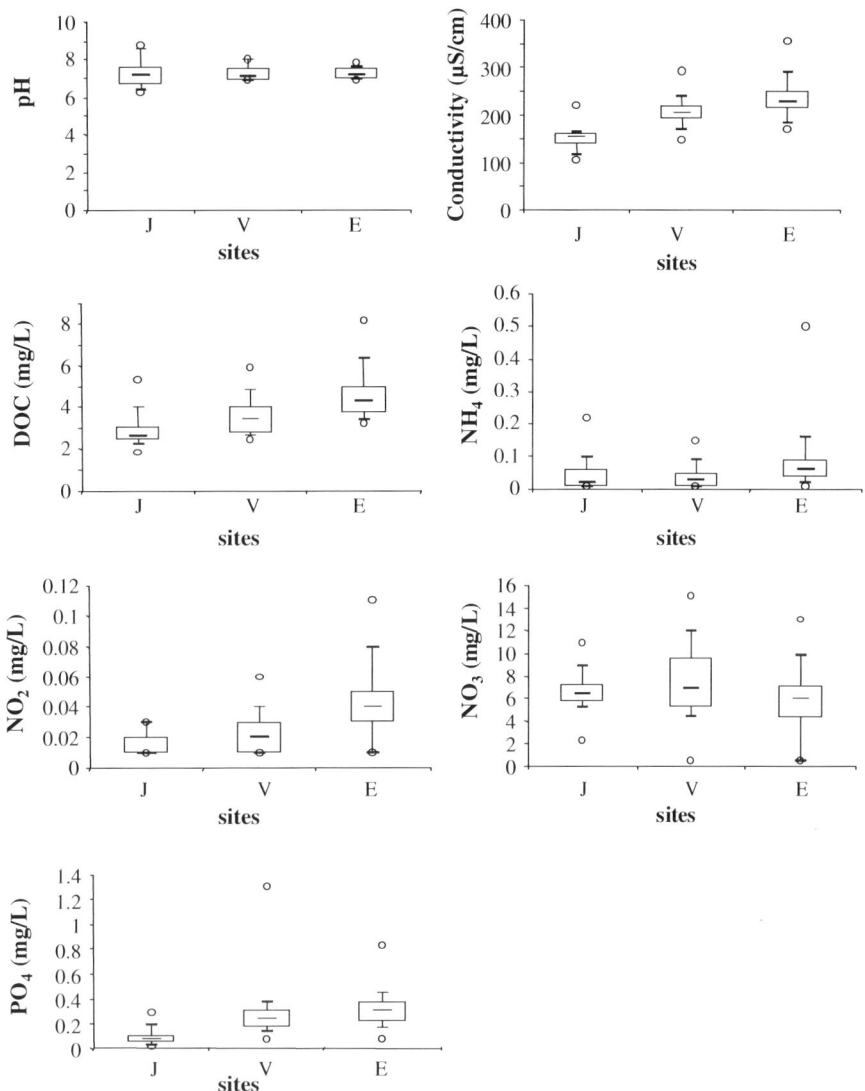

in freshwater for these two metals of 1.6 and 4.4 µg/l, indicating a good quality upstream and a poorer quality (given the hardness) at St Ennemond, with a potential toxic effect.

On average some 20 different organic pesticides were found in water, both at Versauds and St Ennemond (Collectif, 2008). However, only the products most frequently used in the drainage basin and found in the water were taken into consideration here (Rabiet et al., 2008, 2010). They were grouped into two activity families (herbicides: diuron and its breakdown products DCPMU and DCA; fungicides: azoxystrobin, carbendazim, tebuconazole, procymidone and dimetomorph). St Joseph was relatively free of contamination: no substance was found in winter, while diuron was found only three times and fungicides twice in 22 summer samples (Table 1). Versauds and St Ennemond showed a marked occurrence of both herbicide and fungicide (86–100%).

Biomass

Periphytic biomasses were not different between sites ($n = 12$; $P > 0.05$) and between seasons (Fig. 4) and kept a mean value of 0.8 mg AFDW cm$^{-2}$.

**Fig. 3** Boxplot of the concentrations of the main chemicals in the river Morcille (date 2007; $n = 33$). H: herbicide; F: fungicide—J: St Joseph; V: Versauds; E: St Ennemond. Boxplots indicate mean value, first and third quartile, mini and maxi

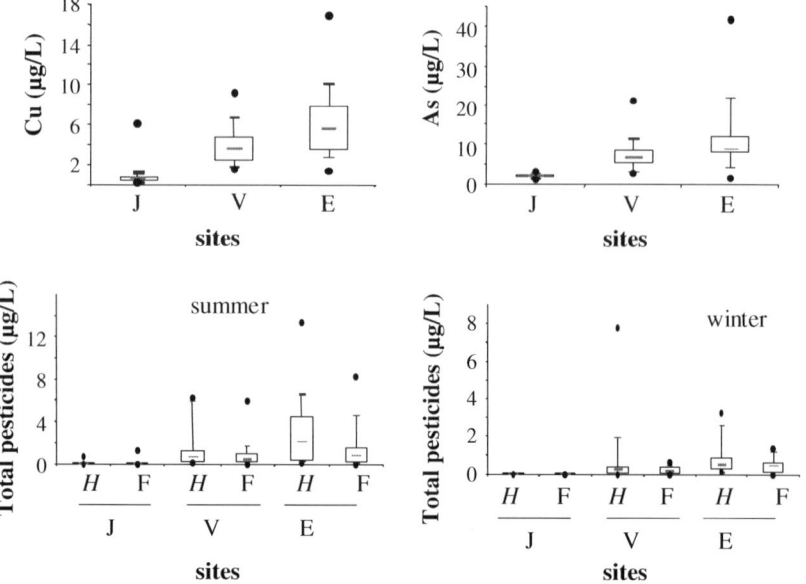

**Table 1** Occurrence of pesticides during 2007 survey: A/B: A: presence (with analytical quantification); B: number of analyses

|  | St Joseph | Versauds | St Ennemond |
|---|---|---|---|
| Winter | H: 0/11 F: 0/11 | H: 13/14 F: 13/14 | H:11/11 F: 10/11 |
| Summer | H: 3/22 F:2/22 | H: 21/21 F: 18/21 | H: 22/22 F 19/22 |

H herbicides, F fungicides

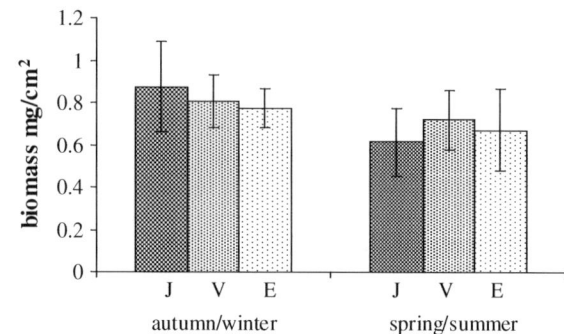

**Fig. 4** Periphytic biomass on the three study sites. Spring/summer: during pesticide treatment on the watershed. Autumn/winter: outside pesticide spreading time (mean values ± SD; $n = 3$). J: St Joseph; V: Versauds; E: St Ennemond

## Community level finger printing

The microbial community was structured along the upstream–downstream gradient (Fig. 5). For the prokaryotic community, the first two axes of each correspondence analysis accounted globally for 88% of the variability in spring and for 87.7% in winter. For the eukaryotic community, the first two axes of each correspondence analysis accounted for 96 and 67% of the variability in spring and in winter, respectively. For both communities, in spring (Fig. 5, left panel) the first axis separates the pristine St Joseph from Versauds and St Ennemond, and the second axis separates these two last stations. In winter for the eukaryotic community (Fig. 5b, right panel) St Joseph and Versauds are very close to each other and separated from St Ennemond by the first axis. In winter, there was a very high dispersion in the eukaryotic diversity between the three plates from the St Ennemond sampling area.

The correspondence analysis based on the relative percentage of the pigments detected in all sampling sites in winter and spring (see Dorigo et al., 2007), depicts the adaptation of the microalgal community pigment structure (Fig. 6). The projection of the plane defined by the first two axes indicates a clear separation between spring and winter samples. The second axis allowed the differentiation in spring of the pristine St Joseph area from the other two, and in winter the separation of the St Ennemond area from Versauds and St Joseph.

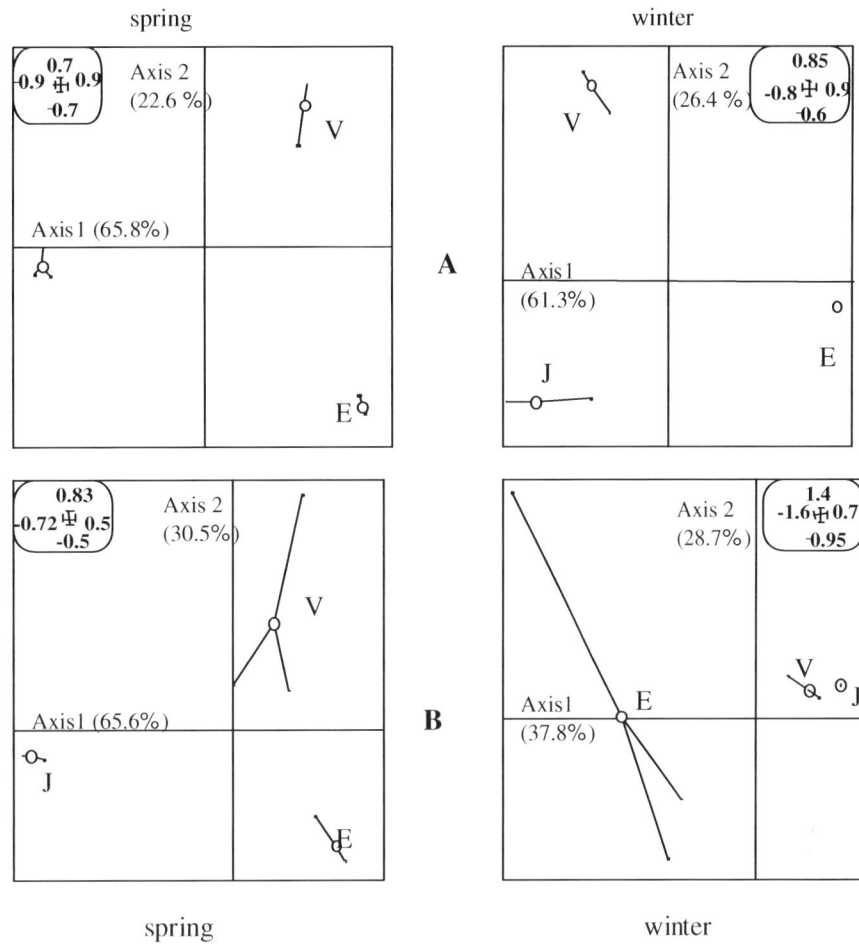

**Fig. 5** Correspondence analysis of prokaryotic DGGE bands (**A**) and eukaryotic DGGE bands (**B**) of each sampling site and area (J, St Joseph; V, Versauds; E, St Ennemond) in spring (*left panel*) and in winter (*right panel*) (from Dorigo et al., 2007). *Dots* are mean value of triplicates and *bars* indicate the variability of triplicates

Community level physiological profiles

*Extracellular enzyme activities*

Extracellular enzyme activities of the river Morcille biofilms were analysed using a two-way ANOVA to test for spatial effect (upstream to downstream), temporal effect, and their interaction (all data were logarithmically transformed to stabilize the variance and when significant differences were detected by using Scheffé post hoc comparisons were used). No clear spatial or temporal pattern was found (Fig. 7). Between-site differences were generally not significant, except for Lap, which significantly but slightly increased from upstream to downstream in June. Temporal effect was significant for $\beta$Xyl and Lap activities (ANOVA2 $P = 0.0115$ and $P = 0.0006$, respectively), with June values higher than May values (Scheffé $P = 0.0135$ and $P = 0.0006$, respectively). The only observed pattern was the activity level of the enzymes, with Lap > $\beta$Glu > $\beta$Xyl.

*Aerobic respiration*

Aerobic respiration was described by MicroResp for bacteria and fungi (heterotrophic microorganisms) as well as the capacity to mineralise different carbon sources. A spatial pattern appears with a marked lowering (mean value: 40%) in the respiration at St Ennemond (Fig. 8), whatever the carbon source (e.g. carbohydrate, amino acids, acids). Finally, substrate-induced respiration is only weakly greater than biofilm respiration on natural DOC (0–15%), whatever the sites.

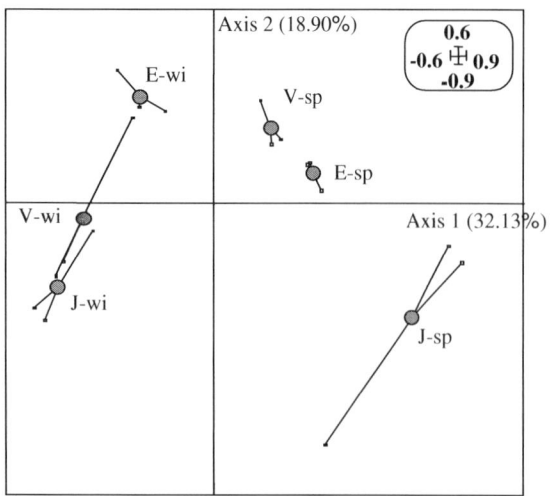

**Fig. 6** Correspondence analysis of the percentage contribution of each pigment of the three biofilm samples at the sampling areas (J, St Joseph; V, Versauds; E, St Ennemond) in spring (sp) and in winter (wi). ADE-4 Software Package (from Dorigo et al., 2007)

*Diuron mineralisation potential*

Diuron mineralisation potential increased significantly from the upstream to the downstream sampling stations (Fig. 8). After a short lag phase of about 10 days, downstream samples exhibited a high biodegradation potential. Diuron mineralisation was then fast (the mean mineralisation rate was about 1.56% per day) and reached a plateau (sixth week) with a mean value of 25% of the initially applied diuron. Conversely, diuron mineralisation obtained with the upstream epilithon was very limited (about 3.3%) and remained close to that observed in autoclaved samples (<4%, data not shown).

*Tolerance of periphyton (Table 2)*

Photosynthesis $EC_{50}$ values increased from upstream to downstream in both spring and winter, with values ranging from 19.71 to 42.23 and from 9.17 to 50.66 µg l$^{-1}$ of diuron, respectively. Except for St Ennemond, $EC_{50}$ values were higher in summer than in winter. At each season, the lowest $EC_{50}$ values (thus lowest tolerance to diuron) were recorded at the upstream area of St Joseph. Copper effects had the same pattern and tolerance of biofilm also increased from upstream to downstream.

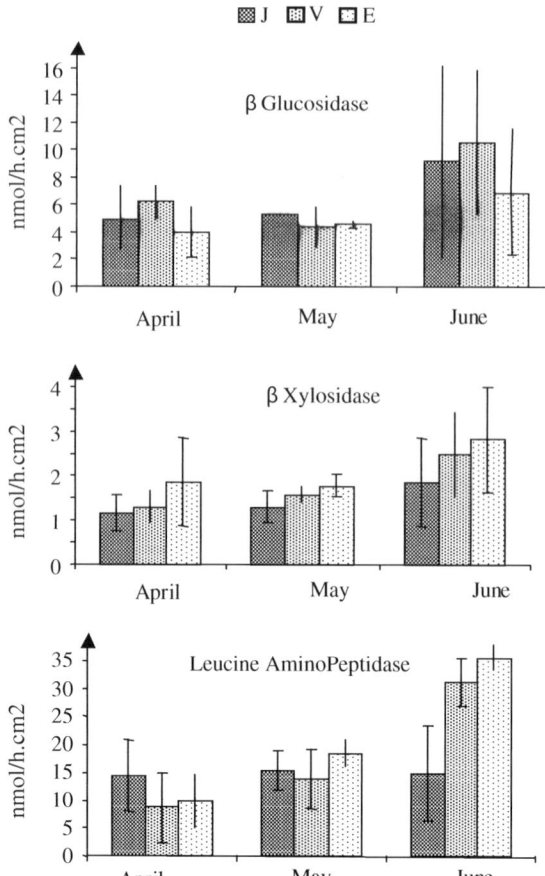

**Fig. 7** Exoenzyme activity gradient in early spring at the three sampling sites. J: St Joseph; V: Versauds; E: St Ennemond (mean values ± SD; $n = 3$)

## Discussion

Relationships between land use and water quality

The characterisation of land use in the Morcille drainage basin identified the possible main causes of the observed nature and dynamics of water quality. One cause is the preponderant use of land for wine growing, but with a very small presence at the first sub-basin (St Joseph). The identification of drainage ditches (respectively, 0.2, 33.4 and 97.1 km for the St Joseph, Versauds and St Ennemond sub-basins) shows the importance of run-off collection (Lagacherie et al., 2006) and its contribution to the pollutant flow entering the river Morcille. Indeed, the land is characterised by a sandy, modified granite shallow soil, easily eroded and poor in organic

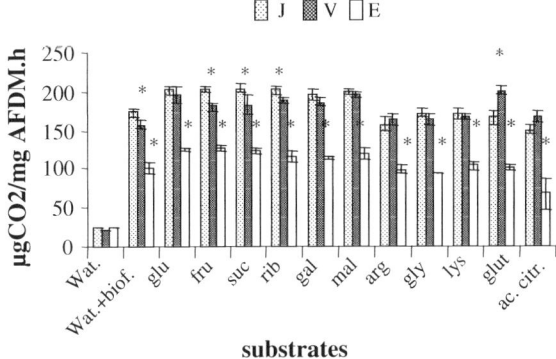

Fig. 8 Potential of carbon substrate mineralisation of biofilms sampled at the three sites. Mean values ± SD; *significant difference with St Joseph (reference site)—Kruskall–Wallis test, $n = 4$, $P = 0.05$. Wat: water alone (blank); Wat + biof: river water + biofilm; glu: glucose; fru: fructose; suc: sucinate; rib: ribose; gal: galactose; mal: maltose; arg: arginin; gly: glycine; glut: glutamic acid; ac citr: citric acid. J: St Joseph; V: Versauds; E: St Ennemond

material. Rainfall intensities as well as sloping relief favour rapid transfer by run-off. This sensitivity to rainfall effects results in high MES levels when the river is in flood (up to 1 g/l, Rabiet et al., 2008). In general, the rainfall pattern is of the 'continental' type, with frequent sudden rainstorms in spring and summer. There can be as many as 3–4 river floods during the summer (Rabiet et al., 2008). The drainage ditches and domestic wastewater flow directly into the river Morcille, shunting the wooded or grassy riverbank areas, whose important role in water self-purification is then limited at this site. This pattern of land use explains the upstream/downstream gradient in levels of both organic chemicals, metals (Cu and As) and nutrients. The surface area covered by farm buildings or dwellings is a good indicator of domestic pressure; its gradual increase along the three sub-basins (respectively, 0.05, 2.2 and 6.9 ha) probably results in the level increase of some organic and inorganic nutrients ($PO_4$ in particular), whatever the season.

These characteristics account for the flows of pollutants into the Morcille, their gradient and their seasonal pattern. The intensive use of pest-control chemicals in vineyards [mostly herbicides (such as diuron, isoproturon or diflufenicanil in early spring) and fungicides (such as tebuconazol, linuron or fenitrothion in spring and summer)] causes rising downstream contamination of the waterway with both organic residues and metals (essentially Cu and As from vineyard treatment). This contamination persists all year, but with a seasonal peak in the spring and part of the summer (generally late March to early August). Its variability is due to (i) differences in the use of herbicides (mostly early in the growing season) and fungicides (mostly in summer), with a predominance of the herbicide diuron and (ii) the pattern of rainfall and run-off, the intensity of which is an important controlling factor in pollutant flow. At peaks of flooding the concentrations of certain phytochemicals can reach several μg/l, or even tens of μg/l (Rabiet et al., 2008), thus profoundly modifying the conditions of exposure of aquatic organisms. As an example, the average concentration of diuron largely exceeds the European environmental quality standard (EQS) of 0.2 μg/l expressed as annual average.

Periphyton biomass, structure and diversity as bioindicators

*Global indicators*, such as periphytic ash-free dry mass or chlorophyll *a*, have been and are still frequently used to characterise the status of a biofilm or a plankton community in response to environmental changes (Rosemond et al., 2000). This global approach does not always respond clearly to levels of

Table 2 Sensitivity of periphyton to copper and diuron (expressed as $EC_{50}$ measured as primary productivity for periphyton samples from the three sampling sites on the river Morcille)

|  | St Joseph | Versauds | St Ennemond | Chemicals |
| --- | --- | --- | --- | --- |
| Survey 1 (winter) | 9.7 ± 3.8 | 20 ± 6.6 | 47 ± 8.4 | Diuron (μg/l$^{-1}$) |
| Survey 2 (summer) | 20 ± 6.5 | 38 ± 10.2 | 42 ± 15.8 | Diuron (μg/l$^{-1}$) |
| Survey 3 (summer) | 4.53 ± 1.54 | 7.8 ± 2.2 | – | Diuron (μg/l$^{-1}$) |
| Survey 4 (summer) | 32 ± 7.5 | 160 ± 40 | – | Copper (μM) |

Mean value ± SD ($n = 3$); –, non determined

nitrogen- and phosphorus-containing nutrients (Bernhard & Likens, 2004). This limitation is seen here for our data on the river Morcille (Fig. 4). Given the intra-site variability of periphytic biomass, and despite a tendency towards a fall in periphyton biomass, no clear-cut upstream/downstream spatial gradient is discernable, in spite of the permanent increase in $PO_4$ levels at Versauds and St Ennemond. Three hypotheses can be advanced: (i) an effect of reduced light (Villeneuve et al., 2010) due to canopy growth in summer, (ii) physical constraints (biofilm abrasion) due to the turbulences of water on small pebbles and (iii) the presence of diuron (and its breakdown product DCMU) and copper, both inhibitors of algal photosynthesis, which antagonise the growth stimulus provided by $PO_4$ (Guasch et al., 2007). These authors showed that $PO_4$ did not modify the inhibiting effect of atrazine (at 100 µg/l) on the periphyton biomass, and did not modify the tolerance of the biofilm to this herbicide. The biomass (expressed in AFDM or chlorophyll) does not therefore seem to be a very reliable indicator of the chemical quality of the ecosystem, especially as this depends also on hydraulic and light conditions (Battin et al., 2003a, b; Villeneuve, 2008). Accordingly, some authors have proposed growth rate rather than biomass as an indicator of eutrophication (Othoniel, 2007).

Most works on biofilms take into account algal communities and their taxonomic diversity. Because of their physiological characteristics, photoautotrophic biofilm communities (microalgae and cyanobacteria) and the taxa that compose them can be good indicators of environmental changes. They also represent a potential target for herbicide residues in the aquatic ecosystems. Herbicides acting on the photosynthetic system (such as Photosystem II) have already been identified as a cause of impairment of these communities (Guasch et al., 2003; Dorigo et al., 2004, 2007).

*Taxonomic approaches*, especially those based on diatoms or on microalgae, are regularly used. The predominance of diatoms in periphyton is a very common situation in lotic systems (Stevenson & Pan, 1999) and allows the design of a specific index. The BDI was originally designed to assess alterations in trophic status (Lenoir & Coste, 1996). For the river Morcille, as shown by Morin et al. (2010), the BDI values along the gradient for 1-month-old communities indicated a trophic pollution and changes in water quality classes, as defined by the European Water Framework Directive and with a good/moderate boundary: BDI = 14, for this national type and stream order between St Joseph (BDI = 14.7 ± 0.1, good ecological quality), Versauds (BDI = 13.8 ± 0.2, good to moderate) and St Ennemond (BDI = 13.1 ± 0.1, moderate). Morin et al. (2010) identified larger proportions of pollution-sensitive species at St Joseph (25.4%) than at Les Versauds (18.2%) or St Ennemond (15.9%). In addition to these approaches based on global indices, some of the species that preferentially developed at Les Versauds (especially in May) and St Ennemond have already been recorded under herbicide concentrations >5 µg/l (*Planothidium lanceolatum*, *Planothidium frequentissimum* and *Cocconeis placentula*); Pérès et al., 1996; S. Morin, pers. comm.). Only a few studies have addressed diatom sensitivity or tolerance to pesticides, but this work in favour of diatom use for the assessment of pesticide contamination has yielded important findings. Pérès et al. (1996) and Schmitt-Jansen & Altenburger (2005) found a remarkable decrease in diatom numbers under atrazine and isoproturon contamination. The shift in diatom community diversity observed in the river Morcille was concomitant with increased diuron-induced tolerance (see below), revealing that pesticide contamination was probably one of the major driving factors, even if the trophic level bioindicator BDI expressed the increasing OM and nutrient gradient in the Morcille (Morin et al., 2010).

*Community level finger printings* (*CLFPs*) are now widely used to assess the diversity of microbial communities and their responses to environmental changes. Fingerprinting techniques based on PCR–DGGE or on pigment abundances have often been used to describe the effect of pollutants (Dorigo et al., 2004; Boivin et al., 2007; Pesce et al., 2009b) or spatiotemporal changes (Lyautey et al., 2005), whatever the biases of these methods (see Dorigo et al., 2007; Marzorati et al., 2008). To obtain ecologically relevant information, fingerprint data are often transformed by statistical treatments such as principal component analysis or other multivariate statistical treatment, but some other straightforward processing methods can be used to obtain a relevant ecological interpretation of fingerprinting patterns (Marzorati et al., 2008). For the river Morcille, correspondence analysis performed on prokaryote and eukaryote fingerprints and on pigment analysis revealed a structured pattern depending on the season (Dorigo et al., 2007, 2009).

More specifically, correspondence analysis, relative to pigment analysis, separated the spring from the winter communities, and communities inhabiting less contaminated areas from those in more contaminated ones (Fig. 6). However, these differences were not linear because of in situ confounding factors (see above). DGGE analyses also allowed a differentiation of prokaryotic and eukaryotic communities, separating those inhabiting less contaminated sampling areas from more contaminated ones (Fig. 5). Thus, the two methods used to assess the structure of these communities (DGGE or HPLC) prove to be useful bioindicators of the ecological status of biofilms, allowing seasonal and spatial differentiations. In spring, the St Joseph reference area was differentiated from the two other sampling areas (i) by the structure of their prokaryotic, eukaryotic and photoautotrophic communities and (ii) by its level of pollution, which was significantly reduced in comparison with Versauds and St Ennemond (Fig. 3). In the same way, and only in winter, the St Joseph and Versauds sampling areas were more similar to each other than to St Ennemond, by both their level of pollution and the structure of their microbial communities. In addition, microcosm studies conducted elsewhere have also shown that environmentally realistic diuron exposure levels can affect aquatic bacterial diversity (Pesce et al., 2006, 2008; Ricart et al., 2009) and it appears that chronic long-term exposure (1 µg/l) can lead to changes in bacterial community structure, more than acute (14 µg/l) and short-term exposure, suggesting a progressive adaptation in microbial communities (Tlili et al., 2008).

Calculating an index such as Jaccard's similarity index yields additional, quantitative indications concerning the differentiation of sites which are therefore more easily exploitable as bioindicators. For example, summer monitoring of the diversity of biofilms grown on artificial supports has shown the low degree of similarity among communities of bacteria (16S) or algae (18S) in the three study sites and according to the season (Table 3) (Montuelle, unpublished data). This index can thus be considered as an indicator of disturbance, by quantifying the intensity of a change in composition of a microbenthic community.

However, these structural changes related to environmental variables or to the toxic contamination

**Table 3** Eukaryotes and prokaryotes similarity index (Jaccard Index) between three sampling sites during a summer survey

|  |  | St Joseph | Versauds | St Ennemond |
|---|---|---|---|---|
| 16 S rRNA | St Joseph | 100 | 23 | 15 |
|  | Versauds |  | 100 | 26.7 |
|  | St Ennemond |  |  | 100 |
| 18S rRNA | St Joseph | 100 | 13.9 | 23.4 |
|  | Versauds |  | 100 | 64.7 |
|  | St Ennemond |  |  | 100 |

present in the river Morcille should be associated with function changes to assess the disturbance of ecological (biochemical) processes and identify the relationship between structure and function.

Functional changes along the pollution gradient

Several functional variables were active on the river Morcille for both autotrophic communities (photosynthesis) and heterotrophic communities (biodegradation).

*Photosynthesis and PICT approach*

To verify that structural changes were related to pesticide contamination of the river Morcille, we applied the PICT concept proposed by Blanck et al. (1988) and used since by several authors (see Bérard et al. 2002, for a review). The PICT concept states that the tolerance of a community to a toxicant is related to the previous exposure of that community to the toxicant or to another toxicant so long as both belong to the same toxicant family (same chemical composition and/or similar mode of action). In our studies, diuron and Cu were taken, respectively, as models of organic and inorganic pesticides inhibiting photosynthetic organisms in biofilms. The most diuron- and Cu-sensitive communities were found in the upstream area at St Joseph, in both winter and spring. On the other hand, the most diuron-tolerant communities were found both in spring and in winter at the St Ennemond sampling area, which was also the most severely contaminated one. These results fit well with the predictions of the PICT concept and suggest that pesticide concentrations constitute a selective pressure on the photoautotrophic communities of the Morcille biofilms, resulting in changes in species composition and in pesticide tolerance. However, nutrients increased concomitantly with the in situ pesticide concentrations (Collectif, 2008), and they could also have driven structural and functional changes in eukaryotic and photoautotrophic communities (deNicola et al., 2006). In such a specific situation, the application of the PICT concept enables us to attribute observed changes to pesticides and not directly to nutrients. In the case of multiple toxic contaminations (diuron and copper), a co-tolerance effect may occur, with a risk of error in the estimation of causality of the effect observed in the biofilms. No co-tolerance between copper and a PSII-inhibiting herbicide (Irgarol) was observed on phytoplankton in Lake Leman (Bérard, unpublished data).

It should also be emphasised that P content could also modify the tolerance of biofilm to Cu (Guasch et al., 2004): an increase in P could induce an increase in the tolerance to copper and care must be taken to sample sites where P and Cu increase together (as in the Morcille River). However, a recent field survey performed in the river Morcille confirmed that despite the possible influence of identified co-varying variables (such as nitrates, conductivity and temperature), the main factor explaining spatio-temporal variation in diuron sensitivity within photoautotrophic biofilm communities was the mean in situ diuron exposure level during their colonisation period (Pesce et al., 2010).

*Biodegradation of pesticides*

The increase in tolerance is also accompanied by a greater capacity for the biodegradation of toxicants by biofilms. This type of adaptative response by heterotrophic communities to exposure to toxicants has not been much studied in aquatic environments (Toräng et al., 2003). In the river Morcille, an in situ microbial adaptation to diuron mineralisation following previous diuron exposure was clearly highlighted in aquatic biofilm communities (Fig. 9). For downstream epilithon, a short lag phase was observed, probably reflecting the time required by the diuron-degrading populations to acclimatise to the experimental conditions. Despite a non-negligible variation between replicates, biofilm samples collected at the downstream diuron-impacted station exhibited higher mineralisation percentages than the samples collected at the upstream station, thus evidencing an increasing

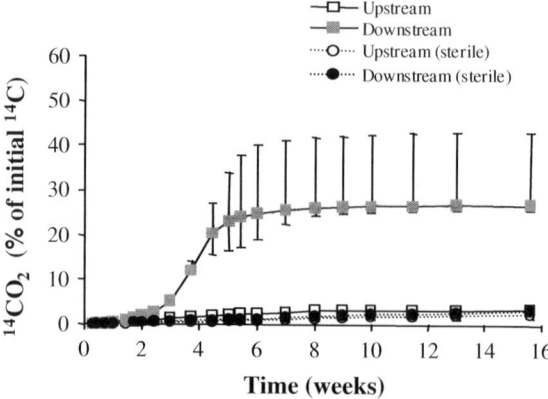

**Fig. 9** Biodegradation potential of $^{14}$C-diuron by periphyton in the river Morcille (modified from Pesce et al., 2009a, b)

natural breakdown potential of diuron along the contamination gradient (Pesce et al., 2009a). From a functional point of view, this potential is quite different from an estimation of the real in-field biodegradation processes of pesticides in biofilms. Given the short residence time of water in the river Morcille and the diffuse input in the watershed, most diuron residues are exported to the downstream rivers. However, this bioassay offers interesting perspectives for assessing the adaptation of heterotrophic microbial communities following pollutant exposure.

*Community level physiological profiles*

Parallel to changes in diversity, the heterotrophic degradation functions exerted by the biofilm also adapted along the gradient. They also provide an integrated response to overall modifications in chemical composition, and can be considered as reflecting the composition of the organic matter in the system (Sinsabaugh et al., 2002; Romani et al., 2004). The cellulases (e.g., glucosidase and xylosidase) that break down plant fibres and peptidases are the key enzymes involved, for example, in the leaf conversion of polymeric compounds into smaller molecules that can be assimilated by microorganisms (Sinsabaugh et al., 2002). No general adaptation pattern was detected in enzymes activity whether temporal or spatial because of a large variability in activity (up to 75%), which prevented characterisation of the increasing tendency of the mean activity values from upstream to downstream as statistically significant.

If increasing concentration in DOC and nutrients could explain such a tendency, many reasons can be offered to explain the high variability. Microbial extracellular enzymes and their activity respond to many environmental variables that can vary rapidly, such as temperature, dissolved oxygen, nycthemeral cycle, microbial biomass and organic matter content (Montuelle & Volat, 1998; Boshker & Cappenberg, 1998). Seasonal or temporal variations in extracellular enzymes activity are well documented (Chappel & Goulder, 1994; Harbott & Grace, 2005), as well as spatial variability (Romani & Sabater, 2000). As expounded by Chrost (1990), the regulation system in extracellular enzymes results from a balance between directly available substrates (low molecular weight molecules inhibit enzyme synthesis and expression), and substrates that are not directly bioavailable but are biodegradable (high molecular weight molecules stimulate enzyme synthesis and activity). Finally, toxicants such as pesticides or heavy metals could have impaired extracellular activity (Lopez et al., 2009; Hussain et al., 2009). On the river Morcille, it is then plausible that these antagonistic factors, linked to the chemistry of water, co-exist and account for the variability of these activities. However, the observed increase in LAP from upstream to downstream in April and May could be a consequence of the increasing housing in the watershed and its greater discharge of protein-rich wastewater.

The degradation of carbon-containing substrates by heterotrophic microbial communities is part of the self-purifying process, in particular pathways that produce $CO_2$ (or $CH_4$ in anaerobiosis), which is then exported from the aqueous ecosystem into the atmosphere. The study of the respiratory capacity of microbial communities based on patterns of use of carbon-containing substrates (SIR), has prompted the development of kits such as Biolog® (Garland, 1996) or MicroResp® (Campbell et al., 2003), which have been used to characterise the status of a community or as a bioindicator of the effect of toxicants (Rutgers et al., 1998; Boivin et al., 2007). The application of this last technique on the river Morcille showed a fall in the overall mineralisation capacity ($CO_2$ production) of downstream biofilms (St Ennemond) (Fig. 8), although DOC levels were higher. Two hypotheses can be advanced: (i) the DOC is less biodegradable downstream than upstream (unlikely, because some hydrolysis activity tends to increase from upstream to downstream) and (ii) the toxicants present modify the expression of the enzymes involved in the mineralisation of the carbon compounds as shown for soil microbial community (Hussain et al., 2009). Enhancement of SIR is likely associated with the high biodegradability level of the C source tested, more important than the natural DOC in the Morcille River. However, the observed lowering of biodegradation capacity reduces the self-purification capacity of the river.

Towards a microbial bioindication tool

The different research carried out on the LTER Morcille River allowed the testing of several parameters describing the periphyton and its different functional or structural responses to a specific land use and to a contamination gradient. Putting into practice periphytic microbial indicators to characterise the ecological state of aquatic environments has progressed greatly during recent years. Their sensitivity to pollution and their early response is now well identified (Sabater et al., 2007). However, much progress is needed to meet the indicator criteria defined by Dale & Beyeler (2001).

> Be easily measured; be sensitive to stresses on the system; respond to stress in a predictable manner; be anticipatory; predict changes that can be averted by management actions; be integrative; have a known response to disturbances, anthropogenic stresses and changes over times; have a low variability in response.

It is not reasonable to assume that only one biological indicator, whatever its integration level, could be enough to capture the complexity of an ecological system. Development of a multimetric index is essential. It should be composed of metrics for structure, diversity and functions (Eisman & Montuelle, 1999) to reflect the multiple dimensions of ecosystem services. In our work on the river Morcille, the use of these three categories of ecosystems response allowed us to take into account the different functions assumed by microbial communities and to identify sensitive biological indicators for ecosystem functions.

The complexity in the choice and in the identification of relevant microbial indicators is a consequence of the action of environmental factors that are likely to interfere with the toxicants actions and then to limit the identification of univocal causality relationships. For example, tolerance of biofilm community to one stressor can be modified by natural abiotic factors, such as light or flow speed (Guasch & Sabater, 1998; Villeneuve, 2008), chemical conditions favouring co-tolerance phenomena (Schmitt-Jansen et al., 2008) or changing exposure to toxicants (chronic or acute; Tlili et al., 2008). Biofilm internal parameters, such initial community composition and species interactions, could also control the biological response of biofilm microbial communities to a contaminant (Guasch et al., 1998).

However, community level approaches are a necessity, since single-species tests are not ecologically relevant. Specific methodology such as PICT may then serve as a diagnostic tool in chemical hazard assessment (McClellan et al., 2008). PICT methodology provides information on the sensitivity of a community to a toxicant, integrating in its response changes in species diversity and cell physiology. Applied to complex environmental contamination (in the field or experimentally tested in 'cosms'), the PICT method allows reducing uncertainty on the causes of an ecological impairment to be reduced. However, care should be taken of the possibility of co-tolerance to different substances (Schmitt et al., 2006).

The technique of dose–response bioassays used in the PICT method allows a rather easy screening of substances, even for emerging molecules such as pharmaceuticals (Franz et al., 2008). The validity domain of PICT methods could then be broadened to the calculation of Environmental Quality Standards (EQS) for the European Priority Substances to obtain more relevant threshold characterisation for risk assessment (McClellan et al., 2008).

New functional parameters suitable for PICT studies and for individualising the responses of autotrophic and heterotrophic communities will strengthen the use of PICT-based bioindicators (e.g. respiration, denitrification and nitrification). However, not all metabolic activities could be used to group CLPP and PICT (for example, extracellular activities seemed to be worse in situ bioindicators, in our case study), but research in this field should continue.

Finally, the analysis of biofilm diversity by the CLFP supplements the information given by the PICT method allowing characterisation of the specific changes of biofilm eukaryote and prokaryote communities exposed to contaminants. Techniques such as DGGE fingerprinting used to assess the structure of eukaryotes and eukaryote communities on the river Morcille have appeared globally in good agreement with pigment analysis and have discriminated the pollution gradient (Dorigo et al., 2007, 2009). In the same way, more classical taxonomic approaches have given relevant insights into the effect of chemicals (Hill et al., 2000; Morin et al., 2010). Other fingerprinting methods (e.g. ARISA and t-RFLP) are also well adapted to such analysis; more powerful molecular methods (sequencing) seem still to be excluded for the construction of bioindication tools because they do not fit with the criteria underlined by Dale & Beyeler (2001). However, it is likely that, in the future, microarray (taxonomic and functional) techniques will give a new insight into bioindication.

The coupling of PICT–CLPPs–CLFPs constitutes a multimetric assessment tool and allows a thorough assessment of biofilm response (function, diversity and tolerance level) to contaminants. The complementarity of such methods allowed in situ characterisation of the individual effects of pollutants in a cocktail of contaminants in a vineyard-dominated watershed. Such multimetric assessment tools give a comprehensive overview of ecosystem impairments and could be applied to different anthropogenic pressures (urban, peri-urban and agricultural). In the perspective of goals defined by the European WFD, our next step will be the use of PICT–CLPPs–CLFPs for studying community resilience and characterising the trajectory of aquatic ecosystem restoration.

**Acknowledgements** The authors thank the two anonymous reviewers for improving the manuscript. They also thank Marjorie Maréchal for Microresp® assays; Laurence Blanc for extracellular enzymes measurements; Bernard Motte for field surveys; Christelle Margoum, Josiane Gahou, Céline Guillemain and Marina Coquery for chemical analysis and for the Morcille river database. The Ardières-Morcille experimental watershed is supported by the LTER Rhône Basin (ZABR). English was checked by ATT Scientific and Technical Translation.

## References

Admiraal, W., H. Blanck, M. Buckert De Jong, H. Guasch, N. Ivorra, V. Lehman, B. A. H. Nyström, M. Paulsson & S. Sabater, 1999. Short term toxicity of zinc to microbenthic algae and bacteria in a metal polluted stream. Water Research 33: 1989–1996.

AFNOR, 2004. Détermination de l'Indice Biotique Global Normalisé (IBGN). NF T 90-350, Association française de normalisation.

Barranguet, C., F. P. van den Ende, M. Rutgers, A. M. Breure, M. Greijdanus, J. J. Sinke & W. Admiraal, 2003. Copper-induced modifications of the trophic relations in riverine algalbacterial biofilms. Environmental Toxicology and Chemistry 22: 1340–1349.

Battin, T. J., L. A. Kaplan, D. Newbold & C. M. E. Hansen, 2003a. Contributions of microbial biofilms to ecosystem processes in stream mesocosms. Nature 426: 439–442.

Battin, T. J., L. A. Kaplan, D. Newbold, J. D. Cheng & C. M. E. Hansen, 2003b. Effects of current velocity on the nascent architecture of stream microbial biofilms. Applied and Environmental Microbiology 69: 5443–5452.

Bérard, A., U. Dorigo, J. F. Humbert, C. Leboulanger & F. Seguin, 2002. La méthode PICT (Pollution-Induced Community Tolerance) appliquée aux communautés algales. Intérêt comme outil de diagnose et d'évaluation du risque écotoxicologique en milieu aquatique. Annales de Limnologie 38: 247–261.

Bérard, A., U. Dorigo, I. Mercier, K. Becker van-Slooten & C. Leboulanger, 2003. Comparison of the ecotoxicological impact of triazines Irgarol 1051 and atrazine on microalgal cultures and natural microalgal communities in Lake Geneva. Chemosphere 53: 935–944.

Bernhard, E. & G. E. Likens, 2004. Controls on periphyton biomass in heterotrophic streams. Freshwater Biology 49: 14–27.

Blanck, H., S. A. Wängberg & S. Molander, 1988. Pollution-induced community tolerance—a new ecotoxicological tool. In Cairns, J. & J. R. Pratt (eds), Functional Testing of Aquatic Biota for Estimating Hazards of Chemicals. ASTM STP, Philadelphia: 219–230.

Boivin, M. E. Y., B. Massieux, A. M. Breure, F. P. van den Ende, G. D. Greve, M. Rutgers & W. Admiraal, 2005. Effects of copper and temperature on aquatic bacterial communities. Aquatic Toxicology 71: 345–356.

Boivin, M. E. Y., G. D. Greve, J. V. Garcia-Meza, B. Massieux, W. Sprenger, M. H. S. Kraak, A. M. Breure, M. Rutgers & W. Admiraal, 2007. Algal-bacterial interactions in metal contaminated floodplain sediments. Environmental Pollution 145(3): 884–894.

Boshker, H. T. S. & T. E. Cappenberg, 1998. Patterns of extracellular enzyme activities in littoral sediments of Lake Gooimeer, The Netherlands. FEMS Microbial Ecology 25(1): 79–86.

Campbell, C. D., S. J. Chapman, C. M. Cameron, M. S. Davidson & J. M. Potts, 2003. A rapid microtiter plate method to measure carbon dioxide evolved from carbon substrate amendments so as to determine the physiological profiles of soil microbial communities by using whole soil. Applied and Environmental Microbiology 69(6): 3593–3599.

Chappel, K. R. & R. Goulder, 1994. Enzymes as river pollutants and the response of native epilithic extracellular-enzyme activity. Environmental Pollution 86: 161–169.

Chrost, R. J., 1990. Microbial ectoenzymes in aquatic environments. In Overbeck, J. (ed.), Aquatic Microbial Ecology, Biochemical and Molecular Approach. Brock/Springer Series in Contemporary BioScience.

Collectif, 2008. Relations entre structures paysagères, transferts hydriques et flux géochimiques, état écologique des milieux aquatiques, Rapport final du programme ECOGER-Papier, INRA-Cemagref, Coord: Grimaldi, C. & B. Montuelle: 33 pp.

Dale, V. H. & H. C. Beyeler, 2001. Challenges in the development and use of biological indicators. Ecological Indicators 1: 3–10.

DCE, 2005. Définition du bon état des eaux, constitution des nouveaux référentiels et des modalités d'évaluation de l'état des eaux douces de surface. Ministère de l'Ecologie et du Développement Durable (ed.): 17 pp.

DeLorenzo, M. E., P. Scott & P. E. Ross, 2001. Toxicity of pesticides to aquatic microorganisms: a review. Environmental Toxicology and Chemistry 20(1): 84–98.

deNicola, D. M., E. de Eyto, A. Wemaere & K. Irvine, 2006. Periphyton response to nutrient addition in 3 lakes of different benthic productivity. Journal of North American Benthological Society 25: 616–631.

Dorigo, U., X. Bourrain, A. Berard & C. Leboulanger, 2004. Seasonal changes in the sensitivity of river microalgae to atrazine and isoproturon along a contamination gradient. Science of Total Environment 318: 101–104.

Dorigo, U., C. Leboulanger, A. Bérard, A. Bouchez, J. F. Humbert & B. Montuelle, 2007. Lotic biofilm community structure and pesticide tolerance along a contamination gradient in a vineyard area. Aquatic Microbial Ecology 50: 91–102.

Dorigo, U., M. Lefranc, C. Leboulanger, B. Montuelle & J. F. Humbert, 2009. Influence of sampling strategy on the assessment of the impact of pesticides on periphytic microbial communities in a small river. FEMS Microbial Ecology 67: 491 501.

Eisman, F. & B. Montuelle, 1999. Microbial methods for contaminants effects assessment in sediment. Reviews of Environmental Contamination and Toxicology 159: 41–93.

Franz, S., R. Altenburger, H. Heilmeier & M. Schmitt-Janssen, 2008. What contributes to the sensitivity of microalgae to triclosan? Aquatic Toxicology 90: 102–108.

Fuerhacker, M., 2009. EU Water Framework Directive and Stockholm Convention: can we reach the targets for priority substances and persistent organic pollutants? Environmental Science and Pollution Research 16: 92–97.

Garland, J. L., 1996. Analytical approaches to the characterization of samples of microbial communities using patterns of potential C source utilization. Soil Biology and Biochemistry 28: 213–221.

Gold, C., A. Feurtet-Mazel, M. Coste & A. Boudou, 2003. Impacts of Cd and Zn on the development of periphytic diatom communities in artificial streams located along a river pollution gradient. Archives of Environmental Contamination and Toxicology 44: 189–197.

Guasch, H. & S. Sabater, 1998. Light history influences the sensitivity to atrazine in periphytic algae. Journal of Phycology 34: 233–241.

Guasch, H., N. Ivorra, V. Lehmann, M. Paulsson, M. Real & S. Sabater, 1998. Community composition and sensitivity of periphyton to atrazine in flowing waters: the role of environmental factors. Journal of Applied Phycology 10: 203–213.

Guasch, H., W. Admiraal & S. Sabater, 2003. Contrasting effects of organic and inorganic toxicants on freshwater periphyton. Aquatic Toxicology 64: 165–175.

Guasch, H., E. Navarro, A. Serra & S. Sabater, 2004. Phosphate limitation influences the sensitivity to copper in periphytic algae. Freshwater Biology 49: 463–473.

Guasch, H., V. Lehmann, B. van Beusekom, S. Sabater & W. Admiraal, 2007. Influence of phosphate on the response of periphyton to atrazine exposure. Archives of Environmental Contamination and Toxicity 52: 32–37.

Harbott, E. L. & M. R. Grace, 2005. Extracellular enzyme response to bioavailability of dissolved organic C in streams of varying catchment urbanization. Journal of the North American Benthological Society 24: 588–601.

Haury, J., M. C. Peltre, M. Trémolières, J. Barbe, G. Thiébaut, I. Bernez, H. Daniel, P. Chatenet, G. Haan-Archipof, S. Muller, A. Dutartre, C. Laplace-Treyture, A. Cazaubon & E. Lambert-Servien, 2006. A new method to assess water trophy and organic pollution—the Macrophytes Biological Index for Rivers (IBMR): its application to different types of river and pollution. Hydrobiologia 570: 153–158.

Hill, B. H., A. T. Herlihy, P. R. Kaufmann, R. J. Stevenson, F. H. Mc Cormick & C. Burch Johnson, 2000. Use of periphyton assemblage data as an index of biotic integrity. Journal of North American Benthological Society 19: 50–67.

Hussain, S., T. Siddique, M. Saalem, M. Arshad & A. Khalid, 2009. Impact of pesticides on soil microbial diversity, enzymes and biochemical reactions. Advances in Agronomy 102: 159–200.

Lagacherie, P., O. Diot, N. Domange, V. Gouy, C. Floure, C. Kao, R. Moussa, J. M. Robbez-Masson & V. Szleper, 2006. An indicator approach for describing the spatial variability of artificial stream networks in regard with herbicide pollution in cultivated watersheds. Ecological Indicators 6: 265–279.

Landry, D., S. Dousset & F. Andreux, 2004. Laboratory leaching studies of oryzalin and diuron through three undisturbed vineyard soil columns. Chemosphere 54: 734–742.

Lenoir, A. & M. Coste, 1996. Development of a practical diatom index of overall water quality applicable to the French national water Board network. In Whitton, B. A. & E. Rott (eds), Use of Algae for Monitoring Rivers II. Studia Student. G.m.b.H, Innsbruck, Austria: 29–43.

Lopez, L., C. Pozo, B. Rodelas, C. Calvo & J. Gonzalez-Lopez, 2009. Influence of pesticides and herbicides presence on phosphatase activity and selected bacterial microbiota of a natural lake system. Ecotoxicology 15: 487–493.

Lyautey, E., S. Tessier, J. Y. Charcosset, J. L. Rols & F. Garabetian, 2003. Bacterial diversity of epilithic biofilm assemblages of an anthropised river section, assesed by DGGE analysis of a 16S rDNA fragment. Aquatic Microbial Ecology 33: 217–224.

Lyautey, E., C. R. Jackson, J. Cayrou, J. L. J. Rols & F. Garabetian, 2005. Bacterial community succession in natural river biofilm assemblages. Microbial Ecology 50: 589–601.

Marzorati, M., L. Wittebolle, N. Boon, D. Dofonchio & W. Verstraete, 2008. How to get more out of molecular fingerprints: practical tools for microbial ecology. Environmental Microbiology 10: 1571–1581.

McClellan, K., R. Altenburger & M. Schmitt-Janssen, 2008. Pollution-induced community tolerance as a measure of species interaction in toxicity assessment. Journal of Applied Ecology 45: 1514–1522.

Montuelle, B. & B. Volat, 1998. Impact of wastewater treatment plant discharge on enzyme activity in sediments. Ecotoxicology and Environmental Safety 40: 154–159.

Morin, S., M. Vivas-Nogues, T. T. Duong, A. Boudou, M. Coste & F. Delmas, 2007. Dynamics of benthic diatom colonization in a cadmium/zinc-polluted river (Riou-Mort, France). Fundamental and Applied Limnology 168: 179–187.

Morin, S., S. Pesce, A. Tlili, M. Coste & B. Montuelle, 2010. Recovery potential of periphytic communities in a river impacted by a vineyard watershed. Ecological Indicators 10: 419–426.

Othoniel, C., 2007. La croissance du biofilm photosynthétique: un indicateur du statut trophique des rivières? PhD Thesis, University of Bordeaux I: 245 pp.

Pérès, F., D. Florin, T. Grollier, A. Feurtet-Mazel, M. Coste, F. Ribeyre, M. Ricard & A. Boudou, 1996. Effects of the phenylurea herbicide isoproturon on periphytic diatom communities in freshwater indoor microcosm. Environmental Pollution 94: 141–152.

Pesce, S., C. Fajon, C. Bardot, F. Bonnemoy, C. Portelli & J. Bohatier, 2006. Effects of the phenylurea herbicide diuron on natural riverine microbial communities in an experimental study. Aquatic Toxicology 78: 303–314.

Pesce, S., C. Bardot, A. C. Lehours, I. Batisson, J. Bohatier & C. Fajon, 2008. Effects of diuron in microcosms on natural riverine bacterial community composition: new insight into phylogenetic approaches using PCR-TTGE analysis. Aquatic Sciences 70: 410–418.

Pesce, S., F. Martin-Laurent, N. Rouard & B. Montuelle, 2009a. Potential for microbial diuron mineralisation in a small wine-growing watershed: from treated plots to lotic receiver hydrosystem. Pest Management Science 65: 651–657.

Pesce, S., I. Batisson, C. Bardot, C. Fajon, C. Portelli, B. Montuelle & J. Bohatier, 2009b. Response of spring and summer riverine microbial communities following glyphosate exposure. Ecotoxicology and Environmental Safety 72: 1905–1912.

Pesce, S., C. Margoum & B. Montuelle, 2010. In situ relationships between spatio-temporal variations in diuron concentrations and phototrophic biofilm tolerance in a contaminated river. Water Research. doi:10.1016/j.watres.2009.11.053.

Rabiet, M., C. Margoum, V. Gouy, N. Carluer & M. Coquery, 2008. Transfert des pesticides et métaux dans un petit bassin versant viticole. Etude préliminaire de l'influence des conditions hydrologiques sur le transport de ces contaminants. Ingénieries EAT«Azote, phosphore et pesticides: stratégies et perspectives de réduction des flux»: 65–76.

Rabiet, M., C. Margoum, V. Gouy, N. Carluer & M. Coquery, 2010. Assessing pesticide concentrations and fluxes in the stream of a small vineyard catchment—effect of sampling frequency. Environmental Pollution 158: 737–748.

Ricart, M., H. Guasch, D. Barcelo, A. Geiszinger, M. Lopez de Alda, A. M. Romani, G. Vidal, M. Villagras & S. Sabater,

2009. Effects of low concentrations of the phenylurea herbicide diuron on biofilm algae and bacteria. Chemosphere 76: 1392–1401.

Rier, S. T. & R. J. Stevenson, 2002. Effects of light, dissolved organic carbon, and inorganic nutrients on the relationship between algae and heterotrophic bacteria in stream periphyton. Hydrobiologia 489: 179–194.

Romani, A. M. & S. Sabater, 2000. Variability of heterotrophic activity in Mediterranean stream biofilms: a multivariate analysis of physical–chemical and biological factors. Aquatic Sciences 6: 205–215.

Romani, A. M., H. Guasch, I. Munoz, J. Ruana, E. Vilalta, T. Schwartz, F. Emtlazi & S. Sabater, 2004. Biofilm structure and function and possible implications for riverine DOC dynamics. Microbial Ecology 47: 316–328.

Rosemond, A. D., P. Mulholland & S. Brawley, 2000. Seasonally shifting limitation of stream periphyton: response of algal populations and assemblage biomass and productivity to variation in light, nutrients and herbivores. Canadian Journal of Fisheries and Aquatic Science 57: 66–75.

Rutgers, M., I. M. Van't Verlaat, B. Wind, L. Posthuma & A. M. Breure, 1998. Rapid method for assessing pollution-induced community tolerance in contaminated soil. Environmental Toxicology and Chemistry 17: 2210–2213.

Sabater, S., 2000. Diatom communities are indicators of environmental stress in the Guadiamar river, S-W Spain, following a major mine tailings spill. Journal of Applied Phycology 12: 113–124.

Sabater, S., H. Guasch, M. Ricart, A. Romani, G. Vidal, C. Klünder & M. Schmitt-Jansen, 2007. Monitoring the effect of chemical on biological communities: the biofilm as an interface. Analytical and Bioanalytical Chemistry 387: 1425–1434.

Schmitt, H., B. Martinali, P. Van Beelen & W. Seinen, 2006. On the limits of toxicant-induced tolerance testing: co tolerance and response variation of antibiotic effects. Environmental Toxicology and Chemistry 25: 1961–1968.

Schmitt-Jansen, M. & R. Altenburger, 2005. Community-level microalgal toxicity assessment by multiwavelength-excitation PAM fluorometry. Aquatic Toxicology 86: 49–58.

Schmitt-Jansen, M., U. Veit, G. Dudel & R. Altenburger, 2008. An ecological perspective in aquatic ecotoxicology: approaches and challenges. Basic and Applied Ecology 9: 337–345.

Sinsabaugh, R. L., M. M. Carreiro & D. A. Repert, 2002. Allocation of extracellular enzymatic activity in relation to litter composition, N deposition, and mass loss. Biogeochemistry 60: 1–24.

Stevenson, R. J. & Y. P. Pan, 1999. Assessing environmental conditions in rivers and streams with diatoms. In Stoermer, E. F. & J. P. Smol (eds), The Diatoms: Applications for the Environmental and Earth Sciences. Cambridge University Press, UK: 11–40.

Teissier, S. & M. Torre, 2002. Simultaneous assessment of nitrification and denitrification on freshwater epilithic biofilms by acetylene block method. Water Research 36: 3803–3811.

Tlili, A., U. Dorigo, B. Montuelle, C. Margoum, N. Carluer, V. Gouy, A. Bouchez & A. Bérard, 2008. Responses of chronically contaminated biofilms to short pulses of diuron. An experimental study simulating flooding events in a small river. Aquatic Toxicology 87: 252–263.

Toräng, L., N. Nyholm & H. J. Albrechtsen, 2003. Shifts in biodegradation kinetics of the herbicide MCPP and 2, 4-D at low concentrations in aerobic aquifer materials. Environmental Science and Technology 37: 3095–3103.

Villeneuve, A., 2008. Effets conjoints de facteurs physiques et chimiques sur la structure et la composition du périphyton: une approche multi échelle, PhD Thesis, U. de Savoie: 223 pp.

Villeneuve, A., B. Montuelle & A. Bouchez, 2010. Effect of minor changes in light intensity, current velocity and turbulence on the structure and function of the periphyton. Aquatic Sciences 72: 33–44.

Vu, S. H., S. Ishihara & K. Watanabe, 2006. Exposure risk assessment and evaluation of the best management practice for controlling pesticide runoff from paddy fields. Part 1: paddy watershed monitoring. Pesticides Management Science 62: 1193–1206.

GLOBAL CHANGE AND RIVER ECOSYSTEMS

# Discharge and the response of biofilms to metal exposure in Mediterranean rivers

Helena Guasch · Güluzar Atli · Berta Bonet · Natàlia Corcoll · Manel Leira · Alexandra Serra

Received: 28 August 2009/Accepted: 22 January 2010/Published online: 25 February 2010
© Springer Science+Business Media B.V. 2010

**Abstract** The expected response of fluvial biofilms to the environment and metal pollution prevailing under different discharge conditions was investigated. The relationship between inter-annual hydrological variability and metal concentration in water and sediments was explored in Mediterranean rivers (Catalonia, NE Spain) affected by low but chronic metal pollution, using monitoring data provided by the Catalan Water Agency (ACA). During the period investigated (2000–2006), metal pollution was characterized by low water concentrations and high concentrations in sediments. The most consistent pattern was observed for sediment cadmium (Cd) concentrations, showing a positive relationship with annual discharge, reaching values of environmental concern (above ecotoxicological benchmarks). A different pattern was observed for Cu, Zn, and As increasing with flow in some sites and decreasing in others. While Cd seems to proceed from diffuse sources being washed by surface runoff, Zn, Pb, and As may proceed from either diffuse or point-sources in the different river sites investigated. The relevance of diffuse metal pollution in the area of study indicates that polluted landfills runoff might be an important source of metals causing repetitive pulses of high metal concentration in the receiving water courses. The experimental results presented demonstrate that metal effects in fluvial biofilms may be accumulative, increasing the toxicity after repetitive pulse exposures. Since draughts and extreme rain events are expected to increase at higher latitudes due to global change, the sources of metal pollution, its final concentration and potential effects on the fluvial ecosystem may also change following the patterns expected for human-impacted Mediterranean rivers.

**Keywords** Antioxidant enzyme activity · Discharge · Fluvial biofilm · Metal · Toxicity · Water scarcity

Guest editors: R. J. Stevenson, S. Sabater / Global Change and River Ecosystems – Implications for Structure, Function and Ecosystem Services

H. Guasch (✉) · B. Bonet · N. Corcoll · A. Serra
Institute of Aquatic Ecology, Universitat de Girona,
Campus de Montilivi, 17071 Girona, Spain
e-mail: helena.guasch@udg.es

G. Atli
Department of Biology, Faculty of Sciences and Letters, University of Çukurova, 01330 Adana, Turkey

M. Leira
Faculty of Sciences, University of A Coruña, Campus da Zapateira, 15071 A Coruña, Spain

## Introduction

Human activity is one of the major causes of elevated concentrations of metals in fluvial ecosystems causing a great concern over potential toxicity and trophic transfer. Metal concentrations are very variable in time and space depending on the source of pollution (diffuse or point-source), the hydrological regime and the processes affecting their transfer from the water phase to other compartments. Furthermore, it is expected that below average surface flow conditions, will also change the fate and effects of metals at ecosystem scales (Guasch et al., 2009b). Based on recent modeling of the fate and transport of metals in streams (Caruso et al., 2008), higher metal concentrations are predicted during low flow (mostly for Zn). This tendency is common in mining areas where metal inputs come from groundwater. In these cases, metal concentration is lowest at highest discharges due to dilution (Audry et al., 2004; Bambic et al., 2006; Armitage et al., 2007). The influence of flow in sites affected by low but chronic metal pollution (i.e., influenced by urban and agricultural activities) has been poorly investigated (i.e., Brown & Peake, 2006). While the differences in metal concentration between the dry and rainy season described by Fianko et al. (2007) were attributed to dilution, other studies report no linkage between discharge and metal pollution (Benson & Etesin, 2008).

The duration and frequency of increased metal concentration episodes is of great relevance since metal bioaccumulation and toxicity are strongly influenced by the time and frequency of exposure (Meylan et al., 2003). The expected linkages between fluvial hydrology and metal exposure of fluvial biofilm communities were addressed in Guasch et al. (2009b). It was indicated that point-sources of metal pollution might cause chronic and variable metal exposure depending on flow conditions (of lower concentration under high-flow conditions due to dilution). On the other hand, diffuse sources of metal pollution such as the urban runoff would cause intermittent metal exposures directly linked with rainfall episodes. It was also concluded that chronic exposure would lead to community adaptation and a decrease in sensitivity, whereas metal toxicity would be maximum if a non-adapted community was suddenly exposed to peak metal concentrations. Community responses to intermittent metal exposures were not directly addressed in this review article.

Reinert et al. (2002) indicate that the long-term effects on non-target organisms of intermittent pesticide exposure can be a function of the damage sustained during exposure, the capacity of the organisms to recover, and the duration of the recovery period between pulses. Furthermore, sequential exposure can lead to an increased or decreased effect during each subsequent exposure, depending on the mode of action of the toxicant, the resilience of the biotic community, and its adaptation capacity (Macinnis-Ng & Ralph, 2002; Hoang & Klaine, 2008). Results of previous studies (Serra et al., 2009), demonstrated that Cu accumulation kinetics and toxicity differed among fluvial biofilms with different Cu-exposure history. Biofilms that had been continuously exposed to Cu, differed from those unexposed and also from those exposed to short Cu-pulses in their species composition (due to the replacement of sensitive algal classes by tolerant ones) and also on their metal content, several orders of magnitude higher. The non pre-exposed and Cu-pulsed communities were more sensitive to Cu than the chronically exposed community showing a slight decrease in the photosynthetic efficiency after the short exposure to higher Cu concentration (Serra et al., 2009). In this study, photosynthesis inhibition was slightly higher in the Cu-pulsed than in the non pre-exposed community indicating that the pulses may enhance Cu toxicity.

Several authors (Pinto et al., 2003; Torres et al., 2008) have proposed the study of "signals of distress" at molecular level as toxicity biomarkers. Studies performed on algal cultures have already shown the sensibility of several enzymatic biomarkers to the presence metals (Tripathi et al., 2006). Antioxidant enzyme activities (AEA) such as superoxide dismutase (SOD), catalase (CAT), ascorbate peroxidase (APX), and glutathione $S$-transferase (GST) activity are of great importance in oxidative stress to cope with free radicals that lead to several disturbances (Geoffroy et al., 2004). Elevated Cu levels induce oxidative stress by generating reactive oxygen species (ROS), such as hydrogen peroxide, superoxide radical, singlet oxygen, and hydroxyl radical, via Haber–Weiss as well as Fenton reactions that can oxidize proteins, lipids and nucleic acids. This often leads to the cell structure being damaged or even cell death (Tripathi & Gaur, 2004; Dewez et al., 2005). AEA characterization may provide early warning systems of detection of toxicity on autotrophic communities at lower exposure time and/or dose than other classical endpoints such as photosynthesis or algal growth (Sabater et al., 2007).

The main objective of this article is to present the expected response of the biota (fluvial biofilms) to the metal pollution prevailing under different discharge conditions. In order to reach this objective the following specific objectives were addressed:

1. To explore the relationship between inter-annual hydrological variability and metal concentration in water and sediments in Mediterranean watersheds (Catalonia, NE Spain) affected by low but chronic metal pollution.
2. To present a specific example dealing with the response of fluvial biofilms to repetitive Cu-pulses using AEA as early warning systems of toxicity detection.
3. To provide a conceptual framework about the influence of temporal dynamics of metal exposure on metal accumulation and toxicity in fluvial biofilms.

In order to reach the first specific objective, monitoring data provided by the Catalan Water Agency (ACA) were selected in order to cover a large environmental gradient and temporal scale (from 2000 to 2006).

In an experimental study, changes in the defense capacity of biofilms were used to evaluate the effects of several pulses of Cu. This complemented a series of investigations dealing with the significance of the time of exposure on the fate and effects of Cu on fluvial biofilms (Serra & Guasch, 2009; Serra et al., 2009).

## Materials and methods

Monitoring

Nine sites located in seven watersheds were included in this study (Table 1). The physical and chemical characterization including nutrient content, water metal concentration, sediment metal concentration, and flow was performed at each site. Furthermore, the Catalan River basins were characterized according to their typology and their impact degree. Since sediment metal concentration was analyzed once a year, average or integrated annual values of the rest of variable were included for comparison. Temporal variability was based on the inter-annual variability over 7 years (2000–2006). Conductivity was determinate by electrical conductivity (ISO 7888:1985), pH by continuous instrument, dissolved oxygen by electrochemical probe method (ISO 5814-1990). Ammonium by flow analysis (continuous flow and flow injection analysis) and spectrometric detection (ISO 11732:2005) and phosphate was determinate by ammonium molybdate spectrometric method (ISO 6878:2004). Metal concentration in water was analyzed by inductively coupled plasma mass spectrometry (ICP-MS) and metal concentration in sediments was determinate by inductively coupled plasma atomic emission spectroscopy (ICP-AS). Flow was obtained from the Catalan network of gauging stations.

*Statistical analysis*

Non parametric correlation analysis (Spearman's rho) was carried out between the relative flow recorded at each sampling station and the metal concentration in water and sediment. The environmental variables (physical, chemical and heavy metal concentration in water and sediments) along the study period were obtained from the Catalan Water Agency (ACA). Some of the variables available were selected on the basis of their relevance for the distribution and abundance of the biofilm community. The variables used were copper, lead, cadmium, arsenic, mercury, chromium, nickel, and zinc concentration in the sediment, and zinc in the water. As metal content and dynamics were likely to be substantially affected by time, the year in which the data were collected was included as independent variables in partial correlation analyses to describe the linear relationship between the variables while controlling for the effects of temporal variations in the metal content in the streams.

Significant differences in metal concentration in streams between flow conditions were examined using one way ANOVA (Winer, 1971). Two possible flow conditions were considered as factors in ANOVA, either high or low flow. Each site was classified in either of both classes according to their deviation from the standard condition. Deviations were calculated as anomalies by subtracting the mean from each observation, then dividing by the standard deviation. The contribution of spatial and temporal random effects to the variance of the sediment metal concentration was assessed by variance components procedure. Maximum likelihood (ML) method was used in the Variance Components procedure. The maximum likelihood method accounts for the

Table 1 Summary of river sites

| River | Sampling site | UTM's | Typology | Impact degree |
|---|---|---|---|---|
| Francolí | Tarragona | 351987X, 4557363Y | Karst feed river | High |
| Foix | Castellet i la Gornal | 385673 X, 4569332Y | Lowland Mediterranean river | High |
| Llobregat | a-Abrera | 409999X, 4595622Y | Large watercourse | High |
|  | b-St.Joan Despí | 416170X, 4588578Y | Large watercourse | High |
| Muga | Castelló d'Empúries | 530429X, 4680997Y | Lowland Mediterranean river | Intermediate |
| Fluvià | L'Armentera | 500869X, 4668447Y | Lowland Mediterranean river | Intermediate |
| Tordera | Fogars de la Selva | 474637X, 4621141Y | Lowland Mediterranean river | Intermediate |
| Besós | a-Montcada i Reixac | 438807X, 45999056Y | Lowland Mediterranean river | Intermediate |
|  | b-Montornès del Vallès | 432124X, 4593424Y | Lowland Mediterranean river | High |

Name of the river (river); name of each sampling site (sampling site) and geographic location (UTM's). Typology according to the Catalan Water Agency (ACA) classification and impact degree according to the Water Framework Directive

temporal and spatial dependence of the observations that characterize our dataset. Using the Variance Components procedure, the year effect's contribution to the random variation in a given variable can be estimated. Multiple comparisons between means were analyzed in all the cases with a Tukey HSD (Honest significant difference) test (Winer, 1971). Statistical analyses were performed using SPSS for Windows (version 13.0; SPSS Inc., Illinois).

Influence of Cu pulses on the structure and function of biofilms

Fluvial biofilms were cultivated and exposed to copper under controlled conditions. The experiments were carried out in an indoor channel system consisting of six Perspex channels (each 170 cm long and 9 cm wide) as described in Serra et al. (2009). Briefly, water was supplied from 10 l carboys located at the end of each channel and was recirculated at a rate of 1 l min$^{-1}$ through centrifuge pumps. Light was provided by halogen lamps (80–100 μmol m$^{-2}$ s$^{-1}$) with a 12 h light/12 h dark cycle and the temperature was kept between 19 and 20°C using a cooling bath. Two consecutive colonization experiments were performed: one with no copper added (No-Cu); and the other with several Cu pulses (referred to as Cu-pulsed). In each colonization experiment, biofilms were allowed to colonize the surface of etched glass substrata (8.5 * 12 cm) placed at the bottom of each channel. In the No-Cu colonization treatment, no copper was added during the whole colonization period (5 weeks). In the Cu-pulsed colonization treatment, biofilms were exposed to three pulses of 20 μg/l Cu (nominal concentration) on the 5th week of colonization. Each pulse lasted for 2.5 h. After each pulse, water from the system was replaced by water without Cu.

At the end of each colonization period, three glass substrata were removed at random from three different channels to characterize the chlorophyll fluorescence of the different algal classes composing the biofilm communities by means of a Phyto-PAM chlorophyll fluorometer (Heinz Walz, Effeltrich, Germany).

These mature communities were thereafter exposed to higher Cu concentration to assess the influence of Cu pre-exposure on Cu sensitivity. Three channels were exposed for 24 h to 100 μg/l Cu (nominal concentration) and the other three channels were used as controls (maintained without Cu in the water) following the procedures described in Serra et al. (2009). In order to follow accumulative effects, biofilms were collected at time 0 and after 6 and 24 h of exposure to study antioxidant enzyme activities. Cu accumulation, the algal biomass (Fo), and quantum yield (effective and optimal quantum yield) were also analyzed following the procedures described in Serra et al. (2009). All fluorescence measurements were performed using the Phyto-PAM. It employs an array of light-emitting diodes (LED) to excite chlorophyll fluorescence at different measuring lights (470, 520, 645, and 665 nm), and to illuminate samples with actinic light and saturation pulses. The deconvolution of the overall fluorescence signal into the contributions of three algal groups is based on the internal 'reference excitation spectra' of a pure culture (Schmitt-Jansen & Altenburger, 2008). The differences in pigment composition of the antenna complexes of photosystem II can be determined because the shapes of the excitation spectra depend on the spectra of three algal groups (Ruser et al., 1999). Reference spectra which have previously been validated for periphyton communities were used (Schmitt-Jansen & Altenburger, 2008). The fluorescence linked to cyanobacteria, referred to as F(Bl), the fluorescence linked to green algae, referred to as F(Gr) and the fluorescence linked to diatoms, referred to as F(Br), were used for evaluating the relative contribution (in percentage) of each algal class to the whole community. The measurements of in vivo chlorophyll fluorescence of PSII were used to estimate F, which corresponds to the steady-state fluorescence in the given actinic irradiance, and F'm, which refers to the maximum fluorescence yield of an actinic-adapted sample. These two parameters were used to calculate the effective and optimal quantum yield (Y) according to Genty et al. (1989). In our study, Y was based on the fluorescence obtained with 665 nm light-emitting diode (F4). Y measurements were used to follow changes in the photosynthetic efficiency of the communities after Cu exposure. The measurements were performed at room temperature (20°C). Saturation pulses were applied in the same actinic light conditions as the ones used for periphyton colonization.

Effective quantum yield (eff QY) is taken as a measure of the photosynthetic efficiency of the community. Yield inhibition indicates that the toxicant is reducing electron flow in the PSII. The optimal quantum yield (opt QY) provides information about the maximum electron flow. It is a potential estimate and is expected to change after longer exposures if the treatment produces alterations in the photosynthetic apparatus (i.e., shade-adapted chloroplasts). Basal fluorescence (Fo) may increase if the toxicant induces fluorescence production (i.e., in the case of herbicides blocking electron flow). In most cases, however, Fo is used to estimate algal biomass since chlorophyll fluorescence is proportional to total chlorophyll content. Therefore, it is expected that Fo will decrease if the treatment causes a reduction in the number of cells due to cell death (structural damage).

*Antioxidant enzyme activities*

Biofilms were removed from the glass substrata with a cell scraper and centrifuged at 2,300*g* for 5 min (+10°C) to remove the excess water. The pellets were frozen immediately in liquid nitrogen and stored at −80°C until the enzyme assays. Samples were homogenized for 3 min on ice by adding 2.5 ml of homogenization buffer, containing 100 mM potassium phosphate buffer (pH 7.4), 100 mM KCl, 1 mM EDTA and 10% (w/v) PPVP (Polyvinylpolypyrrolidone), to the pellet. Homogenates were then centrifuged at 10,000*g* for 30 min at +4°C and the supernatants were used as the enzyme source. The protein contents of the supernatants were determined by the method of Lowry et al. (1951) using bovine serum albumin as a standard.

Catalase activity was measured spectrophotometrically at 240 nm according to Aebi (1984). 750 µl of reaction mixture was contained in a final concentration of 80 mM potassium phosphate buffer (pH 7.0), 20 mM $H_2O_2$ and enzyme extract (∼60 µg protein). The optimum substrate concentration and protein content of enzyme extract were determined by using 5.0, 10, 15, 20, and 25 mM $H_2O_2$ concentrations and approximately 10, 30, 35, and 60 µg protein in the test media, respectively. The decomposition of CAT was determined by measuring the decrease in absorbance at 25°C for 1–5 min during optimization procedures. Enzymatic activity was measured after monitoring for 1 min, at the end of which linearity was shown. CAT activity was calculated as µmol $H_2O_2$/mg protein/min.

Ascorbate peroxidase activity was assessed by monitoring the decrease in absorbance at 290 nm at 25°C for 2 min due to ascorbate oxidation according to Nakano & Asada (1981). Test medium was contained in a final concentration of 80 mM phosphate buffer (pH 7.0), 0.3 mM Na-Ascorbate, 0.5 mM $H_2O_2$ and enzyme extract (∼65 µg protein) in a final volume of 1 ml. The optimum Na-Ascorbate, $H_2O_2$ concentration and protein content of enzyme extract were determined by using 0.1, 0.2, 0.3, 0.5 and 0.1, 0.2, 0.3, 0.5 and approximately 10, 25, 30, 45 µg protein in the test media, respectively. APX activity was calculated as µmol Ascorbate/mg protein/min.

Superoxide dismutase activity was measured by the indirect method involving the inhibition of cytochrome c reduction, which SOD competes with for superoxide radicals, generated by the hypoxanthine/xanthine oxidase system at 550 nm for 1 min (McCord & Fridovich, 1969). The reaction buffer contained 50 mM potassium phosphate buffer (pH 7.8), 0.1 mM EDTA, 10 mM cytochrome c, 0.05 mM hypoxanthine, and enzyme extract (~25 μg protein). The reaction was started by adding 1.87 mU/ml xanthine oxidase in a final volume of 1 ml, which gives a 0.02 absorbance increase at 550 nm. A unit of SOD activity was defined as the amount of enzyme that causes 50% inhibition of cytochrome c reduction and was given as unit/mg protein.

Glutathione S-transferase activity was evaluated by the absorbance increase at 340 nm due to conjugation of reduced glutathione (GSH) and CDNB (1-chloro-2,4-dinitrobenzene) (Habig et al., 1974). The reaction buffer contained 100 mM potassium phosphate buffer (pH 7.4), 1 mM GSH, 1 mM CDNB, and enzyme extract (~250 μg protein) in a final volume of 1 ml. GST activity was calculated as nmol/min/mg protein.

Catalase activity was measured in biofilms for the No-Cu experiment and CAT, APX, SOD, and GST for the Cu-pulsed experiment.

*Statistical analysis*

One way ANOVA test was used to examine differences in the community structure between the no-Cu and the Cu-pulsed treatments and also to compare photosynthesis, and AEA between controls, and Cu exposed communities during acute exposure (100 μg/l Cu). Statistical analyses were performed using SPSS for Windows (version 13.0; SPSS Inc., Illinois).

## Results

Monitoring

The nine selected sites include karsts fed rivers, lowland Mediterranean rivers, and large watercourses from intermediate to high degree of impact (Table 1). Average water discharge values were between 1 and 10 m$^3$/s, ranging from 0 to 65 m$^3$/s (Table 2). Total Organic Carbon (TOC) ranged from low (1 mg/l) to high (20 mg/l) concentration. Dissolved metal concentration was in general low and below detection limits in most cases (data not shown) except for Zn concentration ranging between 13 and 201 μg/l. Sediment metal concentrations (in μg/g) were well above background concentrations and above ecotoxicological benchmarks in many cases (Table 2).

Comparing flow categories, the most consistent pattern was shown by Cd in sediments increasing at higher flow (Table 3). Comparing sites, the most consistent pattern between metals was found in Besós-b (Montornès del Vallès) and Francolí (20–86% and 22–42% increase with water flow, respectively). On the other hand, a negative relationship between flow and metal contents (with the exception of Cd) was found in Besós-a (Montcada i Reixac), 21–39% higher in the lower flow category.

Overall, Cd in sediment varied significantly ($P < 0.001$) across the river flow condition classes. The low flow condition class shows an average Cd concentration in sediment of 0.98 μg/g ranging from 0.41 to 2.31 μg/g while sites within the high flow condition class have Cd concentrations between 0.51 and 4.13 μg/g with mean = 1.57. A significant correlation between Cd sediment values and flow was obtained when temporal variation among sample collection was controlled for (i.e., variables were treated as covariables in partial correlation analyses). The contribution of the different components evaluated to total random variance was 0.128 due to temporal variation (years), 0.003 explained by spatial variation (basin) and 0.007 due to error.

The comparison of Cd concentrations over time shows that Cd values in sediment peaked during 2003 in most of the rivers (Fig. 1a) although no clear correspondence appears to exist between water discharge and Cd sediment concentratrions (Fig. 1b). Another simple way of looking at the relationship between flow and heavy metal concentrations is by grouping water discharge into different classes (Fig. 1c). After grouping water discharge into different classes it can, however, be observed that Cd content in sediment is higher in high flow conditions (Fig. 1c).

Influence of Cu pulses on the structure and function of biofilms

The exposure of mature biofilms to 20 μg/L of Cu for 2.5 h did not alter the effective quantum yield

**Table 2** Summary of variables selected: average, minimum, and maximum discharge (Q, Q Min and Q Max, respectively); conductivity (cond); pH; dissolved oxygen (O₂); nitrate (NO₃); phosphate (PO₄); ammonium (NH₄); Total Organic Carbon (TOC), and water hardness (hardness)

|  | Median | Avg ± SD | Max | Min | Background[a] | Benchmarcks[b] |
|---|---|---|---|---|---|---|
| Q (m$^3$/s) | 1.44 | 3.47 ± 4.52 | 17.57 | 0.00 | | |
| Q Min (m$^3$/s) | 0.16 | 1.14 ± 2.27 | 9.78 | 0.00 | | |
| Q Max (m$^3$/s) | 6.06 | 10.2 ± 12.8 | 64.81 | 0.01 | | |
| Cond (µS/cm) | 1,152 | 1,137 ± 423 | 2,343 | 516 | | |
| pH | 7.99 | 7.99 ± 0.15 | 8.42 | 7.65 | | |
| O$_2$ (mg/l) | 9.08 | 8.96 ± 1.18 | 11.24 | 4.81 | | |
| NO$_3$ (mg/l) | 12.01 | 16.2 ± 10.5 | 51.81 | 4.89 | | |
| PO$_4$ (mg/l P$_2$O$_5$) | 0.54 | 1.67 ± 2.05 | 8.03 | 0.16 | | |
| NH$_4$ (mg/l) | 0.87 | 3.38 ± 5.16 | 19.58 | BDL | | |
| TOC (mg/l) | 4.50 | 5.39 ± 3.26 | 19.87 | 1.15 | | |
| Hardness (mg/l) | 399 | 393 ± 140 | 683 | 159 | | |
| Zn (µg/l) | 32.20 | 50.1 ± 42.1 | 200.70 | 13.00 | | |
| Cd (µg/g) | 1.10 | 1.26 ± 0.79 | 4.10 | 0.15 | 0.2 ± 0.1 | 1 |
| Cr (µg/g) | 26.70 | 31.7 ± 20.7 | 98.70 | 5.90 | 21.6 ± 6.9 | n.l. |
| Cu (µg/g) | 38.00 | 50.3 ± 43.3 | 247.00 | 8.00 | 21 ± 6.4 | 34 |
| Pb (µg/g) | 30.20 | 67.1 ± 128 | 752.00 | 6.00 | 23 ± 3.7 | 47 |
| Ni (µg/g) | 21.20 | 28.4 ± 24.0 | 144.00 | 4.00 | n.l. | 21 |
| Zn (µg/g) | 129.40 | 162 ± 106 | 401.60 | 16.00 | 56 ± 25 | 100 |
| As (µg/g) | 5.3 | 12.9 ± 24.2 | 136.20 | b.d.l. | n.l. | 8 |

Total water metal concentration (Zn) and total sediment metal concentrations (Cd, Cr, Cu, Pb, Ni, Zn, and As). Median, average (±SD), maximum and minimum values of the nine sites over time (2000–2006)

[a] Range of background concentrations (Adamo et al., 2005) and [b] ecotoxicological benchmarks (Peplow & Edmonds, 2005)

*b.d.l.* below detection limits, *n.l.* not in the literature

(Table 4). Furthermore, at the end of the three consecutive pulses (Cu-pulsed community), algal biomass (based on basal fluorescence, Fo), and the community composition (in terms of the percentage of fluorescence of different algal groups) was similar to the non pre-exposed community (no-Cu) (Table 5).

The effects caused by the acute exposure (100 µg/l Cu for 24 h) differed between the no-Cu and the Cu-pulsed communities (Table 6). Effects on photosynthesis (in terms of effective quantum yield) and photosynthetic capacity (in terms of optimal quantum yield) were similar, but the AEA showed a different pattern. While CAT showed a slight increase in the non pre-exposed community (no-Cu), it showed a marked reduction in the Cu-pulsed one (Table 6). SOD and GST were also inhibited after 24 h of exposure (Fig. 2).

## Discussion

### Monitoring

Dissolved metal concentration was low and below detection limits in many cases but sediment metal concentration (in µg/g) was in general high. Sediment metal concentrations were within the values obtained in river sediments which range 0.073–9.5 (Cd); 19.7–72.5 (Cr); 12–131 (Cu); 15–150 (Pb); and 37–303 (Zn) (Huang & Lin, 2003; Jain, 2004; Adamo et al. 2005; Demirak et al., 2006). Cd, Pb, Zn, and Cu values were characteristic of human-impacted rivers (Peplow & Edmonds, 2005; Demirak et al., 2006) well above background concentrations. Previous study has shown that the relationship between metal loads and diatom species composition can be very poor probably due to the fact that metal concentration

Table 3 Average values of discharge (Q), rainfall, conductivity, Zn water concentration and sediment metal concentrations for each flow category: low flow (1) and high flow (2)

| Site | Flow | Q (m³/s) | Rainfall (mm) | Cond (µS/cm) | Zn (µg/l) | Cd (µg/g) | Cr (µg/g) | Cu (µg/g) | Pb (µg/g) | Ni (µg/g) | Zn (µg/g) | As (µg/g) |
|---|---|---|---|---|---|---|---|---|---|---|---|---|
| Besós-a | 1 | 0.76 | 350.7 | 2,343 | 76.00 | 0.80 | 11.10 | 12.20 | 22.00 | 12.00 | 47.00 | 1.50 |
|  | 2 | 0.87 | 450.8 | 1,856 | 56.95 | 1.20* | 55.25** | 167.0*** | 45.40* | 71.95** | 276.1** | 2.65* |
| Besós-b | 1 | 0.39 | 478.4 | 1,187 | 145.1* | 1.10 | 51.73* | 87.73* | 71.33* | 34.13 | 354.9* | 61.00 |
|  | 2 | 0.55 | 494.1 | 1,156 | 93.75 | 2.00* | 33.55 | 53.98 | 30.98 | 28.20 | 187.55 | 43.70 |
| Fluvià | 1 | 4.64 | 708.7 | 818 | 34.67 | 0.90 | 29.40 | 20.03 | 18.53 | 22.90 | 64.20 | 7.47 |
|  | 2 | 8.00 | 733.4 | 777 | 46.60 | 1.55* | 30.60 | 17.90 | 19.85 | 26.35 | 65.15 | 3.85 |
| Foix | 1 | 0.19 | 456.1 | 1,719 | 46.53 | 0.67 | 15.50 | 33.60 | 17.63 | 10.33 | 53.50 | 4.53 |
|  | 2 | 0.57 | 386.5 | 1,553 | 39.85 | 1.60* | 23.05 | 64.05* | 35.30* | 18.75* | 120.8* | 6.70 |
| Francolí | 1 | 0.70 | 462.4 | 1,166 | 20.61 | 0.74 | 20.92 | 40.12 | 39.90 | 17.90 | 114.7 | 4.74 |
|  | 2 | 1.69 | 645.1 | 1,095 | 25.20 | 1.80* | 47.65* | 75.10* | 70.25* | 34.30* | 220.6* | 7.40* |
| Lobregat-a | 1 | 10.71 | 625.4 | 1,486 | 34.15 | 0.85 | 17.75 | 36.70 | 24.25 | 22.80 | 97.00 | 8.70 |
|  | 2 | 15.38 | 534.5 | 1,233 | 27.20 | 1.33* | 32.63* | 48.57 | 32.70 | 32.47 | 158.4* | 9.07 |
| Lobregat-b | 1 | 12.02 | 560.1 | 1,335 | 27.57 | 1.60 | 50.90 | 75.43 | 37.55 | 40.63 | 207.3 | 13.27 |
|  | 2 | 15.52 | 578.2 | 1,310 | 30.10 | 2.20 | 49.80 | 93.40 | 72.10* | 44.60 | 233.5 | 11.70 |
| Muga | 1 | 1.41 | 366.1 | 649 | 32.47 | 1.03 | 22.00 | 22.60 | 17.80 | 17.07 | 124.6* | 4.03 |
|  | 2 | 2.22 | 393.3 | 683 | 46.03 | 0.95 | 18.75 | 19.60 | 15.38 | 13.65 | 68.40 | 4.78 |
| Tordera | 1 | 0.75 | 665.7 | 634 | 92.57* | 1.30 | 22.63 | 33.13 | 91.17 | 28.93 | 200.5 | 4.27 |
|  | 2 | 3.89 | 1,011 | 763 | 61.15 | 2.33* | 35.28* | 34.78 | 390.9** | 47.93* | 258.7 | 2.65 |

For each flow category, asterisks indicate the percentage of deviation from the average: >20% (*); >50% (**); and >75% (***)

in the water was a bad indicator of metal availability (Guasch et al., 2009a). Metal contents in the water phase represented 9.8% of variation in diatom species composition, including only the most polluted sites: lower part of the Besós and Francolí rivers, both affected by urban, industrial, and agricultural activities. In the current study, sediment metal pollution was remarkable in these sites, but also in the lower part of the Llobregat and Tordera rivers where Cd, Cu, Pb, Ni, and Zn concentrations was also high, above ecotoxicological benchmarks in many occasions.

Focusing on temporal variability, the year effect appeared to account for the majority of the variation of the sediment content of cadmium. In fact, temporal variation explained 93% of total random variance, while spatial distribution of the sampling sites and error explained 2 and 5%, respectively. The comparison of Cd concentrations over time shows that Cd values in sediment peaked during 2003 in most of the rivers followed by a subsequent decrease over the next few years (Fig. 1). Remarkably, the river Tordera was the most polluted river in 2003. The average relative water discharge at river Tordera has increased since 2000 showing a sharp decrease since peak values in 2003 (Fig. 1). Although most of the river basins studied are characterized by an increase in sediment Cd concentration coincident with high discharge during 2003, Cd values in sediment do not behave in a predictive manner according to flow in all of them. Precisely, sediment Cd values did not differ under different water discharge conditions within a same basin. For example, sediment Cd concentrations at Francolí were high in 2003 when mean discharge was also above average while Cd values drop to a minimum in 2005 coinciding with the highest discharge for the study period (Fig. 1). However, overall, Cd content in sediment increases as water discharge rises (Fig. 1c). A different pattern was observed for Cu, Zn, and As increasing with flow in some sites and decreasing in others. The differences observed in the total measured heavy metal concentrations can be a consequence of the relative contribution of the different pathways of metal pollution. The relationship between concentration of a substance and discharge exhibits a non-linear

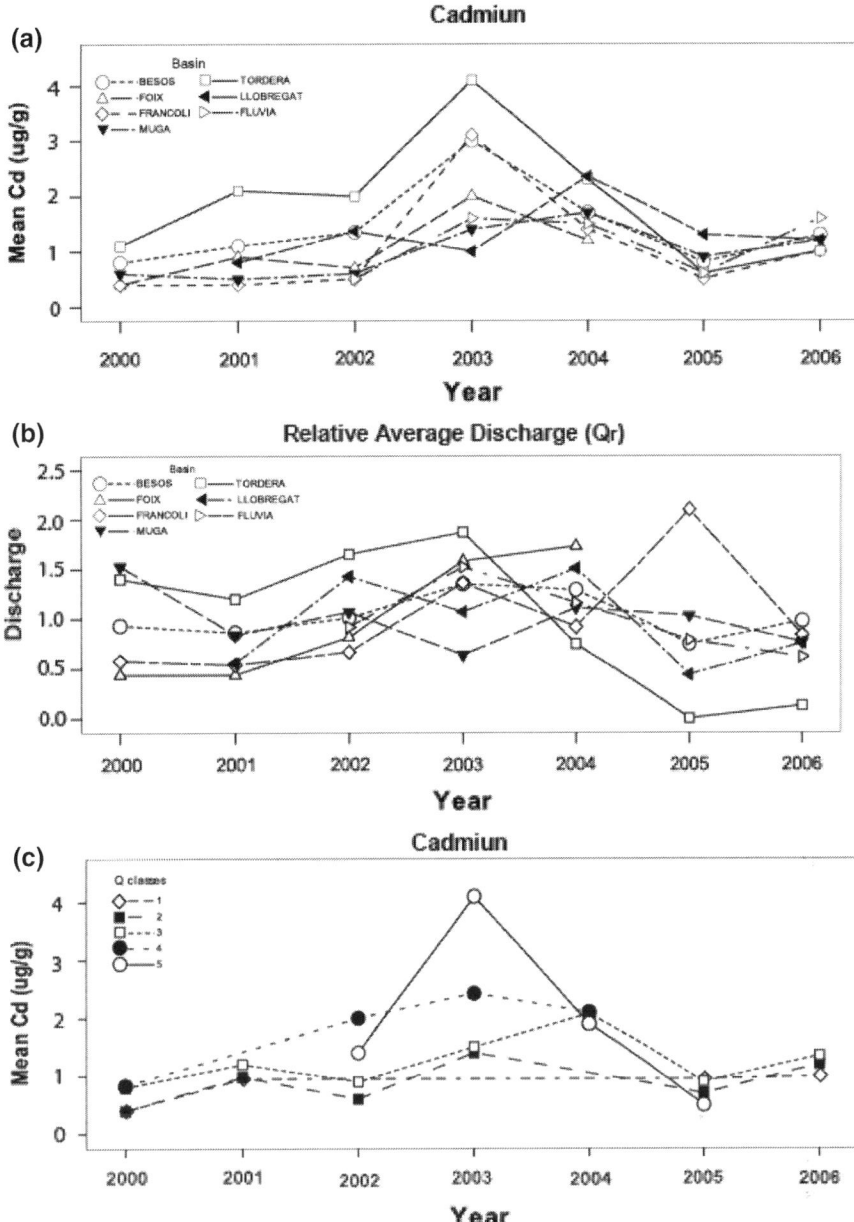

**Fig. 1** Average annual Cd sediment concentrations in the different study rivers (**a**); average annual relative water discharge for all study rivers (**b**); and average annual values for all study rivers grouped in discharge classes (**c**). The water discharge classes were calculated by determining an "upper anchor" and "lower anchor" (the upper and lower 10th percentile) for the average relative annual flow and in doing so the width of the high low water discharge band. The width of the remaining classes was calculated by dividing the interval between the upper and lower anchors equally

**Table 4** Description and effects of the three consecutive Cu-pulses

|  | Pulse 1 | Pulse 2 | Pulse 3 |
| --- | --- | --- | --- |
| Exposure (hours) | 2.5 | 2.5 | 2.5 |
| Cu water (μg/l) | 18.3 | 24.4 | 18.1 |
| Clearance (days) | 3 | 3 | 3 |
| Eff QY (% of pre-exposure) | 96.7 (18.3) | 101.7 (24.4) | 104.4 (18.1) |

Duration of the exposure (exposure), Cu concentration reached at plateau (Cu water), time between pulses (cleaning), and changes in effective quantum yield (eff QY) during exposure

**Table 5** Characterization of the periphytic communities at the end of the respective colonization periods

|  | No-Cu | | Cu-pulsed | |
|---|---|---|---|---|
|  | Avg | SD | Avg | SD |
| Fo | 315 | 260 | 157 | 103 |
| Fo (Bl) | 26.1 | 11.1 | 37.4 | 3.1 |
| Fo (Gr) | 23.9 | 10.9 | 7.0 | 7.3 |
| Fo (Br) | 50.1 | 0.20 | 55.6 | 6.7 |

Chlorophyll *a* fluorescence (Fo) (in relative units of fluorescence) and percentage of chlorophyll fluorescence (F) of each algal class (Bl, blue-green algae; Gr, green algae and Br, brown algae) from the two treatments (Avg ± SD; $n = 3$). No significant differences ($P < 0.05$) among treatments were found for no one of the parameters studied (ANOVA one way test)

dependency. Under this framework the low flow conditions reflect the dominance of point sources from base flow. At higher flow conditions it seems that the input from diffuse sources dominates the transport. While Cd seems to proceed from diffuse sources being washed by surface runoff, Zn, Pb, and As may proceed from either diffuse or point-sources in the different river sites investigated. Although sediment sampling frequency was not optimal (only once a year), sediment characteristics are far more conservative than water metal concentrations and allow an insight within inter-annual variability. Temporal patterns occurring at shorter time scales may also take place. A more frequent sampling is required for a better characterization of metal pollution patterns.

### Influence of Cu pulses on the structure and function of biofilms

Cu pulses were not toxic but caused a progressive Cu accumulation increasing the sensitivity of the Cu-pulsed community in comparison with the non pre-exposed one. Effects on photosynthesis (in terms of effective quantum yield) were similar, but the AEA showed a different pattern. While CAT was activated in the non pre-exposed community (no-Cu), it was clearly inhibited in the Cu-pulsed one. In this later case, SOD and GST were also inhibited indicating that toxicity exceeded the antioxidant cell defenses. The response variability of the antioxidant enzymes may be related to several factors, such as ROS production due to toxic effects of Cu, which can be dependent on the specific toxicity at different cellular sites, organelle localizations of these enzymes, metal concentration and exposure duration (Sauser et al., 1997; Pinto et al., 2003; Li et al., 2006). Decreased CAT and GST activities may be related to the direct binding of Cu on the sulfhydryl groups of the enzyme or elevated levels of ROS leading to deleterious effects on cell structure. In addition, Cu bioaccumulation may cause GSH depletion due to the high affinity of Cu to GSH, which leads to a decrease in GST activity. GST activity in *Fucus* sp. was found to be higher in less contaminated regions than in more contaminated regions of the Portuguese Atlantic coast, which is impacted by complex discharges of contaminants such as petroleum derived products as well as industrial and urban effluents (Cairrao et al., 2004). In the microalgae *Scenedesmus* sp., Cu caused an

**Table 6** Response of non pre-exposed (No-Cu) and pre-exposed biofilms to three consecutive short pulses of 20 µg/l Cu (Cu-pulsed) to an acute Cu exposure (100 µg/l)

|  | No-Cu | Cu-pulsed | No-Cu | Cu-pulsed | No-Cu | Cu-pulsed | No-Cu | Cu-pulsed |
|---|---|---|---|---|---|---|---|---|
| 0 h | 0 | 36.5 | 98.6 | 97.1 | 94.7 | 94.5 | 89 | 100 |
|  | (0) | (16.8) | (12.1) | (13.4) | (4.2) | (8.2) | (23) |  |
| 6 h | 30.0 | 73.0 | 79.4 | 67.5 | 84.2 | 95.2 | 100 | 56 |
|  | (6.2) | (9.5) | (8.9) | (11.6) | (11.5) | (8.3) | (107) |  |
| 24 h | 102.8 | 103.3 | 89.8 | b.d.l. | 94.0 | 85.1 | 107 | 55 |
|  | (47.9) | (35.4) | (5.3) |  | (6.3) | (6.5) | (18) |  |
|  | Cu biofilm (µg/g) | | Eff QY (% control) | | Opt QY (% control) | | CAT (% control) | |

Bioaccumulation (Cu biofilm); effective quantum yield (eff QY); optimum quantum yield (opt QY); and catalase activity (CAT) after 0, 6, and 24 h of exposure

*Average and standard deviation (in parenthesis)

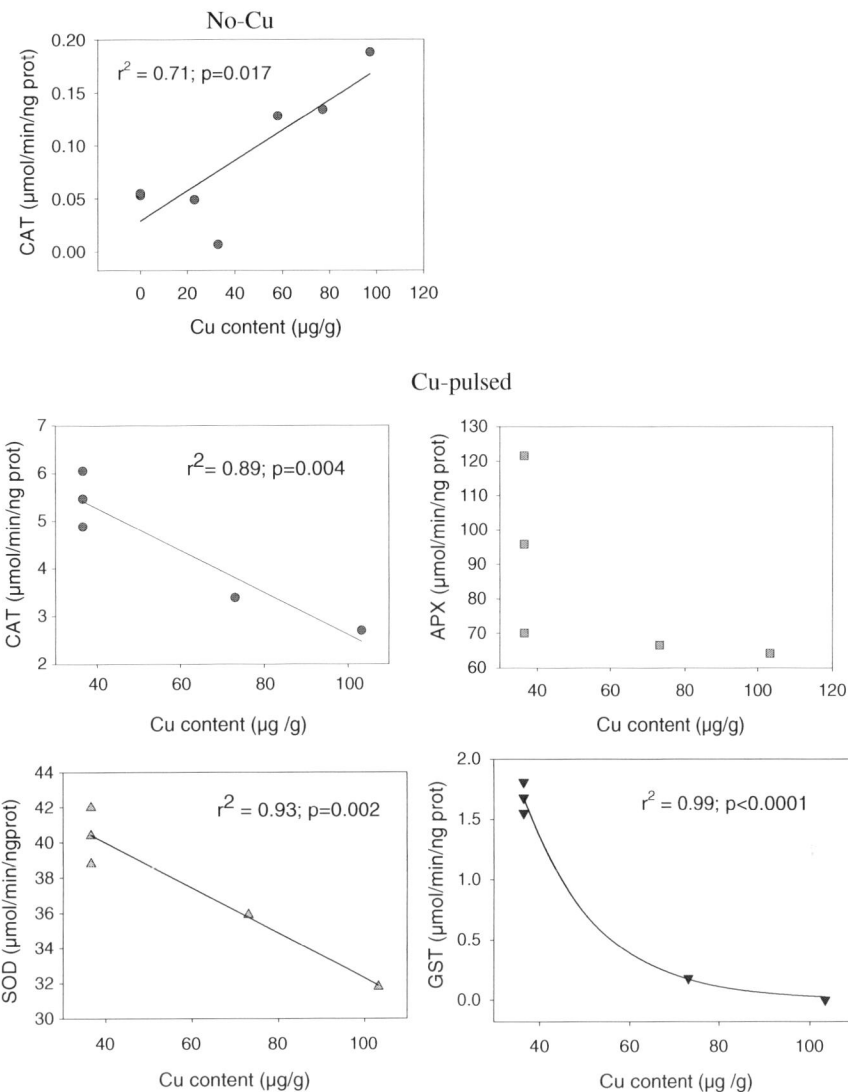

**Fig. 2** Plots of AEA: catalase (CAT); ascorbate peroxidase (APX); superoxide dismutase (SOD); and gutation-S-transferase (GST) versus Cu contents in biofilms after acute Cu exposure on biofilms non pre-exposed (No-Cu) and biofilms pre-exposed to three consecutive short pulses of 20 μg/l Cu (Cu-pulsed). The regression curves (linear for CAT and SOD and exponential decay for GST) and the corresponding regression coefficients and probabilities are also indicated

increase in CAT and APX activity at lower Cu concentrations (up to 5 μM), while at higher concentrations (up to 40 μM) the activities decreased. This was associated with metals binding to sulfhydryl groups on the enzyme or by displacement of an essential element (Tripathi & Gaur, 2004). In conclusion, decreased CAT, SOD, and GST activities appear to be related to high levels of oxidative stress caused by Cu, which is also in agreement with other studies (Sauser et al., 1997; Tripathi & Gaur, 2004; Dewez et al., 2005). It is emphasized that antioxidant responses to toxicants can also depend on the magnitude of stress. Increased antioxidant enzyme activities due to the metal-induced disruption of the oxidative balance can be seen under low dose exposures. However, toxic effects of pollutants may exceed the antioxidant defenses and cell death or damage of cellular mechanisms may follow under acute conditions (Andrade et al., 2006). This is also supported in our data.

The lack of change in the optimal quantum yield (Table 5) indicates that the photosynthetic apparatus was not damaged and that Cu exposure was probably too low to cause cell death in fluvial biofilms after 24 h of exposure. In addition, the effects on the effective quantum yield were slight. In agreement with our results, catalase activity was found to be more sensitive than the photosynthetic activity in

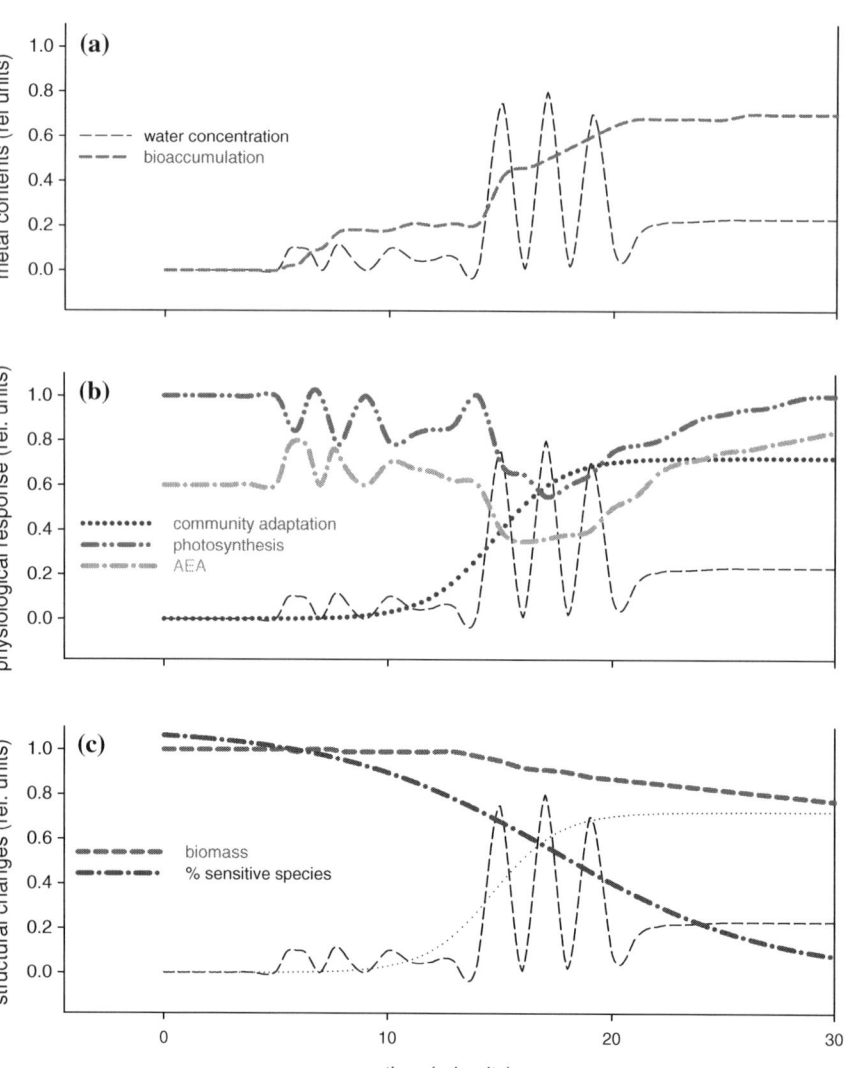

Fig. 3 Conceptual figure showing the expected effects of metal exposure episodes of different duration on fluvial biofilms. Changes in total metal concentration in water and the corresponding pattern of variation of biofilm metal concentration (*upper panel*); expected physiological responses: photosynthesis and antioxidant enzyme activities, AEA (*middle pannel*) and structural changes: biomass and percentage of sensitive species in the community (*lower pannel*)

detecting the effects of the herbicide flumioxazin (Geoffroy et al., 2004).

Time-varying or repeated exposures can have a variety of consequences. The first pulse may select more tolerant species, causing an apparent lessening in toxic response (Allin & Wilson, 2000). Induced community tolerance was observed by Tlili et al. (2008) as a result of the exposure of fluvial biofilms to consecutive diuron pulses. A clearance interval between consecutive pulses may allow time for the community to recover from exposure by depurating or detoxifying the contaminant. However, if the contaminant is not completely depurated or detoxified between transient pollution events, thus the compound could accumulate and result in a cumulative toxic effect over several doses. In a study of the effects of multiple pulses of Cu and Irgarol on the photosynthesis performance of *Zostera*, Macinnis-Ng & Ralph (2002) reported a clear recovery after the first pulse, but a clear damage in the photosynthetic apparatus after the second one, indicating cumulative toxic effects.

Organisms experience toxicity when toxicant accumulation exceeds their tolerance. Fluvial biofilms were not apparently damaged during the short exposure durations in our study. However, marked effects on their AEA did occur after the fourth exposure to higher dose and longer duration. This suggests that Cu accumulation exceeded the tolerance of organisms inhabiting the biofilm after repeated exposures. In contrast with the unexposed biofilm, total Cu content in the Cu-pulsed community was above 30 µg Cu/g DW (36.5 on average) after the three short-pulses suggesting that the biofilm was already exposed and that residual biofilm Cu contents increased the sensitivity of the community. This suggestion is supported by Guan & Wang (2004), who reported that more than about 30% of the accumulated trace metals (Cd, Zn, and Se) remained in *Daphnia magna* after 8 days holding in a clean environment causing retarded lethal effects.

Conceptual model linking temporal variability of metal exposure with fluvial biofilm community responses

The results obtained in our Cu-pulsed experiment contribute to our understanding of the response of fluvial biofilms to metal pulses. Temporal dynamics of total metal concentration in water, the corresponding pattern of variation of biofilm metal concentration and the expected physiological and structural changes are theoretically represented in Fig. 3.

Input to fluvial systems by metals may occur in pulses and by continuous exposure. Biofilm communities exposed to sublethal levels for short periods may lead to progressive internal dosage (bioaccumulation). Once entering the cell, the heavy metal ions may either be detoxified or adversely affect cell processes. Acute exposure may lead to transitory physiological effects. If the community is not adapted, metal exposure is expected to cause oxidative stress, increased activity of the enzymes that may cope with this stress and finally photosynthesis inhibition and toxicity that may damage the cell defense capacity (Fig. 3b). On the other hand, frequent episodes of exposure or chronic exposure may induce an increase in community tolerance driven by the substitution of sensitive species by tolerant ones becoming more resistant to higher concentrations. It is also expected that metal adaptation will be linked to photosynthesis recovery, increase of antioxidant enzyme activities, slight reduction in biomass, and a higher metal accumulation capacity (Fig. 3c).

## Conclusions

Diffuse metal pollution is expected to be high in human-impacted areas, such as the densely populated Mediterranean region. In these areas, pollutants may be accumulated in the terrestrial systems due to low rain, and pollution may be linked to rain events that create urban runoff or industrial spill. The experimental results presented demonstrate that metal effects in fluvial biofilms may be accumulative increasing the toxic effects after repetitive pulse exposures. Longer exposure may allow the community to adapt by the replacement of sensitive species by tolerant ones. Community adaptation is also linked with higher bioaccumulation capacity, in agreement with the results obtained in sediment field samples (monitoring data).

Below average surface flow conditions as well as floods are expected to increase at higher latitudes due to global change. Consequently, the sources of metal pollution, its final concentration and potential effects on the fluvial ecosystem may also change following the patterns expected for human-impacted Mediterranean rivers.

**Acknowledgments** The "Serveis Científics i Tècnics" at the University of Girona provided its facilities and technical help for ICP-MS metal analysis. The research was funded by the Spanish Ministry Science and Education (FLUVIALFITOMARC CGL2006-12785 and FLUVIALMULTISTRESS CTM2009-14111-CO2-01-745), and the EC Sixth Framework Program (MODELKEY 511237-2 GOCE and KEYBIOEFFECTS MRTN-CT-2006-035695). Alexandra Serra and Berta Bonet benefit from a FPI grant of the Spanish Ministry Science and Education. Güluzar Atli benefit from an Erasmus grant between the University of Çukurova (Turkey) and the University of Girona (Spain).

## References

Adamo, P., M. Arienzo, M. Imperato, D. Naimo, G. Nardi & D. Stanzione, 2005. Distribution and partition of heavy metals in surface and sub-surface sediments of Naples city port. Chemosphere 61: 800–809.

Aebi, H., 1984. Catalase in vitro. Methods in Enzymology 105: 121–176.

Allin, C. J. & R. W. Wilson, 2000. Effects of pre-acclimation to aluminum on the physiology and swimming behavior of juvenile rainbow trout (*Oncorhynchus mykiss*) during pulsed exposure. Aquatic Toxicology 51: 213–224.

Andrade, S., L. Contreras, J. W. Moffet & J. A. Correa, 2006. Kinetics of copper accumulation in *Lissonia nigrescens* (Phaeophyceae) under conditions of environmental stress. Aquatic Toxicology 78: 398–401.

Armitage, P. D., M. J. Bowes & H. M. Vincet, 2007. Long-term changes in macroinvertebrate communities of a heavily metal polluted stream: the river Went (Cumbria, U.K.) after 28 years. River Research Applications 23: 997–1015.

Audry, S., G. Blanc & J. Schäfer, 2004. Cadmium transport in the Lot-Garonne River system (France) – temporal variability and a model for flux estimation. The Science of the Total Environment 319: 197–213.

Bambic, D. G., C. N. Alpers, P. G. Green, E. Fanellid & W. K. Silo, 2006. Seasonal and spatial patterns of metals at a restored copper mine site. I. Stream copper and zinc. Environmental Pollution 144: 774–782.

Benson, N. U. & U. M. Etesin, 2008. Metal contamination of surface water, sediment and *Tympanotonus fuscatus* var. radula of Iko River and environmental impact due to Utapete gas flare station, Nigeria. Environmentalist 28: 195–202.

Brown, J. N. & B. M. Peake, 2006. Sources of heavy metals and polycyclic aromatic hydrocarbons in urban stormwater runoff. Science of the Total Environment 359: 145–155.

Cairrao, E., M. Couderchet, A. M. V. M. Soares & L. Guilhermino, 2004. Glutathione-S-transferase activity of *Fucus* sp. as a biomarker of environmental contamination. Aquatic Toxicology 70: 277–286.

Caruso, B. S., L. T. J. Cox, R. L. Runkel, M. L. Velleux, K. E. Bencala, D. K. Nordstrom, P. Y. Julien, B. A. Butler, C. N. Alpers, A. Marion & K. S. Smith, 2008. Metals fate and transport modelling in streams and watersheds: state of the science and SEPA workshop review. Hydrological Processes 22: 4011–4021.

Demirak, A., F. Yilmaz, A. L. Tuna & N. Ozdemir, 2006. Heavy metals in water, sediment and tissues of *Leuciscus cephalus* from a stream in southwestern Turkey. Chemosphere 63: 1451–1458.

Dewez, D., L. Geoffroy, G. Vernet & R. Popovic, 2005. Determination of photosynthetic and enzymatic biomarkers sensitivity used to evaluate toxic effects of copper and fludioxonil in alga *Scenedesmus obliquus*. Aquatic Toxicology 74: 150–159.

Fianko, J. R., S. Osae, D. Adomako, D. K. Adotey & Y. Serfor-Armah, 2007. Assessment of heavy metal pollution of the Iture Estuary in the central region of Ghana. Environmental Monitoring Assessment 131: 467–473.

Genty, B., J. M. Briantais & N. R. Baker, 1989. The relationship between the quantum yield of photosynthetic electron transport and quenching of chlorophyll fluorescence. Biochimica et Biophysica Acta 990: 87–92.

Geoffroy, L., C. Frankart & P. Eullaffroy, 2004. Comparison of different physiological parameter responses in *Lemna minor* and *Scenedesmus obliquus* exposed to herbicide flumioxazin. Environmental Pollution 131: 233–241.

Guan, R. & W. X. Wang, 2004. Cd and Zn uptake kinetics in *Daphnia magna* in relation to exposure history. Environmental Science & Technology 38: 6051–6058.

Guasch, H., M. Leira, B. Montuelle, A. Geiszinger, J. L. Roulier, E. Tornés & A. Serra, 2009a. Use of multivariate analyses to investigate the contribution of metal pollution to diatom species composition. Hydrobiologia 627: 143–158.

Guasch, H., A. Serra, N. Corcoll, B. Bonet & M. Leira, 2009b. Metal ecotoxicology in fluvial biofilms: potential influence of water scarcity. In: The Handbook of Environmental Chemistry review series. Springer.

Habig, W. H., M. J. Pabst & W. B. Jakoby, 1974. Glutathione S-transferases. The first enzymatic step in mercapturic acid formation. Biological Chemistry 249: 7130–7139.

Hoang, T. C. & S. J. Klaine, 2008. Characterizing the toxicity of pulsed selenium exposure to *Daphnia magna*. Chemosphere 71: 429–438.

Huang, K.-M. & S. Lin, 2003. Consequences and implication of heavy metal spatial variations in sediments of the Keelung river drainage basin, Taiwan. Chemosphere 53: 1113–1121.

Jain, C. K., 2004. Metal fractionation study on bed sediments of river Yamuna, India. Water Research 38: 569–578.

Li, M., C. Hu, Q. Zhu, L. Chen, Z. Kong & Z. Liu, 2006. Copper and zinc induction of lipid peroxidation and effects on antioxidant enzyme activities in the microalga *Pavlova viridis* (Prymnesiophyceae). Chemosphere 62: 565–572.

Lowry, O. H., N. J. Rosebrough, N. J. Farr & R. J. Randall, 1951. Protein measurements with the Folin phenol reagent. Journal Biological Chemistry 193: 265–275.

Macinnis-Ng, C. M. O. & P. J. Ralph, 2002. Towards a more ecologically relevant assessment of the impact of heavy metals on the photosynthesis of the seagrass, *Zostera capricorni*. Marine Pollution Bulletin 45: 100–106.

McCord, J. M. & I. Fridovich, 1969. Superoxide dismutase: an enzymatic function for erythrocuprein (hemocuprein). Journal Biological Chemistry 244: 6049–6055.

Meylan, S., R. Behra & L. Sigg, 2003. Accumulation of Cu and Zn in periphyton in response to dynamic variations of metal speciation in freshwater. Environmental Science & Technology 37: 5204–5212.

Nakano, Y. & K. Asada, 1981. Hydrogen peroxide is scavenged by ascorbate-specific peroxidase in spinach chloroplasts. Plant & Cell Physiology 22: 867–880.

Peplow, D. & R. Edmonds, 2005. The effects of mine waste contamination at multiple levels of biological organization. Ecological Engineering 24: 101–119.

Pinto, E., T. C. S. Sigaud-Kutner, M. A. S. Leitao, O. K. Okamoto, D. Morse & P. Colepicolo, 2003. Heavy metal-induced oxidative stress in algae. Journal of Phycology 39: 1008–1018.

Reinert, K. H., J. M. Giddings & L. Judd, 2002. Effects analysis of time-varying or repeated exposures in aquatic ecological risk assessment of agrochemicals. Environmental Toxicology and Chemistry 21(9): 1977–1992.

Ruser, A., P. Popp, J. Kolbowski, M. Reckermann, P. Feuerpfeil, B. Egge, C. Reineke & K. H. Vanselow, 1999. Comparison of chlorophyll-fluorescence-based measuring systems for the detection of algal groups and the determination of chlorophyll-$a$ concenrations. Berichte Forsch.- u. Technologiezentr. Westküste d. Univ. Kiel. 19:27–38.

Sabater, S., H. Guasch, M. Ricart, A. M. Romaní, G. Vidal, C. Klünder & M. Schmitt-Jansen, 2007. Monitoring the effect of chemicals on biological communities. The biofilm as an interface. Analytical Bioanalytical Chemistry 387: 1425–1434.

Sauser, K. R., J. K. Liu & T. Y. Wong, 1997. Identification of a copper-sensitive ascorbate peroxidase in the unicellular green alga *Selenastrum capricornutum*. BioMetals 10: 163–168.

Schmitt-Jansen, M. & R. Altenburger, 2008. Community-level microalgal toxicity assessment by multiwavelength-excitation PAM fluorometry. Aquatic Toxicolology 86: 49–58.

Serra, A. & H. Guasch, 2009. Effects of chronic copper exposure on fluvial systems: linking structural and physiological changes of fluvial biofilms with the in-stream copper retention. Science of the Total Environment 407: 5274–5282.

Serra, A., H. Guasch & N. Corcoll, 2009. Copper accumulation and toxicity in fluvial periphyton: the influence of exposure history. Chemosphere 74(5): 633–641.

Tlili, A., U. Dorigo, B. Montuelle, C. Margoum, N. Carluer, V. Gouy, A. Bouchez & A. Bérard, 2008. Responses of chronically contaminated biofilms to short pulses of diuron. An experimental study simulating flooding events in a small river. Aquatic Toxicology 87: 252–263.

Torres, M. A., M. P. Barros, S. C. G. Campos, E. Pinto, S. Rajamani, R. T. Sayre & P. Colepicolo, 2008. Biochemical biomarkers in algae and marine pollution. Ecotoxicology and Environmental Safety 71: 1–15.

Tripathi, B. N. & J. P. Gaur, 2004. Relationship between copper- and zinc-induced oxidative stress and proline accumulation in *Scenedesmus* sp. Planta 219: 397–404.

Tripathi, B. N., S. K. Mehta, A. Amar & J.-P. Gaur, 2006. Oxidative stress in *Scenedesmus* sp. during short- and long-term exposure to $Cu^{2+}$ and $Zn^{2+}$. Chemosphere 62: 538–544.

Winer, B. J., 1971. Statistical Principles in Experimental Design. McGraw Hill, New York: 907.

GLOBAL CHANGE AND RIVER ECOSYSTEMS

# Effects of eutrophication on the interaction between algae and grazers in an Andean stream

John Ch. Donato-Rondón · Silvia Juliana Morales-Duarte · María Isabel Castro-Rebolledo

Received: 6 August 2009 / Accepted: 17 February 2010 / Published online: 9 March 2010
© Springer Science+Business Media B.V. 2010

**Abstract** Nutrient excess is a common disturbance that affects biological interactions in river ecosystems. The response of nutrient supply on primary producers and *Tricorythodes* sp., a common mayfly grazer, was determined in experimental chambers set in a tropical, high Andean stream. Chambers in an experimentally fertilized reach developed higher amount of both benthic and detached chlorophyll than chambers in an upstream control reach. Fertilization produced a slight increase in grazer biomass, and reduced algal biomass compared to grazer-free chambers. These results show that nutrient excess spread bottom-up effects through the food web, and that relevant top-down effects could also be detected. Eutrophication may produce relevant changes in the food web of tropical high-mountain streams.

**Keywords** Algal biomass · Colombian Andes · *Tricorythodes* sp. · Chambers · Nutrient enrichment

## Introduction

During the 1970s, it was assumed that biomass and productivity of a certain trophic level were defined by the predation and the grazing at higher levels (effects "Top-down"; Steinman, 1996). The predominant control over the distribution and abundance of algae in rivers is given from higher trophic levels (herbivores) to the lower trophic levels (biofilm, benthic algae); however, this control depends upon time, place, and the environmental conditions (Rosemond et al., 1993; Allan & Castillo, 2008). Many experiments established that grazing can significantly affect the structure and dynamics of primary producers' communities (Lamberti & Resh, 1983; Lamberti & Moore, 1984; Hart, 1987; Hill & Knight, 1987; Liess & Hillebrand, 2004; Peters et al., 2007).

Manipulative experiments carried out on the interaction and effects between the herbivorous insects and periphyton in rivers have been limited mostly to caddisflies (Lamberti & Resh, 1983; McAuliffe, 1984). Hill & Knight (1987) carried out a study on the interaction among the mayfly *Ameletus validus* and the periphyton in a small north California stream. All these studies reported a reduction and substantial alteration in periphyton growth, primary production,

Guest editors: R. J. Stevenson, S. Sabater / Global Change and River Ecosystems – Implications for Structure, Function and Ecosystem Services

J. Ch. Donato-Rondón (✉) · M. I. Castro-Rebolledo
Departamento de Biología, Universidad Nacional de Colombia, Av (Cra.) 30 No. 45-03, Bogotá, Colombia
e-mail: jcdonator@unal.edu.co

M. I. Castro-Rebolledo
e-mail: micastror@unal.edu.co

S. J. Morales-Duarte
Escuela de Biología, Universidad Industrial de Santander, Cra. 27 Cll. 9, Bucaramanga, Colombia
e-mail: sjmoralesd@unal.edu.co

and community structure. Other experiments reported the exclusion of herbivorous insects from periphyton (Lamberti & Moore, 1984; Murphy, 1984; Hart, 1985; Hill & Knight, 1987),

Nutrient excess is one of the most common disturbances affecting river ecosystems, through "bottom-up" effects to the whole community structure (Biggs & Smith, 2002). In recent years, both "bottom-up" or "top-down" control of the communities of algae, have been considered important for the structure of trophic webs (McQueen et al., 1986) and the effects of both nutrients and herbivores (Hillebrand & Kahlert, 2001; Hillebrand et al., 2002).

The Andean tropical fluvial ecosystems lodge an enormous biological diversity. The marked seasonal variation in the temperature characteristic of the temperate systems is substituted by altitudinal variation, and the differences in the rainy and drought periods influence the hydrological dynamics (Zapata & Donato, 2005). The incident solar radiation is intense despite of local conditions. In the Andes, the radiation is 50% higher that in the sea level for a regime of equivalent atmospheric humidity (Lewis et al., 1995).

Original forests are fragmented and <30% of their original extension remains (Armenteras et al., 2003). The particular effect nutrient enrichment in the biodiversity and structure of communities in tropical rivers, and, in particular, in the Andean region, is largely unknown. These fluvial systems face increasing population densities and rising eutrophication that add to the seasonal flow variations (Donato & Galvis, 2008).

This article aims to solve the consequences of nutrient enrichment on the relationships between algae and nutrients, as well as the potential implications for the trophic net. The use of controlled experiments has been the main tool to analyze the effect of the nutrient enrichment in the ecological interactions and the biodiversity. This type of experiments may be essential in understanding the response of rivers and streams of tropical fluvial systems. In particular, the main objective of this article was to determine the relationships between nutrient supply, algal biomass, and its link with *Tricorythodes* nymphs grazing in controlled experiments. Results may be indicative of effects of fertilization in Andean streams.

## Methods

### Study area

The Tota stream is located in the Colombian Eastern mountain range (5°35'N and 73°00'W) and its drainage basin has a surface area of 340.6 km$^2$. The average annual temperature and the average annual precipitation are 15°C and 730.5 mm, respectively.

The rainfall regime is bimodal with rainy periods from April to May and from October to November. The driest period ranges from December to February. Discharge is 1.52 m$^3$ s$^{-1}$ during the rainy seasons, and is 10 times lower during the dry period. The natural vegetation of the catchment has been replaced by pasture for cattle rising. The predominant riparian plants trees are alder (*Alnus acuminata*), eucalyptus (*Eucalyptus globulus*), and willow (*Salix humboldtiana*; Castro & Donato, 2008).

The conductivity values in Tota waters are among 27–373 µS cm$^{-1}$, pH has a neutral behavior of 6.67–7.93. The nutrients concentrations have ranges of 0.09–0.87 mg l$^{-1}$ PRS, 0.1–0.5 mg l$^{-1}$ NH$_4^+$, 2.5–11.5 mg l$^{-1}$ silica, 0.06–0.6 mg l$^{-1}$ NT, and 0.05–1.4 mg l$^{-1}$ PT (Zapata & Donato, 2008).

*Tricorythodes* sp. (Ephemeroptera: Leptohyphidae) of the Tota's stream has been not well-characterized despite of being one of the dominant groups in this stream. In tropical areas, most studies of the genus have emphasized its taxonomy (Domínguez et al., 2006; Emmerich, 2007), ecology, and bioindication value (Roldán, 2003; Liévano & Ospina, 2007).

### Experimental design

The study reaches were located in the area of Cuitiva Town in Boyacá State (Colombia). The experiments were performed in a third-order stream (Tota stream). Cobbles were the dominant substrata in the stream. *Alnus acuminata* trees partially shade the stream bed.

The experiment was carried out in a 30 m reach of the stream. The first 15 m of the reach were kept in natural conditions (control reach), while in the other 15 m (fertilized reach) nutrients were added using a 500 l tank. Two commercial grain fertilizers were diluted in the tank (Nitron 26 (26–0–0) and Abocol (NPK) (10–30–10)) to raise at least twice the average

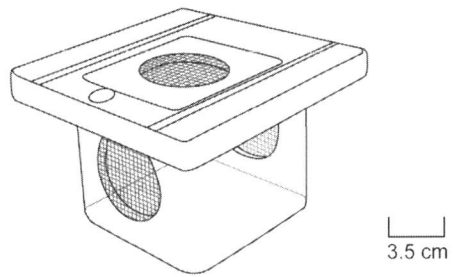

Fig. 1 Chamber used as the artificial substrate in the experiment

basal (natural) phosphates (1.93 μg l$^{-1}$) and ammonium concentrations (16.12 μg l$^{-1}$) in the stream.

Twenty-four chambers were placed in the fertilized reach and in the control reach. These chambers were made of transparent acrylic material, and were 7 cm length, 7 cm height, and 7 cm width. Each chamber had three circular openings of 3.5 cm of diameter covered with a 0.5 mm mesh, to allow continuous flow of water through them (Fig. 1). Fertilization started 15 days before introducing the animals in the chambers. In that period, chambers were placed in the stream to allow periphytic colonization.

In each treatment reach, we randomly choose 12 chambers and introduced 10 *Tricorythodes* sp. nymphs into each chamber. The *Tricorythodes* sp. was collected in the study area. Their average body length was 3.0 mm, and the cephalic capsule was 0.5 mm long by 0.5 mm wide. The herbivores were placed in the compartments at the beginning of the experiment, when the periphyton colonization period was finished.

The sampling was carried out after 3, 10, 17, and 28 days of introducing *Tricorythodes* sp. On each sampling day, six chambers were randomly selected for sampling from both the fertilized and the control reaches. Of the six chambers taken in each treatment reach, three were chambers without grazers and the remaining three contained grazers.

*Tricorythodes* sp. nymphs were taken out of the chambers every sampling day and immediately measured. Extracted chambers were wrapped up in aluminum paper and moved to the laboratory for the estimation of benthic chlorophyll *a* and detached periphyton chlorophyll *a*. Chlorophyll was measured from the detached periphyton and from that suspended in the chamber water using the APHA methods (2005).

Environmental conditions

Water flow (Q) was estimated after daily measurements of the current velocity using a Global digital flowmeter. Temperature (°C) and dissolved oxygen (mg l$^{-1}$ O$_2$) were measured daily with a HACH LDO HQ30d oxygen sensor. Conductivity (μS cm$^{-1}$) was measured with a YSI model 556 MPS multiparametric probe. Light intensity (μmol s$^{-1}$ m$^{-2}$) was measured with a Model LI-COR LI-250$^a$ luxometer. The environmental light was measured thrice each day in study reaches, as well as the light intensity received by each chamber in the water. The pH was measured with a SCHOTT pH 11/SET sensor. The ammonium (μg l$^{-1}$ NH$_4^+$) and the phosphate (μg l$^{-1}$ PO$_4^{3-}$), were measured following the techniques described by Butturini et al. (2009)

Data analysis

Before beginning the experiment, 83 *Tricorythodes* nymphs were collected. In each animal sampled were measured the total body length, length and width of the cephalic capsule. They were oven dried (48 h at 70°C) and weighed with a precision of 0.01 μg. The results were fitted to the equation by Burgherr & Meyer (1997),

$$DM = a * L^b$$

where *a* and *b* are the regression constants, DM dry mass (mg) and *L* total body length (mm). *Tricorythodes* length was related to biomass with the equation:

$$\ln DM = -7.45 + 4.06 \ln(L)$$
$$(n = 83, r^2 = 0.66, P < 0.0001).$$

This equation was used to define the initial biomass of the 240 nymphs of *Tricorythodes* sp. introduced in 24 chambers (12 in the control and 12 in the fertilized reach) and to compare it with the final biomass of the individuals extracted throughout the experiment.

A multivariate analysis of variance (MANOVA) was carried out to detect significant differences between treatments by using the SPSS 15.0 for Windows.

**Results**

Environmental conditions

The reaches showed average values of 0.05 m$^3$ s$^{-1}$ flow, 7.05 pH, 166.69 μS cm$^{-1}$ conductivity,

7.31 mg l$^{-1}$ dissolved oxygen, 16.87°C temperature, 102.48% oxygen saturation, and 907.16 µmol m$^{-2}$ s$^{-1}$ environmental light during the experimental period.

The control and impact reaches showed that significant differences in physical conditions among days (Table 1; Fig. 2), but not between them (Table 2). Only dissolved oxygen was significantly different between days at the two reaches. The dissolved oxygen showed that a similar behavior between reaches during the first day of the experiment (fertilized reach with an average of 7.4 mg l$^{-1}$ and control reach with an average of 7.2 mg l$^{-1}$) until day 25. The oxygen values were slightly higher from day 25 onwards in the fertilized (7.0–7.9 mg l$^{-1}$) than in the control reach (6.9–7.6 mg l$^{-1}$).

The fertilization obviously produced a significant increase in nutrient availability (Fig. 3; Table 1). Concentration of ammonia (NH$_4^+$) rose from 16.12 to 265.11 µg l$^{-1}$, and phosphate (PO$_4^{3-}$) from 1.93 to 68.83 µg l$^{-1}$.

Biota

Significantly higher concentrations of benthic ($n = 24$, $F = 242.543$, $P < 0.0001$) and detached periphyton chlorophyll $a$ ($n = 24$, $F = 52.525$, $P < 0.0001$) were observed due to fertilization (Fig. 4).

In chambers with *Tricorythodes* sp. in the control reach the benthic chlorophyll $a$ showed values of 0.18 µg cm$^{-2}$ (day 19)–0.39 µg cm$^{-2}$ (day 33). Chlorophyll concentrations ranged from 1.00 µg cm$^{-2}$ (day 19) to 2.47 µg cm$^{-2}$ (day 44) in the fertilized reach. The detached periphytic chlorophyll $a$ showed values of 0.07 µg cm$^{-2}$ (day 19)–0.58 µg cm$^{-2}$ (day 33) in chambers of the control reach and from 0.15 µg cm$^{-2}$ (day 19) to 2.79 µg cm$^{-2}$ (day 33) in the fertilized reach.

In the chambers without *Tricorythodes* sp. the benthic chlorophyll $a$ ranged from 0.12 µg cm$^{-2}$ (day 19) to 0.45 µg cm$^{-2}$ (day 44) in the control reach and 0.56 µg cm$^{-2}$ (day 26) to 2.27 µg cm$^{-2}$ (day 44) in the fertilized reach. The detached periphytic chlorophyll $a$ concentration corresponded to 0.10 µg cm$^{-2}$ (day 19)–0.87 µg cm$^{-2}$ (day 44) in the control reach and 0.22 µg cm$^{-2}$ (day 19)–4.78 µg cm$^{-2}$ (day 44) in the impact reach.

Average mayfly biomass was initially 0.06 mg at the control reach and 0.07 mg at the fertilized one. The final biomass did not present any significance difference ($n = 12$, $F = 0.979$, $P = 0.333$) between the two reaches. However, when the mean light value measured in each box was used as covariable differences became statistically significant ($n = 6$, $F = 12.255$, $P = 0.003$). Differences were also significant when biomasses between sampling days were compared ($n = 6$, $F = 42.752$, $P < 0.0001$) due to the rise of the nymphs final biomass in the fertilized reach in relation with the control reach (Fig. 5). Benthic chlorophyll $a$ did not show significant differences regarding the biomass of mayfly nymphs ($n = 12$, $F = 4.985$, $P = 0.037$).

Overall, the detached periphyton chlorophyll $a$ was significantly higher in the chambers with mayfly nymphs ($n = 24$, $F = 22.300$, $P < 0.0001$). We observed that detached periphyton chlorophyll $a$ concentrations were higher in the impact than in control treatments. In the presence of mayfly nymphs, the chlorophyll $a$ concentrations were lower than in the treatments without the herbivore (Fig. 6).

**Table 1** Maximum and minimum values among physical, chemical, and hydrological variables between control and fertilized reach

|  | Control reach | | Fertilized reach | |
| --- | --- | --- | --- | --- |
|  | Min | Max | Min | Max |
| Flow (m$^3$ s$^{-1}$) | 0.01 | 0.32 | 0.01 | 0.37 |
| Temperature (°C) | 14.5 | 20.2 | 14.4 | 20.3 |
| pH | 5.45 | 7.84 | 5.25 | 7.68 |
| Conductivity (µS cm$^{-1}$) | 89 | 224 | 89 | 244 |
| Dissolved Oxygen (mg l$^{-1}$) | 6.65 | 7.56 | 6.77 | 7.98 |
| Ligh (µmol m$^{-2}$ s$^{-1}$) | 235.4 | 1,410.63 | 258.07 | 1,412.53 |
| NH$_4^+$ (µg l$^{-1}$) | 16.12 | 37.68 | 14.22 | 265.11 |
| PO$_4^{3-}$ (µg l$^{-1}$) | 1.93 | 42.7 | 1.37 | 66.37 |

## Discussion

The dissolved oxygen values were significantly higher in the fertilized reach than in the control reach. This change is related to the increase of the primary production as a response to the supply of nutrients (Dodds, 2006), that can be assumed given the higher benthic and detached periphyton chlorophyll $a$ in the fertilized reach. Dodds et al. (2002) also found positive correlations between the detached

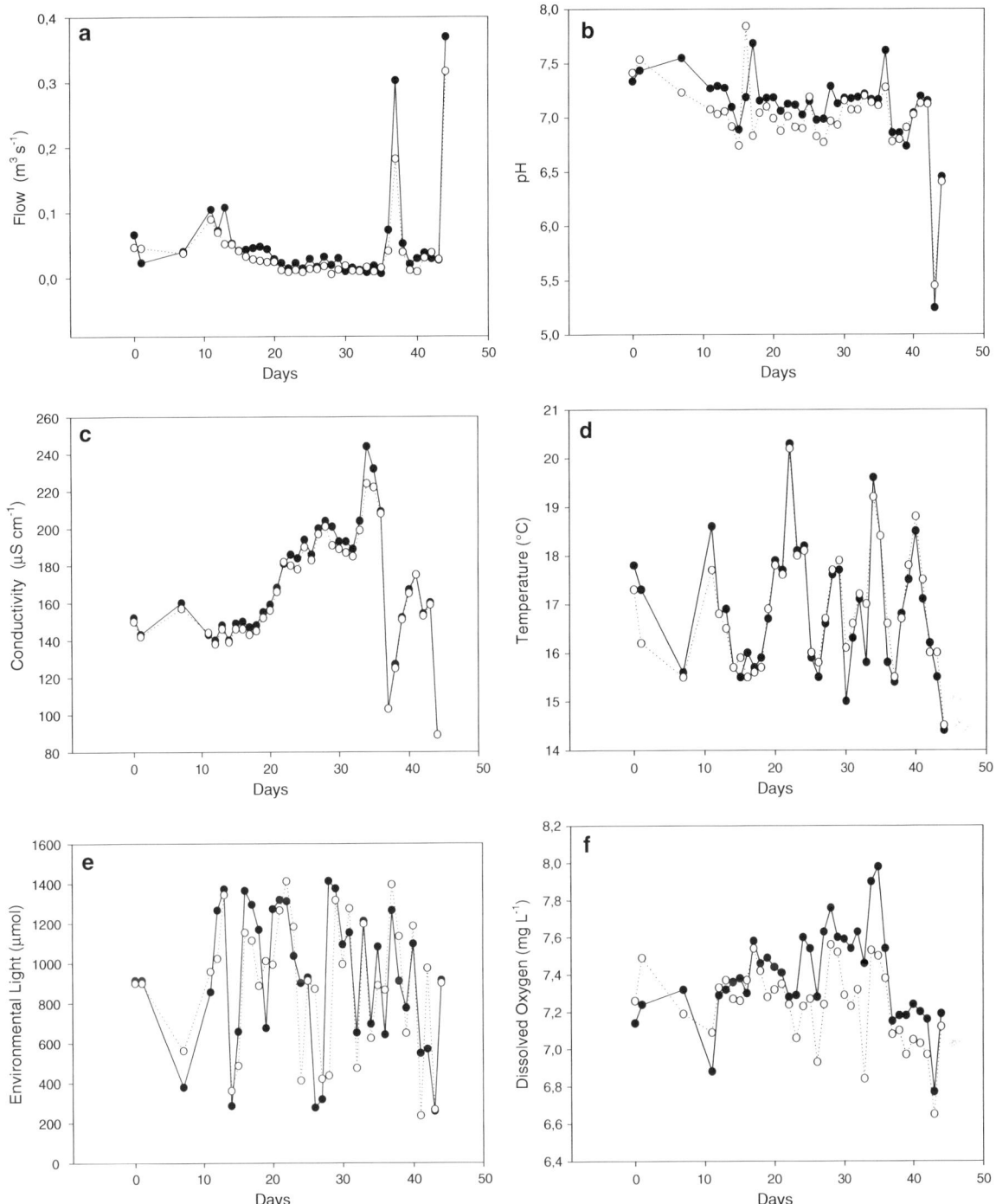

**Fig. 2** **a** Flow, **b** pH, **c** conductivity, **d** temperature, **e** environmental light, and **f** dissolved oxygen values in the sampled days, control (*open circle*) and fertilized (*filled circle*) reach

periphyton chlorophyll *a* and nutrients in the water column. Also, Biggs & Smith (2002) concluded that nutrients supply had strong influence on periphyton despite disturbances, e.g., resulting from hydrological stability. Both the benthic algal biomass and the suspended algae increased as a result of increasing

**Table 2** Using ANOVA analysis, significance values ($P < 0.05$) among the physical, chemical, and hydrological variables between treatments (control–impact) and the sampled days

| | Places (Fertilized–Control) | | Sample Days | |
|---|---|---|---|---|
| | F | P | F | P |
| Flow ($m^3\ s^{-1}$) | 0.736 | 0.394 | 23.291 | <0.0001 |
| pH | 1.409 | 0.239 | 8.924 | <0.0001 |
| Conductivity ($\mu S\ cm^{-1}$) | 0.210 | 0.648 | 156.615 | <0.0001 |
| Dissolved Oxygen ($mg\ l^{-1}$) | 7.326 | 0.008 | 2.915 | 0.001 |
| Temperature (°C) | 0.022 | 0.883 | 31.005 | <0.0001 |
| Light ($\mu mol\ m^{-2}\ s^{-1}$) | 0.160 | 0.691 | 5.784 | <0.0001 |
| $PO_4^{3-}$ ($\mu g\ l^{-1}$) | 21.496 | <0.0001 | 1.389 | 0.162 |
| $NH_4^+$ ($\mu g\ l^{-1}$) | 50.277 | <0.0001 | 0.454 | 0.990 |

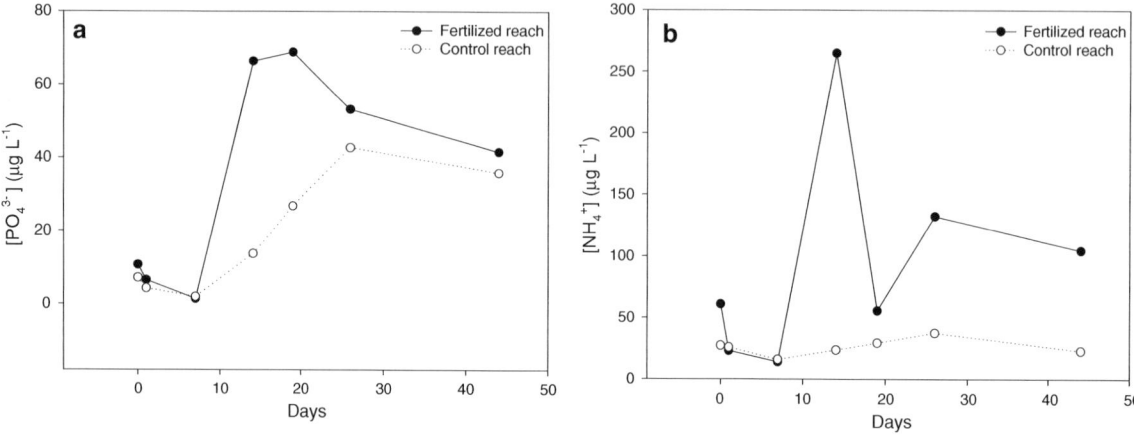

**Fig. 3** Increment of (**a**) phosphate ($PO_4^{3-}$) and (**b**) ammonium ($NH_4^+$) concentrations in the fertilized (*filled circle*) and control (*open circle*) reach during the study days

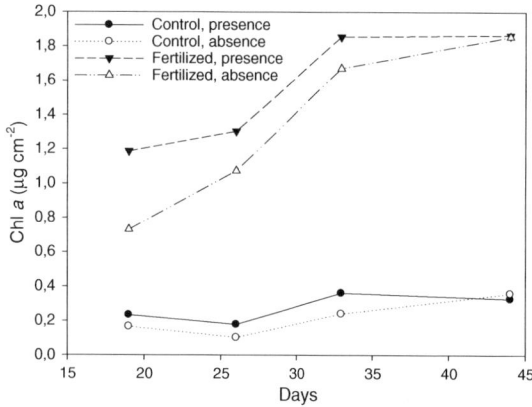

**Fig. 4** Concentration of benthic chlorophyll *a* in fertilized and control boxes in presence and absence of grazers

nutrient supply in Tota's stream. The nutrients concentrations reached during the experiment were within the ranges defined (Rier & Stevenson, 2006; <86 µg DIN $l^{-1}$ and <16 µg P $l^{-1}$) to sustain a maximum algal biomass growth rate.

The higher algal biomass in the fertilized reach had an impact on the nymphs biomass in the impact chambers than in the control ones. The significant higher values rise of final biomass of *Tricorythodes* nymphs in the impact when light factor was a covariable demonstrates that the more intense grazing by the herbivores is due to the influence of light and nutrients in the growth of the periphyton, that in these situations is expressed in greater food availability for the herbivores (Larned & Santos, 2000; Mosisch et al., 2001; Taulbee et al., 2005).

The decrease in the detached periphyton chlorophyll *a* concentrations suggests that *Tricorythodes* nymphs behaved also as collectors according to the descriptions given by Merritt & Cummins (1996). The larvae of many mayfly species have gathering collector feeding structures and tend to feed on the

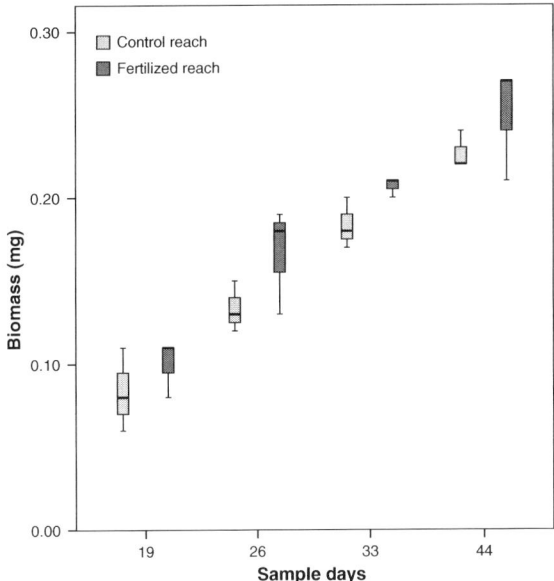

**Fig. 5** Biomass of *Tricorythodes* sp. in fertilized and control reaches

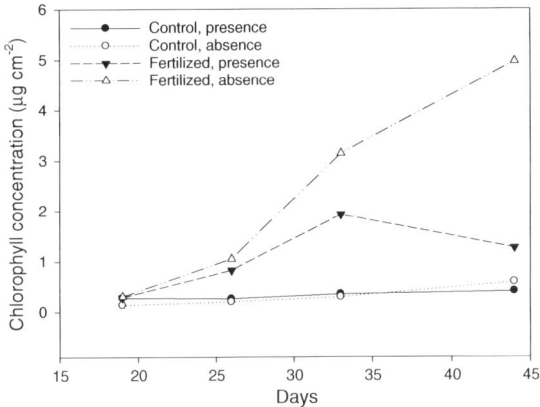

**Fig. 6** Concentration of detached periphyton chlorophyll *a* in fertilized and control boxes in presence and absence of grazers

upper layers, or loosely attached, portions of the periphyton mat (Steinman, 1996). This contrasts with the observations of Rivera et al. (2008) that considered the organisms of this family to be shedders.

The results of this experiment emphasize the ecological relevance of carrying out experiments in situ, where light and nutrients levels are the main restrictive factors of biomass growth in primary producers and vary naturally, than in the fixed conditions in laboratory settings. Different environmental variables can influence organisms' response.

In the present experiment, we showed that the importance of light as a covariable factor that generates significant effects in the increase of the algal biomass and the final biomass of *Tricorythodes*.

Nutrient enhancement in the high Andean stream increased primary productivity ("bottom-up") and generated herbivores responses that regulated periphyton biomass ("top-down"). These results are in agreement with those obtained in temperate rivers (Rosemond et al., 1993; Hillebrand & Kahlert, 2001; Liess & Hillebrand, 2004; Gafner & Robinson, 2007). However, enrichment in tropical systems might be even more effective than in temperate areas, where the base of the stream trophic web depends of allochthonous material and seasonality, while in the high-mountain tropical streams it mostly depends on the contributions of algal biomass (Davies et al., 2008) and the intensity of the physical events of hydrological type (Zapata & Donato, 2005).

**Acknowledgments** The authors thank the financing of the Universidad Nacional de Colombia – Colciencias and the BBVA Foundation, in the framework of the project "Global Changes in fluvial systems: effects on the trophic web, biodiversity and functional aspects" (GLOBRIO). We appreciate the critical reading of the manuscript by Gary Stiles, Universidad Nacional de Colombia.

## References

Allan, J. D. & M. M. Castillo, 2008. Stream ecology: structure and function of running waters. School of Natural Resources and Environment, University of Michigan, USA.

APHA, AWWA & WEF, 2005. Standard Methods for the Examination of Water and Wastewater. American Public Health Association, Washington, DC.

Armenteras, D., F. Gast & H. Villareal, 2003. Andean forest fragmentation and the representativeness of protected natural areas in the eastern Andes, Colombia. Biological Conservation 113: 245–256.

Biggs, B. F. & R. A. Smith, 2002. Taxonomic richness of stream benthic algae: effect of flood disturbance and nutrients. Limnology and Oceanography 47: 1175–1186.

Burgherr, P. & E. I. Meyer, 1997. Regression analysis of linear body dimensions vs. dry mass in stream macroinvertebrates. Archiv für Hydrobiologie 139: 101–112.

Butturini, A., S. Sabater & A. Romaní. 2009. La química de las aguas. Los nutrientes. In A. Elosegi & S. Sabater (eds), Conceptos y técnicas en ecología fluvial. Fundación BBVA, España: 97–116.

Castro, M. I. & J. Donato, 2008. El entorno natural del río Tota. In Donato, J. (ed.), Ecología de un río de montaña de los Andes colombianos (río Tota, Boyacá). Universidad

Nacional de Colombia, Facultad de ciencias, Bogotá, Colombia: 73–79.

Davies, P. M., S. E. Bunn & S. K. Hamilton, 2008. Primary production in tropical streams and rivers. In Dudgeon, D. (ed.), Tropical Stream Ecology. Elsevier, London, UK: 24–37.

Dodds, W. K., 2006. Eutrophication and trophic state in rivers and streams. Limnology and Oceanography 51: 671–680.

Dodds, W. K., V. H. Smith & K. Lohman, 2002. Nitrogen and phosphorous relationships to benthic algal biomass in temperate streams. Canadian Journal of Fisheries and Aquatic Science 59: 865–874.

Domínguez, E., C. Molineri, M. L. Pescador, M. D. Hubbard & C. Nieto, 2006. Ephemeroptera of South America. In Adis, J., R. Arias, G. Rueda-Delgado & K. M. Wantzen (eds), Aquatic Biodiversity of Latin America, Vol. 2. Pensoft, Moscow and Sofia: 1–646.

Donato, J. & G. Galvis, 2008.Tipología de ríos colombianos – Aspectos generales. In Donato, J. (ed.), Ecología de un río de montaña de los Andes colombianos (río Tota, Boyacá). Universidad Nacional de Colombia, Facultad de ciencias, Bogotá, Colombia: 27–52.

Emmerich, D. E., 2007. Two new species of *Tricorythodes* Ulmer (Ephemeroptera: Leptohyphidae) from Colombia. Zootaxa 1561: 63–68.

Gafner, K. & C. T. Robinson, 2007. Nutrient enrichment influences the responses of stream macroinvertebrates to disturbance. Journal of the North American Benthological Society 26: 92–102.

Hart, D. D., 1985. Grazing insects mediate algal interactions in a stream benthic community. Oikos 44: 40–46.

Hart, D. D., 1987. Experimental studies of exploitative competition in a grazing stream insect. Oecologia 73: 41–47.

Hill, W. R. & A. W. Knight, 1987. Experimental analysis of the grazing interaction between a mayfly and stream algae. Ecology 68: 1955–1965.

Hillebrand, H. & M. Kahlert, 2001. Effect of grazing and nutrient supply on periphyton biomass and nutrient stoichiometry habitats of different productivity. Limnology and Oceanography 46: 1881–1898.

Hillebrand, H., M. Kahlert, A. L. Haglund, U. G. Berninger, S. Nagel & S. Wickham, 2002. Control of microbenthic communities by grazing and nutrient supply. Ecology 83: 2205–2219.

Lamberti, G. A. & J. M. Moore, 1984. Aquatic insects as primary consumers. In Resh, V. H. & D. M. Rosenberg (eds), Ecology of Aquatic Insects. Praeger Scientific, New York, USA: 164–195.

Lamberti, G. A. & V. H. Resh, 1983. Stream periphyton and insect herbivores: an experimental study of grazing by a caddisfly population. Ecology 64: 1124–1135.

Larned, S. T. & S. R. Santos, 2000. Light- and nutrient-limited periphyton in low order streams of Oahu, Hawaii. Hydrobiologia 432: 101–111.

Lewis, W. M. Jr., S. K. Hamilton & J. F. Sanders III, 1995. Rivers of northern South America. In Cushing, C. E., K. W. Cummins & G. W. Minshall (eds), River and Stream Ecosystems of the World. University of California Press, Amsterdam: 219–256.

Liess, A. & H. Hillebrand, 2004. Direct and indirect effects in herbivore–periphyton interactions. Archiv Für Hydrobiologie 159: 433–453.

Liévano, A. & R. Ospina, 2007. Guía ilustrada de los macroinvertebrados acuáticos del río Bahamón. Universidad del Bosque, Bogotá, Colombia.

McAuliffe, J. R., 1984. Competition for space, disturbance, and the structure of a benthic stream community. Ecology 65: 894–908.

McQueen, D. L., J. R. Post & E. L. Mills, 1986. Trophic relationships in freshwater pelagic ecosystems. Canadian Journal of Fisheries and Aquatic Science 43: 1571–1581.

Merritt, R. W. & K. W. Cummins, 1996. An Introduction to the Aquatic Insects of North America. Kendall/Hunt Publishing Company, Dubuque, USA.

Mosisch, T. D., S. E. Bunn & P. M. Davies, 2001. The relative importance of shading and nutrients on algal production in subtropical streams. Freshwater Biology 46: 1269–1278.

Murphy, M. L., 1984. Primary production and grazing in freshwater and intertidal reaches of a coastal stream, Southeast Alaska. Limnology and Oceanography 29: 805–815.

Peters, L., H. Hillebrand & W. Traunspurger, 2007. Spatial variation of grazer effects on epilithic meiofauna and algae. Journal of the North American Benthological Society 26: 78–91.

Rier, S. T. & R. J. Stevenson, 2006. Response of periphytic algae to gradients in nitrogen and phosphorus in streamside mesocosms. Hydrobiologia 561: 131–147.

Rivera, C. A., E. Pedraza & A. M. Zapata, 2008. Aproximación preliminar a la dinámica del flujo de la materia orgánica. In Donato, J. (ed.), Ecología de un río de montaña de los Andes colombianos (río Tota, Boyacá). Universidad Nacional de Colombia, Facultad de ciencias, Bogotá, Colombia: 145–162.

Roldán, G. A., 2003. Bioindicación de la calidad del agua en Colombia, uso del método BMWP/Col. Universidad de Antioquia, Medellín, Colombia.

Rosemond, A. D., P. J. Mulholland & J. W. Elwood, 1993. Top-down and Bottom-up control of stream periphyton: effects of nutrients and herbivores. Ecology 74: 1264–1280.

Steinman, A. D., 1996. Effects of grazers on freshwater benthic algae. In Stevenson, J. R., N. Bothwell & R. Lowe (eds), Algal Ecology: Freshwater Benthic Ecosystems. Academic Press, San Diego, CA: 341–373.

Taulbee, W. K., S. D. Cooper & J. M. Melack, 2005. Effects of nutrient enrichment on algal biomass across a natural light gradient. Archiv für Hydrobiologie 164: 449–464.

Zapata, A. & J. Donato, 2005. Cambios diarios de las algas perifíticas y su relación con la velocidad de corriente en un río tropical de montaña (Río Tota, Colombia). Limnética 24: 327–328.

Zapata, A. M. & J. Donato, 2008. Regulación hidrológica de la biomasa algal béntica. In Donato, J. (ed.), Ecología de un río de montaña de los Andes colombianos (río Tota, Boyacá). Universidad Nacional de Colombia, Facultad de ciencias, Bogotá, Colombia: 103–125.

# Comparing fish assemblages and trophic ecology of permanent and intermittent reaches in a Mediterranean stream

Esther Mas-Martí · Emili García-Berthou · Sergi Sabater · Sylvie Tomanova · Isabel Muñoz

Received: 14 July 2009 / Accepted: 28 April 2010 / Published online: 16 May 2010
© Springer Science+Business Media B.V. 2010

**Abstract** Mediterranean streams are characterised by seasonal droughts, the frequency and intensity of which vary spatially and are expected to increase with global change. We studied the potential effects of drought and climate change on the fish assemblage and its trophic ecology in a Mediterranean stream by comparing an intermittent tributary with two more permanent neighbouring reaches. Although the three sites were dominated by the same two fish species, Mediterranean barbel (*Barbus meridionalis*) and chub (*Squalius laietanus*), the intermittent tributary had a lower overall fish density and fewer eel (*Anguilla anguilla*). The intermittent tributary had macroinvertebrates with lower density, smaller taxa and higher diversity. Fish in the intermittent tributary had significantly lower biomasses in their gut contents (adjusted for fish length) and more negative electivities than those in the permanent reaches, as well as significantly lower taxonomic diversity. These results indicate that there was reduced resource availability in the intermittent tributary, which resulted in significantly lower condition and gonadal weight (adjusted for length) of barbel and chub. The data obtained in this Mediterranean stream support the observation that reduced water flow may affect fish at both individual and assemblage levels.

**Keywords** Mediterranean streams · Intermittent stream · Drought · Fish condition · Diet · Resource availability

Guest editors: R. J. Stevenson, S. Sabater / Global Change and River Ecosystems – Implications for Structure, Function and Ecosystem Services

E. Mas-Martí (✉) · I. Muñoz
Department of Ecology, Faculty of Biology, University of Barcelona, Avda. Diagonal 645, 08028 Barcelona, Catalonia, Spain
e-mail: emasmarti@ub.edu

E. García-Berthou · S. Sabater · S. Tomanova
Institute of Aquatic Ecology, University of Girona, Campus de Montilivi, 17071 Girona, Catalonia, Spain

S. Sabater
Catalan Institute for Water Research, Scientific and Technologic Park of the University of Girona, 17003 Girona, Catalonia, Spain

## Introduction

Mediterranean stream ecosystems are characterised by strong variations in stream flow and water temperature throughout the year (Gasith & Resh, 1999; Acuña et al., 2004). These streams also dry out, mostly often during the summer. Therefore, organisms in Mediterranean streams as well as other intermittent streams need to be adapted to highly variable physicochemical factors and biotic interactions (Williams, 1996). The severity of drought is an important determinant of the magnitude of these environmental changes and thus the effects on the local biota (Golladay et al., 2004; Dewson et al., 2007). Flow decrease diminishes both wetted width and depth and consequently reduces habitat availability (Lake, 2003). In addition, habitat suitability may also be altered by reduced water velocity, increased sedimentation, lower resource availability, changes in nutrient concentrations, reduced dissolved oxygen levels or increased water temperatures (e.g. Butturini et al., 2003; Dewson et al., 2007). Drought can affect streams indirectly through effects on connected ecosystems such as riparian zones (Sabater et al., 2001; Acuña et al., 2005). Severe drought can cause local extinctions of the taxa unable to adapt to drought, either through behavioural or life-history strategies (Lytle & Poff, 2004), particularly if water flow is completely interrupted and the stream bed dries out completely.

In current predictions of climate change, both the intensity and frequency of droughts are likely to increase, reducing discharge in many European rivers (Arnell, 1999a, b; IPCC, 2007). Summer precipitation would decrease over Europe, the frequency, severity and duration of heat waves would increase (and consequently evapotranspiration), and therefore runoff would decrease. Overall, a 'mediterranealisation' of many temperate streams would occur (Arnell, 1999a, b; Giorgi et al., 2004; Beniston et al., 2007; IPCC, 2007). Since earlier and longer droughts may occur in the Mediterranean area (Beniston et al., 2007), the seasonal drought effects would be exacerbated, especially in small- and middle-sized streams. In these systems, permanent riffle habitats could dry in summer and intermittent reaches might dry out completely. These effects on riverine ecosystems may be worsened by increasing urbanisation and more intensive irrigation, as both would enhance water abstraction (Arnell, 1999a; IPCC, 2007; Sabater & Tockner, 2010). Temporality would thus extend from Mediterranean ecosystems to other temperate river systems. Therefore, understanding how the stream biota reacts to drought conditions is essential for predicting and mitigating potential effects of increasing water scarcity.

Flow intermittency affects fish populations in several ways (Matthews & Marsh-Matthews, 2003). During dry periods, fish are confined to refugia (i.e. pools), where they reach high densities. In these pools, abiotic (increased water temperature, reduced stream flow and lower oxygen concentration) and biotic interactions (predation and competition for space and food) are intense (Spranza & Stanley, 2000; Lake, 2003; Magoulick & Kobza, 2003). However, the rapid recovery of populations after droughts (Gasith & Resh, 1999; Matthews & Marsh-Matthews, 2003; Vasiliou & Economidis, 2005) points out that many fish species may have developed adaptations to survive in such harsh ecosystems (Labbe & Fausch, 2000; Humphries & Baldwin, 2003). Nevertheless, the existence of behavioural, anatomical and physiological adaptations to small warm-water streams undoubtedly requires some energy investment (Grossman et al., 1998; Spranza & Stanley, 2000). Although adaptations may allow fish survival in these systems, they may not completely prevent fitness consequences (Lake, 2003).

Physiological stress resulting from harsher abiotic conditions or increased competition for resources (either habitat or food) should impact individual fish. Consequently, individuals in intermittent streams might present stress symptoms derived from their higher energy investment, such as a lower condition or reproductive success than others inhabiting permanent river systems. Increased drought frequency and intensity will produce negative effects on habitat quantity (i.e. the occurrence of smaller and shallower pools) and quality (increased temperature, reduced oxygen levels and changes in resource availability) and will thus reduce community stability and persistence (Magoulick, 2000; Oberdorff et al., 2001). Consequently, the effects on fish will increase with drought severity, and the time needed for communities to recover will be longer. Therefore, both the existence of fish adaptations and the severity of drought would determine the necessary time lag for fish to overcome the effects of drought.

A final effect on fish assemblages may be related to the quality and availability of prey under drought conditions. Similarly to fish assemblages, the macroinvertebrate community structure can recover rapidly from drought (Boulton, 2003; Lake, 2003). However, the extension of drought may influence community development during the following year. Surface water disappearance in intermittent streams represents a critical stage that negatively affects survival and next-year recruitment of taxa that have limited mobility, those whose aquatic stage is longer than 1 year, as well as those lacking desiccation-resistant stages (Boulton & Lake, 1992; Boulton, 2003; Acuña et al., 2005;

Bêche et al., 2006; Bêche & Resh, 2007). Compared to permanent rivers, intermittent streams have macroinvertebrate assemblages with lower densities, biomasses and productions, as well as smaller individuals for a given taxon and smaller-bodied taxa with shorter generation times (del Rosario & Resh, 2000; Muñoz, 2003; Halwas et al., 2005; Chadwick & Huryn, 2005, 2007; Bonada et al., 2007). Optimal foraging theory predicts that prey selection is determined by the relative profitability of particular types and sizes of prey (Pyke, 1984). Therefore, the effects of drought on macroinvertebrate fauna may reduce both optimal and overall prey availability, causing shifts in fish diets towards less nutritious prey.

Despite the existence of studies proving the effects of drought on fish populations and their rapid recovery, studies considering delayed effects (such as those arising from changes in food availability) are still rare (Matthews & Marsh-Matthews, 2003), particularly in Mediterranean streams (Magalhães et al., 2007). In this study, we aimed to compare the differences in fish condition and trophic ecology between an intermittent Mediterranean stream and neighbouring permanent reaches. The studied reaches supported the same native fish assemblage typical of the Mediterranean basin (Aparicio et al., 2000; Benejam et al., 2008). Although intermittent and permanent reaches differ in a number of factors, such as basin size, river flow, productivity and connectivity (Nilsson et al., 1994; Taylor et al., 1996; Bonada et al., 2006), comparing them provides large scale, realistic information on how natural or artificial droughts affect overall community structure and functioning. The study was conducted in the spring, a season that allowed maximum recovery from the preceding drought and that is the period prior to the start of the next seasonal drought. We hypothesised that stronger cumulative effects of drought in the intermittent stream would alter macroinvertebrate assemblages and thereby cause changes in fish trophic ecology as well as a reduction in the abundance and individual condition of fish.

## Materials and methods

Study site

The study was performed in La Tordera and one of its tributaries, Fuirosos (NE Iberian Peninsula; Fig. 1).

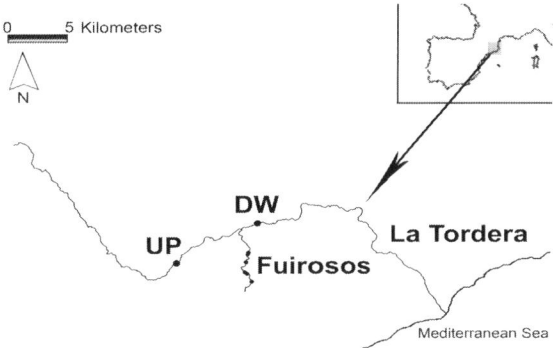

**Fig. 1** Location of sampling sites (*filled circle*) at the mainstream, La Tordera, and its Fuirosos tributary

This catchment is characterised by a typical Mediterranean climate with mild, humid winters and warm, dry summers. Seasonal rainstorms occur during the autumn and spring, and they usually cause spates that increase stream base-flow. Summer drought can highly reduce stream flow, causing streambed drying in the Fuirosos tributary (Acuña et al., 2005; Benejam et al., 2008; Artigas et al., 2009). The study sites are <10 km apart within the Tordera catchment, and thus subject to very similar weather conditions. Therefore, differences in flow among sites were not caused by short term weather-related differences, but by differences in their drainage areas.

La Tordera drains a siliceous catchment with a total area of 895 km$^2$. The studied permanent reaches (UP and DW) are located midstream (Fig. 1). Land uses in the catchment are heterogeneous: while most headwater valleys (including Fuirosos subcatchment) are protected and remain forested with little human activity, the main valley plain is also occupied by both agricultural and urban areas. These activities diminish the water quality in the mainstream, mainly with respect to the high nutrient concentrations (Table 1).

During the study period, both streams were similar in habitat. All reaches alternated between riffles and pools and had similar substrate composition. Reaches were flanked by riparian vegetation, mainly hazel-nut (*Corylus avellana* L.), plane (*Platanus acerifolia* Aiton-Willd.) and cottonwood (*Populus* sp.). The mean stream width was 5.5–8 m in La Tordera and 3–5 m in Fuirosos. During the study period, none of the reaches were light limited since leaf emergence had not yet occurred. Differences in hydrology were remarkable, and the effect of summer drought

**Table 1** Physical and chemical characteristics and fork length values from the fish captured at the sampling sites

|  | Fuirosos (F) | Tordera upstream (UP) | Tordera downstream (DW) |
|---|---|---|---|
| Sub-drainage area (km$^2$) | 15.6 | 152.54 | 315.04 |
| Flow (l s$^{-1}$) | | | |
|   Mean/median (range) | 49.57/11.49 (0–5,000) | 350/60 (0.005–9,690) | – |
| Physical and chemical characteristics | | | |
|   Temperature (°C) | 8.4 | 12.8 | 14.5 |
|   $O_2$ (mg l$^{-1}$) (%) | 10.1 (90.6) | 7.5 (87) | 8.25 (80) |
|   Conductivity (μS cm$^{-1}$) | 223 | 308 | 340 |
|   Phosphate (μg l$^{-1}$) | 3.76 | 340 | 340 |
|   $NH_4^+$ (μg l$^{-1}$) | 2.84 | 360 | 530 |
| Mean fork length (mm) (range) | | | |
|   *Barbus meridionalis* | 86.81 (45–159) | 92.17 (39–197) | 82.83 (60–114) |
|   *Squalius laietanus* | 152.21 (86–246) | 107.52 (54–252) | 114.80 (71–154) |

Physico-chemical data from Fuirosos corresponds to the sampling date, while data from La Tordera is the mean value from Spring 2007. The flow data from both streams correspond to the period of 2000–2009. The flow and chemical data from La Tordera were obtained from the Catalan Water Agency (http://aca.gencat.net:2020/sdim/fillForm.do)

on the remaining flow varied between the streams. Base-flow in the study sites in La Tordera ranged from 350 to 660 l s$^{-1}$ but could be highly reduced due to summer drought. Pool isolation does not occur within the reaches studied in La Tordera (Benejam et al., 2010), whereas the effects of summer drought on the Fuirosos hydrology were more severe. Base-flow in the study reaches (5–20 l s$^{-1}$) could be reduced to complete streambed drying in the Fuirosos, which can last from July/August to September/October (Sabater et al., 2001).

Field study

In spring 2007, we sampled two 100-m reaches in La Tordera, one upstream (UP) and another downstream (DW) from the confluence with Fuirosos (Fig. 1). Fish were sampled by electrofishing with a Smith-Root backpack engine (200–350 V, 2–3 A fully rectified triphasic DC). Due to low fish occurrence, six reaches (885 m in total) were sampled in Fuirosos (F) (Fig. 1) and they were used as replicates to estimate fish abundance but were pooled as a single sample for other statistical analyses. In the field, fish were preserved on ice to avoid digestion of the gut contents and were later frozen in the laboratory. Subsequently, they were measured (fork and total lengths to the nearest 0.5 mm), eviscerated, sexed and weighed (total, eviscerated and gonadal weight to the nearest 0.1 mg), and the entire gut was preserved in 70% alcohol until analysis. Gut contents were examined under a dissecting microscope. All the prey present in the gut were sorted, usually to the family level, counted, and a minimum of 30 individuals (if available) of each prey taxon were measured with an ocular micrometer. Measurements were converted to dry mass according to published length–dry mass relationships (mainly Meyer, 1989; Stead et al., 2003). Other food categories (e.g. detritus and plant debris) were dried until constant weight and weighed to the nearest 0.1 mg.

Quantitative samples of benthic macroinvertebrates were also taken from the different sites. We took three replicate samples from riffles and pools at the two sites in La Tordera using a 0.1-m$^2$ and 250-μm mesh size surber. In Fuirosos, five samples were taken from the dominant substrates, leaves (which accumulated in pools) and rock substrata (dominant in riffles) in two of the study reaches. A cylinder of 115 cm$^2$ was used to sample leaf substrata. Entire cobbles were sampled, and the surface area was calculated after an empirical relationship using the three maximum dimensions of the cobbles (Acuña et al., 2005). All samples were sieved through a 250-μm sieve and immediately fixed in 4% formalin. The differences in macroinvertebrate sampling methods were motivated by the need to maximise sampling efficiency given the different hydraulic conditions of the streams. In the laboratory, organisms were sorted and usually identified to the genus or species level under a dissecting microscope. Benthic macroinvertebrate biomass was obtained following prey biomass measurements (see above).

Data analysis

Differences between sites in total macroinvertebrate density, biomass and individual weights were analysed by means of one-way analyses of variance (ANOVA) and differences between sites controlling for taxon with a two-way ANOVA (with taxon and site as factors). Quantitative variables were log-transformed for the analyses to improve the homoscedasticity and linearity of the data. Macroinvertebrate diversity was assessed using Simpson's index ($D$):

$$D = \sum_i \frac{n_i(n_i - 1)}{N(N - 1)},$$

where $n_i$ is the number of individuals of macroinvertebrate type $i$ and $N$ is the total number of macroinvertebrates (Hurlbert, 1971).

Differences in fish assemblage composition among the three sites were tested by a $\chi^2$ test of independence. Differences in overall density and mean length of the fish species between sites were also analysed by one-way ANOVA. Analysis of covariance (ANCOVA) was used to compare fish weight, gonadal weight and ingested gut biomass between the sites, using fish length as a covariate. Differences in mean prey weight between the sites were also analysed with ANCOVA, using fish length as a covariate. We started with the most complex model, introducing all possible interactions (including interactions of covariates and factors, following García-Berthou & Moreno-Amich, 1993). The general linear model was then simplified by removing non-significant interactions ($P > 0.10$). When the covariate was not significant, it was also removed from the model and ANOVA was used. All factors were considered as fixed effects.

We also measured diet diversity (for each fish) using Simpson's index. In order to compare diet composition versus resource availability, we used Vanderploeg & Scavia's (1979) relativised electivity index ($E^*$):

$$E_i^* = \frac{W_i - (1/N)}{W_i + (1/N)}; \quad \text{where} \quad W_i = \frac{r_i/p_i}{\sum r_i/p_i},$$

where $r_i$ is the relative abundance of prey $i$ in the diet, $p_i$ is the relative abundance of prey $i$ in the environment and $N$ is the number of prey types included in the analysis. This index ranges from $+1$ (positive selection or preference for a certain prey type in relation to its abundance or availability in the environment) to $-1$ (negative selection or avoidance of a certain prey consumption); values near zero indicate neutral electivity. The $E^*$ index was arcsine transformed (arcsine $\sqrt{[(E_i^* + 1)/2]}$) for statistical analysis, as the homoscedasticity and normality were clearly improved. To test whether electivity significantly deviated from 0, a one-sample Student's $t$ test was used. Both electivity and diet diversity were further analysed with ANCOVA (see above), using fish length as a covariate.

The statistical analyses of fish data follow our previous work (e.g. García-Berthou & Moreno-Amich, 2000; Alcaraz & García-Berthou, 2007) and focused on the two most abundant species, *Barbus meridionalis* (Risso) and *Squalius laietanus (*Doadrio, Kottelat & de Sostoa), which are also dominant in middle reaches throughout NE Catalonia (Aparicio et al., 2000). All statistical analyses were performed with SPSS 15.

## Results

Environmental conditions in streams

The different location of the studied reaches within the catchment resulted in large differences in their drainage areas (10 and 20 times larger in UP and DW than F, respectively). Consequently stream flow was 86% lower in F than in UP. Different predominant land uses at the studied sites were probably related to physico-chemical parameters in the reaches. Higher oxygen concentration, as well as lower temperature and conductivity characterised the intermittent stream, which is forested and located in a natural reserve (Table 1). The largest differences between sites corresponded to their nutrient concentrations, phosphorus being 90 times lower, and ammonia 120–185 times lower in Fuirosos than in the others.

Structure of the fish assemblage

The fish assemblages in all the sites were dominated by the Mediterranean barbel (*Barbus meridionalis*) and chub (*Squalius laietanus*). However, there was a significant variation in species composition ($\chi^2_6 = 50.3$, $P < 0.0005$) between the sites. Eels (*Anguilla anguilla* L.) were less abundant in the intermittent stream, and minnows (*Phoxinus* sp.) were only present in the UP site (Fig. 2). Single individuals of common carp (*Cyprinus carpio* L.) and largemouth bass (*Micropterus salmoides* (Lacepède) were detected in the intermittent stream. The total fish density was significantly greater in permanent sites than in the intermittent site (ANOVA using the different transects in Fuirosos as replicates; $F_{2, 9} = 9.41$, $P = 0.006$, Fig. 2).

The mean lengths for most of the fish species were similar between the sites (ANOVAs, $P > 0.05$), except chub which were larger in the intermittent stream ($F_{2, 87} = 14.36$, $P < 0.0005$) (Table 1). The condition (ANCOVA of total weight with length as a covariate) of the barbel was significantly lower in the intermittent stream for both male and female individuals (Table 2). The condition of the chub was significantly lower in intermittent stream (Fig. 3). The same patterns in condition were observed when eviscerated weight was used instead of total weight.

The gonadal weight (adjusted for length with ANCOVA) of the chub was the lowest in individuals

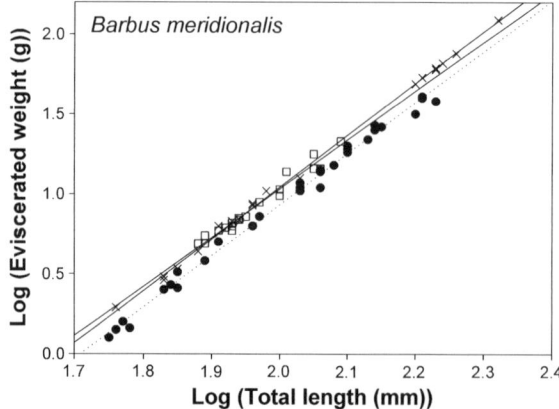

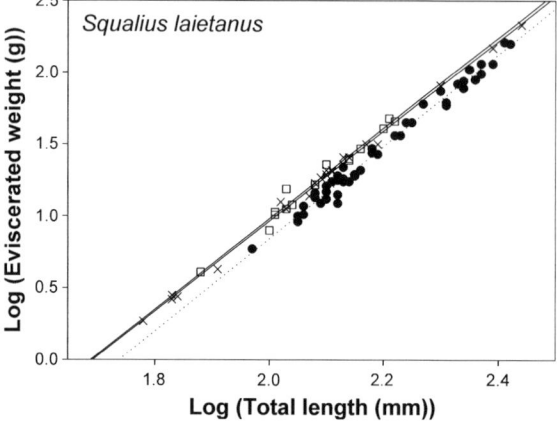

**Fig. 3** Eviscerated weight–length relationship (log scales) of barbel (*top*) and chub (*bottom*) for the F (*filled circle*), DW (*open square*) and UP (*times*) sites. Linear regressions are shown for each site: for barbel, the $r^2$ were 0.994, 0.966 and 0.997 at F (*dotted line*), DW and UP, respectively; for chub, the respective $r^2$ were 0.987, 0.978 and 0.995

from the intermittent stream, especially in the females (Table 2, Fig. 4). Lower gonadal weights were also found in male barbels from Fuirosos (Table 2, Fig. 4). Barbel females were too scarce for analysis. The total biomass of material in guts (adjusted for fish length with ANCOVA) was significantly lower in fish from Fuirosos than in those from La Tordera. This was true both for barbel and chub (Table 2, Fig. 5).

Resource availability

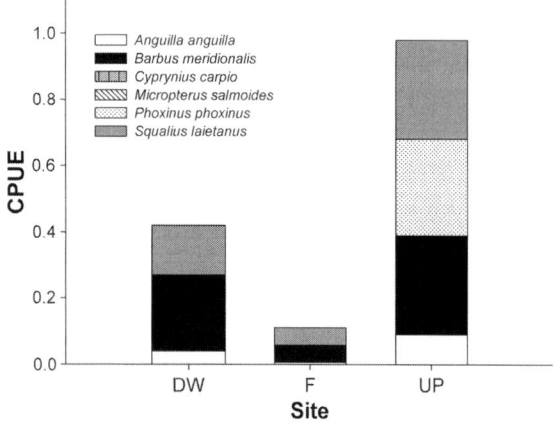

**Fig. 2** Catch per unit effort (CPUE, individuals per meter of electrofishing transect) of fish at the three sampling sites

Although the total macroinvertebrate density did not differ significantly between the sampling sites (ANOVA, $F_{2, 29} = 2.16$, $P = 0.13$), the overall densities at the UP and DW sites were, respectively, 39.0

Table 2 Analyses of covariance of total, eviscerated and gonadal weight and total gut biomass with site and sex (factors) and fish length (covariate)

| | Site | | | Sex | | | Site × sex | | | Log (total length) | | |
|---|---|---|---|---|---|---|---|---|---|---|---|---|
| | df | F | P | df | F | P | df | F | P | df | F | P |
| *Barbus meridionalis* | | | | | | | | | | | | |
| Total weight | 2, 63 | 3.218 | 0.047 | 1, 63 | 5.584 | 0.021 | 1, 63 | 11.688 | 0.001 | 1, 63 | 1784.83 | <0.001 |
| Eviscerated weight | 2, 62 | 1.831 | 0.169 | 1, 62 | 4.65 | 0.035 | 1, 62 | 6.796 | 0.011 | 1, 62 | 2199.62 | <0.001 |
| Female gonadal weight | 1, 16 | 1.309 | 0.269 | – | – | – | – | – | – | 1, 16 | 16.116 | 0.001 |
| Male gonadal weight | 2, 47 | 10.61 | <0.001 | – | – | – | – | – | – | 1, 47 | 27.126 | <0.001 |
| Total gut biomass | 2, 472 | 25.026 | <0.001 | – | – | – | – | – | – | 1, 472 | 62.97 | <0.001 |
| *Squalius laietanus* | | | | | | | | | | | | |
| Total weight | 2, 76 | 72.812 | <0.001 | 1, 76 | 0.311 | 0.578 | 1, 76 | 0.986 | 0.986 | 1, 76 | 5124.776 | <0.001 |
| Eviscerated weight | 2, 77 | 73.038 | <0.001 | 1, 77 | 2.384 | 0.127 | 1, 77 | 0.035 | 0.852 | 1, 77 | 6026.533 | <0.001 |
| Female gonadal weight | 1, 35 | 5.097 | 0.03 | – | – | – | – | – | – | 1, 35 | 257.497 | <0.001 |
| Male gonadal weight | 2, 38 | 3.082 | 0.058 | – | – | – | – | – | – | 1, 38 | 145.666 | <0.001 |
| Total gut biomass | 2, 437 | 81.598 | <0.001 | – | – | – | – | – | – | 1, 437 | 55.961 | <0.001 |

All quantitative variables were $\log_{10}$ transformed

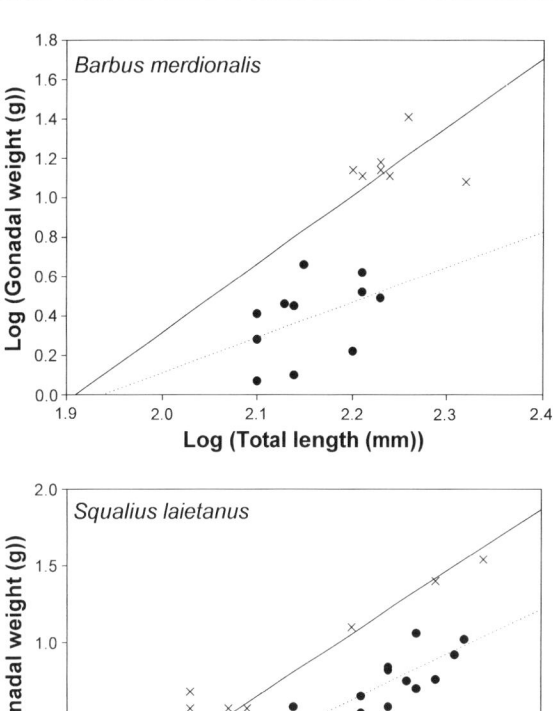

**Fig. 4** Gonadal weight–length relationship (log scales) of female barbel (*top*) and chub (*bottom*) for the F (*filled circle*) and UP (*times*) sites. No female individuals for these species were found in DW. Linear regressions are shown for each site: for barbel, the $r^2$ were 0.183 and 0.855 at F (*dotted line*) and UP, respectively; for chub, the respective $r^2$ were 0.913 and 0.925

and 50.6% higher than those in the intermittent stream (Fig. 6). Similarly, the macroinvertebrate biomass was lower in the intermittent stream, despite differences only being significant between the F and DW sites (Fig. 6, ANOVA, $F_{2, 29} = 12.1$, $P < 0.0005$). Moreover, an analysis of taxonomic groups showed significant differences in both density and biomass among the sites (two-way ANOVA, $F_{100, 1473} = 13.0$, $P < 0.0005$ and $F_{90, 1334} = 12.0$, $P < 0.0005$, respectively). Oligochaeta and especially baetid ephemeropterans were more abundant in the permanent reaches, while chironomid larvae dominated in Fuirosos.

The highest contribution to the biomass in the permanent sites was derived either from ephemeropteran larvae or from the crustacean isopod

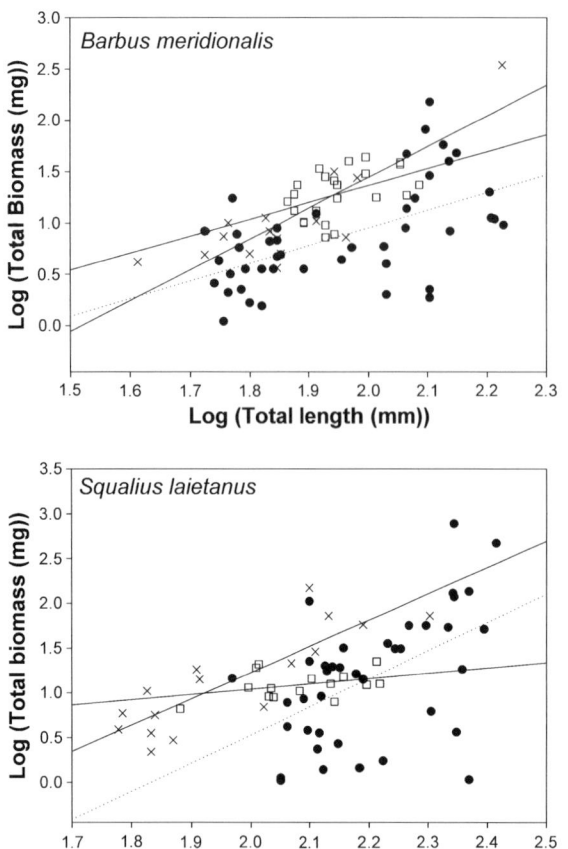

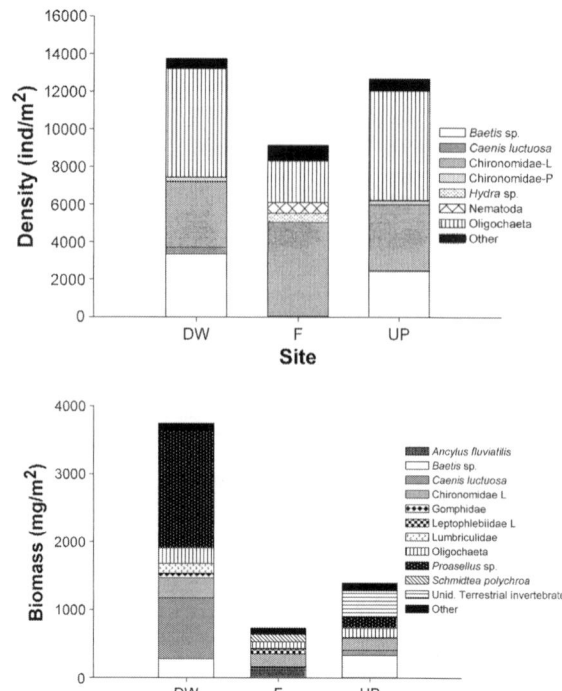

**Fig. 5** Total biomass in the gut–length relationship (log scales) of barbel (*top*) and chub (*bottom*) for the F (*filled circle*), DW (*open square*) and UP (*times*) sites. Linear regressions are shown for each site: for barbel, the $r^2$ were 0.315, 0.213 and 0.709 at F (*dotted line*), DW and UP, respectively; for chub, the respective $r^2$ were 0.259, 0.139 and 0.720

*Proasellus* sp., which was only present in the permanent sites. Terrestrial invertebrates were also important in terms of biomass in the permanent site, UP. Despite the lower mean biomass per macroinvertebrate individual in Fuirosos (Table 3), differences were not significant between the sites

**Fig. 6** Mean macroinvertebrate density (*top*) and biomass (*bottom*) at the three sampling sites. L = larvae, P = pupae, A = adult, Unid = unidentified. Taxa that accounted for <1% of the total density or biomass have been pooled into 'Other'

(ANOVA, $F_{2, 403} = 0.199$, $P = 0.820$). Small-sized individuals were present in the two streams, although the largest individuals were found in La Tordera (Table 3). The macroinvertebrate diversity (Simpson's index) was significantly higher in Fuirosos than in the other sampling sites ($F_{2, 8} = 5.32$, $P = 0.034$; Table 3).

Trophic ecology

Overall, the diet diversity in both barbel and chub did not significantly depend on fish length (ANCOVA,

**Table 3** Weight of macroinvertebrate individuals and overall macroinvertebrate diversity (Simpson's index) found at the three sampling sites

| Site | Macroinvertebrate weight (mg) | | | Macroinvertebrate diversity (*D*) | |
|------|------|------|------|------|------|
| | Mean | SEM | Range (minimum–maximum) | Mean | SEM |
| F | 0.87 | 0.18 | ($2.13 \times 10^{-5}$–14.9) | 0.86 | 0.01 |
| Dw | 2.80 | 0.63 | ($1.50 \times 10^{-6}$–40.5) | 0.84 | 0.01 |
| Up | 1.41 | 0.43 | ($9.85 \times 10^{-4}$–24.9) | 0.79 | 0.03 |

$F_{1, 67} = 3.21, P = 0.078$) and was significantly lower in the intermittent stream (ANOVA, $F_{2, 79} = 7.10$, $P = 0.001$ and $F_{2, 67} = 8.45, P = 0.001$ for barbel and chub, respectively; Fig. 7). The mean prey electivity was not significantly related to barbel length (ANCOVA, $F_{1, 1710} = 1.26, P = 0.261$) and did not differ significantly between the sites (ANOVA, $F_{2, 1711} = 1.96, P = 0.14$), although the mean value was lowest in the intermittent stream (Fig. 8). Similarly, the electivity in chub showed no significant relation to fish length (ANCOVA, $F_{1, 1516} = 0.85, P = 0.357$) and was significantly lower in Fuirosos (ANOVA, $F_{2, 1517} = 22.40, P < 0.0005$; Fig. 8). At all the sites, fish showed significantly negative mean electivities ($t$ tests, $P < 0.001$) for benthic macroinvertebrate species, indicating that the consumption of most of the potential prey was avoided. Barbel and chub mainly fed on detritus and chironomid larvae in both streams. However, a higher proportion of ephemeropterans, *Proasellus* and prey from terrestrial origin were also found in the guts of the permanent stream fish.

Despite the lower mean electivity in Fuirosos, the mean prey weight for prey consumed by the barbel was higher in Fuirosos (ANCOVA, $F_{2, 282} = 4.07$, $P = 0.018$) (1.42 mg ± SD = 6.38) than in La Tordera (0.72 mg ± SD = 1.21 and 0.62 mg ± SD = 1.87 for the DW and UP sites, respectively). However, these differences were not statistically significant for chub (ANCOVA, $F_{2, 187} = 0.54, P = 0.58$), probably due to the high variability in the mean prey weight caused by the occasional presence of large prey such as salamanders or Oligochaeta (Lumbricidae). The mean prey weight increased significantly with fish size in both barbel and chub at all sites (ANCOVA, $F_{1, 282} = 4.51, P = 0.035$ and $F_{1, 187} = 6.87, P = 0.009$ for barbel and chub, respectively).

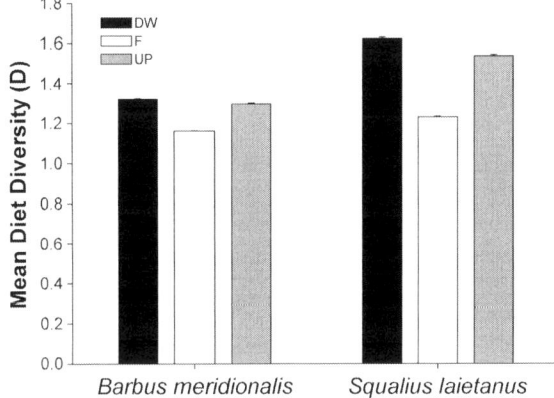

Fig. 7 Mean diet diversity (Simpson's index) for barbel and chub at the three sites. Error bars show the standard error

## Discussion

Comparing intermittent and permanent reaches

The effects of natural- or human-induced reduction on water flow and fish communities were analysed by comparing two permanent reaches and a neighbouring intermittent tributary. The analysis assumed that a permanent reach would turn into intermittent because of hydrological alterations as a result of climate change or water withdrawal. All the reaches supported the same autochthonous Mediterranean fish assemblage, and were close enough to minimise climatic and geologic differences between them. There were substantial differences in fish condition and trophic ecology between the fish communities in the intermittent Mediterranean stream and those in the neighbouring permanent reaches. The intermittent stream had lower water flow and dried up in summer with only a few pools remaining. This stream also had lower drainage area, temperature or nutrient concentrations (Table 1). These differences result from the covariance of many abiotic factors along environmental gradients in rivers. Intermittent tributaries occur at higher altitude, show lower temperature, conductivity, and nutrients, and have lower

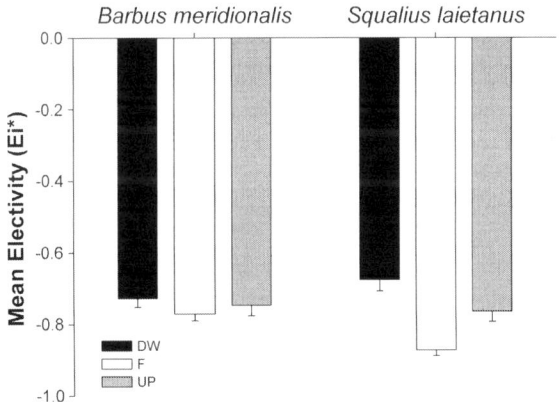

Fig. 8 Mean electivity values (Vanderploeg and Scavia's index) by number for barbel and chub at the three sampling sites. Error bars show the standard error

hydraulic connectivity than perennial streams (e.g. Nilsson et al., 1994; Taylor et al., 1996; Ostrand & Wilde, 2002; Bonada et al., 2006).

Although differences in fish community could be related to drought intensity, alternative hypothesis such as the relevance of decreased habitat or nutrient availability for fish in the intermittent reaches cannot be ruled out as regulating factors. Even though the specific mechanisms that cause the observed differences between the permanent and intermittent reaches are ambiguous and would require of an experimental approach, the comparative results benefit from the large scale and realism of observational studies (Keddy, 1989) and results are consistent with changes expected from drought stress.

Differences in the fish assemblage

Fish density was significantly lower in the intermittent stream, even well after drought had ceased. Davey & Kelly (2007) and Magalhães et al. (2007) found that fish density declined with decreasing flow permanence. Fish population recovery depends on the rate and extent of recolonisation from refugia upon rewetting, the distance from refugia, and the species-specific behaviour (Magoulick & Kobza, 2003; Davey & Kelly, 2007; Albanese et al., 2009). We could not isolate water intermittency as the only cause of diminished fish density in the intermittent stream. However, a previous study in Fuirosos (Aparicio & Vargas, 2004) found that despite the availability of long reaches that were suitable for colonisation, there was no significant expansion of the fish population during the wet period after drought. That study highlighted that although the fish population started to recover during the flow period, the recovery was interrupted by the arrival of a new summer drought. Periods between droughts were insufficiently long for the fish population to recover from disturbance, causing the fish density to remain low. It might be assumed that greater effects on fish populations can be expected (Nicola et al., 2009) if more intense and frequent droughts occur in the future. Fish are long-lived organisms that are sensitive to variations in the intensity, frequency and timing of drying. Fish may show a decrease in survival rates due to mismatches between life-history adaptations (e.g. spawning and hatching periods) and the intervals in water flow.

Differences in species composition between the intermittent and the permanent sites were not due to the most abundant species (Mediterranean barbel and chub), but were due to the presence of minnows in one of the permanent sites and, secondarily, to the lower density of eels in the intermittent stream. These results contrast with others, which found similar fish species composition during and after a drought period (Matthews & Marsh-Matthews, 2003; Aparicio & Vargas, 2004; Magalhães et al., 2007). Our results may be partly explained because the minnow is an introduced species in La Tordera that has expanded in recent years (Benejam et al., 2010) but has not yet reached Fuirosos. The long-lasting disconnection between the intermittent and the permanent streams would have impeded eel migration upstream towards Fuirosos.

We observed lower fish condition (chub, barbel) in the intermittent stream than in the permanent reaches. These results are consistent with others that observed that fish condition decreased with either lower flow or intermittent waters (Torralva et al., 1997; Vila-Gispert & Moreno-Amich, 2001; Oliva-Paterna et al., 2003; but see Spranza & Stanley (2000) for the opposite pattern). Our results also show that the minimum summer flow values do not only affect fish autumnal condition (Oliva-Paterna et al., 2003; Vasiliou & Economidis, 2005), but also effects of seasonal drought prevail until the following spring. Therefore, despite the initial recovery in fish condition in autumn (Oliva-Paterna et al., 2003; Vasiliou & Economidis, 2005), the repeated seasonal drought in Mediterranean regions prevents the complete recovery of fish condition in the favourable wet period that precedes the subsequent drought.

Even though the major resource allocation for reproductive strategies is a plausible cause for the reduction in fish condition (Aparicio & de Sostoa 1998; Oliva-Paterna et al., 2003), as the lower gonadal weight–length relationship obtained in the intermittent stream suggests, reproduction investment may not be the only cause of lower fish condition. Low flow periods lead to decreased fecundity and shorter reproductive period in barbel (Aparicio & de Sostoa 1998). Similarly, recruitment can fail in dry years (Magalhães et al., 2003, 2007; Matthews & Marsh-Matthews, 2003) even if high levels of reproduction occur during an unusually high-flow year (Labbe & Fausch, 2000; Lake, 2003). Drought may cause delayed reproductive effects on fish assemblages, leading to lower abundance in the following year (Matthews & Marsh-Matthews, 2003). Moreover, if drying occurs annually (which is the case for Mediterranean streams), droughts may have cumulative effects expressed in low fish densities and limited population recovery. Because a higher investment in reproduction was not causing the lower fish condition in the intermittent stream, indirect mechanisms must be responsible, at least to some extent, for the low condition.

Stream size differed among permanent and intermittent reaches, the former being ~1.7 times wider than the latter. Although such a difference did not result in different habitat availability among reaches during the study period, it could have an indirect effect through increased drought intensity in the intermittent stream. Smaller streams are more prone to flow fluctuations (Medici et al., 2008), and susceptible to suffer more severe droughts (Sabater & Tockner, 2010). Fuirosos experienced pool disconnection and almost complete bed drying (Acuña et al., 2005), while a decreased water flow during the dry season was never intermittent in La Tordera reaches. Intermittent streams experience reduced habitat availability and suitability (Lake, 2003; Dewson et al., 2007) under drought, which might have negative effects on fish condition.

The increased fish condition in the permanent reaches could also respond to the higher nutrient availability in these reaches. Whole stream fertilization resulted in increased fish and insect growth via autochthonous primary production (Deegan & Peterson, 1992; Peterson et al., 1993). However, effects on insect secondary production were only important for some grazers (the caddisfly *Brachycentrus* and the mayfly *Baetis*) and were not general to the overall insect community (Peterson et al., 1993). Deegan et al. (1997) also observed that stream nutrient addition caused a fivefold increase in chlorophyll levels but not to total insect abundance, fish growth being more related with per capita insect availability than with per capita algal standing stock. Neither whole stream fertilization experiments conducted in Fuirosos intermittent stream had bottom-up effects in its detritus based food web (Sabater et al., 2005). Although higher macroinvertebrate density and biomass could result from flow permanence (del Rosario & Resh, 2000; Muñoz, 2003; Halwas et al., 2005) and nutrient availability (Cross et al., 2006), macroinvertebrates from intermittent streams possess life traits that enable them to cope with streambed drying (Bonada et al., 2007).

Trophic ecology: effects of drought through food availability

There was a lower macroinvertebrate density and biomass in the intermittent stream, as well as a lower abundance of the more frequently consumed prey items in the permanent river. These patterns of lower macroinvertebrate densities and biomass in intermittent streams and after severe droughts have been described elsewhere (Feminella, 1996; Gasith & Resh, 1999; del Rosario & Resh, 2000; Muñoz, 2003, Acuña et al., 2005; Halwas et al., 2005; Bêche et al., 2006). In addition, macroinvertebrate taxa were generally smaller in the intermittent reach (e.g. *Oulimnius* and chironomid larvae, nematodes and *Hydra* sp.) than in the permanent ones (e.g. baetid larvae, *Proasellus* and lumbriculids). This finding is consistent with the observations of Chadwick & Huryn (2007). The combination of the small size and the lower density of invertebrates likely indicate resource limitation for predators (Zaret & Rand, 1971; Oliva-Paterna et al., 2003; Nunn et al., 2007). This possibility was confirmed by the lower ingested biomass in guts of both barbel and chub from the intermittent stream.

*Proasellus* individuals were only found in the permanent reaches, and their absence in the intermittent stream might be due to the life-history traits. *Proasellus* lacks aerial stages, and its life-cycle lasts for more than a year (Tachet et al., 2000); streambed drying would have excluded them from the intermittent stream. This taxon was an important prey item as indicated by their high biomass in the fish gut contents. Confounding effects derived from increased nutrient availability in the permanent reaches on macroinvertebrate structure cannot be ruled out. However, the indirect effects of drought (i.e. elimination of certain prey) are certainly likely determinants for the diminished fish condition in the intermittent reach.

Although macroinvertebrates had the highest diversity in the intermittent stream, fish diets were the least diverse, suggesting that many of the potential trophic resources available were not actually used by the fish. Despite mean electivity indices being negative in the permanent and intermittent streams, the lower values in the latter indicate that fish either rejected more food categories or that they had a lower ability to select the available prey.

Rapid changes in macroinvertebrate abundance follow changes in habitat, food or hydrologic disturbance (Bêche & Resh, 2007). However, seasonal droughts will also affect assemblage composition. Increased consumer effects with longer drought duration (Ludlam & Magoulick, 2009) suggest that the increased drought duration and intensity resulting from global change may increase pool isolation and streambed drying. There is therefore a hydrological threshold for some invertebrate species, either directly, through a mismatch between species lifecycle and hydrology (Lytle & Poff, 2004; Statzner et al., 2004), or indirectly, through increased competition for resources or predation.

In the predicted scenarios of climate change, with reduced summer precipitation and increased temperatures over Europe, droughts in Mediterranean regions are expected to start earlier in season and increase in duration (Beniston et al., 2007). Direct pressures of climate change on river intermittency will probably be exacerbated by human water demand (Arnell, 1999a; Benejam et al., 2010), and therefore many permanent streams could become more intermittent. The observed differences between intermittent and permanent streams provide insight about larger-scale impacts of global change on the inhabiting biota. In this study, the intermittent reach exhibited a significantly lower fish density, as well as reduced condition and reproductive investment, than the permanent river reaches. Further, there was lower prey availability and significantly lower food biomass in the gut of fishes inhabiting the intermittent stream. Diminished prey availability may be responsible for the lower fish condition in the intermittent stream. Therefore, not only would the direct effects of drought on fish condition be important, but also the indirect bottom-up effects resulting from changes in resource availability and suitability.

**Acknowledgements** We thank C. Alcaraz, L. Benejam, J. Benito and L. Zamora for their help in the field and laboratory work and editor R. Jan Stevenson and two anonymous reviewers for helpful comments on the manuscript. This study was financially supported by projects CGL2007-65549/BOS, CGL2008-05618-C02-02, CGL2009-12877-C02-01 and Consolider-Ingenio 2010 CSD2009-00065 of the Spanish Ministry of Science. Additional funding was provided by the Catalan Water Agency, the Government of Catalonia and the Barcelona Provincial Council ('Observatori de la Tordera'). EMM holds a doctoral fellowship (FI 2009-2012) from the Government of Catalonia.

## References

Acuña, V., A. Giorgi, I. Muñoz, U. Uehlinger & S. Sabater, 2004. Flow extremes and benthic organic matter shape the metabolism of a headwater Mediterranean stream. Freshwater Biology 49: 960–971.

Acuña, V., I. Muñoz, A. Giorgi, M. Omella, F. Sabater & S. Sabater, 2005. Drought and postdrought recovery cycles in an intermittent Mediterranean stream: structural and functional aspects. Journal of the North American Benthological Society 24: 919–933.

Albanese, B., P. L. Angermeier & J. T. Peterson, 2009. Does mobility explain variation in colonisation and population recovery among stream fishes? Freshwater Biology 54: 1444–1460.

Alcaraz, C. & E. García-Berthou, 2007. Food of an endangered cyprinodont (*Aphanius iberus*): ontogenetic diet shift and prey electivity. Environmental Biology of Fishes 78: 193–207.

Aparicio, E. & A. de Sostoa, 1998. Reproduction and growth of *Barbus haasi* in a small stream in the N.E. of the Iberian peninsula. Archiv für Hydrobiologie 142: 95–110.

Aparicio, E. & M. J. Vargas, 2004. Influence of hydrological variability on fish populations in Fuirosos stream. IV Trobada d'estudiosos del Montnegre i el Corredor. Barcelona: Diputació de Barcelona. Monografies, núm 4: 119–122 (in Catalan, abstract in English).

Aparicio, E., M. J. Vargas, J. M. Olmo & A. de Sostoa, 2000. Decline of native freshwater fishes in a Mediterranean watershed on the Iberian Peninsula: a quantitative assessment. Environmental Biology of Fishes 59: 11–19.

Arnell, N. W., 1999a. Climate change and global water resources. Global Environmental Change-Human and Policy Dimensions 9: S31–S49.

Arnell, N. W., 1999b. The effect of climate change on hydrological regimes in Europe: a continental perspective. Global Environmental Change-Human and Policy Dimensions 9: 5–23.

Artigas, J., A. M. Romaní, A. Gaudes, I. Muñoz & S. Sabater, 2009. Organic matter availability structures microbial biomass and activity in a Mediterranean stream. Freshwater Biology 54: 2025–2036.

Bêche, L. A. & V. H. Resh, 2007. Short-term climatic trends affect the temporal variability of macroinvertebrates in California 'Mediterranean' streams. Freshwater Biology 52: 2317–2339.

Bêche, L. A., E. P. McElravy & V. H. Resh, 2006. Long-term seasonal variation in the biological traits of benthic-macroinvertebrates in two Mediterranean-climate streams in California, USA. Freshwater Biology 51: 56–75.

Benejam, L., E. Aparicio, M. J. Vargas, A. Vila-Gispert & E. García-Berthou, 2008. Assessing fish metrics and biotic indices in a Mediterranean stream: effects of uncertain native status of fish. Hydrobiologia 603: 197–210.

Benejam, L., P. L. Angermeier, A. Munné & E. García-Berthou, 2010. Assessing effects of water abstraction on fish assemblages in Mediterranean streams. Freshwater Biology 55: 628–642.

Beniston, M., D. B. Stephenson, O. B. Christensen, C. A. T. Ferro, C. Frei, S. Goyette, K. Halsnaes, T. Holt, K. Jylha, B. Koffi, J. Palutikof, R. Scholl, T. Semmler & K. Woth, 2007. Future extreme events in European climate: an exploration of regional climate model projections. Climatic Change 81: 71–95.

Bonada, N., M. Rieradevall, N. Prat & V. H. Resh, 2006. Benthic macroinvertebrate assemblages and macrohabitat connectivity in Mediterranean-climate streams of northern California. Journal of the North American Benthological Society 25: 32–43.

Bonada, N., S. Dolédec & B. Statzner, 2007. Taxonomic and biological trait differences of stream macroinvertebrate communities between mediterranean and temperate regions: implications for future climatic scenarios. Global Change Biology 13: 1658–1671.

Boulton, A. J., 2003. Parallels and contrasts in the effects of drought on stream macroinvertebrate assemblages. Freshwater Biology 48: 1173–1185.

Boulton, A. J. & P. S. Lake, 1992. The ecology of two intermittent streams in Victoria, Australia. II. Comparisons of faunal composition between habitats, rivers and years. Freshwater Biology 27: 99–121.

Butturini, A., S. Bernal, E. Nin, C. Hellin, L. Rivero, S. Sabater & F. Sabater, 2003. Influences of the stream groundwater hydrology on nitrate concentration in unsaturated riparian area bounded by an intermittent Mediterranean stream. Water Resources Research 39: 1–13.

Chadwick, M. A. & A. D. Huryn, 2005. Response of stream macroinvertebrate production to atmospheric nitrogen deposition and channel drying. Limnology and Oceanography 50: 228–236.

Chadwick, M. A. & A. D. Huryn, 2007. Role of habitat in determining macroinvertebrate production in an intermittent-stream system. Freshwater Biology 52: 240–251.

Cross, W. F., J. B. Wallace, A. D. Rosemond & S. L. Eggert, 2006. Whole-system nutrient enrichment increases secondary production in a detritus-based ecosystem. Ecology 87: 1556–1565.

Davey, A. J. H. & D. J. Kelly, 2007. Fish community responses to drying disturbances in an intermittent stream: a landscape perspective. Freshwater Biology 52: 1719–1733.

Deegan, L. A. & B. J. Peterson, 1992. Whole-river fertilization stimulates fish production in an arctic tundra river. Canadian Journal of Fisheries and Aquatic Sciences 49: 1890–1901.

Deegan, L. A., B. J. Peterson, H. Golden, C. C. McIvor & M. C. Miller, 1997. Effects of fish density and river fertilization on algal standing stocks, invertebrate communities, and fish production in an arctic river. Canadian Journal of Fisheries and Aquatic Sciences 54: 269–283.

del Rosario, R. B. & V. H. Resh, 2000. Invertebrates in intermittent and perennial streams: is the hyporheic zone a refuge from drying? Journal of the North American Benthological Society 19: 680–696.

Dewson, Z. S., A. B. W. James & R. G. Death, 2007. Invertebrate community responses to experimentally reduced discharge in small streams of different water quality. Journal of the North American Benthological Society 26: 754–766.

Feminella, J. W., 1996. Comparison of benthic macroinvertebrate assemblages in small streams along a gradient of flow permanence. Journal of the North American Benthological Society 15: 651–669.

García-Berthou, E. & R. Moreno-Amich, 1993. Multivariate analisis of covariance in morphometric studies of reproductive cycle. Canadian Journal of Fisheries and Aquatic Sciences 50: 1394–1399.

García-Berthou, E. & R. Moreno-Amich, 2000. Rudd (*Scardinius erythrophthalmus*) introduced to the Iberian peninsula: feeding ecology in Lake Banyoles. Hydrobiologia 436: 159–164.

Gasith, A. & V. H. Resh, 1999. Streams in Mediterranean climate regions: abiotic influences and biotic responses to predictable seasonal events. Annual Review of Ecology and Systematics 30: 51–81.

Giorgi, F., X. Q. Bi & J. Pal, 2004. Mean, interannual variability, trends in a regional climate change experiment over Europe. II: climate change scenarios (2071–2100). Climate Dynamics 23: 839–858.

Golladay, S. W., P. Gagnon, M. Kearns, J. M. Battle & D. W. Hicks, 2004. Response of freshwater mussel assemblages (Bivalvia: Unionidae) to a record drought in the Gulf Coastal Plain of southwestern Georgia. Journal of the North American Benthological Society 23: 494–506.

Grossman, G. D., R. E. Ratajczak, M. Crawford & M. C. Freeman, 1998. Assemblage organization in stream fishes: effects of environmental variation and interspecific interactions. Ecological Monographs 68: 395–420.

Halwas, K. L., M. Church & J. S. Richardson, 2005. Benthic assemblage variation among channel units in high-gradient streams on Vancouver Island, British Columbia. Journal of the North American Benthological Society 24: 478–494.

Humphries, P. & D. S. Baldwin, 2003. Drought and aquatic ecosystems: an introduction. Freshwater Biology 48: 1141–1146.

Hurlbert, S. H., 1971. The nonconcept of species diversity: a critique and alternative parameters. Ecology 52: 577–586.

IPCC, 2007. Climate Change 2007: Synthesis Report. Intergovernmental Panel on Climate Change. Fourth Assessment Report.

Keddy, P. A., 1989. Competition. Chapman and Hall, London.

Labbe, T. R. & K. D. Fausch, 2000. Dynamics of intermittent stream habitat regulate persistence of a threatened fish at multiple scales. Ecological Applications 10: 1774–1791.

Lake, P. S., 2003. Ecological effects of perturbation by drought in flowing waters. Freshwater Biology 48: 1161–1172.

Ludlam, J. P. & D. D. Magoulick, 2009. Spatial and temporal variation in the effects of fish and crayfish on benthic communities during stream drying. Journal of the North American Benthological Society 28: 371–382.

Lytle, D. A. & N. L. Poff, 2004. Adaptation to natural flow regimes. Trends in Ecology & Evolution 19: 94–100.

Magalhães, M. F., I. J. Schlosser & M. J. Collares-Pereira, 2003. The role of life history in the relationship between population dynamics and environmental variability in two Mediterranean stream fishes. Journal of Fish Biology 63: 300–317.

Magalhães, M. F., P. Beja, I. J. Schlosser & M. J. Collares-Pereira, 2007. Effects of multi-year droughts on fish assemblages of seasonally drying Mediterranean streams. Freshwater Biology 52: 1494–1510.

Magoulick, D. D., 2000. Spatial and temporal variation in fish assemblages of drying stream pools: the role of abiotic and biotic factors. Aquatic Ecology 34: 29–41.

Magoulick, D. D. & R. M. Kobza, 2003. The role of refugia for fishes during drought: a review and synthesis. Freshwater Biology 48: 1186–1198.

Matthews, W. J. & E. Marsh-Matthews, 2003. Effects of drought on fish across axes of space, time and ecological complexity. Freshwater Biology 48: 1232–1253.

Medici, C., A. Butturini, S. Bernal, E. Vazquez, F. Sabater, J. I. Velez & F. Frances, 2008. Modelling the non-linear hydrological behaviour of a small Mediterranean forested catchment. Hydrological Processes 22: 3814–3828.

Meyer, E., 1989. The relationship between body length parameters and dry mass in running water invertebrates. Archiv für Hydrobiologie 117: 191–203.

Muñoz, I., 2003. Macroinvertebrate community structure in an intermittent and a permanent Mediterranean streams (NE Spain). Limnetica 22: 107–116.

Nicola, G. G., A. Almodóvar & B. Elvira, 2009. Influence of hydrologic attributes on brown trout recruitment in low-latitude range margins. Oecologia 160: 515–524.

Nilsson, C., A. Ekblad, M. Dynesius, S. Backe, M. Gardfjell, B. Carlberg, S. Hellqviist & R. Jansson, 1994. A comparison of species richness and traits of riparian plants between a main river channel and its tributaries. Journal of Ecology 82: 281–295.

Nunn, A. D., J. P. Harvey & I. G. Cowx, 2007. The food and feeding relationships of larval and 0+ year juvenile fishes in lowland rivers and connected waterbodies. I. Ontogenetic shifts and interspecific diet similarity. Journal of Fish Biology 70: 726–742.

Oberdorff, T., B. Hugueny & T. Vigneron, 2001. Is assemblage variability related to environmental variability? An answer for riverine fish. Oikos 93: 419–428.

Oliva-Paterna, F. J., P. A. Miñano & M. Torralva, 2003. Habitat quality affects the condition of *Barbus sclateri* in Mediterranean semi-arid streams. Environmental Biology of Fishes 67: 13–22.

Ostrand, K. G. & G. R. Wilde, 2002. Seasonal and spatial variation in a prairie stream-fish assemblage. Ecology of Freshwater Fish 11: 137–149.

Peterson, B. J., L. Deegan, J. Helfrich, J. E. Hobbie, M. Hullar, B. Moller, T. E. Ford, A. Hershey, A. Hiltner, G. Kipphut, M. A. Lock, D. M. Fiebig, V. McKinley, M. C. Miller, J. R. Vestal, R. Ventullo & G. Volk, 1993. Biological responses of a tundra river to fertilization. Ecology 74: 653–672.

Pyke, G. H., 1984. Optimal foraging theory: a critical review. Annual Review of Ecology and Systematics 15: 523–575.

Sabater, S. & K. Tockner, 2010. Effects of hydrologic alterations on the ecological quality of river ecosystems. In Sabater, S. & D. Barceló (eds), Water Scarcity in the Mediterranean. Springer, Berlin.

Sabater, S., S. Bernal, A. Butturini, E. Nin & F. Sabater, 2001. Wood and leaf debris input in a Mediterranean stream: the influence of riparian vegetation. Archiv für Hydrobiologie 153: 91–102.

Sabater, S., V. Acuña, A. Giorgi, E. Guerra, I. Muñoz & A. M. Romaní, 2005. Effects of nutrient inputs in a forested Mediterranean stream under moderate light availability. Archiv für Hydrobiologie 163: 479–496.

Spranza, J. J. & E. H. Stanley, 2000. Condition, growth, and reproductive styles of fishes exposed to different environmental regimes in a prairie drainage. Environmental Biology of Fishes 59: 99–109.

Statzner, B., S. Dolédec & B. Hugueny, 2004. Biological trait composition of European stream invertebrate communities: assessing the effects of various trait filter types. Ecography 27: 470–488.

Stead, T. K., J. M. Schmid-Araya & A. G. Hildrew, 2003. All creatures great and small: patterns in the stream benthos across a wide range of metazoan body size. Freshwater Biology 48: 532–547.

Tachet, H., P. Richoux, M. Bournaud & P. Usseglio-Polatera, 2000. Invertébrés d'eau douce. CNRS éditions, Paris.

Taylor, C. M., M. R. Winston & W. J. Matthews, 1996. Temporal variation in tributary and mainstem fish assemblages in a Great Plains stream system. Copeia 1996: 280–289.

Torralva, M. D. M., M. A. Puig & C. Fernández-Delgado, 1997. Effect of river regulation on the life-history patterns of *Barbus sclateri* in the Segura river basin (south-east Spain). Journal of Fish Biology 51: 300–311.

Vanderploeg, H. A. & D. Scavia, 1979. Calculation and use of selectivity coefficients of feeding: zooplankton grazing. Ecological Modelling 7: 135–149.

Vasiliou, A. & P. S. Economidis, 2005. On the life-history of *Barbus peloponnesius* and *Barbus cyclolepis* in Macedonia, Greece. Folia Zoologica 54: 316–336.

Vila-Gispert, A. & R. Moreno-Amich, 2001. Mass–length relationship of Mediterranean barbel as an indicator of environmental status in South-west European stream ecosystems. Journal of Fish Biology 59: 824–832.

Williams, D. D., 1996. Environmental constraints in temporary fresh waters and their consequences for the insect fauna. Journal of the North American Benthological Society 15: 634–650.

Zaret, T. M. & A. S. Rand, 1971. Competition in tropical stream fishes: support for competitive exclusion principle. Ecology 52: 336–342.

Hydrobiologia (2010) 657:181–198
DOI 10.1007/s10750-009-0080-7

GLOBAL CHANGE AND RIVER ECOSYSTEMS

# Global change and food webs in running waters

Daniel M. Perkins · Julia Reiss ·
Gabriel Yvon-Durocher · Guy Woodward

Received: 20 July 2009 / Accepted: 29 December 2009 / Published online: 26 January 2010
© Springer Science+Business Media B.V. 2010

**Abstract** Riverine habitats are vulnerable to a host of environmental stressors, many of which are increasing in frequency and intensity across the globe. Climate change is arguably the greatest threat on the horizon, with serious implications for freshwater food webs via alterations in thermal regimes, resource quality and availability, and hydrology. This will induce radical restructuring of many food webs, by altering the identity of nodes, the strength and patterning of interactions and consequently the dynamics and architecture of the trophic network as a whole. Although such effects are likely to be apparent globally, they are predicted to be especially rapid and dramatic in high altitude and latitude ecosystems, which represent 'sentinel systems'. The complex and subtle connections between members of a food web and potential synergistic interactions with other environmental stressors can lead to seemingly counterintuitive responses to perturbations that cannot be predicted from the traditional focus of studying individual species in isolation. In this review, we highlight the need for developing new network-based approaches to understand and predict the consequences of global change in running waters.

**Keywords** Climate change · Ecological networks · Freshwater · Food webs · Global warming · Invasive species · Hydrology · Stoichiometry · Streams · Rivers

## Introduction

Running waters are among the most heavily impacted of all natural ecosystems (Ricciardi & Rasmussen, 1999; Sala et al., 2000; Malmqvist & Rundle, 2002). Eutrophication, pollution, acidification, overharvesting, introductions of exotic species, overabstraction, and habitat destruction are all well documented threats to the inhabitants of running waters, and the goods and services they provide (Carpenter et al., 1992). Each of these perturbations occurs worldwide, but often in relatively discrete locales (Malmqvist & Rundle, 2002). In contrast, the unprecedented changes in the climate system predicted for the coming century will have far reaching impacts on freshwaters on a truly global scale (Parmesan & Yohe, 2003; Parmesan, 2006), as well as inducing potentially dangerous synergies with more localised stressors. For instance, anoxic events in organically polluted waters will increase in intensity, duration and frequency as temperatures rise (Matzinger et al., 2007). Climate change is itself a complex amalgam of

Guest editors: R. J. Stevenson, S. Sabater / Global Change and River Ecosystems – Implications for Structure, Function and Ecosystem Services

D. M. Perkins · J. Reiss · G. Yvon-Durocher ·
G. Woodward (✉)
School of Biological & Chemical Sciences, Queen Mary University of London, London E1 4NS, UK
e-mail: g.woodward@qmul.ac.uk

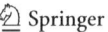

stressors that include temperature rises, altered atmospheric and hydrological conditions, and species invasions (IPCC, 2002, 2007).

Mean global temperatures have increased by 0.3–0.6°C over the past century, due largely to increased emissions of carbon dioxide ($CO_2$), methane ($CH_4$) and other greenhouse gases (IPCC, 2002, 2007). The biological effects of recent warming on the composition of riverine communities are becoming increasingly evident through species range shifts and altered phenologies (Parmesan & Yohe, 2003; Parmesan, 2006; Burgmer et al., 2007), as well as alterations in the body size distribution within food webs (McKee et al., 2002; Daufresne & Boet, 2007; Daufresne et al., 2009).

Mean global temperatures are predicted to rise by 2.8°C (IPCC, 2007 A1B scenario) within this century, but these effects will not be evenly distributed, with rates of warming predicted to be especially fast at high latitudes and altitudes: in Arctic regions, warming of as much as 4.9°C is predicted over the same time frame (IPCC, 2007). Glacial systems at high altitudes are likely to be disproportionately affected, and cold water stenotherms have already been lost from some of these areas (Brown et al., 2007). In some locations reduced and elevated flows are also predicted during summer and winter, respectively (Carpenter et al., 1992; Malmqvist & Rundle, 2002), with summer droughts exacerbating habitat shrinkage and fragmentation, thereby disrupting food web linkages and further isolating already vulnerable populations. In addition, elevated $CO_2$ levels are likely to alter the quality, quantity and timing of leaf-litter inputs, which underpin secondary production in many running waters. These key interlocking components of warming, hydrological and atmospheric changes, range shifts, and species invasions form the focus of this review.

Running waters are particularly vulnerable to global change and its associated stressors because they are relatively isolated and fragmented within terrestrial landscapes and are already heavily exploited for the goods and services they provide. Only 0.006% of the world's surface is represented by running waters, yet freshwater ecosystems contain approximately 6% of all described species (Dudgeon et al., 2006), many of which have restricted thermal and hydrological tolerances (Giller & Malmqvist, 2005). Indeed, reported extinction rates of freshwater fauna in some regions are comparable to those in tropical rain forests (Ricciardi & Rasmussen, 1999) and many important components of ecosystem functioning are also threatened (Cardinale et al., 2006).

It is still unclear how the impacts of stressors will be manifested across different levels of biological organization, but food webs provide a useful bridge between individuals (where trophic interactions occur) and the ecosystems within which they operate. Few studies have considered the higher levels of organisation, or the links between levels, hindering our current ability to predict system-level responses (Woodward, 2009).

Here, we review the array of effects and potential consequences of climate change over increasing levels of biological complexity within a food web context. We discuss the effects of global warming, alterations in hydrology and changes in atmospheric $CO_2$, in turn. Within each of these three topics, we first discuss the nodes (species populations), then links (trophic interactions) and finally the topology and dynamics (community structure and ecosystem functioning) of the trophic network. We also comment on other drivers of change, including species invasions and the potential for synergies to arise between them and other components of global change. Finally, we propose that allometric scaling (Peters, 1983; Brown et al., 2004) coupled with ecological stoichiometry (Sterner & Elser, 2002) can provide new integrative theoretical frameworks that link multiple levels of organisation and that can provide a more mechanistic and hence predictive approach.

## Global warming

An individual's metabolism is determined by its metabolic strategy (e.g. endothermic vs. ectothermic) and scales allometrically with body size and exponentially with environmental temperature (Peters, 1983; Brown et al., 2004; Fig. 1a). Since ectotherms represent approximately 99% of all freshwater species (Parmesan & Yohe, 2003) global warming will have particularly profound impacts on individual metabolism and, by extension, the physiology, bioenergetics, behaviour, abundance and distribution of species populations, with consequences for food web structure and dynamics (Ings et al., 2009; Woodward, 2009).

Migrate, adapt or perish: the role of body size in a warming climate

The ability of ectotherms to operate at different temperatures is described by thermal performance curves (Huey & Stevenson, 1979), which are optimised within a particular temperature range for each species (Vannote & Sweeney, 1980). The proximity of a species to its thermal optimum will determine the direction and magnitude of its response to warming (Thomas et al., 2004; Deutsch et al., 2008), which will force many species to live for periods in sub-optimal conditions, thus impairing individual physiological performance and compromising the population's longer term viability (Thomas et al., 2004). Each species within a community will have to migrate, adapt or face the consequences of warming, with species that are large, rare and high in the food web being especially vulnerable to local extinction (Petchey et al., 1999; Raffaelli, 2004). Tropical species might be especially susceptible in this context, because they tend to have narrower temperature envelopes; an adaptation to lower seasonal and inter-annual variability (Deutsch et al., 2008).

The ability to adapt to warmer conditions tends to be greatest among smaller species, which have large population sizes and fast generation times (Reiss & Schmid-Araya, 2008). For example, a flagellate population with a generation time of 4 days can undergo more than 4,500 generations in just 50 years. Many micro-organisms have resting stages (cysts [Protozoa] or spores [diatoms]) which can lie dormant for many years, in addition to having large population sizes and high dispersal ability (via water, air or animal vectors), leading some to speculate that most species <1 mm in length can be found almost anywhere on the planet (Finlay, 2002; Finlay & Esteban, 2007), if local conditions are suitable (but see Mann & Droop, 1996; Foissner, 1999; Hillebrand et al., 2001). Clear differences in large-scale biogeographical patterns are evident between small and large species (Hillebrand & Azovsky, 2001), but whether the distribution of small species is truly global or not, it is highly likely that many of the 'cryptic taxa' that are usually hidden within the 'seed bank' will manifest themselves as local environmental conditions approach their thermal optima (Finlay & Esteban, 2007).

For many small organisms vector-mediated dispersal enables them to traverse huge distances, as noted by Darwin, who observed that newly hatched molluscs could survive on ducks' feet in damp air for up to 24 h, 'in this length of time a duck or heron might fly at least six or seven hundred miles…[and] would be sure to alight on a pool or rivulet' (Darwin, 1920, p. 324).

Higher in the food web, the insects that dominate the macrofauna of most running waters have an active aerial dispersal phase as adults. Whilst only a few gravid female insects might be required to populate a water body (Hildrew et al., 2004; Petersen et al., 2004), physical barriers such as mountain ranges or straits between islands might be sufficient to prevent this (Milner et al., 2008) if rates of warming are sufficiently fast.

Some local food webs have the potential to at least partially compensate for species loss via the gain of new species, as species track suitable thermal conditions through range shifts. However, on a global scale warming is likely to lead to an overall loss of biodiversity, as these 'replacements' will be largely absent in tropical locations and at higher latitudes and altitudes endemic cold-stenothermal specialists close to their physiological limits will be squeezed out as temperatures continue to rise (Brown et al., 2007). Such effects might be particularly marked for large species at the top of food webs such as fishes, which have limited dispersal ability relative to micro-organisms and flying insects. Cold stenotherms, such as Arctic Charr (*Salvelinus alpines*), that inhabit isolated waterbodies are in a particularly precarious position (Rouse et al., 1997). Sweeney et al. (1992) suggest a 4°C increase in stream temperatures requires a 680 km northward range shift to maintain populations within their current thermal regime, a huge challenge for many freshwater species.

Body size, temperature and interactions in space and time

The differential abilities of species to persist within changing ecosystems suggest that many local food webs will undergo radical restructuring. In addition to range shifts, food web linkages may be disrupted by phenological mismatches between consumer and resources (Durant et al., 2007). In Lake Washington, Winder & Schindler (2004) found spring thermal stratification occurring approximately 21 days earlier in 2002 than in the 1960s, with the phytoplankton bloom shifting accordingly, whereas the consumers *Daphnia* spp., who did not respond similarly, became increasingly unable to exploit peak food availability. In running waters asynchronous phenological responses to warming could be just as dramatic, given that most macrofaunal taxa are insects with an ephemeral adult phase and emergence is driven by photoperiod and temperature.

Trophic interactions and hence the structure of food webs in freshwater systems are largely determined by the respective sizes of both consumers and resources, as many consumers are gape limited and thus consume resources smaller then themselves (Woodward et al., 2005b; Beckerman et al., 2006; Petchey et al., 2008). Interaction strengths scale allometrically with the ratio of predator/prey body mass (Emmerson et al., 2005a; Brose et al., 2006; Berlow et al., 2009) because ingestion rates, attack rates and handling times scale with predator size. Since the metabolism of ectotherms is also determined by temperature, warming might have marked influences on both the occurence and strength of consumer–resource interactions (Fig. 1a). Since many of the components of foraging behaviour are temperature dependent (Persson, 1986; Woodward & Hildrew, 2002), marked changes in the patterning and strength of interactions in these food webs are likely to arise, with potentially important consequences for the dynamic stability of these food webs (McCann, 2000; Neutel et al., 2002, 2007).

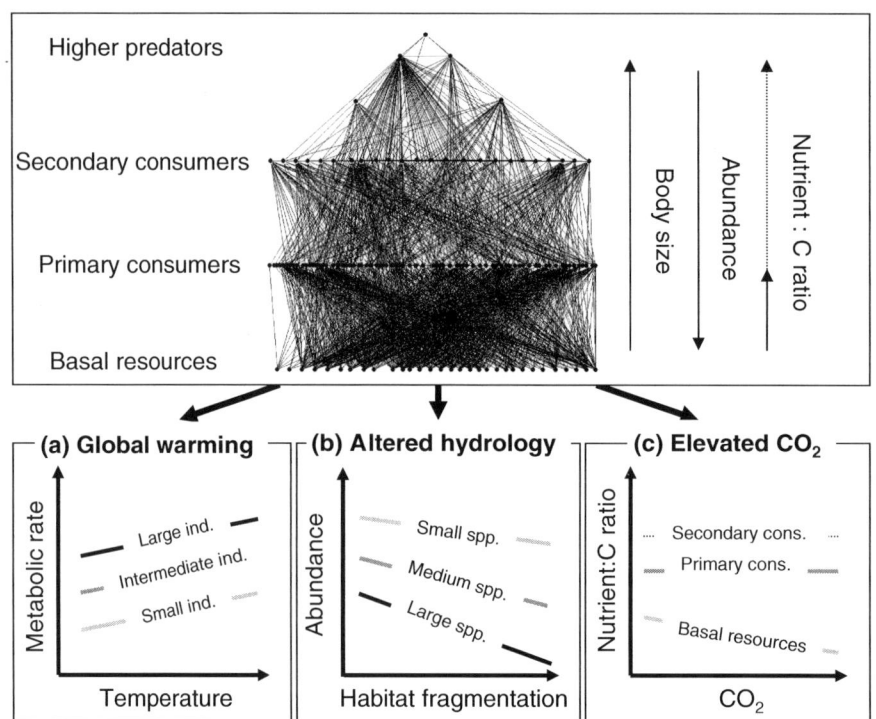

**Fig. 1** Conceptual figure highlighting common characteristics of freshwater food webs and how components of climate change may affect these attributes. Top panel: an example of a highly-resolved stream food web, redrawn after Woodward et al. (2008). *Arrows* highlight hypothetical changes in body size, abundance and nutritional quality with trophic position. **a** The effects of global warming on the basal metabolic rates for large, intermediate and small-sized individuals, on log–log scales. Metabolic demands increase with temperature and are higher for larger individuals in the food web. **b** The effects of habitat fragmentation on the abundance of different-sized species. Large species are disproportionately affected due to greater home ranges, limited dispersal ability, and reduced scope for adaptation. **c** Increasing levels of $CO_2$ and nutrient:carbon ratios for different trophic levels. Imbalances between consumers and resources arise via decreased nutrient:carbon ratios among basal resources and strict homeostasis by consumers

Biodiversity, food web stability and ecosystem functioning

A few recent studies have examined the effects of warming on the higher levels of organisation, including its potential to modulate biodiversity–ecosystem functioning relations. Petchey and co-workers (Petchey et al., 1999; Petchey, 2000), who pioneered much of this field using experimental protist assemblages in microcosms to study trajectories of community change and food web responses to warming, found that large, rare species (i.e. top predators and herbivores) were most prone to extinction, and assemblages became dominated by smaller, more abundant autotrophs and bacterivores. They also found that more diverse communities were more resilient, lending support to the insurance hypothesis (Yachi & Loreau, 1999).

The use of long-term data series over thermal gradients, provides a complementary approach to studying whole system responses. Burgmer et al. (2007) for instance found significant compositional changes in macroinvertebrate communities in response to warming in Swedish lakes and rivers since the 1990s. Durance & Ormerod (2007) detected declines in macroinvertebrate species richness and spring abundances as temperatures have risen over a 25 year period, and similar trends have also been reported in both surveys and field experiments (Mouthon & Daufresne, 2006; Hogg & Williams, 1996). These studies suggest that surprisingly dramatic biotic responses to warming can arise even over relatively small temperature gradients of just a few degrees.

Changes in the abundance, composition and diversity of species have implications for food web stability and the fluxes of resources through the web as a whole (i.e. ecosystem processes). Many theoretical studies have highlighted how weak interactions in food webs dampen the effects of destabilising, strong interactions (McCann, 2000; Neutel et al., 2002, 2007) and as interaction strength is related to body size of consumers and resources, any alterations in the size structure of the food web due to warming will have consequences for its stability. Species loss (or gain) has further implications for the overall functioning of the ecosystem, as species richness is positively correlated with many key ecosystem processes (Doak et al., 1998; Yachi & Loreau, 1999; Cardinale et al., 2006).

Biogeographical patterns: the distribution of body size in freshwaters

A common approach used to anticipate the effects of warming and resultant changes in freshwater communities is to infer patterns from large scale natural gradients in temperature. The effects of environmental temperature on body size have been studied for over 100 years (Bergmann, 1847), with individuals tending to be smaller in warmer climates (i.e. at lower latitudes). For ectotherms this has been attributed to the temperature-size rule, where individuals at warm temperatures typically mature at a smaller maximum size due to differential temperature dependencies of catabolic and anabolic processes (Atkinson, 1994). McKee et al. (2002) showed experimentally that the maximum adult body size of two cladoceran species was reduced by a 3°C year-round temperature rise, and a long-term study in three French rivers revealed a reduction in the mean body size within fish populations over the last 25 years as temperatures have risen (Daufresne & Boet, 2007). In the latter study, there were additional increases in the proportion of 'southern' species, which also led to reduced mean body size in the fish assemblage as a whole. This phenomenon, if widespread, could represent a fundamental biological response to warming (Daufresne et al., 2009) that may prove critical to understanding (and predicting) the impacts of global warming on local communities (Millien et al., 2006). If elevated temperatures benefit the small, as appears to be the case (Daufresne et al., 2009), resultant shifts in the distribution of body size at the community (i.e. the food web) level has serious implications for the dynamic stability and maintenance of ecosystem functioning in these systems (Fig. 1a).

Model systems at high latitudes and altitudes for studying warming effects on food webs

Warming is predicted to be especially pronounced in high altitude and/or latitude systems (IPCC, 2007), where it is expected to have particularly marked effects on the biota (Wrona et al., 2006; Heino et al., 2009), for several reasons: first, these areas often have low organismal density, low species diversity and organisms with slow generation times, so there is less scope for adaptation; second, many organisms are close to the edge of their physiological tolerance; third, these habitats will shrink and become increasingly fragmented as warming proceeds, reducing the regional pool of cold-water refugia. The latter point is highlighted by research carried out in glacial-fed streams in Alaska (Milner et al., 2000; Milner et al., 2008). Long-term datasets (c. 40 years) of community assembly in a few model systems, in addition to snapshots of more extensive space-for-time surveys across multiple sites, have revealed clear patterns of food web assembly and the increasing importance of cross-system subsidies, such as the supply of marine-derived nutrients to streams and ultimately their neighbouring terrestrial ecosystems, as glaciers have retreated (Milner et al., 2000; Allan et al., 2003; Wipfli et al., 2003).

Another approach is to make use of 'natural experiments', such as the temperature gradients that can be found in geothermally or tectonically heated areas at high latitudes. One example comes from recent study in a set of streams in a single geothermal catchment in Iceland, which range from about 5–25°C but which are all connected to the same mainstem and have no confounding differences in water chemistry across the thermal gradient (Friberg et al., 2009; Woodward et al., 2009). In the warmer streams the herbivore assemblage is dominated by large-bodied and very efficient grazers, in the form of the snail *Radix peregra*, whereas in the colder streams this niche is occupied by small chironomid midge larvae, some of which are specialist cold stenotherms (Woodward et al., 2009). Decomposition of terrestrial-derived detritus and algal production were faster (Fig. 2a–c, respectively) and top-down control of algae by grazers (i.e. interaction strength) was stronger in the warmed streams (Friberg et al., 2009). Fish were also more abundant and larger and food chain length (measured via stable isotope analysis) increased with temperature (Fig. 3; Woodward et al., 2009). Essentially, the key process rates and fluxes of energy between trophic levels ran faster in the warmer streams, with taxonomic community changes across the thermal gradient also representing marked 'functional shifts' in the food web.

### Altered hydrology and habitat fragmentation

The effects of drought and consequent habitat fragmentation are additional, but still poorly understood, stressors associated with climate change in many fresh waters (Carpenter et al., 1992). For example, small headwater streams in Europe with a warm, rainy climate are predicted to experience reduced flows and thus greater habitat fragmentation in summer over the coming decades (e.g. examples in Malmqvist & Rundle, 2002; Giller & Malmqvist, 2005).

Longer term oscillations in the climate also play a role here, such as the decreases in magnitude, duration and frequency of critical flows observed in a floodplain in Montana that tracked the warm phases of the Pacific decadal oscillation (PDO, warm and cold surface waters of the Pacific Ocean oscillating in a 20–30 year pattern) (Whited et al., 2007). Similarly, the El Nino-Southern Oscillation (ENSO) and the North Atlantic Oscillation (NAO) influence the hydrology of many running waters (see Malmqvist & Rundle, 2002). Since the hydrology of a stream or river depends on the climate, basin geomorphology and geology (Giller & Malmqvist, 2005) and the source of its base flow (i.e. from a lake, groundwater, rainfall and snow or ice melt) the effects of climate change will be somewhat system-specific. Reduced snow and earlier melting in glacier fed rivers, could shift the peak runoff into a different season (Barnett et al., 2005), whereas in other systems dominated by rainfall inputs, reduced precipitation and increased evaporation can result in drought and temporary fragmentation of the stream or river, with concomitant shifts in community structure (Ledger & Hildrew, 2001).

Isolated nodes in fragmented habitats

Many freshwater species have mechanisms to cope with at least some degree of hydrological change and recovery from temporary droughts can be rapid because some species remain in the dry stream surface and others recolonise the rewetted stream from refugia within or linked to the main channel (Ledger & Hildrew, 2001; Rundle & Robertson, 2002). The rate of recovery is thus coupled to the number, size and connectivity of these refugia within

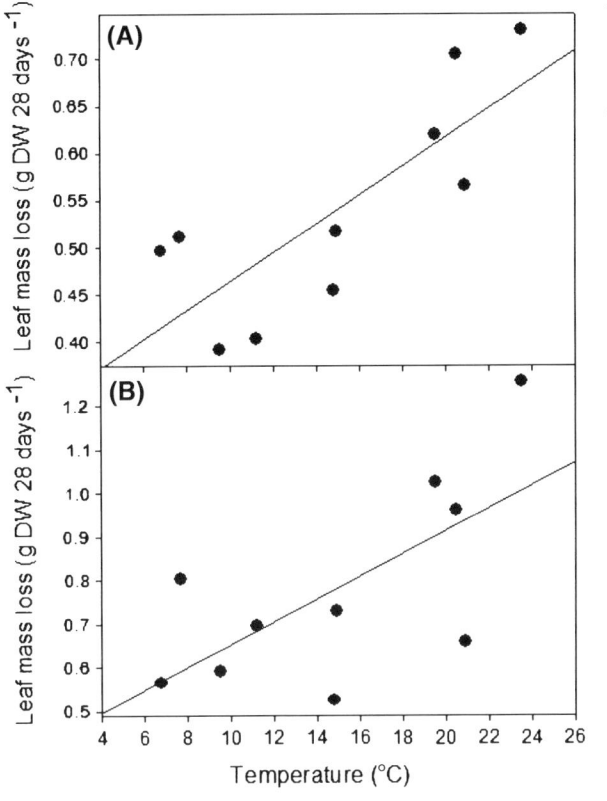
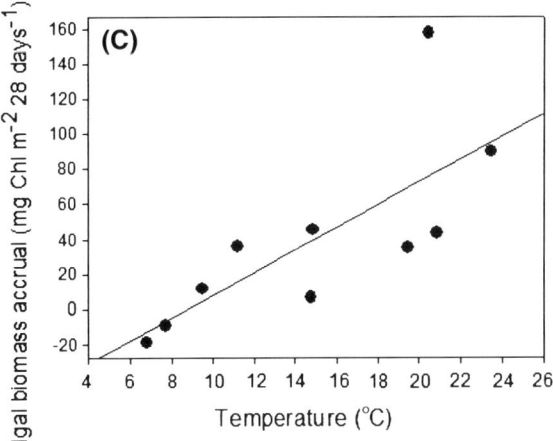

**Fig. 2** Rates of ecosystem processes in 10 Icelandic streams varying in temperature, during a 28-day experimental period. **a** Leaf litter breakdown rates in fine mesh, **b** accompanying coarse mesh bags and **c** algal biomass accrual on nutrient diffusion substrates (between different nutrient substrate treatments) as a function of temperature. Figure redrawn from Friberg (2009)

a system, in addition to the mobility or resistance of potential recolonists (Boulton, 2003). The organisms that can best withstand dry conditions are the smaller taxa with resting phases: mostly bacteria, protists, meiofauna or mesozooplankton at the base of the food web (Rundle & Robertson, 2002). Ledger & Hildrew (2001) investigated the algal assemblage in a small headwater stream after a drought and found that viable cells remained attached to the dry substratum despite 9 weeks of drying and algal biomass increased rapidly after rewetting. Macroinvertebrates can colonise new stream sections via drift from upstream sections, or by active dispersal upstream of the flying adults (Milner et al., 2000, 2008). This potential for rapid recovery is illustrated clearly by two studies, in which artificial stream channels connected to a lowland chalk stream were colonised by 127 macrofaunal taxa within 2 years (Harris et al., 2007) and in the second case the total number of taxa in the Little Stour in England increased from 60 to 80 over 3 years following a severe drought (Boulton, 2003).

However, certain taxa, including the vast majority of fish species, depend on a connected network of streams for migration and droughts represent a considerable threat to their ability to maintain viable populations. Although fish might typically migrate downstream when drought occurs in the headwaters (Giller & Malmqvist, 2005), protracted droughts could severely threaten reproductive success and the availability of habitat. Fish migration within river systems is also further restricted by habitat fragmentation caused by man-made obstacles (Zwick, 1992), and this could be exacerbated if localised droughts are induced by global climate change. When streams become fragmented into separate pools as result of drought, these might allow the survival of some of the more 'lentic' species (Boulton, 2003), but not of those that require a larger home range or fast flowing water (Fig. 1b).

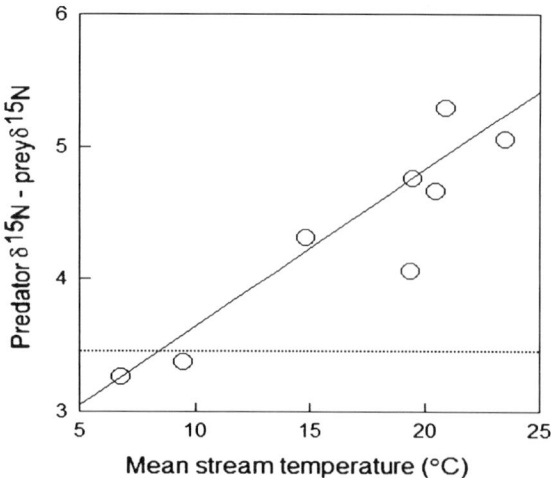

Fig. 3 Shifts in trophic height of the top predator (brown trout, *Salmo trutta*) across a temperature gradient in Icelandic geothermal streams, expressed as the difference in its $\delta 15N$ isotopic signature from that of a prey species (*Simulium vittatum*) common to each of the eight streams within a single catchment where fish were present. The *dashed line* represents a difference of one 'trophic level', assuming $\delta 15N$ fractionation of 3.4‰. Redrawn from Woodward et al. (2009)

Hydrological disturbances: effects on links, chains and food webs

Changes in hydrology can alter the relative abundance of species and promote shifts in dominance. These responses will depend on the initial species pool, because species exhibit different resistance or resilience strategies. Ledger et al. (2008) induced droughts of different length and frequency on artificial stream channels and found that in the absence of disturbance the algae biofilm was dominated by green encrusting algae. However, droughts reduced the dominance of these crust-forming species by opening space for an assemblage of mat-forming diatoms, although the establishment of the latter depended to a large extent on priority effects: i.e. which species were present before the experiment. These findings imply that trophic links between basal resources and primary consumers could be altered profoundly and that changes in hydrology will alter food chains and energy fluxes to the higher trophic levels.

There will be winners and losers associated with low flows and drought: large species at the top of food webs appear most vulnerable and are likely to be lost first, whereas small cryptic species with resting stages might even benefit from changed conditions (Fig. 1b). However, predictions to changes within communities are likely to be complicated by changes in biotic interactions within the wider food web (Hannah et al., 2007).

In addition to a general increase in summer droughts in temperate regions, climate change is also predicted to increase the frequency and magnitude of winter spates and floods (IPCC, 2007; Malmqvist and Rundle, 2002). An increase in flows during these periods provides an additional selection pressure on riverine biota that is diametrically opposed to the adaptations required to survive summer droughts. High flows could exceed the ability of many organisms to persist, especially if they lack traits to resist such conditions (e.g. extreme streamlining, hooks and grapples; Giller & Malmqvist, 2005) or if the system has few flow refugia (Gjerløv et al., 2003).

Attempting to withstand the current is one tactic for coping with increased flows; escaping or avoiding the current is another. Flow refugia may provide protection from high intensity but short-term disturbances, with dispersal in space among and within various habitats types reducing the risk of population extinctions (Hildrew & Giller, 1994; Gjerløv et al., 2003). Thus, flow refugia are likely to become increasingly important for the resilience of many species in a more variable climate, but their provision is dependant upon the system's physical and hydrological heterogeneity, both of which have often been compromised by additional anthropogenic influences, such as river engineering (Zwick, 1992). Under extreme flow events habitat refugia can be completely wiped out, such as the loss of whole riffle habitats, which can reduce consumer populations, thereby weakening top-down control within the food web (Riseng et al., 2004).

Although some level of disturbance may even enhance biodiversity (and ecosystem functioning) on a local scale, in line with the intermediate disturbance hypothesis, the key question is whether these mechanisms can still operate when the frequency of hydrological change exceeds the tolerances of key taxa within the food web.

### Elevated $CO_2$

Atmospheric $CO_2$ concentrations are predicted to nearly double to $\sim 740$ ppm by the end of the century

(IPCC [A1B scenario]), and because its aqueous concentration is regulated by complex chemical, geological and biological processes, in addition to physical gas exchange at the water surface (Wetzel, 2001), a wide range of effects are likely to be manifested in running waters.

Altered carbon:nutrient ratios in basal resources

The ecotone between the stream and the riparian zone is an area of intense biogeochemical activity (Carpenter et al., 1992). Elevated $CO_2$ levels may stimulate terrestrial primary production (Cotrufo et al., 1998; Norby et al., 2001), thereby increasing the quantity of detrital inputs to running waters and hence the degree of heterotrophy. The nutritional quality of these resources should decline, however, as the accumulation of carbon under elevated $CO_2$ conditions dilutes concentrations of nitrogen in plant tissues, thereby increasing foliar C:N ratios (Norby et al., 2001; Graca et al., 2005; Fig. 1c), which are important drivers of decomposition rates in streams (Hladyz et al., 2009).

Although many plant species can acclimatize to elevated $CO_2$ relatively quickly, others may maintain elevated growth rates (Cotrufo et al., 1998). Some species are therefore likely to gain a relative competitive advantage as the atmosphere changes, resulting in shifts in the composition of the riparian vegetation that fuels many stream food webs over and above those predicted due to biogeographical species range shifts due to warming (Fig. 4).

Elevated $CO_2$ could also induce similar changes in autochthonous basal resources, because like terrestrial plants, freshwater algae increase photosynthetic rates in response to dissolved $CO_2$ (Urabe et al., 2003). Algal nutrient content is strongly affected by the balance of light and nutrient supplies and, as photosynthesis and nutrient uptake are not perfectly coupled, altered $CO_2$ concentrations could influence the balance of algal carbon fixation relative to N and P (Fig. 1c). Some recent evidence supports this notion: elevated $CO_2$ resulted in increased algal C assimilation relative to P absorption resulting in elevated high C:P ratios (Urabe et al., 2003). Further investigation into the generality of this effect is required, however, because algal primary producers can differ markedly in their mechanisms of carbon accumulation.

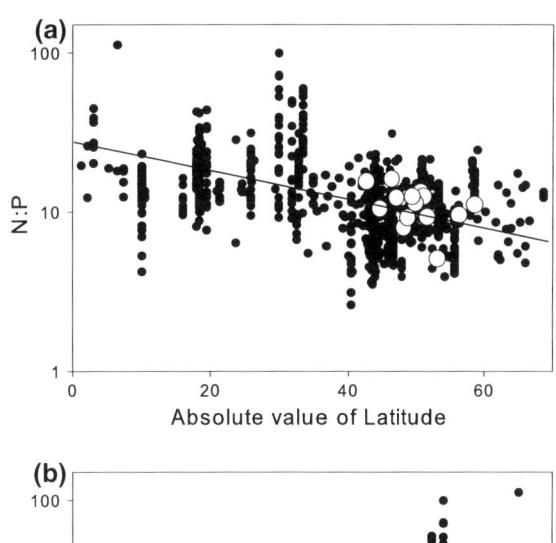

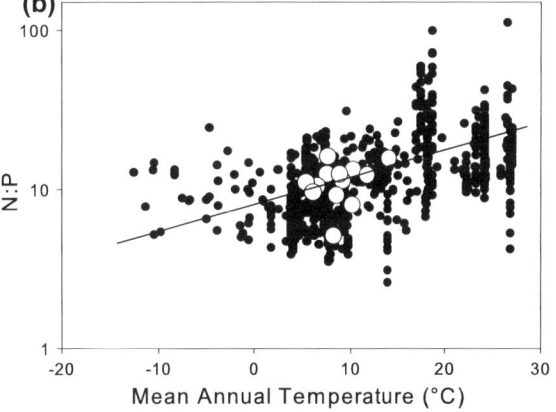

Fig. 4 Averaged leaf N:P ratio in relation to **a** absolute latitude and **b** mean annual temperature, for over 875 terrestrial plant species. *Open circles* correspond to common riparian vegetation to central Europe and North America. *Solid lines* represents regression fits **a** $r^2$ 0.24, $P < 0.0001$, $n$ 878, and **b** $r^2$ 0.31, $P < 0.0001$, $n = 894$). Redrawn with kind permission from Reich & Oleksyn (2004)

Consumer–resource stoichiometric imbalances

Primary consumers typically contain greater ratios of C to N (or P) and exhibit less variation in their elemental composition of CNP ratios than do the basal resources (Cross et al., 2006; Hladyz et al., 2009; Fig. 1c). Stoichiometric differences among consumers species from the same functional group are small (Fig. 5) and nutrient content tends to vary with trophic position and taxonomy: predators are typically the most nutrient rich (vertebrate followed by invertebrate) then primary consumers (herbivores followed by detritivores), and finally basal resources (Cross et al., 2003, 2006; Frost et al., 2006).

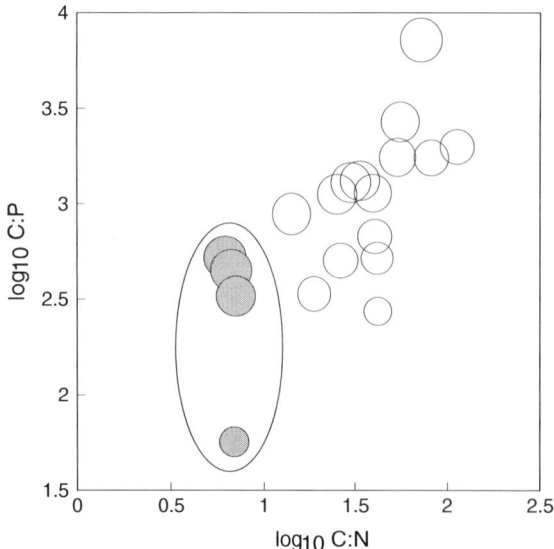

**Fig. 5** CNP (C:N, C:P and N:P) ratios in leaf-litter and consumer body tissues, illustrating stoichiometric consumer–resource imbalances among 15 leaf types (species or cultivar varieties) of varying resource quality and the four dominant shredder species (*grey bubbles*, within *dotted ellipse*). Note axes are log-transformed, and the area of each circle is proportional to $\log_{10}$ N:P. Redrawn from Hladyz et al. (2009)

Stoichiometric imbalances (i.e. where consumer ratios differ from those of their resources) suggest that consumer growth and production is tightly constrained by the nutrient content of their food, in particular by the most scarce element (Fig. 5). The consequences of these imbalances have been addressed extensively in freshwaters, but predominately in phytoplankton–zooplankton lake communities (Sterner & Elser, 2002). More recently, a few studies have addressed mismatches in benthic communities of streams and rivers (Cross et al., 2003; Frost et al., 2006; Hladyz et al., 2009) where the effects of reduced resource quality of terrestrially derived resources could be especially pervasive. Hladyz et al. (2009) showed that breakdown rates of leaf litter by stream detritivores declined as imbalances widened and resource quality fell (Fig. 5). The authors concluded that breakdown was driven by the biochemical traits (rather than taxonomic identity per se) of the terrestrial flora. Similar experiments have also emerged addressing changes in nutrient content of leaf litter following $CO_2$ enrichment. Higher C:N, total phenolic compounds and lignin content of *Populus tremuloides* grown under doubled $CO_2$ conditions resulted in 59% lower bacterial production (Tuchman et al., 2002). Reduced quality of litter and bacterial biomass greatly impacted the primary consumers (craneflies), which consumed and assimilated less, and grew 12 times slower than on ambient-grown leaf litter. In a similar experiment mosquito larvae showed increased mortality and delayed larval development on leaf litter grown under elevated $CO_2$ (Tuchman et al., 2003). Such findings question the ability of species to display homeostatic regulation, whereby elevated C:N and C:P ratios among basal resources force consumers to feed faster to extract the same amount of nutriment (Sterner & Elser, 2002). Aquatic detritivores thus appear unable to elevate consumption rates to meet homeostatic demands, at least in the time scales of short-term laboratory studies (Tuchman et al., 2002, 2003).

Similar effects might also arise in autochthonous-based food chains: Urabe et al. (2003) showed the growth of planktonic herbivore *Daphnia pulicaria*, slowed when fed with algae grown under elevated $pCO_2$. Similarly, although not manipulating changes in $CO_2$ conditions directly, Stelzer & Lamberti (2002) found that the snail *Elimia livescens* grew 40–66% faster on periphyton with high P content but that they did not actively compensate for the lower quality food by increasing ingestion rates. When faced with resources varying in quality, generalist consumers may therefore display adaptive foraging. Kominoski et al. (2007) showed that crayfish (*Orconectes virilise*) fed faster on the higher-quality periphyton that grew using DOC from trees (*Populus tremuloides*) reared under 'ambient' conditions, than on periphyton reared under elevated $CO_2$.

Dynamic stability of allochthonous
and autochthonous pathways

Since increases in productivity and decreases in the quality of basal resources has profound implications for the nodes (e.g. population production and growth rates) and the pattering and strength of interactions (consumer-resource stoichiometric imbalances and ingestion rates) such changes will influence the dynamic stability and fluxes of energy through the food web. Despite the current scarcity of literature, we can make some cursory assessments and tentative predictions of the effects of elevated $CO_2$ at these higher levels of organisation.

Whilst few studies have addressed greater amounts of terrestrial inputs expected with increasing $CO_2$ directly, Wallace et al. (1997) conducted a 3-year study of the effects of litter exclusion on a benthic stream food web, which induced a strong bottom up suppression of consumer populations. Consequently, increased litter inputs due to $CO_2$ enrichment might increase secondary production, diversity, and hence stability, by favouring 'slow' detrital-based pathways within the food web (Rooney et al., 2006). If, however, elevated $CO_2$ favours the relative importance of autochthonous algal-based pathways in the food web, due to greater carbon availability, and if facultative herbivore–detritivores switch to the more nutritious algal resources, this could potentially weaken the stabilising effect of detrital-based pathways (Rooney et al., 2006). This could create an indirect feedback onto the algal-based pathways, as detrital subsidies sustain consumers at higher densities than would otherwise be possible if they fed exclusively on algae.

Emmerson et al. (2005a) used a modelling approach to simulate a scenario of $CO_2$ enrichment on simple food chains. They assumed that elevated $CO_2$ increases productivity yet decreases the food quality of autotrophs. Consumption rates of primary consumers were fixed to resemble homeostasis, i.e. consumption rates increase to compensate for a decrease in resource quality. Finally, primary consumers were eaten by an omnivorous predator species. The authors found that these assumptions had implications for the persistence of food chains. Herbivores increased their consumption rates to compensate for reduced resource quality, and the rise in interaction strengths destabilised the system, in line with the predictions of food web theory (cf. Neutel et al., 2002).

An alternative scenario could arise if consumers do not elevate feeding rates in response to lower quality food, in which case they must either adapt to lower nutrient levels or face increased metabolic costs of nutrient acquisition (Fig. 1c). Whilst some evidence suggests certain taxa might be capable of deviating from strict homeostatis (Cross et al., 2003; Rothlisberger et al., 2008), the overriding view (Stelzer & Lamberti, 2002; Sterner & Elser, 2002; Tuchman et al., 2002, 2003; Urabe et al., 2003) is that consumers maintain fairly rigid inherent stoichiometric ratios: i.e. they should experience elevated metabolic costs associated with elevated $CO_2$ conditions, resulting in slower growth rates. However, further research is required to test the generality of this observation.

**Range shifts and species invasions**

Species range shifts and invasions associated with global change will be overlain on changes in the physicochemical environment, imposing additional biotic stressors upon freshwater food webs. For instance, in addition to any potential changes to stoichiometric imbalances resulting from atmospheric change, differential range shifts of consumers and resources are likely to lead to altered elemental fluxes through stream food webs. C:N:P ratios differ among terrestrial vegetation types over large spatial scales, and this subsidy fuels much of the secondary production in headwater streams (Wallace et al., 1997; Woodward et al., 2005b). In a large scale meta-analysis, Reich & Oleskyn (2004) highlighted that concentrations and ratios of key nutrients, nitrogen (N) and phosphorous (P) in leaf material tend to decrease with increasing latitude (Fig. 4a) and thus increase with temperature (Fig. 4b). The mechanistic basis of this has been ascribed to geographical gradients in soil substrate age and physiological stoichiometry, whereby elevated N and P in plants at higher latitudes might result from the requirements for rapid growth during the short growing season. N and P are associated with proteins and ribosomal RNA, both of which are fundamental apparatus for somatic growth (Sterner & Elser, 2002). Therefore, elevated N and/or P relative to C in plants adapted to high latitude conditions might result from increased allocations to these subcellular structures to achieve rapid growth in a short time. Thus, range expansions by 'southern species' in response to warming will be constrained to some extent by predetermined soil conditions and physiological adaptations to local climate. A change in the identity of riparian vegetation and/or nutritional quality (Fig. 4) has potentially far reaching implications for riverine communities. Reduced quality of basal resources could impair in-stream rates of decomposition and secondary production across the food web as a whole. In addition, consumer populations might also be pre-adapted to the taxonomic identity and quality of local terrestrial inputs (Carpenter et al., 1992), further inhibiting the efficiency of energy transfer through the food web if the basal resources change.

New consumer species that are able to exploit the predicted changes in temperature, shorter duration of ice cover and altered hydrology could become increasingly familiar colonists of riverine food webs in the coming decades, as has been predicted for warmwater fishes in North America and Europe, which are expected to expand their ranges northwards as cold waterbodies warm (Rahel & Olden, 2008). Within a relatively short space of time we could witness a weakening, or even breaking, of many contemporary food web linkages, as colonisation rates will differ among species (and trophic position), in a manner that is somewhat analogous to phenological mismatches—except that in this instance the mismatch arises from an interaction between temporal and spatial dimensions.

The differential dispersal abilities of species and source-sink dynamics come into play here. If microscopic taxa are effectively a large global diaspora, as suggested by Finlay & Esteban (2007), then we would expect rapid changes in community composition at the lower trophic levels within food webs as the regional species pool responds to changing local conditions and new species manifest themselves within the food web. At the other extreme of the body size spectrum, large, mobile top predators may be able to 'island-hop' to keep track of changing conditions, thereby linking together otherwise relatively isolated food webs.

## Additional drivers, synergies and feedbacks

Altered thermal, atmospheric and hydrological conditions, and species invasions associated with a changing climate will not be operating in isolation. These drivers of change are interconnected in a complex manner and as such are likely to cause higher order synergistic (i.e. non-additive) interactions and feedbacks (IPCC, 2007), making predictions to the overall effects of climate change in running waters problematic. Further, freshwaters worldwide are already exposed to a multitude of localised human impacts, some of which are not necessarily directly linked to climate change (e.g. eutrophication, acidification, overexploitation of fisheries). Some synergies among these components of change have already been recognised (Malmqvist & Rundle, 2002; Giller et al., 2004; Woodward, 2009), but overall, the potential effects of interactive components of global change on biotic communities remains poorly understood (Woodward et al., 2010).

Arguably the greatest combined threats for many riverine systems are the reduced summer flows and elevated temperatures predicted for many areas (IPCC, 2007), as the biota will be subjected to entirely new thermal conditions and greater habitat fragmentation. Species populations that are susceptible to elevated temperatures (e.g. due to localised anoxia) may also be vulnerable to reduced flows (e.g. due to habitat fragmentation), resulting in a non-additive net effect when both stressors are operating simultaneously. Such effects are likely to be amplified by other human activities, such as water abstraction (Carpenter et al., 1992; Malmqvist & Rundle, 2002). Additional synergies may occur as the quality of basal resources is altered in response to elevated $CO_2$. Metabolic requirements will rise with temperature (unless the initial conditions are under cold-stress), but they might be constrained by increased stoichiometric imbalances, which could suppress biomass production across the food web as a whole. This could result in a shortening of food chains, compromising the supply of valuable ecosystem services such as the maintenance of commercial fisheries.

Species invasions are likely to create additional synergies, as polewards range shifts in vegetation could also lead to poorer or higher quality litter inputs in temperate systems (due to colonisation by Mediterranean/chapparal species) and boreal systems (due to replacement of conifers by broadleaf species), respectively. Overlain on this is the human induced introduction of 'alien' (i.e. non-native) species, which can infiltrate contemporary food webs and which could alter ecosystem process rates at the base of the web (Hladyz et al., 2009), or induce top down effects if the invaders enter higher in the food chain (Guan & Wiles, 1997; Rahel & Olden, 2008).

Although we have highlighted how multiple stressors might combine to amplify the effects of change, in some instances the opposite may be true (Malmqvist & Rundle, 2002; Giller et al., 2004;

Woodward, 2009). Considerably, more research is now required in this area, especially in heavily impacted running waters, in which a wide range of stressors are already operating.

## Future directions

We have highlighted how different components of climate change have had, and will continue to have, significant effects on species populations, interactions, and the structure and functioning of food webs in running waters. Although some of these topics have already received substantial research interest, many important questions remain unanswered and we need to develop a more integrated approach to assess the consequences of global change. In particular, we need to move away from focussing too heavily on contingent case studies to identify more general responses if we are to predict future impacts in riverine food webs and the goods and services they provide. The question, then, is: how might we progress towards this goal?

Quantitative scaling from individuals to food webs

By considering what constrains each organism, we should be able to make links to the more complex, higher levels of organisation, since these are ultimately composed of numerous interacting individuals. We have seen how body size is a strong determinant of a plethora of physiological, biological and ecological attributes of an individual, and also of key ecosystem properties (Woodward et al., 2005a). Given that individual metabolism represents the fundamental energetic and resource acquisition constraints acting upon an individual (Fig. 1a), the body size–metabolism relationship provides one mechanistic link between individuals and the behaviour of entire systems, and it is a useful predictor of food web structure and dynamics (Emmerson et al., 2005b; Woodward et al., 2005a). Recent metabolic scaling theories (Brown et al., 2004) have sought to couple long-established body size allometries (Peters, 1983) and temperature scaling (Gillooly, 2000), which offers a powerful means of linking individuals to populations, communities and ultimately to food webs and ecosystems in a warming climate

(Woodward & Warren, 2007; Berlow et al., 2009). Similarly, ecological stoichiometry offers an alternative approach with which to understand consumer–resource interactions, based upon principles of mass balance for multiple chemical elements (Sterner & Elser, 2002), which offers great promise for predicting system-level responses to climate change.

Whilst a suite of theoretical frameworks are now starting to emerge, the testing of predictions and further development of mechanistic models is often constrained by the data collected (Brown & Gillooly, 2003). This is particularly true of food web ecology. We can see how allometric, metabolic and foraging theories related to individual body size and metabolism can be linked to the higher levels of organization, but the empirical and experimental data with which to test and validate models are in short supply. Part of the reason for this is that traditionally the nodes within food webs have been described by 'species averaging', rather than being constructed from the individual level (Ings et al., 2009). However, when constructed from individual body sizes (i.e. where nodes represent an individual's body size and prey size of the individuals it has consumed), the patterning and structure of food web can alter markedly (Ings et al., 2009). To the best of our knowledge at present only a single stream food web, that of Broadstone Stream, has been constructed in this fashion (Woodward et al., 2005b; Woodward & Warren, 2007), which restricts our ability to generalise. Thus, a clear research priority must be to collect additional food web data at the appropriate level of resolution, to facilitate the testing and development of theoretical models (Brown & Gillooly, 2003), and aquatic systems appear to provide an ideal opportunity in which to achieve this (Ings et al., 2009).

Beyond food webs

To date, still relatively few studies have addressed biodiversity–ecosystem functioning (B–EF) relationships in freshwaters in respect to climate and global change (but see Petchey et al., 1999). An additional area for future research lies in unravelling the mechanistic link between food web structure and ecosystem functioning. Food webs exhibit structural properties which affect their stability, but their relationship to ecosystem processes, goods and services remains poorly understood. For instance, species richness (i.e. food web size) is the most commonly used measure of biodiversity but it is not the only one, and if we consider functional diversity this could be related to the distribution of (individual) body sizes within the food web (Reiss et al., 2009). Individual body size is an easily measured functional 'trait' that determines an individual's (and ultimately a population's) contribution to many critical process rates (e.g. nutrient cycling, biomass production), via allometric relations and foraging constraints. In addition, interactions (i.e. links) within a food web can be just as important as species richness (i.e. nodes) for determining ecosystem functioning (Woodward, 2009). Future research aimed at examining the connections between the architecture of a network and function has the potential to inform management strategies and prioritize the protection of network structures that preserve ecosystem stability and functioning. For instance, large, mobile organisms that link together local food webs might promote the maintenance of a diverse and variable assemblage of organisms in the larger regional network, which in turn could buffer the potential impacts of environmental change (McCann, 2000). Whilst the integration of food web and biodiversity-ecosystem functioning research has started to emerge (Thebault & Loreau, 2003; Reiss et al., 2009; Woodward, 2009) there remain many fruitful areas for future research, particularly with respect to investigating biological responses to climate change.

Identifying suitable approaches and model systems

High latitude/altitude systems can be viewed as particularly useful 'sentinel systems' by providing early warnings of wider scale change, and studying these relatively simple food webs will also enable us to gain better insight into climate change impacts on the more complex (i.e. species rich) systems in warmer climes. We can usefully apply multiple approaches here, via assessing long-term community assembly during glacier retreat (Milner et al., 2000, 2008) and by using space-for-time substitutions, including assessment of altitudinal gradients as a proxy for temperature. 'Natural experiments' in high latitude geothermal areas can also be very powerful model systems, as confounding biogeographical and other physicochemical gradients can be removed (Friberg et al., 2009). We also need to combine correlational survey-based studies with experiments to identify the underlying causal relationships that produce the patterns we see in natural systems (Woodward, 2009). Manipulations of whole food webs are still vanishingly rare, but these are now needed if we are to achieve a more mechanistic, and hence predictive understanding, and also if we are to gain insight into how temperature acts as a dynamic stressor, rather than as a 'static' environmental descriptor. We might expect to see rather different food web responses in a cold stream that is warmed experimentally, as the resident cold-adapted fauna and flora are placed under an unfamiliar stress, than inferences drawn from comparing cold and warm streams at their respective equilibrial conditions. This issue relates to both the approach and the timescale of the study, and there is a potential mismatch between the phenomena manifested in experimental manipulations (i.e. transient dynamics under stress) and those in space-for-time surveys (i.e. equilibrial conditions under different optima). This critical point needs to be considered carefully when extrapolating to predict future scenarios. Clearly, we still have a long way to go, but the rate of progress is accelerating rapidly—by using a combination of approaches, from surveys to experiments to models, and a range of carefully selected model systems we can hope to gain much greater insight into how riverine food webs will respond to the dramatic changes that they will face in the coming decades.

**Acknowledgements** We would like to thank the Natural Environment Research Council for financial support awarded to GW (grant reference: NE/D013305/1) which funded DMP and JR and to Dr Jose Montoya, Dr Mark Trimmer and GW (NER/S/A2006/14029), which funded GY-D.

## References

Allan, J. D., M. S. Wipfli, J. P. Caouette, A. Prussian & J. Rodgers, 2003. Influence of streamside vegetation on inputs of terrestrial invertebrates to salmonid food webs. Canadian Journal of Fisheries and Aquatic Sciences 60: 309–320.

Atkinson, D., 1994. Temperature and organism size – a biological law for ectotherms. Advances in Ecological Research 25: 1–58.

Barnett, T. P., J. C. Adam & D. P. Lettenmaier, 2005. Potential impacts of a warming climate on water availability in snow-dominated regions. Nature 438: 303–309.

Beckerman, A. P., O. L. Petchey & P. H. Warren, 2006. Foraging biology predicts food web complexity. Proceedings of the National Academy of Sciences of the United States of America 103: 13745–13749.

Bergmann, C., 1847. Über die Verhältnisse der Wärmeökonomie der Thiere zu ihrer Grösse. Göttinger Studien 3: 595–708.

Berlow, E. L., J. A. Dunne, N. D. Martinez, P. B. Stark, R. J. Williams & U. Brose, 2009. Simple prediction of interaction strengths in complex food webs. Proceedings of the National Academy of Sciences of the United States of America 106: 187–191.

Boulton, A. J., 2003. Parallels and contrasts in the effects of drought on stream macroinvertebrate assemblages. Freshwater Biology 48: 1173–1185.

Brose, U., T. Jonsson, E. L. Berlow, P. Warren, C. Banasek-Richter, L. F. Bersier, J. L. Blanchard, T. Brey, S. R. Carpenter, M. F. C. Blandenier, L. Cushing, H. A. Dawah, T. Dell, F. Edwards, S. Harper-Smith, U. Jacob, M. E. Ledger, N. D. Martinez, J. Memmott, K. Mintenbeck, J. K. Pinnegar, B. C. Rall, T. S. Rayner, D. C. Reuman, L. Ruess, W. Ulrich, R. J. Williams, G. Woodward & J. E. Cohen, 2006. Consumer–resource body-size relationships in natural food webs. Ecology 87: 2411–2417.

Brown, J. H. & J. F. Gillooly, 2003. Ecological food webs: high-quality data facilitate theoretical unification. Proceedings of the National Academy of Sciences of the United States of America 100: 1467–1468.

Brown, J. H., J. F. Gillooly, A. P. Allen, V. M. Savage & G. B. West, 2004. Toward a metabolic theory of ecology. Ecology 85: 1771–1789.

Brown, L. E., D. M. Hannah & A. M. Milner, 2007. Vulnerability of alpine stream biodiversity to shrinking glaciers and snowpacks. Global Change Biology 13: 958–966.

Burgmer, T., H. Hillebrand & M. Pfenninger, 2007. Effects of climate-driven temperature changes on the diversity of freshwater macroinvertebrates. Oecologia 151: 93–103.

Cardinale, B. J., D. S. Srivastava, J. E. Duffy, J. P. Wright, A. L. Downing, M. Sankaran & C. Jouseau, 2006. Effects of biodiversity on the functioning of trophic groups and ecosystems. Nature 443: 989–992.

Carpenter, S. R., S. G. Fisher, N. B. Grimm & J. F. Kitchell, 1992. Global change and freshwater ecosystems. Annual Review of Ecology and Systematics 23: 119–139.

Cotrufo, M. F., P. Ineson & A. Scott, 1998. Elevated $CO_2$ reduces the nitrogen concentration of plant tissues. Global Change Biology 4: 43–54.

Cross, W. F., J. P. Benstead, A. D. Rosemond & J. B. Wallace, 2003. Consumer–resource stoichiometry in detritus-based streams. Ecology Letters 6: 721–732.

Cross, W. F., J. P. Benstead, P. C. Frost & S. A. Thomas, 2006. Ecological stoichiometry in freshwater benthic systems: recent progress and perspectives (vol 50, pg 1895, 2005). Freshwater Biology 51: 986–987.

Darwin, C., 1920. The origin of species by means of natural selection, or the preservation of favoured races in the struggle for life. 6th ed. John Murray, London.

Daufresne, M. & P. Boet, 2007. Climate change impacts on structure and diversity of fish communities in rivers. Global Change Biology 13: 2467–2478.

Daufresne, M. K., K. Lengfellner & U. Sommer, 2009. Global warming benefits the small in aquatic ecosystems. Proceedings of the National Academy of Sciences of the United States of America 106: 12788–12793.

Deutsch, C. A., J. J. Tewksbury, R. B. Huey, K. S. Sheldon, C. K. Ghalambor, D. C. Haak & P. R. Martin, 2008. Impacts of climate warming on terrestrial ectotherms across latitude. Proceedings of the National Academy of Sciences of the United States of America 105: 6668–6672.

Doak, D. F., D. Bigger, E. K. Harding, M. A. Marvier, R. E. O'Malley & D. Thomson, 1998. The statistical inevitability of stability–diversity relationships in community ecology. American Naturalist 151: 264–276.

Dudgeon, D., A. H. Arthington, M. O. Gessner, Z. I. Kawabata, D. J. Knowler, C. Leveque, R. J. Naiman, A. H. Prieur-Richard, D. Soto, M. L. J. Stiassny & C. A. Sullivan, 2006. Freshwater biodiversity: importance, threats, status and conservation challenges. Biological Reviews 81: 163–182.

Durance, I. & S. J. Ormerod, 2007. Effects of climatic variation on upland stream invertebrates over a 25 year period. Global Change Biology 13: 942–957.

Durant, J. M., D. O. Hjermann, G. Ottersen & N. C. Stenseth, 2007. Climate and the match or mismatch between predator requirements and resource availability. Climate Research 33: 271–283.

Emmerson, M., T. M. Bezemer, M. D. Hunter & T. H. Jones, 2005a. Global change alters the stability of food webs. Global Change Biology 11: 490–501.

Emmerson, M. C., J. M. Montoya, G. Woodward, 2005b. Body size, interaction strength, and food web dynamics. In de Ruiter, P., V. Wolters & J. C. Moore (eds), Dynamic Food Webs: Multispecies Assemblages, Ecosystem Development and Environmental Change. Elsevier: 167–178.

Finlay, B. J., 2002. Global dispersal of free-living microbial eukaryote species. Science 296: 1061–1063.

Finlay, B. J. & G. F. Esteban, 2007. Body size and biogeography. In Hildrew, A. G., D. Raffaelli & R. Edmonds-Brown (eds), Body Size: The Structure and Function of Aquatic Ecosystems. Cambridge University Press, Cambridge: 167–185.

Foissner, W., 1999. Protist diversity: estimates of the near-imponderable. Protist 150: 363–368.

Friberg, N., J. B. Dybkjaer, J. S. Olafsson, G. M. Gislason, S. E. Larsen, & T. Lauridsen, 2009. Relationships between structure and function in streams contrasting in temperature. Freshwater Biology 54: 2051–2068.

Frost, P. C., J. P. Benstead, W. F. Cross, H. Hillebrand, J. H. Larson, M. A. Xenopoulos & T. Yoshida, 2006. Threshold elemental ratios of carbon and phosphorus in aquatic consumers. Ecology Letters 9(7): 774–779.

Giller, P. S. & B. Malmqvist, 2005. The Biology of Streams and Rivers. Oxford University Press, New York.

Giller, P. S., H. Hillebrand, U.-G. Berninger, M. O. Gessner, S. Hawkins, P. Inchausti, C. Inglis, H. Leslie, B. Malmqvist, M. T. Monaghan, P. J. Morin & G. O'Mullan, 2004. Biodiversity effects on ecosystem functioning: emerging issues and their experimental test in aquatic environments. Oikos 104: 423–436.

Gillooly, J. F., 2000. Effect of body size and temperature on generation time in zooplankton. Journal of Plankton Research 22: 241–251.

Gjerløv, C., A. G. Hildrew & J. I. Jones, 2003. Mobility of stream invertebrates in relation to disturbance and refugia: a test of habitat templet theory. Journal of the North American Benthological Society 22: 207–223.

Graca, M. A. S., F. Barlocher & M. O. Gessner. 2005. Methods to Study Litter Decomposition: A Practical Guide. Springer.

Guan, R. Z. & P. R. Wiles, 1997. Ecological impact of introduced crayfish on benthic fishes in a British lowland river. Conservation Biology 11: 641–647.

Hannah, D. M., L. E. Brown, A. M. Milner, A. M. Gurnell, G. R. McGregord, G. E. Petts, B. P. G. Smith & D. L. Snook, 2007. Integrating climate–hydrology–ecology for alpine river systems. Aquatic Conservation – Marine and Freshwater Ecosystems 17: 636–656.

Harris, R. M. L., P. D. Armitage, A. M. Milner & M. E. Ledger, 2007. Replicability of physicochemistry and macroinvertebrate assemblages in stream mesocosms: implications for experimental research. Freshwater Biology 52: 2434–2443.

Heino, J., R. Virkkala & H. Toivonen, 2009. Climate change and freshwater biodiversity: detected patterns, future trends and adaptations in northern regions. Biological Reviews 84: 39–54.

Hildrew, A. G. & Giller, P. S., 1994. Patchiness, species interactions and disturbance in the stream benthos. In Giller, P., A. Hildrew & D. Raffaelli (eds), Ecology: Scale Pattern and Process. Blackwell, Oxford: 21–62.

Hildrew, A. G., G. Woodward, J. H. Winterbottom & S. Orton, 2004. Strong density dependence in a predatory insect: large-scale experiments in a stream. Journal of Animal Ecology 73: 448–458.

Hillebrand, H. & A. I. Azovsky, 2001. Body size determines the strength of the latitudinal diversity gradient. Ecography 24: 251–256.

Hillebrand, H., F. Watermann, R. Karez & U. G. Berninger, 2001. Differences in species richness patterns between unicellular and multicellular organisms. Oecologia 126: 114–124.

Hladyz, S., M. O. Gessner, P. S. Giller, J. Pozo & G. Woodward, 2009. Resource quality and stoichiometric constraints on stream ecosystem functioning. Freshwater Biology 54: 957–970.

Hogg, I. D. & D. D. Williams, 1996. Response of stream invertebrates to a global-warming thermal regime: an ecosystem-level manipulation. Ecology 77: 395–407.

Huey, R. B. & R. D. Stevenson, 1979. Integrating thermal physiology and ecology of ectotherms – discussion of approaches. American Zoologist 19: 357–366.

Ings, T. C., J. M. Montoya, J. Bascompte, N. Bluthgen, L. Brown, C. F. Dormann, F. Edwards, D. Figueroa, U. Jacob, J. I. Jones, R. B. Lauridsen, M. E. Ledger, H. M. Lewis, J. M. Olesen, F. J. F. Van Veen, P. H. Warren & G. Woodward, 2009. Ecological networks – beyond food webs. Journal of Animal Ecology 78(1): 253–269.

IPCC, 2002. Climate Change 2002: The Scientific Basis. Cambridge University Press, Cambridge.

IPCC, 2007. Climate Change 2007: The Physical Sciences Basis. In Parry, M. L., O. F. Canziani, J. P. Palutikof, P. J. van der Linden & C. E. Hanson, (ed.), Contribution of Working Group I to the Fourth Assessment Report of the Intergovernmental Panel on Climate Change. Cambridge University Press, Cambridge.

Kominoski, J. S., P. A. Moore, R. G. Wetzel & N. C. Tuchman, 2007. Elevated $CO_2$ alters leaf-litter-derived dissolved organic carbon: effects on stream periphyton and crayfish feeding preference. Journal of the North American Benthological Society 26: 663–672.

Ledger, M. E. & A. G. Hildrew, 2001. Recolonization by the benthos of an acid stream following a drought. Archiv Fur Hydrobiologie 152: 1–17.

Ledger, M. E., R. M. L. Harris, P. D. Armitage & A. M. Milner, 2008. Disturbance frequency influences patch dynamics in stream benthic algal communities. Oecologia 155: 809–819.

Malmqvist, B. & S. Rundle, 2002. Threats to the running water ecosystems of the world. Environmental Conservation 29: 134–153.

Mann, D. G. & S. J. M. Droop, 1996. Biodiversity, biogeography and conservation of diatoms. Hydrobiologia 336: 19–32.

Matzinger, A., M. Schmid, E. Veljanoska-Sarafiloska, S. Patceva, D. Guseska, B. Wagner, B. Muller, M. Sturm & A. Wuest, 2007. Eutrophication of ancient Lake Ohrid: global warming amplifies detrimental effects of increased nutrient inputs. Limnology and Oceanography 52: 338–353.

McCann, K. S., 2000. The diversity–stability debate. Nature 405: 228–233.

McKee, D., D. Atkinson, S. Collings, J. Eaton, I. Harvey, T. Heyes, K. Hatton, D. Wilson & B. Moss, 2002. Macrozooplankter responses to simulated climate warming in experimental freshwater microcosms. Freshwater Biology 47: 1557–1570.

Millien, V., S. K. Lyons, L. Olson, F. A. Smith, A. B. Wilson & Y. Yom-Tov, 2006. Ecotypic variation in the context of global climate change: revisiting the rules. Ecology Letters 9: 853–869.

Milner, A. M., E. E. Knudsen, C. Soiseth, A. L. Robertson, D. Schell, I. T. Phillips & K. Magnusson, 2000. Colonization and development of stream communities across a 200-year gradient in Glacier Bay National Park, Alaska. U.S.A. Canadian Journal of Fisheries and Aquatic Sciences 57: 2319–2335.

Milner, A. M., A. E. Robertson, K. Monaghan, A. J. Veal & E. A. Flory, 2008. Colonization and development of a stream community over 28 years, Wolf Point Creek in Glacier Bay, Alaska. Frontiers in Ecology and Environment 6: 413–419.

Mouthon, J. & M. Daufresne, 2006. Effects of the 2003 heatwave and climatic warming on mollusc communities of the Saone: a large lowland river and of its two main tributaries (France). Global Change Biology 12: 441–449.

Neutel, A. M., J. A. P. Heesterbeek & P. C. de Ruiter, 2002. Stability in real food webs: weak links in long loops. Science 296: 1120–1123.

Neutel, A. M., J. A. P. Heesterbeek, J. van de Koppel, G. Hoenderboom, A. Vos, C. Kaldeway, F. Berendse & P. C.

de Ruiter, 2007. Reconciling complexity with stability in naturally assembling food webs. Nature 449: 599-U511.

Norby, R. J., M. F. Cotrufo, P. Ineson, E. G. O'Neill & J. G. Canadell, 2001. Elevated $CO_2$, litter chemistry, and decomposition: a synthesis. Oecologia 127: 153–165.

Parmesan, C., 2006. Ecological and evolutionary responses to recent climate change. Annual Review of Ecology Evolution and Systematics 37: 637–669.

Parmesan, C. & G. Yohe, 2003. A globally coherent fingerprint of climate change impacts across natural systems. Nature 421: 37–42.

Persson, L., 1986. Temperature-induced shift in foraging ability in 2 fish species roach (Rutilus-rutilus) and perch (Perca-fluviatilis) – implications for coexistsence between poikliotherms. Journal of Animal Ecology 55: 829–839.

Petchey, O. L., 2000. Prey diversity, prey composition, and predator population dynamics in experimental microcosms. Journal of Animal Ecology 69: 874–882.

Petchey, O. L., P. T. McPhearson, T. M. Casey & P. J. Morin, 1999. Environmental warming alters food-web structure and ecosystem function. Nature 402: 69–72.

Petchey, O. L., A. P. Beckerman, J. O. Riede & P. H. Warren, 2008. Size, foraging, and food web structure. Proceedings of the National Academy of Sciences of the United States of America 105: 4191–4196.

Peters, R. H., 1983. The Ecological Implications of Body Size. Cambridge University Press, Cambridge.

Petersen, I., Z. Masters, A. G. Hildrew & S. J. Ormerod, 2004. Dispersal of adult aquatic insects in catchments of differing land use. Journal of Applied Ecology 41: 934–950.

Raffaelli, D., 2004. How extinction patterns affect ecosystems. Science 306: 1141–1142.

Rahel, F. J. & J. D. Olden, 2008. Assessing the effects of climate change on aquatic invasive species. Conservation Biology 22: 521–533.

Reich, P. B. & J. Oleksyn, 2004. Global patterns of plant leaf N and P in relation to temperature and latitude. Proceedings of the National Academy of Sciences of the United States of America 101: 11001–11006.

Reiss, J. & J. M. Schmid-Araya, 2008. Existing in plenty: abundance, biomass and diversity of ciliates and meiofauna in small streams. Freshwater Biology 53: 652–668.

Reiss, J., J. R. Bridle, J. M. Montoya & G. Woodward, 2009. Emerging horizons in biodiversity and ecosystem functioning research. Trends in Ecology and Evolution 24: 505–514.

Ricciardi, A. & J. B. Rasmussen, 1999. Extinction rates of North American freshwater fauna. Conservation Biology 13(5): 1220–1222.

Riseng, C. M., M. J. Wiley & R. J. Stevenson, 2004. Hydrologic disturbance and nutrient effects on benthic community, structure in midwestern US streams: a covariance structure analysis. Journal of the North American Benthological Society 23(2): 309–326.

Rooney, N., K. McCann, G. Gellner & J. C. Moore, 2006. Structural asymmetry and the stability of diverse food webs. Nature 442: 265–269.

Rothlisberger, J. D., M. A. Baker & P. C. Frost, 2008. Effects of periphyton stoichiometry on mayfly excretion rates and nutrient ratios. Journal of the North American Benthological Society 27: 497–508.

Rouse, W. R., M. S. V. Douglas, R. E. Hecky, A. E. Hershey, G. W. Kling, L. Lesack, P. Marsh, M. McDonald, B. J. Nicholson, N. T. Roulet & J. P. Smol, 1997. Effects of climate change on the freshwaters of arctic and subarctic North America. Hydrological Process 11: 873–902.

Rundle, S., A. L. Robertson & J. M. Schmid-Araya, 2002. Freshwater Meiofauna: Biology and Ecology. Blackhuys, Leiden.

Sala, O. E., F. S. Chapin, J. J. Armesto, E. Berlow, J. Bloomfield, R. Dirzo, E. Huber-Sanwald, L. F. Huenneke, R. B. Jackson, A. Kinzig, R. Leemans, D. M. Lodge, H. A. Mooney, M. Oesterheld, N. L. Poff, M. T. Sykes, B. H. Walker, M. Walker & D. H. Wall, 2000. Biodiversity – Global biodiversity scenarios for the year 2100. Science 287: 1770–1774.

Stelzer, R. S. & G. A. Lamberti, 2002. Ecological stoichiometry in running waters: periphyton chemical composition and snail growth. Ecology 83: 1039–1051.

Sterner, R. W. & J. J. Elser, 2002. Ecological Stoichiometry: The Biology of Elements from Molecules to the Biosphere. Princeton University Press, Princeton, NJ.

Sweeney, B. W., J. K. Jackson, J. D. Newbold, & D. H. Funk, 1992. Climate change and the life histories and biogeography of aquatic insects in eastern North-America. In Firth, P. & S. G. Fisher (eds), Global Climate Change and Freshwater Ecosystems. Springer-Verlag, New York: 143–176.

Thebault, E. & M. Loreau, 2003. Food-web constraints on biodiversity–ecosystem functioning relationships. Proceedings of the National Academy of Sciences of the United States of America 100(25): 14949–14954.

Thomas, C. D., A. Cameron, R. E. Green, M. Bakkenes, L. J. Beaumont, Y. C. Collingham, B. F. N. Erasmus, M. F. de Siqueira, A. Grainger, L. Hannah, L. Hughes, B. Huntley, A. S. van Jaarsveld, G. F. Midgley, L. Miles, M. A. Ortega-Huerta, A. T. Peterson, O. L. Phillips & S. E. Williams, 2004. Extinction risk from climate change. Nature 427: 145–148.

Tuchman, N. C., R. G. Wetzel, S. T. Rier, K. A. Wahtera & J. A. Teeri, 2002. Elevated atmospheric $CO_2$ lowers leaf litter nutritional quality for stream ecosystem food webs. Global Change Biology 8: 163–170.

Tuchman, N. C., K. A. Wahtera, R. G. Wetzel, N. M. Russo, G. M. Kilbane, L. M. Sasso & J. A. Teeri, 2003. Nutritional quality of leaf detritus altered by elevated atmospheric $CO_2$: effects on development of mosquito larvae. Freshwater Biology 48: 1432–1439.

Urabe, J., J. Togari & J. J. Elser, 2003. Stoichiometric impacts of increased carbon dioxide on a planktonic herbivore. Global Change Biology 9: 818–825.

Vannote, R. L. & B. W. Sweeney, 1980. Geographic analysis of thermal equiliberia – a conceptual–model for evaluating the effects of natural and modified thermal regimes on aquatic insect communities. American Naturalist 115: 667–695.

Wallace, J. B., S. L. Eggert, J. L. Meyer & J. R. Webster, 1997. Multiple trophic levels of a forest stream linked to terrestrial litter inputs. Science 277: 102–104.

Wetzel, R., 2001. Limnology. Lake and River Ecosystems, 3rd ed. Academic Press, San Diego.

Whited, D. C., M. S. Lorang, M. J. Harner, F. R. Hauer, J. S. Kimball & J. A. Stanford, 2007. Climate, hydrologic disturbance, and succession: drivers of floodplain pattern. Ecology 88: 940–953.

Winder, M. & D. E. Schindler, 2004. Climate change uncouples trophic interactions in an aquatic ecosystem. Ecology 85: 2100–2106.

Wipfli, M. S., J. P. Hudson, J. P. Caouette & D. T. Chaloner, 2003. Marine subsidies in freshwater ecosystems: salmon carcasses increase the growth rates of stream-resident salmonids. Transactions of the American Fisheries Society 132: 371–381.

Woodward, G., 2009. Biodiversity, ecosystem functioning and food webs in freshwaters: assembling the jigsaw puzzle. Freshwater Biology 54: 2171–2187.

Woodward, G. & A. G. Hildrew, 2002. Differential vulnerability of prey to an invading top predator: integrating field surveys and laboratory experiments. Ecological Entomology 27: 732–744.

Woodward, G. & P. Warren, 2007. Body size and predatory interactions in freshwaters: scaling from individuals to communities. In Hildrew, A. G., D. Raffaelli & R. Edmonds-Brown (eds), Body Size: The Structure and Function of Aquatic Ecosystems. Cambridge University Press, Cambridge: 179–197.

Woodward, G., B. Ebenman, M. Ernmerson, J. M. Montoya, J. M. Olesen, A. Valido & P. H. Warren, 2005a. Body size in ecological networks. Trends in Ecology & Evolution 20: 402–409.

Woodward, G., D. C. Speirs & A. G. Hildrew, 2005b. Quantification and resolution of a complex, size-structured food web. Advances in Ecological Research 36: 85–135.

Woodward, G., G. Papantoniou, F. Edwards & R. B. Lauridsen, 2008. Trophic trickles and cascades in a complex food web: impacts of a keystone predator on stream community structure and ecosystem processes. Oikos 117(5): 683–692.

Woodward, G., J. B. Christensen, J. S. Olafsson, G. M. Gislason, E. R. Hannesdottir & N. Friberg, 2009. Sentinel systems on the razor's edge: effects of warming on Arctic stream ecosystems. Global Change Biology. doi:10.1111/j.1365-2486.2009.02052.x.

Woodward, G. D., M. Perkins & L. Brown, 2010. Climate change in freshwater ecosystems: impacts across multiple levels of organisation. Philosophical Transactions of the Royal Society of London Series B – Biological Sciences.

Wrona, F. J., T. D. Prowse, J. D. Reist, J. E. Hobbie, L. M. J. Levesque & W. F. Vincent, 2006. Climate change effects on aquatic biota, ecosystem structure and function. Ambio 35: 359–369.

Yachi, S. & M. Loreau, 1999. Biodiversity and ecosystem productivity in a fluctuating environment: The insurance hypothesis. Proceedings of the National Academy of Sciences of the United States of America 96: 1463–1468.

Zwick, P., 1992. Stream habitat fragmentation – a threat to biodiversity. Biodiversity and Conservation 1: 80–97.

GLOBAL CHANGE AND RIVER ECOSYSTEMS

# Effects of hydromorphological integrity on biodiversity and functioning of river ecosystems

Arturo Elosegi · Joserra Díez · Michael Mutz

Received: 5 August 2009 / Accepted: 29 December 2009 / Published online: 22 January 2010
© Springer Science+Business Media B.V. 2010

**Abstract** River channels tend to a dynamic equilibrium driven by the dynamics of water and sediment discharge. The resulting fluctuating pattern of channel form is affected by the slope, the substrate erodibility, and the vegetation in the river corridor and in the catchment. Geomorphology is basic to river biodiversity and ecosystem functioning since the channel pattern provides habitat for the biota and physical framework for ecosystem processes. Human activities increasingly change the natural drivers of channel morphology on a global scale (e.g. urbanization increases hydrological extremes, and clearing of forests for agriculture increases sediment yield). In addition, human actions common along world rivers impact channel dynamics directly, e.g. river regulation simplifies and fossilizes channel form. River conservation and restoration must incorporate mechanisms of channel formation and ecological consequences of channel form and dynamics. This article (1) summarizes the role of channel form on biodiversity and functioning of river ecosystems, (2) describes spatial complexity, connectivity and dynamism as three key hydromorphological attributes, (3) identifies prevalent human activities that impact these key components and (4) analyzes gaps in current knowledge and identifies future research topics.

**Keywords** River ecosystem · Hydromorphology · Biodiversity · Functioning

Guest editors: R. J. Stevenson, S. Sabater / Global Change and River Ecosystems – Implications for Structure, Function and Ecosystem Services

A. Elosegi (✉) · J. Díez
Faculty of Science and Technology, University of the Basque Country, PO Box 644, 48080 Bilbao, Spain
e-mail: arturo.elosegi@ehu.es

M. Mutz
Department of Freshwater Conservation, Brandenburg University of Technology, Seestr. 45, 15526 Bad Saarow, Germany

## Introduction

Lotic ecosystems are integral elements of landscapes, shaped by the transport of water and materials from their drainage basins (Hynes, 1975). Because the unidirectional transport occurs in a dendritic network and is highly episodic, river channels are spatially complex and temporally variable (Rosgen, 1996). As river ecologists discovered the relevance of transport, flood dynamics, channel complexity, parafluvial and floodplain areas, and other key characteristics of river ecosystems, different concepts dominated the field of river ecology. Pioneering works (Hawkes, 1975) described rivers as composed of discrete biological zones set downstream in a predictable order. The River Continuum Concept (RCC, Vannote et al., 1980), on the other hand, stressed the fact that zones are not discrete, but change in a rather continuous way along the river, driven primarily by changes in channel morphology. The RCC was criticized by some authors (Winterbourn et al., 1981; Statzner & Higler, 1985), and concepts were advanced regarding morphological discontinuity. For instance, the Serial Discontinuity Concept (Ward & Stanford, 1983) made predictions on the effect of dams, largely based on the RCC, and the network dynamics hypothesis (Benda et al., 2004) analyzed the effect of tributary confluences. On the other hand, the Flood Pulse Concept (Junk et al., 1989; Junk & Wantzen, 2003) stressed the role of the flood regime on the ecology of large rivers, and the Riverine Productivity Model (Thorp & Delong, 1994) challenged some of the tenets of the RCC regarding the source of organic matter. More recently, the Riverine Ecosystem Synthesis (Thorp et al., 2006) depicted rivers as an array of large 'hydrogeomorphic patches' (equivalent to constrained, braided or meandering sections), and stressed the fact that there is no simple way to predict the position of these patches along a river, but they are associated to distinct functional zones. Thus, biological communities are shaped by a hierarchy of environmental factors, from ones affecting regional-scale distribution of organisms, down to factors affecting communities at the reach-scale, and even at the scale of individual riffles (Parsons & Thoms, 2007).

Although scientists have long explored the role of channel morphology on river ecology, managers only recently recognized hydromorphology as an important element of streams and rivers, often focusing mainly on fish and benthic invertebrate habitat, neglecting other aspects of biodiversity and ecosystem functioning. This probably occurred because river morphology is highly variable, depending on a hierarchy of controlling factors in the catchment, including upstream and, to a lesser extent, downstream parts of the river network. Natural constraints such as climate and geology and human activities such as land use and flow regulation, determine the main drivers—hydrological regime, sediment regime and riparian vegetation—that shape the local morphology.

Because channel form and hydraulics provide a structural template that shapes ecological processes, many authors stressed their importance for river biodiversity and functioning. Different terms have been coined to stress this interplay: hydromorphology (EU, 2000), eco-geomorphology (Thoms & Parsons, 2002), functional ecomorphology (Fisher et al., 2007) and others. Despite this recognition there is still a lack of understanding of complex responses of biological processes to hydromorphology, which often shows shifting patterns rather than static balances (Lenders et al., 1998). Even when acknowledged, hydromorphology is mostly seen as static (e.g. riffle-pool sequences), neglecting the dynamic aspect of channel form, which is central to hydromorphological quality.

Newson & Large (2006) defined natural river channels as those whose geometry and features represent the full interplay of unmanaged water and sediment fluxes with local boundary conditions. Such channels are free to adjust by aggradation, degradation or by lateral interaction with the floodplain or valley floor in response to unmanaged flows and sediment supplies (short term) or in response to long-term changes in system or local drivers. They are not wilderness channels but may inspire a holistic perception of being 'intact', a popular human perception of reference conditions deriving mainly from landscape aesthetics. 'Natural' channels require minimum management intervention to offer resilience and a diversity of physical habitat, though neither of these 'natural services' is universal or perpetual.

Here we analyze the importance of hydromorphological integrity (i.e. of natural river channels), on biodiversity and functioning of river ecosystems, focusing in three key attributes for stream ecosystem functioning: spatial complexity, connectivity and dynamism. We also discuss the effects of human activities on the hydromorphological integrity of rivers and identify current gaps in knowledge, and topics for future research.

## Hydromorphological attributes of importance to stream ecology

The term hydromorphology, which was recently made popular by the European Water Framework Directive (EU, 2000), reflects the inseparable association of channel form and flow. Depending on river type, hydromorphology is expressed by a different array of morphological elements and hydrodynamic features. For instance, oxbow lakes are important constituents of lowland rivers meandering in wide valleys, but not in v-shaped constrained ones. It is beyond the objectives of the present article to give a comprehensive list of all elements constituting channel form and dynamics and to specify their characteristics for different rivers. We instead give ecologically significant examples that illustrate three attributes of key importance for river ecosystems—spatial complexity, connectivity and dynamism (Table 1). Spatial complexity is created and maintained by a series of processes acting from microhabitat to catchment scale and driven by periodic and stochastic events in time, such as discharge and sediment input. Sequences of complex reaches along a river network further increase dynamics and spatial complexity. Hence channel complexity, connectivity, and dynamism are closely linked.

Spatial complexity of channel and river corridor

Channel complexity results from processes occurring at a wide variety of mutually dependent scales (Frisell et al., 1986; Thorp et al., 2006). At the microhabitat scale, smaller than the width of the stream, sediments are usually sorted by grain size, resulting in patches that provide habitat for different organisms. Sorting processes depend on the local flow pattern that is generated at the reach scale, and largely modified by local flow obstructions such as boulders, macrophytes or large wood (Fig. 1). In the vertical and lateral dimension, variability of sediments and hydraulic conductivity often reflect a series of succeeding historic channel forms, former flood events and sediment deposition. Legacies of historical channel processes increase the complexity of current morphological pattern at all scales (Frisell et al., 1986; Gregory et al. 1991). At the reach scale (at least 1–2 magnitudes of the stream width), physical complexity is expressed as changes in slope, cross-section and plan form. Stream ecologists often categorize distinct visible morphological elements like riffles, various forms of pools, bars, banks and so on. These elements differ in width, depth, water velocity and grain size, sorting and packing of the sediments, and hence control the microhabitat complexity. At the catchment scale, channel form differs along the drainage network, according to constraints such as geology and lithology, increasing discharge and changes in sediment load or slope and valley form. On this large scale, there is a general sequence of straight to braided to meandering morphology from the source to the mouth of a river, and this sequence affects complexity at both the reach and microhabitat scales. Another general trend along the course of rivers is the increasing width of the river corridor and increasing significance of the floodplain, which itself can be expressed in various floodplain-specific water bodies. However, these general patterns are often broken by stream confluences, lakes, regional geology and changes of valley form (Ward & Stanford, 1983).

Connectivity

The concept of connectivity was first applied to river systems by Amoros & Roux (1988). Although mainly recognized as an important ecosystem aspect, connectivity in rivers is primarily a hydromorphological attribute. Pringle (2001) defined hydrologic connectivity as water-mediated transfer of matter, energy and organisms within or between elements of the hydrologic cycle. In rivers, connectivity works in three dimensions (Kondolf et al., 2006). Longitudinal connectivity controls the downstream flux of water and sediments along the river network, and thus the basic processes shaping channel form. To a limited degree there is even downstream to upstream connectivity, which can be seen in case of backward erosion or sediment aggradation caused by shifts in downstream erosion base level. Additionally, water and sediments can be transported in the lateral and vertical dimensions. Lateral connectivity between the channel and the floodplain is also important; sediments deposited during floods form the floodplains and can return to the main channel when the channel migrates laterally (Junk et al., 1989). Hydraulic connectivity between the stream channel and shallow groundwater aquifers can extend for considerable distances laterally into the banks and below floodplains (Standford & Ward, 1988), though the major exchange in most streams and rivers is in vertical dimension across the stream bed. Vertical connectivity is created by the exchange of water between the water column and the hyporheic zone as well as the vertical accretion of sediment deposits. Water flux across the bed and banks is driven by hydraulic pressure variability and depends on the porosity of the sediments. Vertical connectivity is hence closely related to

Table 1 Attributes of hydromorphology of particular relevance to stream and river ecology, scales at which they are important, examples of variables affected, ecological significance and main human impacts

| Attribute | Scale | Features related | Significance for biodiversity | Significance for functioning | Main impacts |
|---|---|---|---|---|---|
| Complexity | Microhabitat | Sediment grain size, sorting and packing; wood, CPOM | Functional habitat diversity; refuge; food availability; spawning areas | Diversity/function links; hydraulic, nutrient and OM retention; redox gradients; metabolism | Siltation; flow detraction and regulation; snagging; canalization |
| | Reach | Channel slope, cross-section & plan form; riffle/pools sequences; bars & undercut banks | Macrohabitat diversity; resource partitioning for organisms | Functional zones; metabolism | Gravel mining, bank and bed revetments; vegetation removal |
| | Catchment | Channel typology along drainage network | Regional biodiversity | Seedling establishment; exchange of nutrients and OM | Extensive land use change; urbanization |
| Connectivity | Longitudinal | Continuity | Large-scale mobility; metapopulation dynamics; resilience; community persistence | Barrier effect; sediment tranport; population structure | Dams, engineered reaches; flow regulation, flood control, and drainage |
| | Lateral | Channel plan form; hydraulic connectivity; flooding | Refuge (floods); breeding in floodplains; terrestrial and aquatic organism interactions | Sediment and OM retention; enhances vertical connectivity | Canalization; dikes; flow regulation; vegetation removal |
| | Vertical | Variability of channel slope and cross-section; hyporheos; wood jams | Refuge (floods, droughts); life cycles; biodiversity hotspots | Diversity/function links; nutrient retention; river metabolism | Siltation; canalization; dikes |
| Dynamism | Hours/days | Storm runoff; floods; droughts | Disturbance; opportunity for colonists; avoidance of competitive exclusion | Resets succession | Regulation; dredging, bank fixation; flood control |
| | Months/years | Hydrologic regime; sediment budget; channel migration | Disturbance regime shapes life cycles; habitat, refuge | Connects river with floodplain | Floodplain forestry, urbanization; regulation |
| | Centuries | Island formation; planform adjustment by biogeomorphic interaction | Key habitats (oxbows...), LW input; important for evolution; habitat, refuge | Hydrogeomorphic patches; LW hotspot for metabolism | Floodplain occupation; channel artificialization |

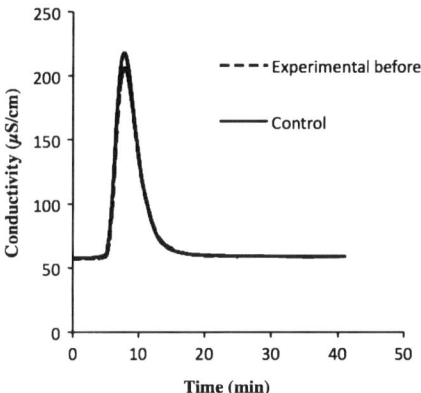
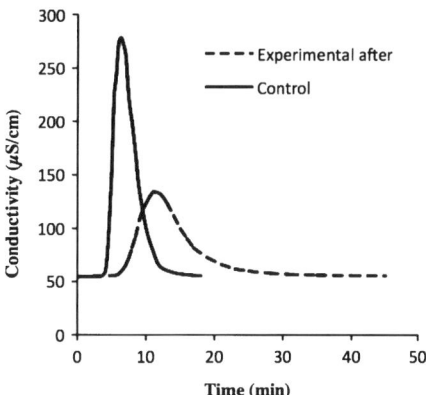

**Fig. 1** Changes in hydraulics as a result of experimental introduction of large wood into Latxe stream, Basque Country, Spain. Figures show break through, measured 100 m downstream from the slug addition of salt, at a control and an experimental reach, before (*left*) and after (*right*) introduction of 144 m³ of large wood per hectare of streambed. Note that addition of wood increased the travel time of salt as a result of decreased water velocity, and that peak conductivity decreased as a result of enhanced dispersion and transient storage

channel complexity, since the variability of hydraulic pressure at the bed results from interaction of flow with bed forms or slope discontinuities at reach scale (Savant et al., 1987; Harvey & Bencala, 1993; Elliott & Brooks, 1997; Kasahara & Hill, 2006) and the porosity is determined by sorting and packing of the sediments (Freeze & Cherry, 1979). On all scales vertical exchange is highly variable in space and time. For instance, in a reach-scale flume study, wood input caused sudden sediment deposition and increase of the vertical water flux through the bed (Mutz et al., 2007). Vertical water flux increased again after bed forms extended due to elevated flow velocity (Fig. 2).

Hence, rivers exhibit shifting mosaics of patches of hyporheic exchange produced by the variable arrangement of morphological features such as wood, riffles, pools and bars. Vertical water flux through the bed sediments also causes physical retention of fine sediments and colloids, which can form clogging layers and reduce vertical connectivity (Brunke, 1999). Sustainability of vertical connectivity and morphological dynamism are coupled because flushing floods and sediment redepostion are needed to periodically remove clogged layers and regenerate vertical connectivity (Schälchli, 1992).

Dynamism

River hydromorphology is far from static because river channels are shaped by the transport of water and sediments that varies from periodic to highly

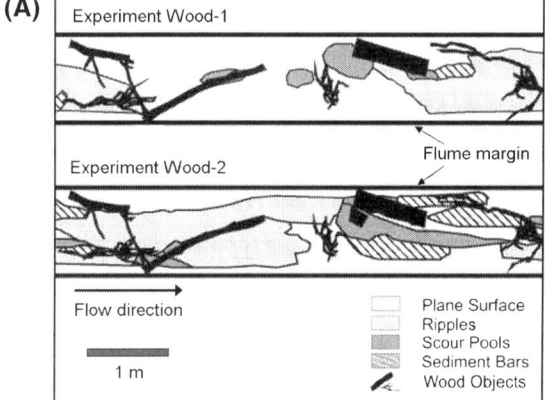
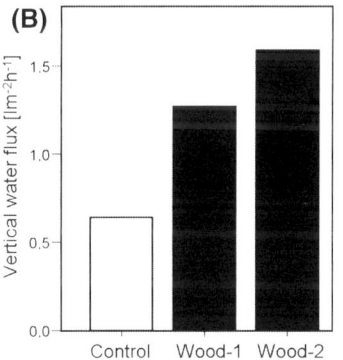

**Fig. 2** Short-term alteration of bed forms and vertical connectivity by experimental wood addition to plane sand bed. **A** Wood distribution and resulting bed forms in experiment Wood-1 with bed form generated by flow of 0.13 ms⁻¹ and Wood-2 with bed forms generated by 0.20 ms⁻¹; **B** Vertical water flux across the bed for the control without wood and the experiments with wood (Mutz et al., 2007)

episodic. Channels are constantly shifting and adjusting to changes in stream power, sediment yield and valley features, making rivers highly dynamic ecosystems. Dynamism is an intrinsic characteristic of rivers that shapes both channel form and connectivity, processes often modified by human activities.

Dynamism is linked to channel complexity and connectivity, as for instance, when meander migration results on formation of oxbow lakes or when floods reconnect them to the main channel. Channel dynamics depend on many basin and reach characteristics, but flow is of paramount importance. Such dynamism is important for river ecology at a broad range of temporal scales, from days to centuries (Frisell et al., 1986). At a scale of hours to days, flood events resulting from storm runoff produce extensive movement and rearrangement of sediments and litter packs. They also cause substantial changes in lateral and vertical connectivity, driving transport of water and sediments in both dimensions. At a scale of months to years, channel form is shaped by seasonal and interannual variations in discharge. Bankfull discharge, which occurs on average once every 2 years, is extremely important shaping channel dimensions. Additionally, landslides, severe storms and other events produce mass failures in the riparian forest, resulting in sporadic inputs of sediments and large wood, crucial to channel morphology (May & Gresswell, 2004). At a scale of centuries, lateral migration of river channels leads to meander cutoff, formation of oxbow lakes, and terrestrialization of abandoned paleochannels, and there is a cycle of formation, growth and decay of islands in braided channels (Gurnell et al., 2001).

**Relevance for biodiversity and functioning**

Physical habitat shapes biological communities and ecosystem functioning. Smith & Powell (1971) depict its effect as a series of environmental 'screens' (from large-scale, biogeographic factors to fine physiological and biotic interactions) that eventually shapes the composition of local communities. Southwood (1977) stressed the idea that the environment forms the template to which organisms have adapted through natural selection, and therefore, affects life traits. Several authors (Tomanova & Usseglio-Polatera, 2007) showed that habitat characteristics at the mesoscale determine invertebrate biological traits, and this idea is in the core of the concept of functional habitats (Harper et al., 1992), that is still in use (Harvey & Clifford, 2008). Nevertheless, attempts to identify one-to-one connections between surface flow types, units of channel morphology and functional habitats oversimplify a complex and dynamic hydraulic environment, and some authors (Harvey et al., 2008) proposed to use instead a nested hierarchy of reach-scale physical and ecological habitat structures, characterized by transferable assemblages of functional units.

Complexity

The relationship between physical complexity of ecosystems and diversity of biological communities leads to the concept of ecological niche, borrowed from the field of architecture and reminiscent of the old saying that the more niches in a church, the more saints could fit in. In a classical paper, Hutchinson (1959) already stressed the role of what he called 'the mosaic nature of the environment' in maintaining a large number of species, and thereafter, the role of architectural complexity on community diversity has become more and more evident (Oldeman, 1983). Less evident but also important is the effect of physical complexity on ecosystem functioning.

At the microhabitat scale, high diversity of sediment size is clearly linked to biodiversity, because patches differing in grain size often constitute different functional habitats (*sensu* Harper et al., 1992). Many invertebrate taxa are tightly linked to grain size (Fig. 3). For example, in Basque streams *Capnioneura* and Limnephilinae are found almost exclusively in organic matter and *Rithrogena* mainly on sand and stones. Many studies report dependencies of macroinvertebrate diversity, abundance, traits or productivity with substrate diversity or surface-perimeter ratio (Beisel et al., 1998, 2000; Lancaster, 2000). Among the elements of the stream bed, organic material plays a key role as a substratum for ecosystem functioning (Aldridge et al., 2009). The physical complexity of dead wood affects abundance and diversity of fish and macroinvertebrates (Crook & Robertson, 1999; Scealy et al., 2007). Invertebrate taxa use wood directly as habitat and for food (Dudley & Anderson, 1982; Hoffman &

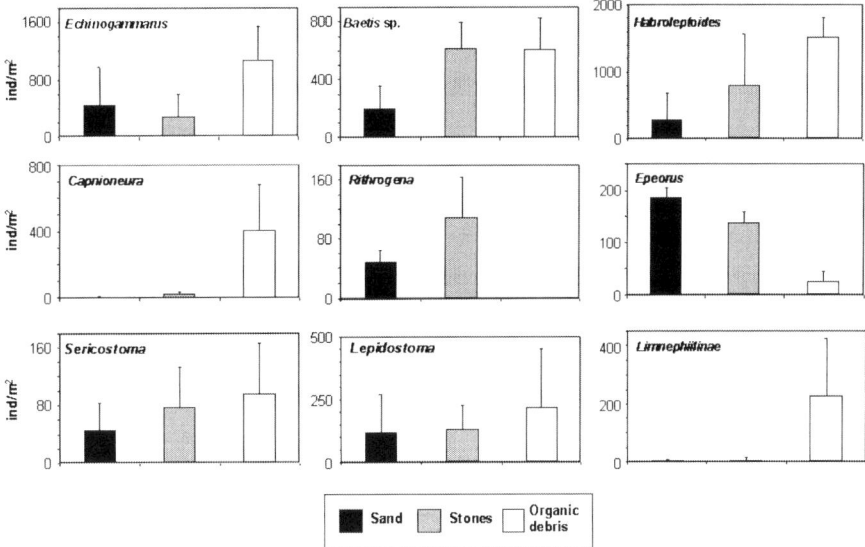

**Fig. 3** Densities of selected invertebrate taxa on microhabitats at the headwater of the Añarbe stream, Basque Country, Spain. *Vertical bars* are standard errors

Hering, 2000), while many fish seek refuge and cover.

At the reach level, channel complexity is related with the diversity of riffle/pool sequences or with the abundance and diversity of gravel bars and features such as undercut banks. Again, this level of complexity is related to biodiversity and functioning, and loss of these features can result in local extinctions. As an example, many fish species require different habitats through their life cycle. Brown trout spawn in gravel-bed riffles or runs, young fry concentrate in shallow areas, and adults prefer deep pools and wood accumulations. Because predatory fish species have large effects on invertebrate food chains, the absence of key habitats can decrease fish populations, with effects cascading through the food webs (Katano et al., 2006).

At the catchment scale, natural rivers usually show large differences in physical structure, species associations and ecological functions both from headwaters to lower river, and among different tributaries of similar order. Dependence of biotic communities on local hydromorphological setting lead scientists to propose different schemes of zonation, based for instance on invertebrates (Ilies & Botosaneanu, 1963), or fish (Huet, 1962), or associated changes in ecosystem function (Vannote et al., 1980). Therefore, at the catchment scale, biodiversity is linked to diverse hydrogeomorphic patterns throughout the river network.

Connectivity

Connectivity is important in all communities and ecosystems for many reasons, from maintaining gene pools in populations to recolonizing an area after a major disturbance. Since disturbance, mainly in form of floods or droughts, is so common, connectivity might be especially important in river ecosystems. All three dimensions of connectivity, longitudinal, lateral and vertical, are important for communities and ecosystem processes.

Longitudinal connectivity with the entire drainage net is essential for migratory species that live in different reaches along their life cycles, and for organisms in general to recolonize a reach after disturbance. Colonists from adjacent reaches influence the success or failure of river restoration projects (Kail & Hering, 2009). The dendritic characteristics of river networks determine the location of refuges (Meyer et al., 2007), and thus, affect recolonization trajectories. Furthermore, because river communities are always subject to a continuous supply of propagules from upstream reaches, changes in connectivity can affect local communities. Not all natural rivers have high longitudinal connectivity. Indeed, large waterfalls can block entire sections of the catchment and act as barriers to organism dispersal for millennia, resulting in different communities up and downstream. Longitudinal connectivity also can affect ecosystem functioning. For instance, Taylor et al. (2006) showed that decreases in abundance of a detrital-feeding fish reduced downstream transport of organic carbon and increased primary production and respiration in a South American river. Upstream migration of salmon periodically carries energy and nutrients from the ocean into reaches where carcasses fertilize the stream and, mediated by predation and lateral transport by bears, provide N influx to riparian forests (Helfield & Naiman, 2006; Quinn et al., 2009). Characteristics that make a reach a suitable corridor are highly species-dependent and not necessarily limited to channel variables. For instance, Roberts & Angermeier (2007) showed experimentally that riparian cover significantly affects the movement of fish along a reach.

Lateral connectivity, or the ability for materials and organisms to cross the border between river channel and the riparian areas and floodplains, is often linked to hydrological connectivity, and to the physical characteristics of banks and riparian areas. Steep, undercut banks form more difficult barriers than smooth ones, and the transition between aquatic and terrestrial habitats is probably easiest in complex channels with dead arms and fallen logs. Lateral sloughs, alcoves, and side channels increase the interface between channel and floodplain and act as preferential paths for organisms between these two habitats. Log jams and other large wood structures (including beaver dams) increase water stage and make flooding more common. Lateral connectivity is important for instream biodiversity, at least in large rivers (Paillex et al., 2007), increasing the survival of species that spend a part of their life cycle on the floodplain and another in the channel, like tropical fish species reproducing on floodplains (Welcomme, 1985; Agostinho et al., 2004). It is equally important for ecosystem function because it regulates the transport of nutrients and organic matter between floodplain and channel. For instance, experimental cover of stream channel eliminated input of terrestrial insects into streams, changed fish and macroinvertebrate populations, and also affected terrestrial spider and bat populations because it decreased aquatic insect emergence, which is subsidy to terrestrial predators (Baxter et al., 2005). Additionally, lateral connectivity may increase the chances of finding refuge during disturbances, as when fish move into low velocity river margins and invertebrates crawl to floodplain trees during flood time. Through these complex interactions, lateral connectivity influences biodiversity at the floodplain scale (Ward & Tockner, 2001).

Vertical connectivity across the bed and the banks occurs through the hyporheic zone, the dynamic ecotone between the surface stream and the shallow groundwater aquifer (Gibert et al., 1990) in which both waters mix (White, 1993). Quantity and quality of water exchange control the magnitude of the hyporheic pore space and surfaces with the associated micro organisms and habitat of the hyporheos, which contribute significantly to stream biodiversity (Williams & Hynes, 1974; Bretschko, 1991; Boulton et al., 1998). It is also a temporary habitat for early instars, and pupae of many benthic invertebrates (Boulton, 2000), as well as for the development of eggs and embryos of many fish species (Vaux, 1962; Malcolm et al., 2005). Further, the hyporheic zone can strongly influence metabolism at the reach scale (Grimm & Fisher, 1984; Naegeli & Uehlinger, 1997). Organic carbon, dissolved oxygen and other electron acceptors, enter the hyporheic zone through downwelling surface water or upwelling groundwater (Jones et al., 1995). Additionally, organic particles can be stored within the subsurface zone during sediment deposition (Metzler & Smock, 1990; Fuss & Smock, 1996). Because hyporheic metabolism may be limited by supply of organic carbon or exhaustion of terminal electron acceptors from the surface stream, vertical connectivity may limit stream metabolism and regulate biogeochemical transformations. Vertical connectivity is spatially complex; flow direction, water residence time, and hyporheic flow path length vary with channel form and sediment structure. Abundant and short flow paths may result in higher overall metabolic rates (Malard et al., 2002). On the other hand, longer hyporheic flow paths and small scale heterogeneity in vertical connectivity create contrasting conditions that permit the coexistence of oxidation and reduction processes within small spatial scales and maintain lower water temperatures. Bacterial production can be enhanced in downwelling zones whereas in upwelling water anaerobic conditions can promote ammonification, denitrification and sulphate reduction (Hendricks, 1993). Thus, upwelling water containing nutrients released by hyporheic metabolism can promote patches of high benthic primary production (Grimm & Fisher, 1989). The proportion of total discharge in vertical exchange with the hyporheic zone influences the quality of the water discharged to downstream reaches (Findlay, 1995). However, the effect of vertical connectivity on water quality depends on the spatial arrangement and the dynamics of the hydromorphological elements regulating vertical connectivity (Fisher et al., 1998) and varies greatly in different river networks.

Dynamism

Lotic organisms have adapted through evolution to dynamic channels, to ecosystems that expand and contract, and to physical habitat that changes dramatically with the rise and fall of water stage (Poff, 1997). Different components of the flow dynamics affect river ecosystems. Floods below bankfull discharge are essential for the transport of nutrients, organic matter and sediments, and for organisms dispersal. When they disturb small patches of sediments they may promote higher biodiversity by creating patchiness and avoiding competitive exclusion. Small floods can also slough algal mats, flush fine sediments, and thus, promote nutrient retention by encouraging algal regrowth (Elósegui et al., 1995). Floods above bankfull promote lateral exchange of matter and organisms with the floodplain, and in large rivers with long and predictable floods riverine animal biomass relates to production in the floodplain (Junk et al., 1989). The frequency, duration and magnitude of such floods control composition and spatial distribution of floodplain vegetation (Hupp & Osterkamp, 1996). Duration and timing of low flows control biodiversity on exposed banks and alluvial gravel bars (Nilsson et al., 1993) and also the efficiency of terrestrial predators feeding on aquatic insects (Paetzold et al., 2005). Hence, intensity and ecological effects of lateral connectivity are controlled largely by the extremes of hydrological dynamism.

When floods exceed thresholds of sediment mobility, they function as disturbances, displacing organisms, causing mortality, destroying and reorganizing habitat structure and resetting succession. Disturbance regime can be considered the dominant factor organizing stream ecology (Resh et al., 1988), and recovery trajectories that can take several months to century can be measured at both the community and the ecosystem level (Fisher et al., 1982; Grimm, 1987; Corenblit et al., 2007). Moreover, vegetation and channel form are linked reciprocally, e.g. island evolution promoted by drift wood resulting from bank erosion in braided rivers (Gurnell et al., 2001). This ecosystem engineering function of woody riparian vegetation causes long-term shifts of river plan form (Murray et al., 2008) and hence such 'biogeomorphic' feedback mechanisms can occur at scales of regional landscapes and timeframes of centuries (Corenblit et al., 2007).

Severe disturbances can also be caused by drought (Lake, 2000). During drought and desiccation, riffle habitats disappear and competition and predation increase largely in the remaining isolated pools. Once the flow is interrupted, physicochemical conditions change abruptly (Acuña et al., 2005). With re-aeration greatly reduced, microbial processing of benthic organic matter causes hypoxia and subsequent shifts of the invertebrate community. Loss of species can alter stream functioning, such as reduced leaf decomposition during stream fragmentation (Schlief & Mutz, 2009). If streams dry up completely, most species will be eliminated and subsequent recovery of the system can be slow and somewhat unpredictable, since recolonization depends on source of colonists and their dispersal (Bond et al., 2006).

Ecological effects of extreme floods and droughts depend on their predictability, what determines their potential to trigger organism adaptation and evolution (Poff, 1997). Organisms show complex adaptations to floods and droughts, from morphology to behaviour and life history (Ward, 1992; Brock et al., 2003). When mortality by flood or drought is highly seasonal and thus, predictable, organisms develop risk-avoidance mechanisms, like emergence or diapause. Less predictable environments favour risk-spreading strategies, like asynchronous egg hatching in many Plectopera (Hynes, 1976), or rapid growth rates (Huryn & Wallace, 2000). Thus, a community at a given site represents the balance among benefits and costs of adaptations to the environmental dynamics, and survival is strongly influenced by the level of dynamism.

## Human impacts on river ecosystem through changes in channel hydromorphological integrity

Human activities are changing deeply entire landscapes, making them more homogenous, what lead to decreased diversity (at the ß scale). Examples of these changes include deforestation in agricultural areas or afforestation in prairie regions (Baldi & Paruelo, 2008). Because rivers integrate the characteristics of their drainage basins, the array of human activities in a catchment can affect river ecosystems. Human activities vary widely depending on the physical, biological and societal context, but some generalities can be drawn from how different human civilizations have modified during millennia catchments, river channels and flows.

Direct occupation and modification of floodplain areas and river corridors is one of the main sources of impact on river hydromorphology, and one that is becoming more prevalent as population growth pushes people to settle in flood-prone areas. Effects of floodplain alteration on river channels depend on the type of occupation, agricultural, urban or industrial, and thus, can be site-specific (Hupp et al., 2009; Kail et al., 2009). Nevertheless, human development in floodplains always decreases the naturalness of these areas, which are integral parts of the riparian corridors, and constrain and fossilize the channels. Most significant, occupation of floodplain creates flooding hazard, which leads to every sort of flood defences, from lateral dikes to flood-control dams (Hudson et al., 2008).

One main source of impact from developed floodplains but also from catchment-wide change of land use is changes in sediment yield that cause the river to adjust and change channel form. These floodplain alterations are widespread; in many areas of the world clearing of forest for grazing or for agriculture increased erosion rates and created siltation problems. Indeed, suspended sediments are considered among the most prevalent contaminants in developed countries (USEPA, 2000). Siltation decreases vertical connectivity, can result in hyporheic anoxia, and affects the relative proportion of functional feeding groups in the benthos (Rabeni et al., 2005). On the other hand, recovery of forest cover in much of Europe has reduced sediment yield and leading to channel incision and to changes in planform from braided to single channel (Surian et al., 2009). Similar effects are produced by large dams that block the transport of sediments and result in subsidence of delta areas (Heine & Lant, 2009). Reduction of sediment supply decreases sediment patch mobility, leading to more fixed patches that are immobile through localized bed armouring (Nelson et al., 2009). In cases of severe incision, the water table falls and the floodplain becomes disconnected from the channel except during exceptionally large floods, thus eliminating many of the characteristics that make these areas uniquely productive and diverse (Scott et al., 2004).

Similar modifications of channel form can be caused by changes in stream power, either from river regulation or increased soil imperviousness. Ecological effects of water abstraction are multiple, and can be either direct (decreased velocity, depth and wetted habitat surface) or indirect (increased temperature and conductivity). Indirect effects affect especially to invertebrate communities (Miller et al., 2007), and become important when a large fraction of the flow is being diverted. Altered flow regimes are one of the most prevalent human impacts for river ecosystems. Flow regimes are a critical component of the habitat template to which organism life histories have adapted. They control connectivity and affect biological invasions (Bunn & Arthington, 2002). Apart from the obvious effects on habitat abundance, flow regime changes also affect the disturbance regime, a major determinant of community structure. Water abstraction also reduces longitudinal connectivity, because decreased flow may make some reaches impassable by fish, and in extreme cases, even dewater entire reaches (Galat et al., 1998; Kondolf et al., 2006).

Dams are prevalent impacts on hydromorphology in rivers across the world, fragmenting river ecosystems and reducing catchment scale connectivity (Nilsson et al., 2005). Even low dams affect channel morphology and sediment dynamics in their vicinity, and most detrimental, create barriers for dispersal of organisms. Unlike natural wood accumulations that are permeable to fish and other organisms (Harmon et al., 1986), man-made dams are largely impassable without passage structures and reduce numbers of upstream migrating fish species. Even when they do not form totally closed barriers, large numbers of dams, as they are present in many developed regions (Fig. 4), can have large effects on fish, and have historically been related to declines in migrating fish stocks, e.g. salmon (Mills, 1989). Apart from barriers to upstream migration, dams can divert downstream-migrating fish and other organisms to canals where their survival can be reduced. Large dams are even more detrimental for stream and river ecosystems,

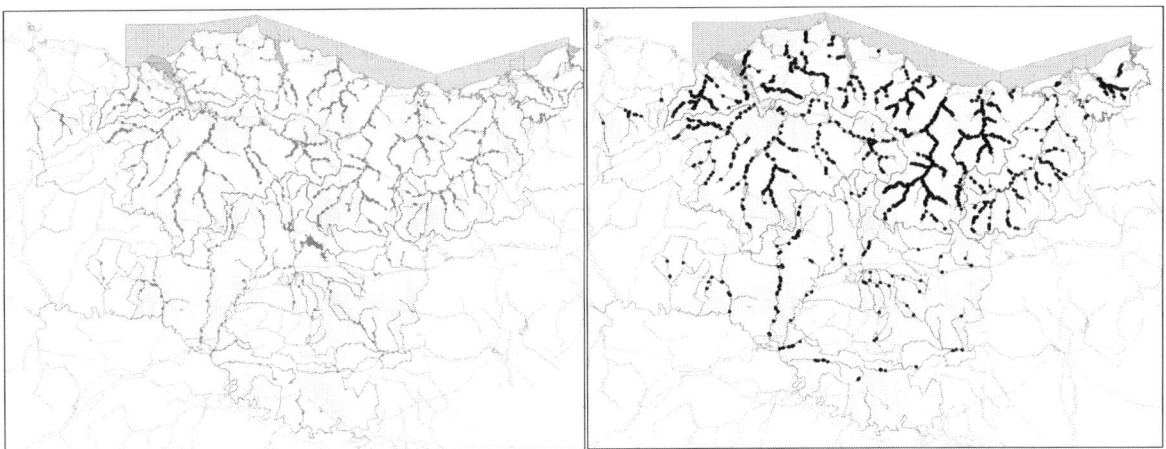

**Fig. 4** Distribution of low dams (*left*) and flood defences (*right*) in the Basque Autonomic Region, Spain. The number of low dams is 1,035, 10% of the 2,028 km of rivers are modified by flood defences, and a further 1.7% of the channels is buried under concrete. Source, Basque Government (2005)

because, in addition to the barrier effect, they have large impacts on sediment yields, and can also profoundly affect the discharge, hydrologic regime, and water temperature and quality. Presently, over half the large river systems in the world are affected by dams (Nilsson et al., 2005), resulting in threefold increase of water residence time, reduced supply of sediments to coastal areas (Vörösmarty et al., 2003), alteration of the thermal regime, and severe habitat fragmentation (Revenga et al., 2000). Although the rate at which large dams are being built is decreasing (Gleick, 2003), global climate change scenarios suggest that pressure to build more dams will be strong in many regions in the future (Oki & Kanae, 2006). The combined effect of these barriers is a likely decrease in biodiversity at the catchment level, as has been reported for fish (McLaughlin et al., 2006) and mussels (Williams et al., 1993).

Streams and rivers across the world are subject to direct modification of river channels for a multitude of purposes, from navigation to water diversion to flood control. The most common activities are reinforcement of banks, building lateral dikes, channel resectioning and straightening, dredging, snagging and so on. In general, these activities result in more homogenous flow, decreased channel complexity, reduced river dynamism, decreased longitudinal, vertical and lateral connectivity (Kondolf et al., 2006), and loss of important habitats. Even subtle changes in lateral connectivity can have profound impacts on channel morphology and ecosystem function. For instance, an effect of reduced lateral connectivity is the impact of channelization on riparian spider populations, that decrease as a consequence of reduced prey densities (Laeser et al., 2005). Similarly, livestock have been described as geomorphic agents, because they trample banks, consume riparian vegetation, and contribute animal wastes in river channels (Trimble & Mendel, 1995). Simply fencing banks can have substantial consequences on rivers. The role of other large mammals, like hippopotamuses or elephants in African rivers, is probably much larger (Butler, 1995).

Many articles report the effects of snagging or removal of in-stream wood in rivers, mainly in Australian and North American rivers throughout the last 200 years (Gippel et al., 1992; Maser & Sedell, 1994). This research revealed the significance of wood as a control of channel hydraulics, morphology and biological responses. Thus, rivers without large wood tend to be wider, straighter, less biodiverse and less productive (Gregory et al., 2003).

Often, it is not possible to discern which human alteration is most detrimental for river ecosystems, as most reaches are affected by multiple stressors (UNEP, 2007). Multiple stressors also explain why biological communities do not recover as rapidly as water quality in reaches where sewage treatment plants and clean production practices have been implemented. This is the case, for instance, of the Pyrenean desman, an endangered water mole endemic from the Iberian Peninsula, which is still declining despite large improvements in water quality during the last decades (Fernandes et al., 2008). The most likely explanation for this decline is that many remnant populations are too small to be viable, and that habitat modifications prevent contact among subpopulations of this highly selective mammal (Nores, 2007).

**Present uncertainties and need for future research**

In this article, we stress the importance of spatial complexity, connectivity and dynamism on river biodiversity and functioning. Though the relevance of these three attributes is well recognized by river ecologists and is starting to be considered in river restoration projects, large uncertainties remain about appropriate future conditions, thresholds that should not be crossed, and the degree to which we can mimic nature under the increasing human pressures.

In the case of river ecosystem complexity, it is necessary to define which channel form and features are key to maintain natural biodiversity and important ecosystem functions, to make clear at which spatial scales are they relevant, and then to define the geomorphic processes that are necessary to keep these features with minimal maintenance. Because the society cannot give up some of the present uses of natural resources (e.g. drinking water supply or use of the space), river ecosystems cannot simply be reverted to a 'natural' status, however, one might define it, and we must seek ways to combine use of resources with ecosystem functioning in a sustainable way. Examples of improved river management are new strategies to reduce the effect of large reservoirs on sediment yield (Rovira & Ibañez, 2007) or techniques to recover a more natural channel geometry under reduced discharges (Thorne et al., 1996). As is often the case in nature conservation, there will be no simple answers to the question of which channel features to maintain, as each spatial configuration will benefit some species at the expense of others and increase some functions at the detriment of others. Therefore, clear goals must be set for biodiversity and functioning, and these general goals must be translated into specific, reach-level goals.

Similarly, it is necessary to define the appropriate level of connectivity to maintain biodiversity and functioning and to develop standard methods to measure it. For instance, hydrological connectivity can be measured by experiments of solute additions, but most experiments performed so far focus mainly on the longitudinal dimension over short timeframes. We lack a clear understanding of how water moves in the three dimensions (longitudinal, lateral and vertical) and how hydromorphology affects all three. Responses of aquatic organisms are even more complex because they differ greatly in their dispersal abilities in and out of the water. Therefore, it is important to conduct studies on mobility of different groups of organisms, at least at two scales: short-term, as is usually performed by tracking tagged animals, and long term, more often approached from the point of view of genetic variability at the metapopulation level. Development of techniques in molecular biology has triggered intense research in population genetics, but with the exception of some particularly well studied species, like the Atlantic salmon (Verspoor et al., 2007), our understanding of individual mobility and genetic flow is limited or non-existent for most populations. A major concern is the effect of connectivity on biological invasions. Humans are acting as the most important vector for transport of organisms through the world, and exotic species are now one of the main global pressures on biodiversity (Millennium Ecosystem Assessment, 2005). Management of river ecosystems and their connectivity must address both metapopulation dynamics and biological invasions.

Thirdly, the natural flow regime for streams and rivers should be defined in each regional hydro-climatic context, especially in terms of timing, intensity and recurrence of floods and droughts. In this context, the focus on allocation of water for the environment should shift from in-stream perspectives and well established 'environmental flows' to 'environmental flow regimes' and consideration of dynamics along the river corridor (Arthington et al., 2006). The significance of extreme and uncommon floods and droughts for long-term morphological and biological development is still unclear. They are suggested to reset biological processes, e.g. growth of riparian trees, which in turn function as environmental engineers on fluvial morphology (Naiman et al., 2008). More research is necessary to define how much departure from the natural regime can occur for specific river morphology and associated habitats without affecting biodiversity and ecosystem services.

A word of caution is necessary to interpret the concept 'natural' above. We don't think management goals should be static; on the contrary, our endpoint should be a dynamic river constantly adjusting to a changing landscape. Humans have and will always change the landscape. It is time we start doing so in a conscious, planned manner, not as a byproduct of multiple, parallel but separate, narrow-minded decisions. This will not be easy for rivers because hydromorphology often reflects historical changes in temporal scales much longer than the scale at which new human impacts arise. Even if it is difficult, these issues in river management cannot be ignored if our societal goal is to ensure long-term sustainability in a healthy environment.

**Acknowledgements** This paper was supported by the Project 'Complextream: effects of channel complexity on stream communities and ecosystem functioning', funded by the Spanish Ministry of Science and Innovation (project CGL2007-65176/HID). Stan Gregory (Oregon State University) checked the English style.

# References

Acuña, V., I. Muñoz, A. Giorgi, M. Omella, F. Sabater & S. Sabater, 2005. Drought and postdrought recovery cycles in an intermittent Mediterranean stream: structural and functional aspects. Journal of the North American Benthological Society 24: 919–933.

Agostinho, A. A., L. C. Gomes, S. Veríssimo & E. K. Okada, 2004. Flood regime, dam regulation and fish in the Upper Paraná River: effects on assemblage attributes, reproduction and recruitment. Reviews in Fish Biology and Fisheries 14: 11–19.

Aldridge, K. T., J. D. Brookes & G. G. Ganf, 2009. Rehabilitation of stream ecosystem functions through the reintroduction of coarse particulate organic matter. Restoration Ecology 17: 97–106.

Amoros, C. & A. L. Roux, 1988. Interaction between water bodies within the floodplains of large rivers: function and development of connectivity. Münstersche Geographische Arbeiten 29: 125–130.

Arthington, A. H., S. E. Bunn, N. L. Poff & R. J. Naiman, 2006. The challenge of providing environmental flow rules to sustain river ecosystems. Ecological Applications 16: 1311–1318.

Baldi, G. & Paruelo, J. M., 2008. Land-use and land cover dynamics in Saouth American temperate grasslands. Ecology and Society 13: 6. [online] URL: http://www.ecologyandsociety.org/vol13/iss2/art6/.

Basque Government, 2005. Confrontación de la situación administrativa de presas y azudes de la Comunidad Autónoma del País Vasco: 390 pp.

Baxter, C. V., K. D. Fausch & W. C. Saunders, 2005. Tangled webs: reciprocal flows of invertebrate prey link streams and riparian zones. Freshwater Biology 50: 201–220.

Beisel, J.-N., P. Usseglio-Polatera, S. Thomas & J.-C. Moreteau, 1998. stream community structure in relation to spatial variation: the influence of mesohabitat characteristics. Hydrobiologia 389: 73–88.

Beisel, J.-N., P. Usseglio-Polatera & J.-C. Moreteau, 2000. The spatial heterogeneity of a river bottom: a key factor determining macroinvertebrate communities. Hydrobiologia 422(423): 163–171.

Benda, L., N. L. Poff, D. Miller, T. Dunne, G. Reeves, M. Pollock & G. Pess, 2004. Network dynamics hypothesis: spatial and temporal organization of physical heterogeneity in rivers. BioScience 54: 413–427.

Bond, N. R., S. Sabater, A. Glaister, S. Roberts & K. Vanderkruk, 2006. Colonisation of introduced timber by algae and invertebrates, and its potential role in aquatic ecosystem restoration. Hydrobiologia 556: 303–316.

Boulton, A. J., 2000. The subsurface macrofauna. In Jones, J. B. & P. J. Mulholland (eds), Streams and Ground Waters. Academic Press, San Diego, California: 337–362.

Boulton, A. J., S. Findlay, P. Marmonier, E. H. Stanley & H. M. Vallet, 1998. The functional significance of the hyporheic zone in streams and rivers. Annual Review of Ecology and Systematics 29: 59–81.

Bretschko, G., 1991. The limnology of a low order alpine gravel bed stream (Ritrodat-Lunz study area, Austria). Verhandlungen der Internationalen Vereinigung für theoretische und angewandte Limnologie 24: 1333–1339.

Brock, M. A., D. L. Nielsen, R. J. Shiel, J. D. Green & J. D. Langley, 2003. Drought and aquatic community resilience: the role of eggs and seeds in sediment of temporary wetlands. Freshwater Biology 48: 1207–1218.

Brunke, M., 1999. Colmation and depth filtration within streambeds: retention of particles in hyporheic interstices. International Review of Hydrobiology 84: 99–117.

Bunn, S. E. & A. H. Arthington, 2002. Basic principles and ecological consequences of altered flow regimes for aquatic biodiversity. Environmental Management 30: 492–507.

Butler, D. R., 1995. Zoogeomorphology: Animals as Geomorphic Agents. Cambridge University Press, Cambridge.

Corenblit, D., E. Tabacchi, J. Steiger & A. M. Gurnell, 2007. Reciprocal interactions and adjustments between fluvial landforms and vegetation dynamics in river corridors: a review of complementary approaches. Earth-Science Reviews 84(1/2): 56–86.

Crook, D. A. & A. I. Robertson, 1999. Relationship between riverine fish and woody debris: implications for lowland rivers. Marine Freshwater Research 50: 941–953.

Dudley, T. & N. H. Anderson, 1982. A survey of invertebrates associated with wood debris in aquatic habitats. Melanderia 39: 1–21.

Elliott, A. H. & A. H. Brooks, 1997. Transfer of non-sorbing solutes to a stream bed with bed forms: laboratory experiments. Water Resources Research 33: 137–151.

Elósegui, A., X. Arana, A. Basaguren & J. Pozo, 1995. Self-purification processes in a medium-sized stream. Environmental Management 19: 931–939.

European Union Water Framework Directive, 2000. Directive 2000/60/EC of the European Parliament and of the Council of 23 October 2000 establishing a framework for Community action in the field of water policy.

Fernandes, M., Herrero, J., Aulagnier, S. & Amori, G., 2008. *Galemys pyrenaicus*. In IUCN 2009. IUCN Red List of Threatened Species. Version 2009.1. www.iucnredlist.org.

Findlay, S., 1995. Importance of surface-subsurface exchange in stream ecosystems: the hyporheic zone. Limnology and Oceanography 40: 159–164.

Fisher, S. G., L. J. Gray, N. B. Grimm & D. E. Busch, 1982. Temporal succession in a desert ecosystem following flash flooding. Ecological Monographs 52: 93–110.

Fisher, S. G., N. B. Grimm, E. Martí & R. Gómez, 1998. Hierarchy, spatial configuration and nutrient cycling in a desert stream. Australian Journal of Ecology 23: 41–52.

Fisher, S. G., J. B. Heffernan, R. A. Sponseller & J. R. Welter, 2007. Functional ecomorphology: feedbacks between form and function in fluvial landscape ecosystems. Geomorphology 89: 84–96.

Freeze, R. A. & J. A. Cherry, 1979. Groundwater. Prentice-Hall, Inc., Englewood Cliffs, New Jersey.

Frisell, C. A., W. J. Liss, C. E. Warren & M. D. Hurley, 1986. A hierarchical framework for stream classification: viewing streams in a watershed context. Environmental Management 10: 199–214.

Fuss, C. L. & L. A. Smock, 1996. Spatial and temporal variation of microbial respiration rates in a blackwater stream. Freshwater Biology 36: 339–349.

Galat, D. L., L. H. Fredrickson, D. D. Humburg, K. J. Bataille, J. R. Bodie, J. Dohrenwend, G. T. Gelwicks, J. E. Havel, D. L. Helmers, J. B. Hooker, J. R. Jones, M. F. Knowlton, J. Kubisiak, J. Mazourek, A. C. McColpin, R. B. Renken & R. D. Semlistsch, 1998. Flooding to restore connectivity of regulated, large-river wetlands. BioScience 48: 721–733.

Gibert, J., M.-J. Dole-Olivier, P. Marmonier & P. Vervier, 1990. Surface water-groundwater ecotones. In Naiman R. J. & H. Décamps (eds), The ecology and management of aquatic-terrestrial ecotones. United Nations Educational, Scientific, and Cultural Organization, Paris and Parthenon Publishers, Carnforth, UK: 199–226.

Gippel, C. J., I. C. O'Neill & B. L. Finlayson, 1992. The Hydraulic Basis for Snag Management. Centre for Environmental Applied Hydrology. Department of Civil and Agricultural Engineering, University of Melbourne, Victoria, Australia.

Gleick, P. H., 2003. Global freshwater resources: soft-path solutions for the 21st century. Science 302: 1524–1528.

Gregory, S. V., F. J. Swanson, W. A. McKee & K. W. Cummins, 1991. An ecosystem perspective of riparian zones. Focus on links between land and water. BioScience 41: 540–551.

Gregory, S. V., Boyer, K. L. & Gurnell, A. M, (eds), 2003. The ecology and management of wood in world rivers. American Fisheries Society, Bethesda.

Grimm, N. B., 1987. Nitrogen dynamics during succession in a desert stream. Ecology 68: 1157–1170.

Grimm, N. B. & S. G. Fisher, 1984. Exchange between interstitial and surface water: implications for stream metabolism and nutrient cycling. Hydrobiologia 111: 219–228.

Grimm, N. B. & S. G. Fisher, 1989. Stability of periphyton and macroinvertebrates to disturbance by flash floods in a desert stream. Journal of the North American Benthological Society 8: 293–307.

Gurnell, A. M., G. E. Petts, D. M. Hannah, B. P. G. Smith, P. J. Edwards, J. Kollmann, J. V. Ward & K. Tockner, 2001. Riparian vegetation and island formation along the gravel-bed Fiume Tagliamento, Italy. Earth Surface Processes and Landforms 26: 31–62.

Harmon, M. E., J. F. Franklin, F. J. Swanson, P. Sollins, S. V. Gregory, J. D. Lattin, N. H. Anderson, S. P. Cline, N. G. Aumen, J. R. Sedell, G. W. Lienkaemper, K. C. Cromack & K. W. Cummins, 1986. Ecology of coarse woody debris in temperate ecosystems. Advances in Ecological Research 15: 133–302.

Harper, D. M., C. D. Smith & P. J. Barham, 1992. Habitat as the building blocks for river conservation assessment. In Boon, P. J., P. Calow & G. E. Petts (eds), River Conservation and Management. John Wiley & Sons Ltd., Chichester: 311–319.

Harvey, J. W. & K. E. Bencala, 1993. The effect of stream bed topography on surface–subsurface water exchange in mountain catchments. Water Resource Research 29: 89–98.

Harvey, G. L. & N. J. Clifford, 2008. Distribution of biologically functional habitats within a lowland river, United Kingdom. Aquatic Ecosystem Health and Management 11: 465–473.

Harvey, G. L., N. J. Clifford & A. M. Gurnell, 2008. Towards and ecologically meaningful classification of the flow biotope for river inventory, rehabilitation, design and appraisal purposes. Journal of Environmental Management 88: 638–650.

Hawkes, H. A., 1975. River zonation and classification. In Whitton, B. A. (ed.), River Ecology. Blackwell, Oxford: 312–374.

Heine, R. A. & C. L. Lant, 2009. Spatial and temporal patterns of stream channel incision in the Loess Region of the Missouri River. Annals of the Association of American Geographers 99: 231–253.

Helfield, J. M. & R. J. Naiman, 2006. Keystone interactions: salmon and bear in riparian forests of Alaska. Ecosystems 9: 167–180.

Hendricks, S. P. 1993. Microbial ecology of the hyporheic zone: a perspective integrating hydrology and biology. Journal of the North American Benthological Society 12: 70–78.

Hoffman, A. & D. Hering, 2000. Wood-associated macroinvertebrate fauna in central European streams. International Review of Hydrobiology 85: 25–48.

Hudson, P. F., H. Middelkoop & E. Stouthamer, 2008. Flood management along the Lower Mississippi and Rhine Rivers (The Netherlands) and the continuum of geomorphic adjustment. Geomorphology 101: 209–236.

Huet, M., 1962. Influence du courant sur la distribution des poissons dans les eaux courantes. Aquatic Sciences-Research Across Boundaries 24: 412–432.

Hupp, C. R. & W. R. Osterkamp, 1996. Riparian vegetation and fluvial geomorphic processes. Geomorphology 14: 277–295.

Hupp, C. R., A. R. Pierce & G. B. Noe, 2009. Floodplain geomorphic processes and environmental impacts of

human alteration along coastal plain rivers, USA. Wetlands 29(2): 413–429.
Huryn, A. D. & J. B. Wallace, 2000. Life history and production of aquatic insects. Annual Review of Entomology 45: 83–110.
Hutchinson, G. E., 1959. Homage to Santa Rosalia or why are there so many kinds of animals? American Naturalist 93: 145–159.
Hynes, H. B. N., 1975. The stream and its valley. Verhandlungen der Internationalen Vereinigung für theoretische und angewandte Limnologie 19: 1–15.
Hynes, H. B. N., 1976. Biology of Plecoptera. Annual Review of Entomology 21: 135–153.
Ilies, J. & L. Botosaneanu, 1963. Problèmes et méthodes de la classification et de la zonation écologique des eaux courantes, considerées surtout du point de vue faunistique. Schweizerbart'sche Verlagsbuchhandlung, Stuttgart.
Jones, J. B., S. G. Fisher & N. B. Grimm, 1995. Vertical hydrologic exchange and ecosystem metabolism in a Sonoran Desert stream. Ecology 76: 942–952.
Junk, W. J. & Wantzen, K. M., 2003. The flood pulse concept: new aspects, approaches and applications, an update. In Welcomme R. L. & T. Petr (eds), Proceedings of the Second International Symposium on the Management of Large Rivers for Fisheries, Vol. 1. Food and Agriculture Organization of the United Nations & Mekong River Commission. FAO Regional Office for Asia and the Pacific, Bangkok. RAP Publication 2004/16: 117–140.
Junk, W. J., Bayley, P. B. & Sparks, R. E., 1989. The floodpulse concept in river-floodplain systems. In Dodge D. P. (ed), Proceedings of the International Large River Symposium. Canadian Special Publication in Fisheries and Aquatic Sciences.
Kail, J. & D. Hering, 2009. The influence of adjacent stream reaches on the local ecological status of Central European mountain streams. River Research and Applications 25: 537–550.
Kail, J., S. C. Jähnig & D. Hering, 2009. Relation between floodplain land use and river hydromorphology on different spatial scales: a case study from two lower-mountain catchments in Germany. Fundamental and Applied Limnology 164: 63–73.
Kasahara, T. & A. R. Hill, 2006. Hyporheic exchange flows induced by constructed riffles and steps in lowland streams in southern Ontario, Canada. Hydrological Processes 20: 4278–4305.
Katano, O., T. Nakamura & S. Yamamoto, 2006. Intraguild indirect effects through trophic cascades between stream-dwelling fishes. Journal of Animal Ecology 75: 167–175.
Kondolf, G. M., Boulton, A. J., O'Daniel, S., Poole, G. C., Rahel, F. J., Stanley, E. H., Wohl, E., Bång, A., Carlstrom, J., Cristoni, C., Huber, H., Koljonen, S., Louhi, P. & Nakamura, K., 2006. Process-based ecological river restoration: visualizing three-dimensional connectivity and dynamic vectors to recover lost linkages. Ecology and Society 11(2). [online] URL: http://www.ecologyandsociety.org/vol11/iss2/art5/.
Laeser, S. R., C. V. Baxter & K. D. Fausch, 2005. Riparian vegetation loss, stream channelization, and web-weaving spiders in northern Japan. Ecological Research 20: 646–651.
Lake, P. S., 2000. Disturbance, patchiness and diversity in streams. Journal of the North American Benthological Society 19: 573–592.
Lancaster, J., 2000. Geometric scaling of microhabitat patches and their efficacy as refugia during disturbance. Journal of Animal Ecology 69: 442–457.
Lenders, H. J. R., B. G. W. Aarts, H. Strijbosch & G. Van der Velde, 1998. The role of reference and target images in ecological recovery of river systems: lines of thought in the Netherlands. In Nienhuis, P. H., R. S. E. Leuven & A. M. J. Ragas (eds), New Concepts for Sustainable Management of River Basins. Backhuys, The Netherlands: 35–52.
Malard, F., K. Tockner, M.-J. Dole-Olivier & J. V. Ward, 2002. A landscape perspective of surface-subsurface hydrological exchanges in river corridors. Freshwater Biology 47: 621–640.
Malcolm, I. A., C. Soulsby, A. F. Youngson & D. Hannah, 2005. Catchment-scale controls on groundwater-surface water interactions in the hyporheic zone: implications for salmon embryo survival. River Research and Applications 21: 977–989.
Maser, C. & J. R. Sedell, 1994. From the Forest to the Sea. The Ecology of Wood in Streams, Rivers, Estuaries and Oceans. St. Lucie Press, Delray Beach, Florida.
May, C. L. & R. E. Gresswell, 2004. Spatial and temporal patterns of debris-flow deposition in the Oregon Coast Range, USA. Geomorphology 57: 135–149.
McLaughlin, R. L., L. Porto, D. L. G. Noakes, J. R. Baylis, L. M. Carl, H. R. Dodd, J. D. Goldstein, D. B. Hayes & R. G. Randall, 2006. Effects of low-head barriers on stream fishes: taxonomic affiliations and morphological correlates of sensitive species. Canndian Journal Fisheries and Aquatic Sciences 63: 766–779.
Metzler, G. M. & L. A. Smock, 1990. Storage and dynamics of subsurface detritus in a sand-bottomed stream. Canadian Journal of Fisheries and Aquatic Sciences 47: 588–594.
Meyer, J. L., D. L. Strayer, J. B. Wallace, S. L. Eggert, G. S. Helfman & N. E. Leonard, 2007. The contribution of headwater streams to biodiversity in river networks. Journal of the American Water Resources 43: 86–103.
Millennium Ecosystem Assessment, 2005. Ecosystems and Human Well-Being: Biodiversity Synthesis. World Resources Institute, Washington, DC. Indicators. Leaflet. ECNC.
Miller, S. W., D. Wooster & J. Li, 2007. Resistance and resilience of macroinvertebrates to irrigation water withdrawals. Freshwater Biology 52: 2494–2510.
Mills, D., 1989. Ecology and Management of Atlantic Salmon. Chapman & Hall, London.
Murray, A. B., M. A. F. Knaapen, M. Tal & M. L. Kirwan, 2008. Biomorphodynamics: physical–biological feedbacks that shape landscapes. Water Resources Research 44: W11301. doi:10.1029/2007WR006410.
Mutz, M., E. Kalbus & S. Meinecke, 2007. Effect of instream wood on vertical water flux in low-energy sand bed flume experiments. Water Resources Research 43: W10424. doi:10.1029/2006WR005676.
Naegeli, M. W. & U. Uehlinger, 1997. Contribution of the hyporheic zone to ecosystem metabolism in a prealpine gravel-bed river. Journal of the North American Benthological Society 16: 794–804.

Naiman, R. J., J. J. Latterell, N. E. Pettit & J. D. Olden, 2008. Flow variability and the biophysical vitality of river systems. Comptes Rendus Geoscience 340: 629–643.

Nelson, P. A., J. G. Venditti, W. E. Dietrich, J. W. Kirchner, H. Ikeda, F. Iseya & L. S. Sklar, 2009. Response of bed surface patchiness to reductions in sediment supply. Journal of Geophysical Research-Earth Surface 114: F02005. doi:10.1029/2008JF001144.

Newson, M. D. & R. G. Large, 2006. "Natural" rivers, "hydromorphological quality" and river restoration: a challenging new agenda for applied fluvial geomorphology. Earth Surface Processes and Landforms 31: 1606–1624.

Nilsson, C., E. Nilsson, M. E. Johansson, M. Dynesius, G. Grelsson, S. Xiong, R. Jansson & M. Danving, 1993. Processes structuring riparian vegetation. Current Topics in Botanical Research 1: 419–431.

Nilsson, C., C. A. Reidi, M. Dynesius & C. Revenga, 2005. Fragmentation and flow regulation of the world's large river systems. Science 308: 405–408.

Nores, C., 2007. *Galemys pyrenaicus* (E. Geoffroy Saint-Hilaire, 1811). In Palomo, L. J., J. Gisbert & J. C. Blanco (eds), Atlas y Libro Rojo de los Mamíferos Terrestres de España. Dirección General para la Biodiversidad. SECEM - SECEMU, Madrid: 96–98.

Oki, T. & S. Kanae, 2006. Global hydrological cycles and world water resources. Science 313: 1068–1072.

Oldeman, R. A. A., 1983. Tropical rain forest, architecture, silvigenesis and diversity. In Sutton, S. L., T. C. Whitmore & A. C. Chadwick (eds), Tropical Rain Forests: Ecology and Management. Blackwell, Oxford: 139–150.

Paetzold, A., C. J. Shubert & K. Tockner, 2005. Aquatic terrestrial linkages along a braided-river: riparian arthropods feeding on aquatic insects. Ecosystems 8: 748–759.

Paillex, A., E. Castella & G. Carron, 2007. Aquatic macroinvertebrate response along a gradient of lateral connectivity in river floodplain channels. Journal of the North American Benthological Society 26: 779–796.

Parsons, M. & M. C. Thoms, 2007. Hierarchical patterns of physical-biological associations in river ecosystems. Geomorphology 89: 127–146.

Poff, N. L., 1997. Landscape filters and species traits: toward mechanistic understanding and prediction in stream ecology. Journal of the North American Benthological Society 16: 391–409.

Pringle, C. M., 2001. What is hydrologic connectivity and why is it ecologically important? Hydrological Processes 17: 2685–2689.

Quinn, T. P., S. M. Carlson, S. M. Gende & H. B. Rich, 2009. Transportation of Pacific salmon carcasses from streams to riparian forests by bears. Canadian Journal of Zoology-Revue Canadienne de Zoologie 87(3): 195–203.

Rabeni, C. F., K. E. Doisy & L. D. Zweig, 2005. Stream invertebrate community functional responses to deposited sediment. Aquatic Sciences 67: 395–402.

Resh, V. H., A. V. Brown, A. P. Covich, M. E. Gurtz, H. W. Li, W. G. Minshall, S. R. Reice, A. L. Sheldon, J. B. Wallace & R. Wissmar, 1988. The role of disturbance in stream ecology. Journal of the North American Benthological Society 7: 433–455.

Revenga, C., J. Brunner, N. Henninger, K. Kassem & R. Payne, 2000. Pilot Analysis of Global Ecosystems: Freshwater Systems. World Resources Institute, Washington, DC.

Roberts, J. H. & P. L. Angermeier, 2007. Movement responses of stream fishes to introduced corridors of complex cover. Transactions of the American Fisheries Society 136: 971–978.

Rosgen, D. L., 1996. Applied River Morphology. Wildland Hydrology, Pagosa Springs, Colorado (USA).

Rovira, A. & C. Ibañez, 2007. Sediment management options for the lower Ebro River and its delta. Journal of Soils and Sediments 7: 285–295.

Savant, S. A., D. D. Reible & L. J. Thibodeaux, 1987. Convective transport within stable river sediments. Water Resources Research 23: 1763–1768.

Scealy, J. A., S. J. Mika & A. J. Boulton, 2007. Aquatic macroinvertebrate communities on wood in an Australian lowland river: experimental assessment of the interactions of habitat, substrate complexity and retained organic matter. Marine and Freshwater Research 58: 153–165.

Schälchli, U., 1992. The clogging of coarse gravel river beds by fine sediment. Hydrobiologia 235(236): 189–197.

Schlief, J. & M. Mutz, 2009. Effect of sudden flow reduction on the decomposition of alder leaves (*Alnus glutinosa* [L.] Gaertn.) in a temperate lowland stream: a mesocosm study. Hydrobiologia 624: 205–217.

Scott, M. L., G. C. Lines & G. T. Auble, 2004. Channel incision and patterns of cottonwood stress and mortality along the Mojave River, California. Journal of Arid Environments 44: 99–414.

Smith, C. L. & Powell, C. R., 1971. The Summer Fish Communities of Brier Creek. Marshall County, Oklahoma. American Museum Novitates No 2458, 30 pp.

Southwood, T. R. E., 1977. Habitat, the templet for ecological strategies? Journal of Animal Ecology 46: 337–365.

Standford, J. A. & J. V. Ward, 1988. The hyporheic habitat of river ecosystems. Nature 335: 64–66.

Statzner, B. & B. Higler, 1985. Questions and comments on the river continuum concept. Canadian Journal of Fisheries and Aquatic Sciences 42: 1038–1044.

Surian, N., Rinaldi, M., Pellegrini, L., Audisio, C., Maraga, F., Teruggi, L. B., Turitto, O. & Ziliani, L., 2009. Channel adjustments in northern and central Italy over the last 200 years. Geological Society of America Special Paper 451: 83–95. doi:10.1130/2009.2451(05).

Taylor, B. W., A. S. Flecker & R. O. Hall, 2006. Loss of a harvested fish species disrupts carbon flow in a diverse tropical river. Science 313: 833–836.

Thoms, M. C. & M. E. Parsons, 2002. Ecogeomorphology: an interdisciplinary approach to river science. International Association of Hydrological Sciences 227: 113–119.

Thorne, C. R., R. G. Allen & S. Andrew, 1996. Geomorphological river channel reconnaissance for river analysis, engineering and management. Transactions of the Institute of British Geographers 21: 4969–4983.

Thorp, J. H. & M. D. Delong, 1994. The riverine productivity model: a heuristic view of carbon sources and organic processing in large river ecosystems. Oikos 70: 305–308.

Thorp, J. H., M. C. Thoms & M. D. Delong, 2006. The riverine ecosystem synthesis: biocomplexity in river networks across space and time. River Research and Applications 22: 123–147.

Tomanova, S. & P. Usseglio-Polatera, 2007. Patterns of benthic community traits in neotropical streams: relationships to mesoscale spatial variability. Fundamental and Applied Limnology 170: 243–255.

Trimble, S. W. & A. C. Mendel, 1995. The cow as a geomorphic agent – a critical review. Geomorphology 13: 233–253.

UNEP, 2007. Global environment outlook 4. Environment for development. United Nations Environment Programme. United Nations, Valletta, Malta.

USEPA, 2000, Nutrient criteria technical guidance manual: lakes and Reservoirs. Office of Water, Office of Science and Technology, Report EPA-822-B00-001, Washington, DC.

Vannote, R. L., G. W. Minshall, K. W. Cummins, J. R. Sedell & C. E. Cushing, 1980. The river continuum concept. Canadian Journal of Fisheries and Aquatic Sciences 37: 130–137.

Vaux, W. G., 1962. Interchange of stream and intergravel water in a salmon spawning riffle. United States Fisheries and Wildlife Service Research Reports: Fisheries 405: 1–11.

Verspoor, E., L. Stradmeyer & J. Nielsen (eds), 2007. The Atlantic Salmon. Genetics, Conservation and Management. Blackwell, Oxford.

Vörösmarty, C. J., M. Meybeck, B. Fekete, K. Sharma, P. Green & J. Syvitski, 2003. Anthropogenic sediment retention: major global impact from registered river impoundments. Global Planetary Change 39: 169–190.

Ward, J. V., 1992. Aquatic Insect Ecology. 1. Biology and Habitat. Wiley, New York.

Ward, J. V. & J. A. Stanford, 1983. The serial discontinuity concept of lotic ecosystems. In Fontaine, T. D. & S. M. Bartell (eds), Dynamics of Lotic Ecosystems. Ann Arbor Science, Ann Arbor, Michigan.

Ward, J. V. & K. Tockner, 2001. Biodiversity: towards a unifying theme for river ecology. Freshwater Biology 46: 807–819.

Welcomme, R. L., 1985. River fisheries. FAO Fisheries Technical Paper 262. Rome, Food and Agricultural Organization of the United Nations: 330 pp.

White, D. S., 1993. Perspectives on defining and delineating hyporheic zones. Journal of the North American Benthological Society 12: 61–69.

Williams, D. D. & H. B. N. Hynes, 1974. The occurrence of benthos deep in the substratum of a stream. Freshwater Biology 4: 233–256.

Williams, J. D., M. L. Warren, K. S. Cummings, J. L. Harris & R. J. Neves, 1993. Conservation status of fresh-water mussels of the United States and Canada. Fisheries 18: 6–22.

Winterbourn, M. J., J. S. Rounick & B. Cowie, 1981. Are New Zealand stream ecosystems really different? New Zealand Journal of Marine and Freshwater Research 15: 321–328.

GLOBAL CHANGE AND RIVER ECOSYSTEMS

# Organic matter availability during pre- and post-drought periods in a Mediterranean stream

Irene Ylla · Isis Sanpera-Calbet · Eusebi Vázquez · Anna M. Romaní · Isabel Muñoz · Andrea Butturini · Sergi Sabater

Received: 5 August 2009 / Accepted: 17 February 2010 / Published online: 7 March 2010
© Springer Science+Business Media B.V. 2010

**Abstract** Mediterranean streams are characterized by water flow changes caused by floods and droughts. When intermittency occurs in river ecosystems, hydrologic connectivity is interrupted and this affects benthic, hyporheic and flowing water compartments. Organic matter use and transport can be particularly affected during the transition from wet to dry and dry to wet conditions. In order to characterize the changes in benthic organic matter quantity and quality throughout a drying and rewetting process, organic matter, and enzyme activities were analyzed in the benthic accumulated material (biofilms growing on rocks and cobbles, leaves, and sand) and in flowing water (dissolved and particulate fractions). The total polysaccharide, amino acid, and lipid content in the benthic organic matter were on average higher in the drying period than in the rewetting period. However, during the drying period, peptide availability decreased, as indicated by decreases in leucine aminopeptidase activity, as well as amino acid content in the water and benthic material, except leaves; while polysaccharides were actively used, as indicated by an increase in $\beta$-glucosidase activity in the benthic substrata and an increase in polysaccharide content of the particulate water fraction and in leaf material. During this process, microbial heterotrophs were constrained to use the organic matter source of the lowest quality (polysaccharides, providing only C), since peptides (providing N and C) were no longer available. During the flow recovery phase, the microbial community rapidly recovered, suggesting the use of refuges and/or adaptation to desiccation during the previous drought period. The scouring during rewetting was responsible for the mobilization of the streambed and loss of benthic material, and the increase in high quality organic matter in transport (at that moment, polysaccharides and amino acids accounted for 30% of the total DOC). The dynamics of progressive and gradual drought effects, as well as the fast recovery after rewetting, might be affected by the interaction of the individual dynamics of each benthic substratum: sand sediments and leaves providing refuge for microorganisms and organic matter storage, while on cobbles, an active bacterial community is developed in the rewetting. Since global climate change may favor a higher intensity and frequency of droughts in

Guest editors: R. J. Stevenson & S. Sabater / Global Change and River Ecosystems – Implications for Structure, Function and Ecosystem Services

I. Ylla (✉) · A. M. Romaní · S. Sabater
Institute of Aquatic Ecology, University of Girona, Campus Montilivi, 17071 Girona, Spain
e-mail: irene.ylla@udg.edu

I. Sanpera-Calbet · E. Vázquez · I. Muñoz · A. Butturini
Department of Ecology, University of Barcelona, Av. Diagonal, 645, 08028 Barcelona, Spain

S. Sabater
Catalan Institute for Water Research (ICRA), 17071 Girona, Spain

streams, understanding the effects of these disturbances on the materials and biota could contribute to reliable resource management. The maintenance of benthic substrata heterogeneity within the stream may be important for stream recovery after droughts.

**Keywords** Mediterranean stream · Drought · Organic matter · Lipids · Polysaccharides · Amino acids · Extracellular enzyme activities · Benthic substrata

## Introduction

Both natural and anthropogenic factors regulate water intermittency (Lehner et al., 2006) and water flow alterations. Climate and global change will produce shifts in the precipitation and discharge patterns of rivers and streams in temperate regions (Schröter et al., 2005). These are expected to cause an increase in the frequency and intensity of floods and droughts (Arnell et al., 1996), and a substantial change in organic matter (OM) accumulation and processing (Lake, 2000; Acuña et al., 2007). Mediterranean stream ecosystems are characterized by a high hydrologic variability (Acuña et al., 2005). This variability is likely to increase (Mariotti et al., 2002), determining the effects of extreme hydrologic episodes on the biochemistry of OM and on carbon cycling in streams. This effect is expected to be increasingly relevant, not only in Mediterranean streams but also in many other climates (Sabater & Tockner, 2010).

The transport and recycling of OM are two major river ecosystem functions (Fisher & Likens, 1973; Cummins, 1974), in which the velocity and efficiency of OM processing, microbial use, and biogeochemical transformations are highly dependent on river hydrology (Butturini et al., 2003; Acuña et al., 2005). All these processes can be affected during the transitions from wet to dry (summer drought) and from dry to wet (flood or rewetting events) conditions that occur under water intermittency. Mediterranean streams are physically, chemically and biologically shaped by predictable floods and droughts through the annual cycle (Gasith & Resh, 1999). Nowadays, the influence of floods and droughts on river and stream catchments have been relatively well studied (e.g. Fisher & Grimm, 1991; Poff et al., 1997; Caramujo et al., 2008), but little is known about their effects on the OM.

Drought severity determines the degree of loss of the hydrologic connectivity, which is often sequential during the drought period (Butturini et al., 2003). In the final stages of the drying process, the fluvial network may be converted into a fragmented landscape of isolated water pools where sediments and organic detritus accumulate and cannot be exported (Lake, 2003). As a result, the drying process is gradual in time and heterogeneous in space. Carbon limitation for the microorganisms during these dry periods benefits autotrophic production (Humphries & Baldwin, 2003). The leaf fall dynamics are also affected by flow cessation. In dry years, a longer leaf fall period is related to hydric stress, causing progressive accumulation of OM in the streambed (Sabater et al., 2001; Acuña et al., 2007). Plausibly, the quality of materials in transport after the first rains is affected by processes occurring in the OM accumulated in dry conditions (Langhans & Tockner, 2006). Photodegradation is one of these processes since solar radiation causes chemical oxidation reactions in the OM accumulated in the streambed (Wetzel et al., 1995; Moran & Zepp, 1997). The importance of photochemical reactions to the microbial communities in aquatic systems will depend on the sources of the dissolved organic matter (DOM) and on its initial bioavailability (Howitt et al., 2008).

Severe drought periods can be followed by intense rainfall episodes (punctuated changes), often leading to floods, and, therefore, to the mobility of materials downstream and between compartments. Infiltration/exfiltration processes at the surface as well as in the groundwater affect the accumulation and biochemistry of OM (Dahm et al., 2003). Nitrate mobilization occurs in the transition from dry to wet interfaces (Butturini et al., 2003), and dissolved organic carbon (DOC) drastically changes in content and composition during rewetting (Vázquez et al., 2007). Since the mobilized DOM is a heterogeneous mixture of carbohydrates, proteins, lignins, organic acids, and humic substances (Thurman, 1985), the relative increase in biodegradable DOC (BDOC) increases microbial activity during rewetting (Romaní et al., 2006). A high amount of carbon is derived from floodplain sources in arid rivers after flooding, having a significant impact on river productivity (Burford et al., 2008). Microbial activity may positively correlate with sediment moisture content and respond to the existence of senescent algae, which can restrain water loss from surface sediments in the moist habitat (Claret & Boulton, 2003). Thus, persistence of microbial communities in drought conditions may be dependent on available wet refuges in the ecosystem (Amalfitano et al., 2008).

This study aimed to characterize the changes in benthic OM (quantity and quality) during the transition from wet to dry and dry to wet conditions in an intermittent Mediterranean stream. We hypothesized that abrupt water flow alterations (water disappearance and return) in headwater streams will impact on the OM concentration and composition in the flowing water as well as in the benthic substrata. A different effect of drought and rewetting was expected to occur on the biofilm benthic substrata of the stream (cobbles and sand) and on leaf material due to their specific mobility and sensitivity to flow regime and specific autotrophic/heterotrophic microbial community associated with different substrata. Both rates and direction of temporal changes in OM content and composition in the flowing water should differ during drought, with gradual changes expected, versus during rewetting when pulsed changes are predicted; and temporal changes should vary among habitat types. Therefore, special attention has been given within and between habitat types during the pre- and post-drought periods.

# Methods

Study site

Fuirosos is a third-order stream located in Montnegre-Corredor Natural Park, a forested range close to the Mediterranean Sea (50 km north of Barcelona, NE Spain). The climate is typically Mediterranean; precipitation is distributed irregularly, and mostly falls in autumn and spring with occasional summer storms. A long dry period (2–3 months) in summer is followed by a short but intense (80 l m$^{-2}$ month$^{-1}$; Acuña et al., 2007) stream recharge period in late summer–early autumn.

The studied stream reach is 3–5 m wide and 10 m in length. The riparian vegetation is made up of alder (*Alnus glutinosa*), hazelnut (*Corylus avellana*), poplar (*Populus nigra*) and plane trees (*Platanus acerifolia*). Most of the leaf input into the river channel occurs in summer due to hydric stress, and from autumn (especially *Alnus* leaves) to late winter (especially *Platanus* leaves). The DOC concentrations in the stream water during basal and storm discharge conditions range between 2–4 and 5–10 mg l$^{-1}$, respectively (Vázquez et al., 2007). The DOC concentration can increase up to 10–20 mg l$^{-1}$ during the hydrologic transition between the dry and wet periods (Butturini et al., 2008). The stream channel morphology of the studied reach included a riffle (with boulders and cobbles) and a large pool (with accumulated leaf material and sand). The superficial water flow was progressively interrupted, and the hydrologic connectivity between the stream habitats was lost since June 19, 2006 and remained completely dry until September 13, 2006 (Fig. 1). The pool (maximum depth of 45 cm) was the last stretch to dry out completely.

Sampling strategy

The first sampling period was before drought (pre-drought period) and the second one after drought (post-drought period). The pre-drought period (with a total of six sampling occasions) started on May 8, 2006, when the stream flow was at basal levels (7.7 l s$^{-1}$). During this period of the pre-drought, the stream discharge decreased progressively until June 19, 2006, when there was no surface water. At that time, two samplings were done for benthic material. The post-drought period with six sampling occasions started on September 13, 2006 when water flow started after 2 months of drought. The post-drought period ended on December 4, 2006, when basal hydrologic conditions were resumed (Fig. 1). Four transects (3 m apart) were defined in the studied reach, and the relative cover (%) of each streambed substratum was identified every 20 cm in each sampling occasion. Dissolved oxygen, water temperature (Hach DO meters), and conductivity (WTW conductivity meter) were measured in the field.

Organic matter (quantity and quality) was analyzed in three types of benthic accumulated material, leaves, and particulate material, biofilms growing on rocks and cobbles, and biofilm and fine material accumulated in sand, as well as in the flowing water (dissolved and particulate fractions). On each sampling occasion, one sample of each benthic substratum type was

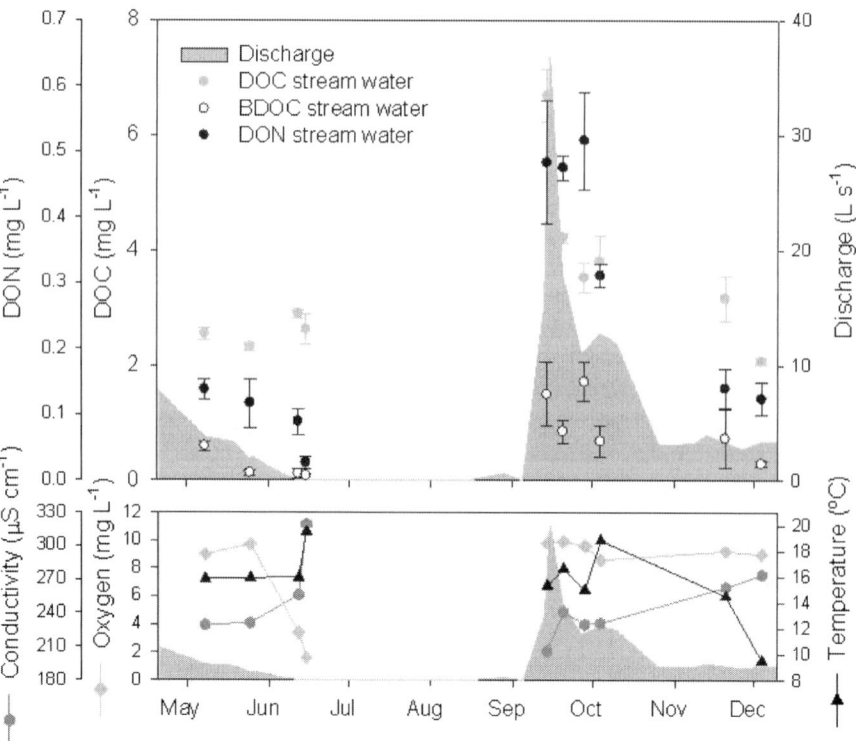

**Fig. 1** Temporal variations in discharge dissolved organic carbon (DOC) and dissolved organic nitrogen (DON) in the Fuirosos stream from May to December 2006. The lower graph corresponds to the temporal variation in temperature, conductivity and oxygen during the same period. Each sampling campaign is represented in a temporal scale; each dot corresponds to one sampling day (the two sampling campaigns done when there was no surface water were not represented)

randomly collected in the same site of each of the four transects. Sand and leaf materials were sampled by coring an area (4.3 cm$^2$) between 5 and 10 cm in depth; cobbles were taken directly from the streambed. Stream water (approx. 8 l) was also collected for analysis. Moreover, the water discharge was measured in each sampling day, and water samples to analyze DOC, BDOC, and dissolved organic nitrogen (DON) were collected. All the samples were refrigerated and transported to the laboratory (1-h travel time). Once inside the laboratory, water was immediately filtered through precombusted GF/F filters to separate the DOM from the particulate organic matter (POM). The OM from the different benthic substrata was detached by immersing the sand samples in 120 ml of distilled water and then sonicating them (3 min, Selecta sonication bath at 40 W and 40 kHz). Leaf material was also immersed in 120-ml distilled water, sonicated (3 min, Selecta sonication bath at 40 W and 40 kHz) and then homogenized with a mixer (kitchen mixer). Leaf material, therefore, included both the leaves as well as their microbial colonizers. Cobbles were immersed in 60 ml of distilled water, scraped with a toothbrush and sonicated (3 min, Selecta sonication bath at 40 W and 40 kHz) to obtain the epilithic material. The extracted and homogenized benthic material (which was separated into subsamples) and the particulate and dissolved water fractions, were analyzed for chlorophyll (excluding the water fractions) and bacterial biomass; polysaccharide, lipid and protein content; and extracellular enzyme activities.

DOC, BDOC, and DON analysis

Water samples to determine DOC, BDOC, and DON were filtered (precombusted GF/F glass-fiber filters; Whatman) before analysis. DOC and total nitrogen (TN) concentrations were determined using a Shimadzu TOC-VCS with a coupled TN analyzer unit. The BDOC was measured (Servais et al., 1989) in samples incubated for 28 days at room temperature in the dark. Glassware was previously heated at 450°C for 4 h to insure complete organic carbon release. All DOC samples were acidified with 2 M HCl (2%) and preserved at 4°C until analysis.

Dissolved organic nitrogen was determined as the difference between the total nitrogen and the total inorganic forms (the sum of nitrate, nitrite, and ammonia). Inorganic forms were determined colorimetrically using a Technicon autoanalyzer. Nitrate was determined by the Griess–Ilosvay method (Keeney & Nelson, 1982) after reduction by percolation through a copperized cadmium column. Ammonia was determined after oxidation by salicylate using sodium nitroprusside as the catalyst (Hach Company, 1992).

Microbial biomass and extracellular enzyme activities in water and benthic material

*Benthic chlorophyll concentration*

Chlorophyll *a* was measured in sand and cobble biofilms and in leaf material (four replicates). For each sample, between 5 and 10 ml of the extracted and homogenized benthic material were filtered (GF/C Whatman) and then chlorophyll was extracted in 90% acetone for 12 h in the dark at 4°C. Samples were further sonicated (2 min, Selecta sonication bath at 40 W and 40 kHz) to insure complete chlorophyll extraction. After filtration (GF/C Whatman) of the extract, the chlorophyll concentration was determined spectrophotometrically (Lambda UV/VIS spectrophotometer, Hitachi) following Jeffrey & Humphrey (1975).

*Bacterial density*

Bacterial density was estimated in sand and cobble biofilms, in leaf material and in water samples (four replicates per substratum). Live and dead bacteria were counted using the Live/Dead Baclight bacterial viability kit, which contains a mixture of SYTO® 9 and propidium iodide. The live cells (with intact cell membranes) appeared green after excitation with blue light, whereas dead cells (with damaged cell membranes) appeared red (Freese et al., 2006). Fifty microliters of the extracted benthic material of the sand samples were diluted in 2 ml of sterilized river water. Aliquots of 200 µl of cobbles and leaves extracts were also diluted with 2 ml of sterilized river water. For water samples 2 ml were taken directly (no dilution). After appropriate dilution, a 1:1 mixture of SYTO® 9 and propidium iodide was added (3 µl), and samples were incubated for 15 min. Samples were then filtered through 0.2 µm black polycarbonate filters (Nucleopore, Whatman). The filters were dried, placed on a slide with mounting oil and examined by epifluorescence microscopy (Nikon E600). At least 20 random fields were examined on each slide for a minimum of 300 bacteria cells as a compromise between observational effort and reliability. The fraction of live bacteria was calculated as the abundance of live cells divided by the total count obtained with the live/dead method.

*Extracellular enzyme activities*

Sand and cobble biofilms, leaf material and water samples (POM and DOM, four replicates) were analyzed to determine the activity of the enzymes $\beta$-D-1,4-glucosidase (EC 3.2.1.21), lipase (EC 3.1.1.3) and leucine-aminopeptidase (EC 3.4.11.1). Each sample from sand, cobbles and leaves consisted of 1 ml of the extracted and homogenized benthic material. Four ml of river water and GF/F filtered river water were considered, respectively, for the total and dissolved activity in water. Extracellular enzyme activities were determined spectrofluorometrically using the artificial substrates 4-methylumbelliferyl-$\beta$-D-glucopyranoside for $\beta$-glucosidase, 4-methylumbelliferyl palmitate for lipase activity and L-leucine-7-amido-4-methylcoumarin hydrochloride for peptidase activity (Sigma-Aldrich), as the respective substrate-analogs.

All the samples were incubated with 0.3 mM substrate (saturated conditions; Romaní & Sabater, 2001) in the dark under continuous shaking for 1 h at 18°C. Blanks and standards of methylumbelliferone (MUF) and AMC (aminomethylcoumarin) were also incubated. At the end of the incubation glycine buffer (pH 10.4) was added (1/1 vol/vol), and the fluorescence was measured at 365/455 nm excitation/emission for MUF and 364/445 nm excitation/emission for AMC.

Polysaccharide, protein, and lipid content in water and benthic material

Polysaccharide, protein, and lipid content was analyzed in water (dissolved and particulate fraction), in the biofilm benthic materials collected from sand and cobbles and in leaf material. Four replicates were considered for each sample type.

*Polysaccharide content*

Polysaccharide content was measured following the 3-methyl-2-benzothiazolinone hydrochloride (MBTH) method with modifications (Pakulski & Benner, 1992; Chanudet & Filella, 2006). A total of 50 ml of DOM, filters for POM (Whatman GF/F), and 10 ml of the previously obtained extracted material from the samples (sand and cobble biofilms; leaf material) were freeze-dried. The dried samples were then acidified with 1 ml of 12 M $H_2SO_4$ for 2 h at room temperature. Then, the samples were diluted with 4 ml Milli-Q water, sonicated (2 min) and hydrolyzed at 100°C for 3 h. After cooling, the pH of the hydrolysis solution was neutralized with NaOH. Next, monosaccharides were reduced to alditols by the addition of potassium borohydride. The reduction reaction was terminated by the addition of 2 M HCl. The samples were left overnight at 4°C. The following day, triplicate aliquots of hydrolysis products (and duplicate blanks) were placed in test tubes and were oxidized to formaldehyde by the addition of 0.025 M periodic acid. The oxidation reaction was terminated by the addition of 0.25 M sodium metaarsenite. After the addition of 2 M HCl, the aldehyde was reacted with MBTH reagent, ferric chloride solution and acetone. Absorbance was measured at 635 nm with a spectrophotometer (Spectronic® 20 Genesys). Absorbance of the blanks was subtracted from all samples. Glucose standard curves were generated concurrently.

*Protein content*

Amino acids were analyzed using high performance liquid chromatography (HPLC). All samples were first freeze-dried (including 5 ml of DOM, filters for POM (Whatman GF/F), and 100 µl of the previously extracted material from sand and cobble biofilms as well as from leaf material) and then hydrolyzed in sealed vials with 6 M HCl at 110°C for 20 h. After this step, the remaining HCl was removed using nitrogen flushing steps. The residue was derivatized with a fluorescent reagent (AccQ Fluor reagent, Waters®) following the manufacturer's instructions, and then analyzed on a Waters HPLC amino acid analysis system. The HPLC system included a Waters AccQ Tag column for separation of amino acids, a Waters 2475 fluorescence detector, a 717plus autosampler, and a 1525 Waters binary pump. An internal standard ($\alpha$-aminobutyric acid) was added during the treatment of the samples and standards.

*Lipid content*

Samples were first freeze-dried. Cobbles and sand biofilms as well as leaf material samples were weighed to the nearest 0.1 mg. Samples were homogenized with an ultrasonic homogenizer (200 W, 24 kHz; Hielscher Ultrasonics GmbH, Teltow, Germany). The lipids were extracted with a mixture of chloroform and methanol (2:1) following Bligh & Dyer (1959). The total lipid content was analyzed by the colorimetric sulphophosphovanillin method (Zollner & Kirsch, 1962). The dissolved fraction could not be analyzed because the lipid content was below the detectable level ($<0.01$ mg l$^{-1}$).

Statistical analyses

Differences in chlorophyll, bacterial density, enzyme activities, and total polysaccharide, protein, and lipid content within the pre- and the post-drought periods for the different substrata were analyzed using a 1-way repeated measures analysis of variance (RM-ANOVA). Probabilities within groups (Day and Day × S) were corrected for sphericity using the Greenhouse–Geisser correction. All the probabilities were adjusted by the Dunn–Sidak correction. All the variables included in the analyses were $\log(x + 1)$ transformed to improve the homogeneity and heterogeneity of variance which was checked by visual inspection of the residual distributions.

Differences of bacterial density, total polysaccharide, protein, and lipid content between the last day of the pre-drought period and the first day of the post-drought period were checked using multivariate analysis of variance (MANOVA).

Pearson correlation was performed to determine the potential relationships between the biogeochemical and biologic variables studied. The relationships between the OM composition (proteins, polysaccharide, and lipid content) of the benthic substrata and their enzyme activities were examined by regression analyses. All the statistical analyses were carried out using the SPSS software package for Windows (Ver.14.0.1, SPSS Inc. 1989–2005).

## Results

### Physical and chemical parameters

During the pre-drought period, there was a progressive increase in the stream water conductivity (from 229 to 329 µS cm$^{-1}$). Dissolved oxygen decreased to very low values (1.6 mg $O_2$ l$^{-1}$). The maximum water temperature (19.5°C) was achieved in the pool immediately before stream water cessation (Fig. 1).

The water flow peak on September 13th (36.8 l s$^{-1}$; Fig. 1) was associated with an increase in DOC (6.7 mg l$^{-1}$), BDOC (1.5 mg l$^{-1}$) and DON (0.5 mg l$^{-1}$) in the stream water. DOC, BDOC, and DON were positively correlated with water flow ($r = 0.705$, $P < 0.001$; $r = 0.891$, $P < 0.001$; $r = 0.896$, $P < 0.001$, respectively) throughout the studied period. During the high flow episode dissolved oxygen increased (9.8 mg $O_2$ l$^{-1}$) and conductivity returned to pre-drought basal values (206 µS cm$^{-1}$) (Fig. 1). Peak flow in September caused important changes in the structure of the stream bed. While the reach during the drought period consisted of a riffle (with cobbles and rocks covering 70% of the streambed surface), a pool (with a large accumulation of sandy sediment, 80%) and some patches of deposited leaves (5%), the heavy rains homogenized the stream bed. After the high flow episode, the stream became dominated by cobbles and rocks (70%). Both the fine substrata (sand) and leaves were washed downstream after the episode.

### Microbial biomass and metabolism

#### Benthic chlorophyll and organic matter

Benthic chlorophyll overall decreased significantly during the pre-drought period, but remained unchanged during the post-drought period (Table 1). During the pre-drought period, chlorophyll decreased in sand and cobble biofilms, but remained steady on leaves (Table 1, Day × S interaction). Chlorophyll-$a$ accounted for 182 ± 47.3 mg chl m$^{-2}$ in sand and 23.5 ± 8.9 mg chl m$^{-2}$ in cobbles during the pre-drought period. These values decreased, respectively, to 126 ± 39.7 and 5.8 ± 1.73 mg chl m$^{-2}$ after drought. The chlorophyll content in leaves was most similar before and after the drought (1,489 ± 825 and 1,229 ± 388 mg chl m$^{-2}$, respectively).

The dry weight per cm$^2$ of benthic materials was higher in the pre-drought than in the post-drought period. Sand had 1.27 ± 0.26 g cm$^{-2}$ and cobbles had 0.03 ± 0.008 g cm$^{-2}$ in the pre-drought period, and 0.95 ± 0.26 and 0.01 ± 0.008 g cm$^{-2}$, respectively, after drought. Similar dry weight values were obtained before and after drought (1.66 ± 1.16 and 1.69 ± 0.57 g cm$^{-2}$, respectively) in leaves.

#### Bacterial density

Total bacteria and the percentage of live bacteria progressively decreased during the pre-drought period in the stream water (RM-ANOVA, Day effect, $P = 0.001$) and in the cobble biofilms substrata, but

Table 1 Results of the repeated-measures analysis of variance considering one factor: Substratum type (cobbles and sand biofilms; and leaf material) for biofilm structure and composition, and activity variables

| Source of variation | Chl | Total bacterial density | Life bacteria (%) | Polysaccharide content | Amino acid content | Lipid content | β-Glucosidase | Peptidase | Lipase |
|---|---|---|---|---|---|---|---|---|---|
| Pre-drought period | | | | | | | | | |
| Day | **0.032** | **<0.001** | 0.134 | **<0.001** | **<0.001** | 0.594 | **<0.001** | **<0.001** | **<0.001** |
| Substratum | **<0.001** | **<0.001** | **<0.001** | **<0.001** | **<0.001** | **<0.001** | **<0.001** | **<0.001** | **<0.001** |
| Day × S | **<0.001** | **<0.001** | **0.020** | **<0.001** | **<0.001** | **0.044** | **<0.001** | **<0.001** | 0.430 |
| Post-drought period | | | | | | | | | |
| Day | 0.953 | **0.017** | 0.118 | **0.002** | **<0.001** | 0.527 | **<0.001** | **0.002** | **<0.001** |
| Substratum | **<0.001** | **<0.001** | **<0.001** | **<0.001** | **<0.001** | **0.002** | **<0.001** | **<0.001** | 0.359 |
| Day × S | 0.766 | 0.200 | 0.283 | 0.179 | **0.047** | 0.302 | **<0.001** | **<0.001** | **0.005** |

Probability within groups (Day and Day × S) are corrected for sphericity by the Greenhouse–Geisser correction. All probabilities are adjusted by the Dunn–Sidak correction. Values <0.05 are indicated in boldface type

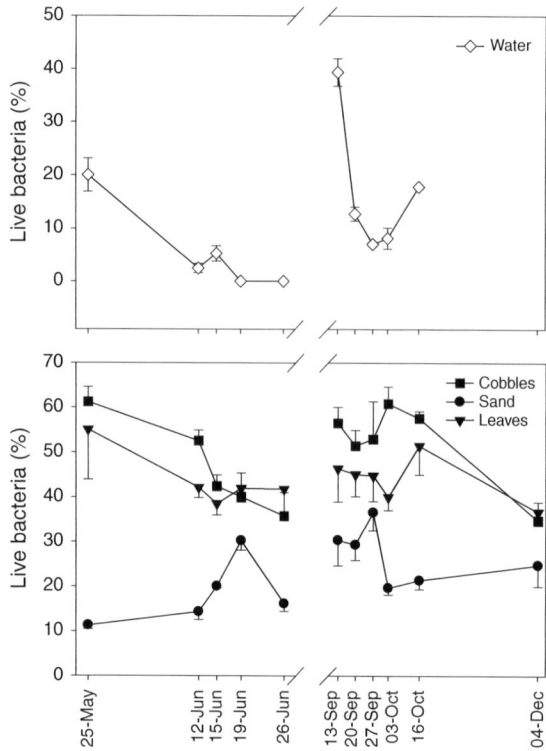

**Fig. 2** Temporal dynamics of live bacteria (%) in water, cobbles, and sand biofilms and leaf material in the Fuirosos stream, May 2006–December 2006. On the $x$ axis the exact sampling days are shown. Mean ± standard error (*vertical bars*), $n = 4$

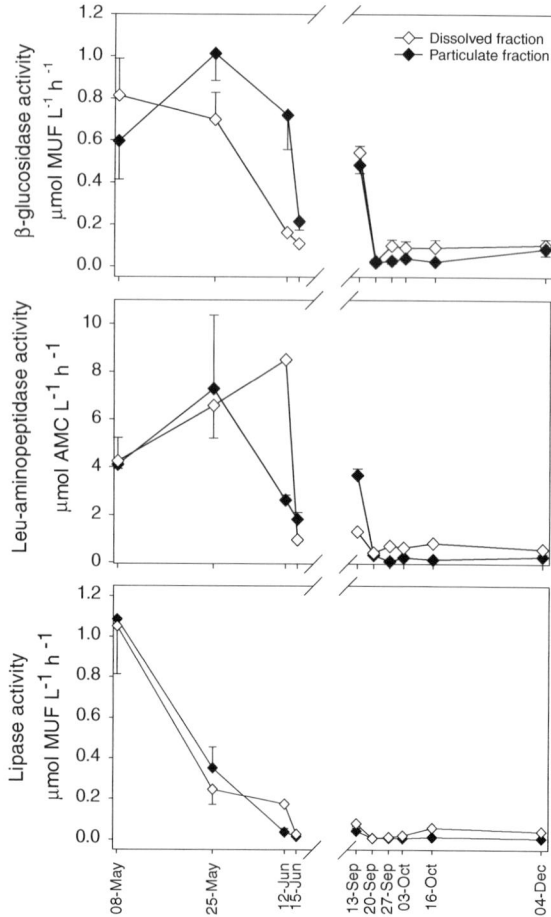

**Fig. 3** Temporal changes in extracellular enzymatic activities in the stream water particulate fraction and dissolved fraction in the Fuirosos stream, May 2006–December 2006. On the $x$ axis, the exact sampling days are shown. Mean ± standard error (*vertical bars*), $n = 4$

increased in the sand biofilms up to the end of June (Table 1, Day × S interaction effects; Fig. 2). The first rains after the drought caused an instantaneous recovery of live bacterial density in water and cobbles (MANOVA, $P < 0.02$).

The total number of bacteria (live plus dead cells) was higher in leaf and sandy material ($3 \times 10^8$ bacteria cm$^{-2}$) than on cobble biofilms ($5 \times 10^6$ bacteria cm$^{-2}$) both in the pre- and post-drought periods. However, the highest number of active bacteria was observed on cobble biofilms, followed by leaf material, sand biofilms, and finally in the stream water.

*Extracellular enzyme activities*

The extracellular enzyme activities measured in the dissolved water fraction followed a similar pattern than in the particulate fraction (Fig. 3). The extracellular enzyme activities progressively decreased when the stream dried out (RM-ANOVA, Day effect $P < 0.001$). Enzymatic activities peaked after the rewetting, coinciding with that in the proportion of active bacteria (Fig. 2).

The enzyme activities in the benthic substrata (Fig. 4) followed similar patterns as those in the water except for $\beta$-glucosidase activity. Leucine aminopeptidase and lipase activities decreased during the drying process while $\beta$-glucosidase activity increased (Table 1, Day effect). After rewetting, enzyme activities recovered and were maintained at similar levels to those before drought. $\beta$-glucosidase activity before and after drought was the lowest in leaf material whereas peptidase activity had the

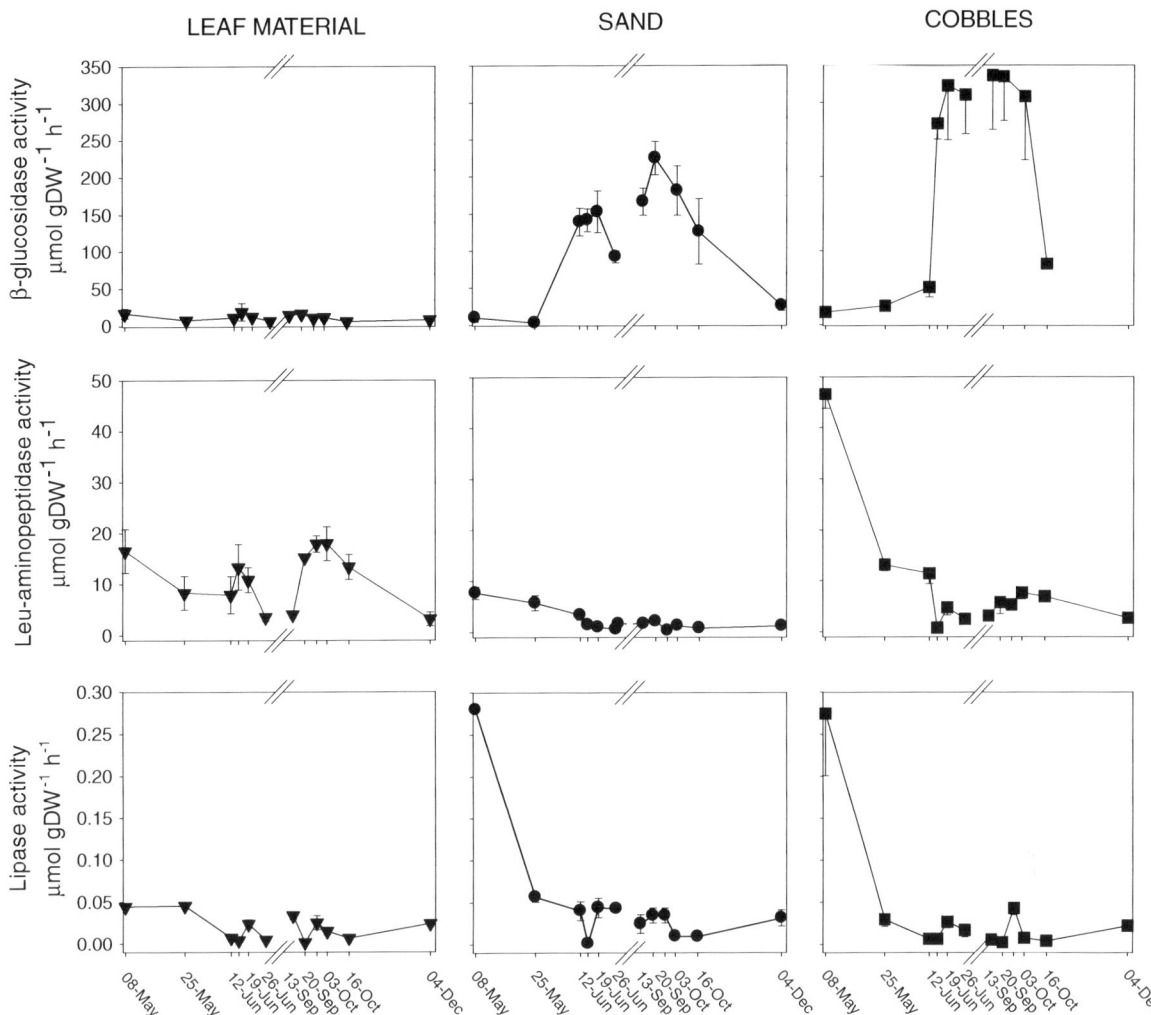

**Fig. 4** Changes in extracellular enzyme activities ($\beta$-glucosidase, leucine amino peptidase, and lipase) in leaf material, and in sand and cobble biofilms in the Fuirosos stream during the study period. On the *x* axis, the exact sampling days are shown. Mean ± standard error (*vertical bars*), $n = 4$

lowest values in the sand biofilm (Table 1, Substratum effect; Fig. 4). Lipase activity behavior was similar during the drought process for the three benthic substrata; however, after drought lipase activity recovered faster on leaves and sand biofilm (Table 1, Day × S effects; Fig. 4).

During the whole study period, $\beta$-glucosidase activity on the benthic substrata was the highest, followed by leucine aminopeptidase, and then, with very low values by lipase. However, in the stream water leucine aminopeptidase activity was the highest followed by $\beta$-glucosidase, and finally by lipase.

Organic matter composition

*Water*

Polysaccharides accumulated in the particulate water fraction while amino acids decreased (RM-ANOVA, Day effect, $P < 0.000$; Fig. 5) in the pre-drought period. Total lipids in water also accumulated during the pre-drought period (RM-ANOVA, Day effect, $P < 0.000$; Fig. 5). During the post-drought period, time differences in OM composition were not detected.

**Fig. 5** Dynamics of the different components of OM (polysaccharides, amino acids, and lipids) in the stream water (dissolved and particulate fraction) and in the different substrata (leaf material, sand, and cobble biofilms). Values are means of four replicates for each sampling date ± standard error. Results of polysaccharide content were expressed as glucose equivalents. In the *x* axis the exact sampling days are shown

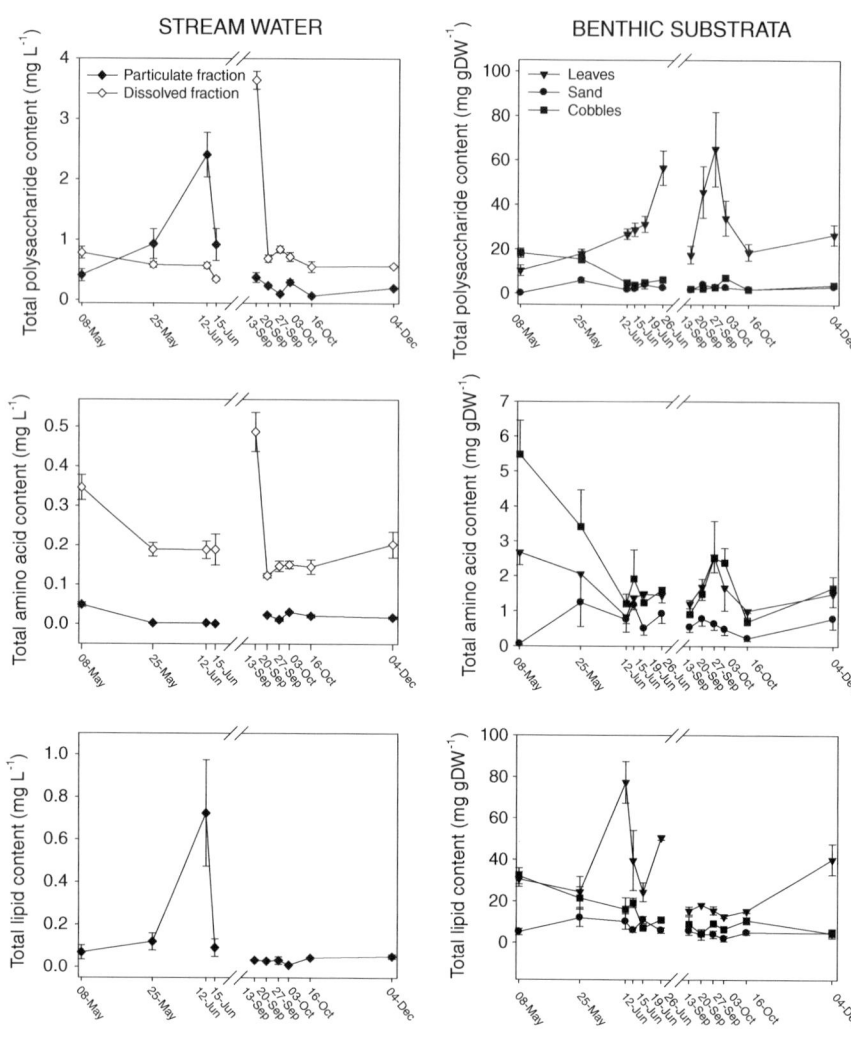

In the dissolved water fraction, polysaccharide and amino acid content decreased (RM-ANOVA, Day effect, $P < 0.000$; Fig. 5) during the pre-drought period. Dissolved polysaccharides reduced from 12% to a 5% of total DOC during the pre-drought period (Fig. 6). Similarly, total dissolved amino acids reduced from 5.6% to 3% of total DOC during the same period (Fig. 6).

A peak of polysaccharides and peptides in the dissolved water fraction occurred immediately after rewetting (Fig. 5). These peaks contributed to the immediate increases of DOC and DON (Fig. 1). At this rewetting moment, polysaccharides accounted for 20% of total DOC, and amino acids were 3% of total DOC (Fig. 6). When basal water flow conditions were re-established (end of post-drought period), amino acids had increased up to 4% of total DOC (Fig. 6).

*Benthic substrata*

Total polysaccharides, amino acid, and lipid content per dry weight showed significant time differences as the stream dried out. The amino acid content (most notably) and the lipid and polysaccharide content decreased in sand and cobbles (Table 1, Day × S interaction; Fig. 5) in parallel with the progressive decrease in benthic chlorophyll content. However, polysaccharide and lipids increased on the leaf material.

In the post-drought period, differences between benthic substrata were evidenced by the highest

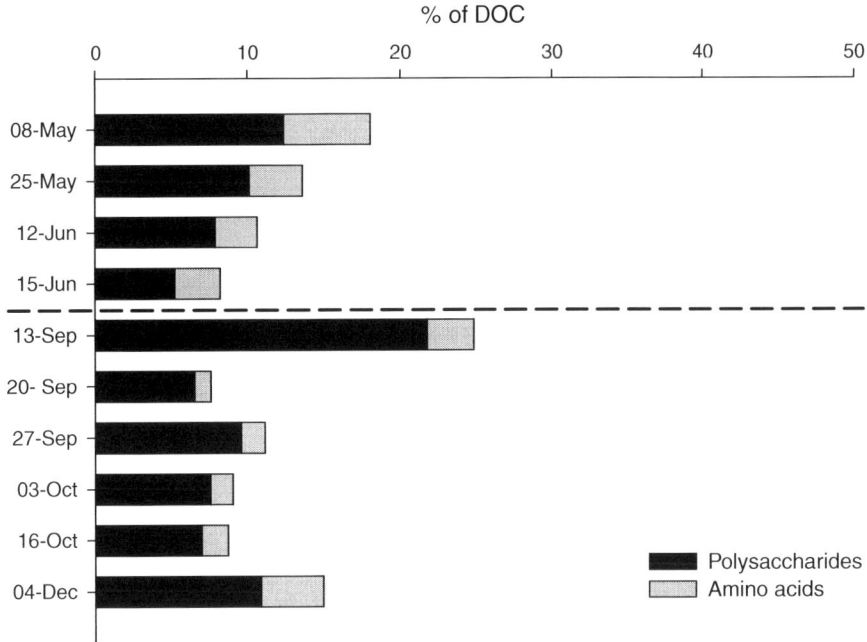

Fig. 6 Contribution of polysaccharides and amino acids (in units of C) to the total DOC. The *dashed line* divides the pre-drought period from the post-drought period

values of polysaccharides (most remarkably) and lipids in leaves, as well as low amino acids in sand biofilms (Table 1, Substratum effect).

Total polysaccharides, amino acid, and lipid content per dry weight on the benthic substrata were higher when the stream was dried out than at the rewetting (MANOVA, $P < 0.05$).

Relationships between extracellular enzyme activities and organic matter

The extracellular enzyme activities in the stream water were higher when the available OM from the benthic substrata was also high. $\beta$-glucosidase in the stream water was correlated with the polysaccharide content in cobble biofilms ($r = 0.674$, $P = 0.033$), and lipase in the stream water was correlated with lipid content in cobble biofilms ($r = 0.766$, $P = 0.010$). Furthermore, in some moments, the available dissolved OM was also related to the water enzyme activities. In the post-drought period, $\beta$-glucosidase activity measured in the dissolved water fraction was correlated to the polysaccharide content in it ($r = 0.978$, $P = 0.001$), while peptidase activity in this same water fraction was associated with the amino acid availability in it ($r = 0.847$, $P = 0.033$).

The enzyme activities on the benthic substrata were significantly related to their OM composition. Peptidase activity on cobbles and sand biofilms and in leaf material were positively related with their amino acid content ($R^2 = 0.663$, $P < 0.001$) according to a linear regression (Fig. 7). In contrast, $\beta$-glucosidase was negatively and exponentially related to substratum-associated polysaccharides ($R^2 = 0.147$, $P = 0.021$; Fig. 7), mainly showing that in the lower range of polysaccharide content (as it occurs in cobbles and sand), there can be high $\beta$-glucosidase activity which decreases rapidly with increasing polysaccharide content; and in the higher range (as it occurs in leaves) changes in polysaccharide content are not affecting the $\beta$-glucosidase activities which are very low (Fig. 7). No significant relationship was found between lipase and the respective lipid composition of benthos (Fig. 7).

Discussion

Drought and rewetting caused mostly two linked effects in the Mediterranean forested stream: (1) changes in the microbial use of available organic matter, and (2) changes in downstream loss of OM dynamics.

Drought affected the availability of OM for microbial use as well as its food quality, in a differential way in the pre- and post-drought periods.

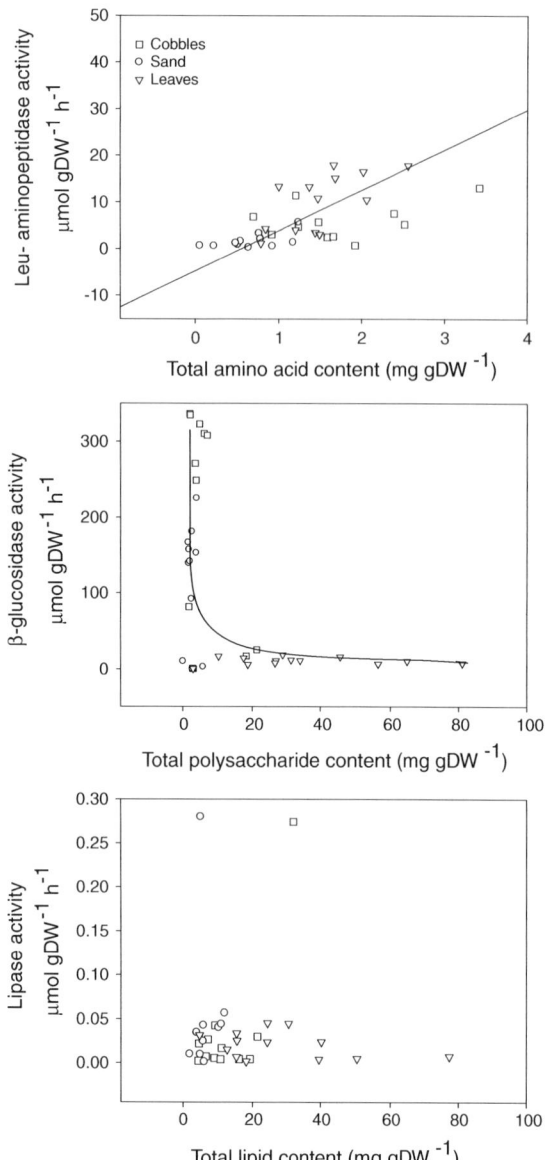

Fig. 7 Regression analyses between the OM composition (amino acid, polysaccharide, and lipid content) of cobbles and sand biofilms as well as of leaf material and their enzyme activities (Leu-aminopeptidase, $\beta$-glucosidase and lipase)

Drought events in Mediterranean river ecosystems have a great effect not only on hydrologic and biogeochemical conditions but also on OM availability, use, and recycling (Sabater & Tockner, 2010). Droughts affect the biota and the stream ecosystem in a degree according to their duration and time of occurrence (Boulton, 2003; Lehner et al., 2006). As drying proceeds, shallow surface habitats such as riffles disappear first, and a series of fragmented pools remain together with a low water flow in the hyporheic zone (Lake, 2003). Increases in water flow may cause a complete resetting of the physical habitat, as well as the downstream drift of many individuals and materials. It has been observed that the biogeochemical response to water flow increases is closely linked to the previous hydrologic "history" of the system (duration of drought) (Vázquez et al., 2007). Our observations during the drying and hydrologic recovery of a Mediterranean stream indicate that the processes occurring during the post-drought period (the fast recovery of active bacteria and most extracellular enzyme activities; the transport of high quality OM) are related to those occurring during the pre-drought period.

In Mediterranean streams, the major drought period takes place in early to mid summer. Prior to this period, we expect the highest accumulation of benthic material derived from in-stream primary producers. In the Fuirosos, the optimal period for the development and growth of the biofilm communities is in the spring. The high availability of light and nutrients, as well as the steady hydrologic conditions in that period, led to very high biofilm biomass as well as to high primary production and microbial metabolism (Ylla et al., 2007). High quality OM (mainly rich in peptides and polysaccharides) is, therefore, available to the stream food web during that period. The abundance of algal material, with equal contributions of amino acids and carbohydrates and lower amounts of lipids (Harvey & Mannino, 2001) probably explain the higher leucine aminopeptidase and $\beta$-glucosidase activity in the benthic substrata, as well as the lower lipase activity.

When discharge declines and the stream dries out, hydric stress on aquatic biota causes a gradual loss (faster in the last days with flowing water) of aquatic habitat, depletion of food resources and a decline in water quality. As the dry season progresses and water flow is even more reduced, habitat conditions become harsher (Gasith & Resh, 1999), transport of OM (detritus, leaves and plant material) and fine sediments declines and high quantities of OM are stored in pools (Cuffney & Wallace, 1989; Boulton & Lake, 1992; Wright & Symes, 1999). In addition, conductivity and water temperature rise, and low oxygen levels lead to facultative aerobic and anaerobic respiration.

In the Fuirosos, these physical and biogeochemical changes were accompanied by variations in the quantity and quality of the OM and in its use by the microbial biota. During the drought process, there was a progressive decrease of the polysaccharide, amino acid, and lipid content in cobbles and sand biofilms. The loss of high quality OM during this period is striking, and is probably due to decaying organisms. In the final stages of the drought, when water ceased to flow, such effects were more substantial on the epilithic material than on the epipsammic and epixylic. At the same time, water was progressively enriched in its particulate fraction, with higher polysaccharide and lipid content, but lower peptide content. Besides, in the dissolved water fraction there was a reduction of polysaccharides and amino acid content. Altogether, the quality and biodegradability of materials in the dissolved fraction progressively decreased during the drought. Polysaccharide plus amino acid content decreased from 18 to 8% of total DOC throughout the drying process (Fig. 6). These changes in the quality of the OM were accompanied by changes in extracellular enzyme activities in the benthic substrata: leucine aminopeptidase and lipase activities decreased, and $\beta$-glucosidase activity increased during the drought period. Therefore, as drought progressed, stream microbial heterotrophs mostly used the lowest quality OM (polysaccharides, providing only C), since peptides (providing N and C) were no longer available. At the same time, the expression of extracellular enzyme activities was modulated by the nature of available OM in each substratum (Artigas et al., 2008). This was evidenced by the positive relationship between amino acid content of benthic substrata and leucine–aminopeptidase activity (Fig. 7). However, the opposite relationship was found for the $\beta$-glucosidase activity and polysaccharide content, eventually resulting in no relationship between high polysaccharide content and $\beta$-glucosidase. This might be related to the use of the available polysaccharides from the particulate water fraction during drought, when polysaccharides in cobbles and sand biofilms decrease.

The harsh conditions for the microbial community during the drying of the stream caused a reduction of both algae (reduction in chlorophyll) and active bacteria. Lower algal density during drought is related to cells breaking under desiccation (Usher & Blinn, 1990; Peterson et al., 2001; Stanley et al., 2004). Hydric stress is also lethal to bacterial cells by damaging membranes, proteins, and nucleic acids (Billi & Potts, 2002). Further, the decrease in DOC inputs during drought may lead to C limitation, and consequently lower heterotrophic production (Humphries & Baldwin, 2003). Harsh conditions were more evident on cobble biofilms than on sand biofilms and leaf material, where moist conditions could be more easily maintained, providing a place in which aquatic bacterial communities could survive (Amalfitano et al., 2008).

The return of water to a previously dry streambed can be seen as a "hot biogeochemical moment" (McClain et al., 2003). Biogeochemical reactions restart or accelerate after long quiescent periods. Aerobic penetration increases in previously dry sediments as once anaerobic zones become aerobic (Baldwin & Mitchell, 2000). Re-wetted sediments liberate phosphorus and nitrogen, as a consequence of death-induced microbial cell lysis, which again may enhance instream primary productivity. Such enormous changes were recorded in the Fuirosos when the first rains occurred in autumn. The September flow peak basically had two effects: the cleaning of the streambed and the downstream transport of materials, and the simultaneous import of OM from upstream. This flow moved and redistributed streambed materials, including the OM accumulated on the streambed during drought. This OM was mostly made up of allochthonous (terrestrial) material coming from the decaying leaves and plant litter. This leaf material produce a fresh and biodegradable high quality lixiviates (Francis & Sheldon, 2002). There is also a noticeable fraction of autochthonous material derived from algal origin. The cleaning effect of the flow peak reduced the total amount of dry weight per $cm^2$ and reduced the polysaccharide, peptide, and amino acid content of the benthic materials. There was a mobilization of OM downstream, and high polysaccharide and amino acid concentrations were detected, especially in the dissolved water fraction. At that moment, these high polysaccharides and amino acids concentrations accounted for 30% of the total DOC, indicating a high loading of high quality OM. The high DOC, BDOC and DON concentrations after rewetting revealed the remarkable transport of dissolved C and N compounds following the first important rains.

The high availability of OM during the peak flow was related to the high metabolic activities (extracellular enzyme activities and active bacteria) recorded in the stream water. This initial response after rewetting lasted for a week, and then values returned to basal levels. Bioavailability of materials returns to basal flow levels at the end of the flow peak (Stepanauskas et al., 2000). During the flow peak there is, therefore, fast and efficient microbial use of the available labile fresh material transported during the rewetting period (Romaní et al., 2006).

In spite of the substantial losses of benthic material during the peak flow, microbial activity recovered immediately after rewetting. The existence of refuges in the stream (for instance in the hyporheic), as well as the capacity of the biota to recover from droughts once these have finished, may explain the fast recovery. Robson (2000) determined that the presence of dry residual biofilms on rocks enhanced recovery and strongly influenced community development. Cobbles, litter, and coarse woody debris in the dry streambed could be used as refuges for microbial organisms (Bond et al., 2008). The physiological plasticity of some taxa makes them able to withstand extreme desiccation and recover rapidly (Stanley et al., 2004). For example, desiccated cyanobacterial mats, started to photosynthesize in the laboratory within 2 h after rewetting (Romaní & Sabater, 1997). The excretion of extracellular polysaccharides by algae and cyanobacteria facilitates cellular water retention in the biofilm, and makes rapid rehydration possible. In our study, algae in sand and leaves were more resilient to the drought than algae on cobble.

In conclusion, the dynamics of progressive and gradual drought effects, as well as the fast and punctuated recover after rewetting, might be affected by the interaction of the individual dynamics of each benthic substratum. As an example, while sediments provide refuge during desiccation, recovery of living bacteria and algal biomass is faster on cobbles; leaf material serves as a refuge and as a source of OM during the whole period. Since global climate change may favor a higher intensity and frequency of droughts in streams (Arnell et al., 1996), understanding the effects of these disturbances on the materials and biota will contribute to reliable resource management (Lake, 2003). The maintenance of benthic substrata heterogeneity within the stream may increase resilience of stream ecosystem processes with increasing frequency and duration of drought-rewetting disturbances.

**Acknowledgments** This study was funded by projects CGL 2007-65549/BOS and CGL2008-05618-C02/BOS of the Spanish Ministry of Science and Innovation and SCARCE (Consolider-Ingenio CSD2009-00065). Andrea Butturini participation was funded by GCL200760144. We thank Francesco Ricciardi for his help with the amino acid analysis, Aitor Larrañaga for his help with the lipid protocol, and Patricia Rodrigo for her help with the lipid analysis.

# References

Acuña, V., I. Muñoz, A. Giorgi, M. Omella, F. Sabater & S. Sabater, 2005. Drought and postdrought recovery cycles in an intermittent Mediterranean stream: structural and functional aspects. Journal of the North American Benthological Society 24: 919–933.

Acuña, V., A. Giorgi, I. Munoz, F. Sabater & S. Sabater, 2007. Meteorological and riparian influences on organic matter dynamics in a forested Mediterranean stream. Journal of the North American Benthological Society 26: 54–69.

Amalfitano, S., S. Fazi, A. M. Zoppini, A. B. Caracciolo, P. Grenni & A. Puddu, 2008. Responses of benthic bacteria to experimental drying in sediments from Mediterranean temporary rivers. Microbial Ecology 55: 270–279.

Arnell, R., B. Bates, H. Land, J. J. Magnusson & P. Mulholland, 1996. Hydrology and freshwater ecology. In Watson, R. T., M. C. Zinyowera, R. H. Moss & D. J. Dokken (eds), Climate Change 1995: Impacts, Adaptations, and Mitigation. Scientific-Technical Analysis. Cambridge University Press, Cambridge, UK: 325–364.

Artigas, J., A. M. Romani & S. Sabater, 2008. Relating nutrient molar ratios of microbial attached communities to organic matter utilization in a forested stream. Fundamental and Applied Limnology 173: 255–264.

Baldwin, D. S. & A. M. Mitchell, 2000. The effects of drying and re-flooding on the sediment and soil nutrient dynamics of lowland river-floodplain systems: a synthesis. Regulated Rivers: Research and Management 16: 457–467.

Billi, D. & M. Potts, 2002. Life and death of dried prokaryotes. Research in Microbiology 153: 7–12.

Bligh, E. G. & W. J. Dyer, 1959. A rapid method of total lipid extraction and purification. Canadian Journal of Biochemistry and Physiology 37: 911–917.

Bond, N. R., P. S. Lake & A. H. Arthington, 2008. The impacts of drought on freshwater ecosystems: an Australian perspective. Hydrobiologia 600: 3–16.

Boulton, A. J., 2003. Parallels and contrasts in the effects of drought on stream macroinvertebrate assemblages. Freshwater Biology 48: 1173–1185.

Boulton, A. J. & P. S. Lake, 1992. Benthic organic matter and detritivorous macroinvertebrates in two intermittent streams in south-east Australia. Hydrobiologia 241: 107–118.

Burford, M. A., A. J. Cook, C. S. Fellows, S. R. Balcombe & S. E. Bunn, 2008. Sources of carbon fuelling production in an arid floodplain river. Marine and Freshwater Research 59: 224–234.

Butturini, A., S. Bernal, E. Nin, C. Hellin, L. Rivero, S. Sabater & F. Sabater, 2003. Influences of the stream groundwater hydrology on nitrate concentration in unsaturated riparian area bounded by an intermittent Mediterranean stream. Water Resources Research 39: 1–13.

Butturini, A., M. Alvarez, S. Bernal, E. Vazquez & F. Sabater, 2008. Diversity and temporal sequences of forms of DOC and $NO_3$-discharge responses in an intermittent stream: predictable or random succession? Journal of Geophysical Research-Biogeosciences 113. G03016. doi: 10.1029/2008JG000721.

Caramujo, M. J., C. R. B. Mendes, P. Cartaxana, V. Brotas & M. J. Boavida, 2008. Influence of drought on algal biofilms and meiofaunal assemblages of temperate reservoirs and rivers. Hydrobiologia 598: 77–94.

Chanudet, V. & M. Filella, 2006. The application of the MBTH method for carbohydrate determination in freshwaters revisited. International Journal of Environmental and Analytical Chemistry 86: 693–712.

Claret, C. & A. J. Boulton, 2003. Diel variation in surface and subsurface microbial activity along a gradient of drying in an Australian sand-bed stream. Freshwater Biology 48: 1739–1755.

Cuffney, T. F. & J. B. Wallace, 1989. Discharge-export relationships in headwater streams: the influence of invertebrate manipulations and drought. Journal of the North American Benthological Society 8: 331–341.

Cummins, K. W., 1974. Structure and function of stream ecosystems. Bioscience 24: 631–641.

Dahm, C. N., M. A. Baker, D. I. Moore & J. R. Thibault, 2003. Coupled biogeochemical and hydrological responses of streams and rivers to drought. Freshwater Biology 48: 1219–1231.

Fisher, S. G. & N. B. Grimm, 1991. Streams and disturbances: are cross-ecosystem comparisons useful? In Cole, J. C., G. M. Lovett & S. E. G. Findlay (eds), Comparative Analyses of Ecosystems: Patterns, Mechanisms, and Theories. Springer-Verlag, New York: 196–221.

Fisher, S. G. & G. E. Likens, 1973. Energy flow in Bear Brook, New Hampshire – integrative approach to stream ecosystem metabolism. Ecological Monographs 43: 421–439.

Francis, C. & F. Sheldon, 2002. River Red Gum (*Eucalyptus camaldulensis* Dehnh.) organic matter as a carbon source in the lower Darling River, Australia. Hydrobiologia 481: 113–124.

Freese, H. M., U. Karsten & R. Schumannn, 2006. Bacterial abundance, activity, and viability in the eutrophic river Warnow, Northeast Germany. Microbial Ecology 51: 117–127.

Gasith, A. & V. H. Resh, 1999. Streams in Mediterranean climate regions: Abiotic influences and biotic responses to predictable seasonal events. Annual Review of Ecology and Systematics 30: 51–81.

Hach Company, 1992. Hach Water Analysis Handbook, 2nd ed. Hach Company, Loveland, CO.

Harvey, H. R. & A. Mannino, 2001. The chemical composition and cycling of particulate and macromolecular dissolved organic matter in temperate estuaries as revealed by molecular organic tracers. Organic Geochemistry 32: 527–542.

Howitt, J. A., D. S. Baldwin, G. N. Rees & B. T. Hart, 2008. Photodegradation, interaction with iron oxides and bioavailability of dissolved organic matter from forested floodplain sources. Marine and Freshwater Research 59: 780–791.

Humphries, P. & D. S. Baldwin, 2003. Drought and aquatic ecosystems: an introduction. Freshwater Biology 48: 1141–1146.

Jeffrey, S. W. & G. F. Humphrey, 1975. New spectrophotometric equations for determining chlorophylls a, b, c1 and c2 in higher-plants, algae and natural phytoplankton. Biochemie und Physiologie der Pflanzen 167: 191–194.

Keeney, D. R. & D. W. Nelson, 1982. Nitrogen – inorganic forms. In Page, A. L., R. H. Miller & D. R. Keeney (eds), Methods of Soil Analysis: Part 2. Chemical and Microbiological Properties, 2nd ed. American Society of Agronomy and Soil Science Society of America, Madison, WI, USA: 643–698.

Lake, P. S., 2000. Disturbances, patchiness, and diversity in streams. Journal of the North American Benthological Society 19: 573–592.

Lake, P. S., 2003. Ecological effects of perturbation by drought in flowing waters. Freshwater Biology 48: 1161–1172.

Langhans, S. D. & K. Tockner, 2006. The role of timing, duration, and frequency of inundation in controlling leaf-litter decomposition in a river-floodplain ecosystem (Tagliamento, NE Italy). Oecologia 147: 501–509.

Lehner, B., P. Döll, J. Alcamo, T. Henrichs & F. Kaspar, 2006. Estimating the impact of global change on flood and drought risks in Europe: a continental, integrated analysis. Climatic Change 75: 273–299.

Mariotti, A., M. V. Struglia, N. Zeng & K. M. Lau, 2002. The hydrological cycle in the Mediterranean region and implications for the water budget of the Mediterranean Sea. Journal of Climate 15: 1674–1690.

McClain, M. E., J. E. Richey & T. P. Pimentel, 2003. Biogeochemical hot spots and hot moments at the interface of terrestrial and aquatic ecosystems. Ecosystems 6: 301–312.

Moran, M. A. & R. G. Zepp, 1997. Role of photoreactions in the formation of biologically labile compounds from dissolved organic matter. Limnology and Oceanography 42: 1307–1316.

Pakulski, J. D. & R. Benner, 1992. An improved method for the hydrolysis and MBTH analysis of dissolved and particulate carbohydrates in seawater. Marine Chemistry 40: 143–160.

Peterson, C. G., H. M. Valett & C. N. Dahm, 2001. Shifts in habitat templates for lotic microalgae linked to interannual variation in snowmelt intensity. Limnology and Oceanography 46: 858–870.

Poff, N. L., J. D. Allan, M. B. Bain, J. R. Karr, K. L. Prestegaard, B. D. Richter, R. E. Sparks & J. C. Stromberg, 1997. The natural flow regime: a paradigm for river conservation and restoration. Bioscience 47: 769–784.

Robson, B. J., 2000. Role of residual biofilm in the recolonization of rocky intermittent streams by benthic algae. Marine and Freshwater Research 51: 724–732.

Romaní, A. M. & S. Sabater, 1997. Metabolism recovery of a stromatolitic biofilm after drought in a Mediterranean stream. Archiv Für Hydrobiologie 140: 261–271.

Romaní, A. M. & S. Sabater, 2001. Structure and activity of rock and sand biofilms in a Mediterranean stream. Ecology 82: 3232–3245.

Romaní, A. M., E. Vázquez & A. Butturini, 2006. Microbial availability and size fractionation of dissolved organic carbon after drought in an intermittent stream: biogeochemical link across the stream-riparian interface. Microbial Ecology 52: 501–512.

Sabater, S. & K. Tockner, 2010. Effects of hydrologic alterations on the ecological quality of river ecosystems. In Sabater, S. & D. Barceló (eds), Water Scarcity in the Mediterranean. Handbook Environmental Chemistry. Springer Verlag. doi:10.1007/698_2009_24.

Sabater, S., S. Bernal, A. Butturini, E. Nin & F. Sabater, 2001. Wood and leaf debris input in a Mediterranean stream: the influence of riparian vegetation. Archiv Für Hydrobiologie 153: 91–102.

Schröter, D., W. Cramer & R. Leemans, 2005. Ecosystem service supply and vulnerability to global change in Europe. Science 310: 1333–1337.

Servais, P., A. Anzil & C. Ventresque, 1989. Simple method for determination of biodegradable dissolved organic-carbon in water. Applied and Environmental Microbiology 55: 2732–2734.

Stanley, E. H., S. G. Fisher & J. B. J. Jones, 2004. Effects of water loss on primary production: a landscape-scale model. Aquatic Sciences 66: 130–138.

Stepanauskas, R., H. Laudon & N. O. G. Jorgensen, 2000. High DON bioavailability in boreal streams during a spring flood. Limnology and Oceanography 45: 1298–1307.

Thurman, E. M., 1985. Organic Geochemistry of Natural Waters. Nijhoff, M & Junk, W Publishers, Dordrecht, The Netherlands.

Usher, H. D. & D. W. Blinn, 1990. Influence of various exposure periods on the biomass and chlorophyll a content of *Cladophora glomerata* (Chlorophyta). Journal of Phycology 26: 244–249.

Vázquez, E., A. M. Romani, F. Sabater & A. Butturini, 2007. Effects of the dry-wet hydrological shift on dissolved organic carbon dynamics and fate across stream-riparian interface in a Mediterranean catchment. Ecosystems 10: 239–251.

Wetzel, R. G., P. G. Hatcher & T. S. Bianchi, 1995. Natural photolysis by ultraviolet irradiance of recalcitrant dissolved organic matter to simple substrates for rapid bacterial metabolism. Limnology and Oceanography 40: 1369–1380.

Wright, J. F. & K. L. Symes, 1999. A nine-year study of the macroinvertebrate fauna of a chalk stream. Hydrological Processes 13: 387–399.

Ylla, I., A. M. Romaní & S. Sabater, 2007. Differential effects of nutrients and light on the primary production of stream algae and mosses. Fundamental and Applied Limnology 170: 1–10.

Zollner, N. & K. Kirsch, 1962. Über die quantitative Bestimmung von Lipoiden (Mikromethode) mittels der vielen natürlichen Lipoiden (allen bekanntem Plasmalipoiden) gemeinsamen Sulphophosphovanillin-Reaktion. Zeitsehrift ffir die gesamte experimentelle Medizin 135: 545–561.

GLOBAL CHANGE AND RIVER ECOSYSTEMS

# Flow regime alteration effects on the organic C dynamics in semiarid stream ecosystems

V. Acuña

Received: 21 September 2009 / Accepted: 29 December 2009 / Published online: 11 January 2010
© Springer Science+Business Media B.V. 2010

**Abstract** There is evidence of an ongoing alteration of the flow regime owing to climate change forcing, which has resulted in substantial increases in the frequency and magnitude of extreme events such as floods and droughts. Such changes in the flow regime may have major implications in freshwater ecosystems and, in particular, in the organic carbon dynamics in semiarid stream ecosystems. Much is known about the role of extreme flow events on structuring stream ecosystems, but few studies explored the effects of extreme flow events magnitude, timing, and sequence on stream ecosystems. To assess the effect of extreme events on stream organic C dynamics, a simple and flexible modeling approach was applied to simulate the organic carbon dynamics in a simplified river reach. The river reach model was initially calibrated and tested using long-term data for stream water velocity and amount of organic carbon in sediment. After that, multiple scenarios differing in the extreme flow events (floods and droughts) sequence and magnitude were used to simulate the effects of possible flow regime changes on the stream organic carbon dynamics. Initial expectations were that: (i) an increase in the magnitude or frequency of extreme flow events would reduce the amount of organic carbon respired within the simulated river reach, and (ii) relationship between the timings of the extreme flow events and of the litterfall input would influence considerably the effects of the extreme flow events. Results pointed out that: (i) the amount of processed carbon respect the amount entering the ecosystem was affected by extreme events such floods and droughts, but the relevance of those events differed along the year, with a maximal effect during the litterfall period; (ii) extreme event timing rather than the magnitude was more relevant to the stream organic carbon dynamics; and (iii) the amount of respired carbon in the ecosystem could be amplified or reduced depending on event sequence. Increasing awareness of the role of inland waters in the global carbon cycle and the shaping role of hydrology on the stream organic carbon dynamics stress the need to better quantify carbon fluxes and the hydrological controls on these fluxes.

**Keywords** Global climate change · Extreme flow events · Freshwater ecosystems · Modeling · Global carbon cycle · Stream metabolism

Guest editors: R. J. Stevenson, S. Sabater / Global Change and River Ecosystems – Implications for Structure, Function and Ecosystem Services

V. Acuña (✉)
Catalan Institute for Water Research (ICRA), Carrer Emili Grahit 101, Edifici H2O, Parc Científic i Tecnològic de la Universitat de Girona, 17003 Girona, Spain
e-mail: vicenc.acuna@icra.cat

## Introduction

Flow fluctuation is an important ecological driver in stream ecosystems. Flow fluctuations are however undergoing major changes owing to climate change forcing, which has resulted in substantial increases in the frequency and magnitude of flow fluctuations in many stream ecosystems. In this direction, several studies reported an increasing frequency and intensity of extreme hydrologic events (e.g., New et al., 2001; Huntington, 2006; Hirabayashi et al., 2008), as well as of events such as ENSO over the past decades (Huntington, 2006; IPCC, 2007). Furthermore, another consequence of climate change might be changes in the timing of river flows (Schindler, 2001). In particular, Mediterranean streams are predicted to undergo severe alterations in the flow regime because of a decrease in the number of precipitation days, and an increase in days with heavy rains (Hirabayashi et al., 2008; Sillmann & Roeckner, 2008). Regional climatic models for southern Europe predict an increase in the frequency and duration of heat waves and heavy rainfall events during summer (Sanchez et al., 2004; Giorgi & Lionello, 2008), and stress that the Mediterranean might be an especially vulnerable region to global change (Giorgi & Lionello, 2008). Therefore, more extreme and unpredictable flow events, such as floods and droughts, are anticipated, thereby creating novel environmental conditions in stream ecosystems.

The effects of gradual climatic trends such as global warming on ecosystems have been studied in much more detail than sudden events (Jentsch et al., 2007). In stream ecology, there are however numerous studies exploring the effects of extreme flow events such as floods and droughts, particularly in arid and semiarid regions (e.g., Grimm & Fisher, 1989; Boulton & Lake, 1992; Stanley et al., 1994; Acuña et al., 2004, 2005). Among those studies, few attempted to characterize the distinct effects of extreme events magnitude and timing or the effects of the sequence of flow events.

Freshwaters receive approximately 1.9 Pg C year$^{-1}$ from the terrestrial system at the global scale, of which approximately 0.9 Pg C year$^{-1}$ finally enters the oceans (Cole et al., 2007). Hence, slightly more than half of the organic C that enters freshwaters is either stored in lakes, reservoirs, and fringing wetlands, or it escapes to the atmosphere as $CO_2$ or $CH_4$ after biological processing (Cole et al., 2007). So far, studies on the role of inland waters in the global C cycle focused on improving the C continental balances, but the role of hydrology has been only addressed by few studies (e.g., Algesten et al., 2004). There are however studies showing that flow extremes may have a strong influence on the stream organic C dynamics (Acuña et al., 2004). Overall, the lack of studies and the indications about the potential relevance of flow regime on the C dynamics stress, therefore, the need to better study the effect of extreme events if aiming to do good predictions of the effects of climate change on the global C cycle.

To address the lack of understanding about the distinct effects of magnitude and timing, as well as sequence of extreme flow events, we modeled the organic C dynamics in a Mediterranean headwater stream ecosystem. This study tested the responses of downstream transport, respiration, and C storage to: individual extreme events of both types of varying magnitude (1), sequences that included floods and droughts at different times (2). Initial hypothesis was that the changes in the timing rather than the magnitude of extreme flow events are relevant for the stream organic C dynamics. The rationale of the hypothesis lies on a previous study by Acuña et al. (2007), which reported differential effects of floods on the organic C dynamics depending on the season.

## Methods

### Study approach

A simple model was used to simulate organic C dynamics in river reaches. The first step was to evaluate different formulations of processes affecting organic C dynamics at the reach level. The respective models were evaluated using an extensive data set from a Mediterranean headwater forested stream (Fuirosos stream, NE Iberian Peninsula). The best-fit model was selected to simulate organic C dynamics under different discharge scenarios. Model simulations were performed with the computer program AQUASIM (version 2.1f), which is designed for the simulation of aquatic systems (Reichert, 1994, 1998).

### Reach-scale models development and selection

Given that there is no "true" model, but a number of adequate simplifying descriptions of the complex real system, we compared 12 competing models with alternative formulations of the processes affecting the

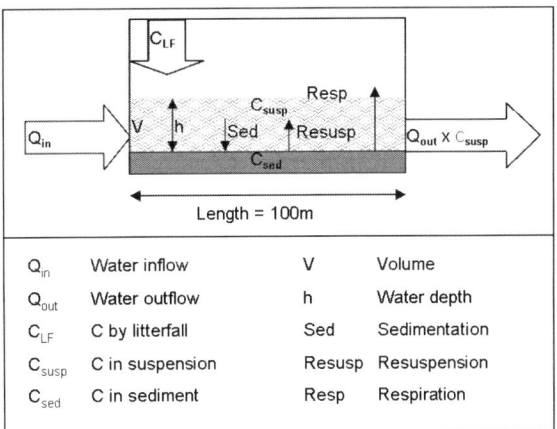

**Fig. 1** Schematic diagram of the used reach-scale model

stream organic C dynamics. We fitted all models to the observed data from the Fuirosos stream (Acuña et al., 2004, 2007) and ranked them based on the goodness of fit (Chatfield, 1995; Johnson & Omland, 2004).

The reach-scale models were designed as mixed reactors of variable water volume, with constant length (100 m) and width (2 m), with water inflow and outflow, and inputs and transformation processes of organic C which could be suspended ($C_{susp}$) or deposited at the sediment ($C_{sed}$) (Fig. 1). Note that both forms of C, $C_{sed}$ and $C_{susp}$ were in particulate form, so that no dissolved forms of C were considered in the model. For water inflow data, we used real discharge data from the Fuirosos stream (from November 2001 to December 2002). Water outflow was calculated as the product of an empirical discharge-volume coefficient ($cQ$) (day$^{-1}$ m$^{-3/2}$) and the power of the water volume ($V$) (m$^3$) by a second empirical discharge-volume coefficient ($cV$), so $cQ \times V^{cV}$. The organic C input to the mixed reactor was modeled as coarse particulate organic C from terrestrial origin, with a seasonal component following the litterfall dynamics of riparian forests dominated by deciduous species (data from the Fuirosos stream; Acuña et al., 2007).

The modeled transformation processes of organic C inside the reactors were sedimentation, resuspension, and respiration. Among these processes, sedimentation was the only one with the same formulation in all models, which was the product of $C_{susp}$ by the sedimentation velocity ($v_{sed}$) (m day$^{-1}$) (Table 1). For resuspension or movement of C from the streambed into the water column, two different processes were involved, namely resuspension and catastrophic removal during floods. Resuspension was, in contrast to catastrophic removal, active during the entire simulations, and was assumed to be proportional to the product of discharge ($Q$) (m$^3$ day$^{-1}$) and the amount of organic C in sediment exceeding a minimum value ($C_{sed,0}$) (Table 1). Differences among resuspension formulations relied on the value of the exponent of $Q$, the exponent of the difference between $C_{sed}$ and $C_{sed,0}$, and the different values of the empirical resuspension coefficient ($c_{resusp}$) (Table 1). Catastrophic removal of C from the streambed occurred only during floods, and was considered as discharges exceeding a critical threshold of discharge determined in a previous study in the Fuirosos stream (50,000 m$^3$ day$^{-1}$; Acuña et al., 2007). Thus, discharge values above this threshold implied the activation of this process, which due to the high value of $k_{flood}$ (= 100 day$^{-1}$), involved rapid falls of $C_{sed}$, which approached then $C_{sed,0}$ (minimum amount of $C_{sed}$ that remains despite floods) (2.1 in Table 1). The rationales for these formulations were: (i) resuspension is assumed to be a process proportional to the bottom shear stress, and based on the Manning-Strickler friction law for a rectangular channel such that simulated the bottom shear stress should vary with $Q^{3/5}$ (Chow, 1959; Henderson, 1966) (3.1 in Table 1); (ii) the concentration and abrasive impact of suspended particles increase with $Q$, and therefore a resuspension rate proportional to $Q^2$ was used (Uehlinger et al., 1996) (3.2 in Table 1); and (iii) resuspension may increase more rapidly with growing thickness of the layer of accumulated

**Table 1** Alternative formulations for the transformation processes of organic C

| Sedimentation | Catastrophic removal | Resuspension | Respiration |
|---|---|---|---|
| 1.1 $C_{susp}v_{sed}$ | 2.1 $k_{flood}(Q)_i(C_{sed} - C_{sed,0})$ | 3.1 $c'_{resusp}Q_i^{3/5}(C_{sed} - C_{sed,0})$ | 4.1 $R_{T_0}\frac{C_{sed}}{C_{sed}+k_{sed}}e^{E(T-T_0)/kTT_0}$ |
|  | 2.2 Inactive | 3.2 $c''_{resusp}Q_i^2(C_{sed} - C_{sed,0})$ | 4.2 $R_{T_0}\frac{C_{sed}}{C_{sed}+k'_{sed}}e^{E(T-T_0)/kTT_0}$ |
|  |  | 3.3 $c'''_{resusp}Q_i(C_{sed} - C_{sed,0})^2$ |  |

organic C at the streambed, and a proportional loss to the square of $X$ was accordingly tried (Uehlinger et al., 1996) (3.3 in Table 1).

Finally, respiration was formulated in two different ways, including or excluding a term accounting for the effects of the amount of $C_{sed}$ on respiration (4.1 and 4.2 in Table 1). Thus, a maximum respiration rate (g C m$^{-2}$ day$^{-1}$) at the reference temperature $T_0$ ($R_{T_0}$) was influenced in both formulations by water temperature following the universal temperature dependence of biological processes (Gillooly et al., 2001). In this term, $T$ and $T_0$ were water temperatures in K, $E$ was the activation energy (0.62 eV), and $k$ was Boltzmann's constant (8.616 × 10$^{-5}$ eV K$^{-1}$). Note that the used reference temperature was 20°C, and that $R_{T_0}$ is expressed as $R_{20}$ in further sections. A Monod-type formulation was used for the term accounting for the effects $C_{sed}$ on respiration, where $k_{sed}$ (g C m$^{-2}$) was the half-saturation coefficient. Note that there was a term accounting for the effect of $C_{sed}$ in both alternative formulations for respiration to avoid active respiration at $C_{sed} < 0$. Furthermore, there was another term in both formulations to avoid active respiration at $V = 0$, so that there was no respiration on dry streambeds. The difference between the formulations relied on $k_{sed}$, as $k_{sed}$ ranged between 2 and 10 g C m$^{-2}$ (4.1 in Table 1), while $k'_{sed}$ was set to 0 (4.2 in Table 1).

A total of 12 alternative models were independently fitted to observed $C_{sed}$ in the Fuirosos stream (data from Acuña et al., 2004). Note that the use of $C_{sed}$ as target variable for the model calibration allowed the determination of all 5 model parameters, as $C_{sed}$ was affected by all active processes in the model (sedimentation, resuspension and respiration). The 12 models arose from the combination of: (i) inclusion or exclusion of catastrophic removal by floods (2.1 or 2.2 in Table 1), (ii) the three formulations for resuspension (3.1, 3.2 or 3.3 in Table 1), and (iii) the two formulations for respiration (4.1 or 4.2 in Table 1). Specifically, a stepwise calibration (two steps) was performed given that the simulated C dynamics strongly depend on the accuracy of the simulated water flows. Therefore, both discharge-volume coefficients were determined after fitting the modeled water velocity to the observed water velocity in the Fuirosos stream ($n = 25$) (data from Acuña et al., 2004). After this hydrological calibration, all alternative models were then evaluated, leading to a series of calculations of different model complexity. The best performing model (lowest $\chi^2$) was then selected and used to simulate stream organic C dynamics under different flow conditions.

Model simulations

Ninety-six annual flow regimes were assembled in order to test for the effect of timing, magnitude, and sequence of extreme flow events, considering magnitude as the temporal length of the extreme flow event rather than the magnitude of the maximum flow. Flood magnitude was simulated as floods exceeding the critical threshold of 100 L/s during 5 (I1), 10 (I2), or 15 (I3) days; while flood timing was explored by simulating floods at each month during the year. Thus, 36 annual flow records were simulated (12 months × 3 magnitudes = 36 simulation scenarios) (Fig. 2a). Similarly, drought magnitude was simulated as zero flows during 5 (I1), 10 (I2), or 15 (I3) days for 12 months (36 simulation scenarios) (Fig. 2b). Finally, the relevance of flow extremes sequence was explored by simulating sequences of flood–drought or drought–flood for the same months used for the drought scenarios (12 months × 2 sequences = 24 simulation scenarios) (Fig. 2c).

Given that there were periods with nonflow for some of the simulated flow regimes, the formulation and stoichiometric coefficients for the C input from the terrestrial systems had to be adapted, so that the C entering the mixed reactor could become either $C_{susp}$ during periods with flow or $C_{sed}$ during periods of no flow, thus assuming C inputs to deposit directly on the sediment during dry conditions. Simulations were made for 500 days, but only the values from days 150 to 500 were used for the analyses in order to avoid dependence on the initial conditions. In order to assess the impact of the flow extremes on the stream organic C dynamics, results are primarily expressed as the amount of C respired within the simulated reach respect all C entering the reach (as ratio of processed to total C).

**Results**

Model evaluation

Hydrologic model coefficients were optimized to obtain the best match between simulated and

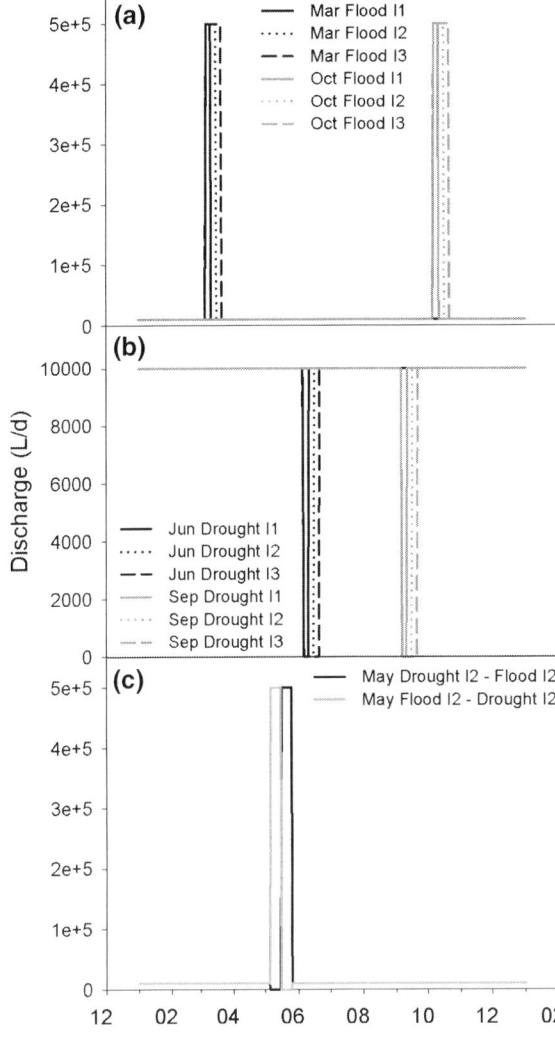

Fig. 2 Selected data series to explore the effects of flood magnitude and timing (a), drought magnitude and timing (b), and sequence of extreme flow events (c). Note that the magnitude of the extreme flow events corresponds to 5 (I1), 10 (I2), and 15 (I3) days

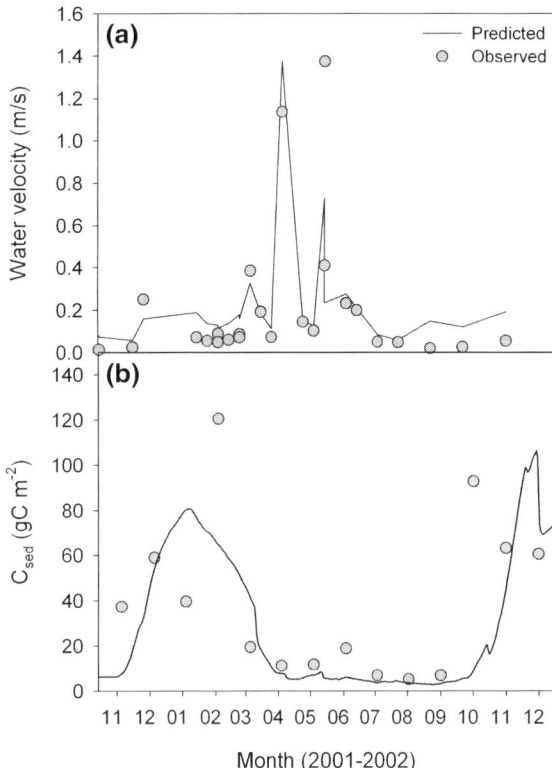

Fig. 3 Temporal changes in observed and predicted data for **a** water velocity and **b** $C_{sed}$. Note that the predicted $C_{sed}$ corresponds to the reach-scale model with the formula showing the best fit (1.1, 2.2, 3.1, 4.1; see Table 1 for details)

measured water velocity. The results of the parameter fit of the reach-scale model are shown in Fig. 3a. The values of $cQ$ (= 1.96 day$^{-1}$ m$^{-3/2}$) and $cV$ (= 1.5) were further used for all alternative models developed to simulate organic C dynamics.

The results of the best three models among the 12 tested alternative reach-scale models are summarized in Table 2. Among these, the model 1.1, 2.2, 3.1, 4.1 was the best supported by the observed data (see Eqs. 1 and 2 for details) (Fig. 3b), and was therefore selected to explore the effect of extreme flow events magnitude, timing, and sequence:

$$\frac{dC_{susp}}{dt} = LF \frac{1}{w_i h_i} + c_{resusp} Q_i^{3/5} (C_{sed} - C_{sed,0}) - C_{susp} v_{sed} \quad (1)$$

$$\frac{dC_{sed}}{dt} = LF \frac{1}{w_i} + C_{susp} v_{sed} - c_{resusp} Q_i^{3/5} (C_{sed} - C_{sed,0}) - R_{T_0} \frac{C_{sed}}{C_{sed} + k_{sed}} e^{E(T-T_0)/kTT_0} \quad (2)$$

where $w$ was the mixed reactor width, $h$ the mixed reactor depth, and LF the litterfall time series (g C m$^{-1}$ day$^{-1}$). In this model, $C_{susp}$ depended on litterfall, resuspension (3.1), and sedimentation, while $C_{sed}$ depended on litterfall (only active during nonflow conditions), sedimentation (1.1), resuspension and respiration with dependence on temperature and on the amount of $C_{sed}$ (4.1).

**Table 2** Results of parameter fit (minimization of $\chi^2$) of the best three alternative models for stream organic C dynamics at the reach scale

| Model codes | $c_{resusp}$ (no units) 0.00005–0.1 | $v_{sed}$ (m day$^{-1}$) 10–500 | $C_{sed,0}$ (g C m$^{-2}$) 0.01–6 | $R_{20}$ (g C m$^{-2}$ day$^{-1}$) 4–20 | $k_{sed}$ (g C m$^{-2}$) 2–10 | $\chi^2$ |
|---|---|---|---|---|---|---|
| 1.1, 2.1, 3.1, 4.1 | 0.001 | 107.2 | 5.53 | 4 | 10 | 21,699 |
| 1.1, 2.2, 3.1, 4.1 | 0.001 | 106.6 | 6 | 4 | 10 | 21,697 |
| 1.1, 2.2, 3.3, 4.1 | 0.00005 | 500 | 6 | 4 | 8 | 28,392 |

The first column contains the codes for the selected formulation (see Table 1); columns 2–6 list the final values for the five parameters. Beneath each parameter (from left to right: resuspension coefficient, sedimentation velocity, minimum $C_{sed}$, maximum respiration rate at 20°C, half-saturation coefficient), the units and boundary values are given

Effects of extreme flow events timing and magnitude

The temporal dynamics of both forms of instream C ($C_{sed}$ and $C_{susp}$) differed considerable among scenarios, reacting very sensitively to extreme flow events. During nonflow periods, C accumulated at the dry streambed due to a combination of physical storage and very low respiration rates. In contrast, simulated floods enhanced resuspension of $C_{sed}$ and a sharp increase in $C_{susp}$, leading in few days to the minimum allowed $C_{sed}$ by the model ($C_{sed,0}$).

Floods and droughts had a different effect on the amount of C respired within the simulated reach, as floods decreased much more than droughts the ratio of processed to total organic C (Table 3). As shown in Table 3, higher magnitudes of both floods and droughts involved a decrease in the ratio, but the decreases were more important for floods. For example, floods of magnitude 2 involved a decrease of 3% (from 0.5026 to 0.4871), while droughts of magnitude 2 involved a decrease of only 0.5% (from 0.5026 to 0.5003).

**Table 3** Effects of extreme flow events magnitude and timing on the amount of C respired within the simulated reach (as ratio of processed to total organic C)

| Magnitude | Floods | | Droughts | |
|---|---|---|---|---|
| | Mean | St. deviation | Mean | St. deviation |
| 5 | 0.4930 | 0.006 | 0.5027 | 0.001 |
| 10 | 0.4871 | 0.010 | 0.5003 | 0.005 |
| 15 | 0.4814 | 0.014 | 0.4977 | 0.009 |

The obtained ratios with the simulations with different flood and drought time and intensity are expressed as mean and standard deviation per each magnitude (5, 10 and 15 days)

The effects of timing and magnitude were different between floods and droughts, as both timing and magnitude of floods were more influential on the stream organic C dynamics. Thus, the major relevance of floods respect floods was reflected in larger temporal variability in the ratios of processed to total organic C between months (Fig. 4). Indeed, there were considerable differences along the annual cycle, so that floods had a different effect on the C balance of the stream depending on the month of the year. For example, floods in March had little effect on the ratio of processed to total C, while the effect on November was the highest (Fig. 4a). The differences in the standard deviations along the year observed in the error bars in Fig. 4a reflected the varying relevance of flood magnitude along the year, so that changes in the flood magnitude were almost irrelevant in March (standard deviation of 0.001), but highly relevant in November (0.012). The effect of droughts on the C balance of the stream also depended on the month of the year, but the differences among months were smaller (Fig. 4b), as well as the variability between magnitudes for each month.

Effects of extreme flow events sequence

The sequence of extreme events seemed to be an important determinant of the stream organic C dynamics, as the sequence of drought–flood appeared to have a stronger effect on the ratio of processed to total C compared with the sequence of flood–drought (Fig. 5). The differences between both sequences in terms of reduction of the mentioned ratio varied along the year, so that the differences between sequences were minimal from December to April (differences between 0.004 and 0.009 in the ratio of processed to total C) and maximal from June to

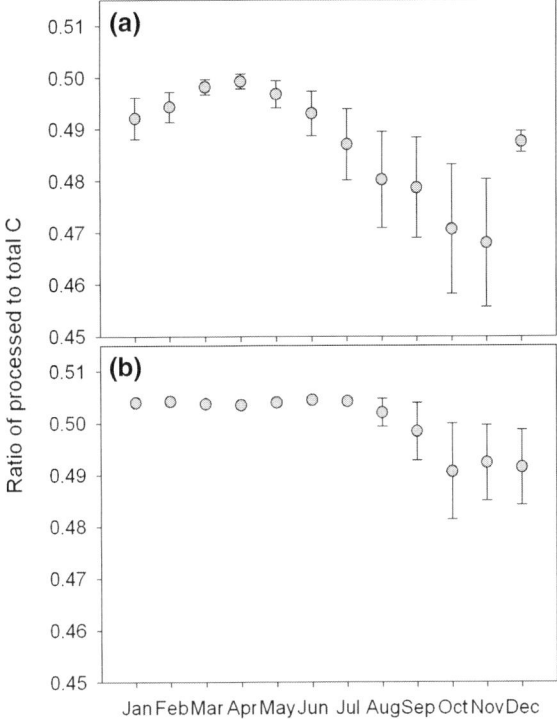

**Fig. 4** Effects of flood (**a**) and droughts (**b**) timing and magnitude on the amount of C respired within the simulated reach (as ratio of processed to total organic C). *Circles* indicate mean and vertical error bars standard deviation of the ratio of processed to total C of the three simulated magnitudes (5, 10, and 15 days) per each month

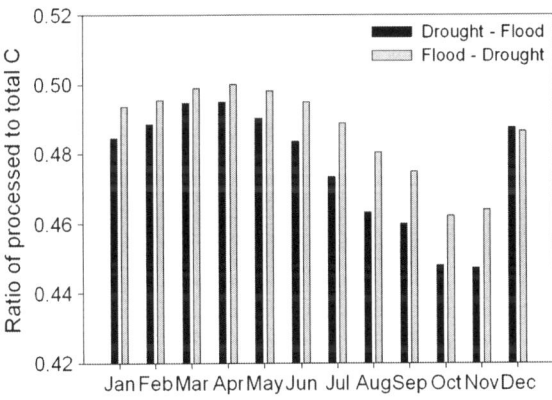

**Fig. 5** Effects of extreme flow sequence on the amount of C respired within the simulated reach (as ratio of processed to total organic C) during the annual cycle. Floods and droughts were simulated as 10 days of discharge above the critical threshold of 100 L/s or 10 days of zero flow respectively

November (differences between 0.011 and 0.017 in the ratio of processed to total C) (Fig. 5). The timing effect was also reflected on the time needed to return

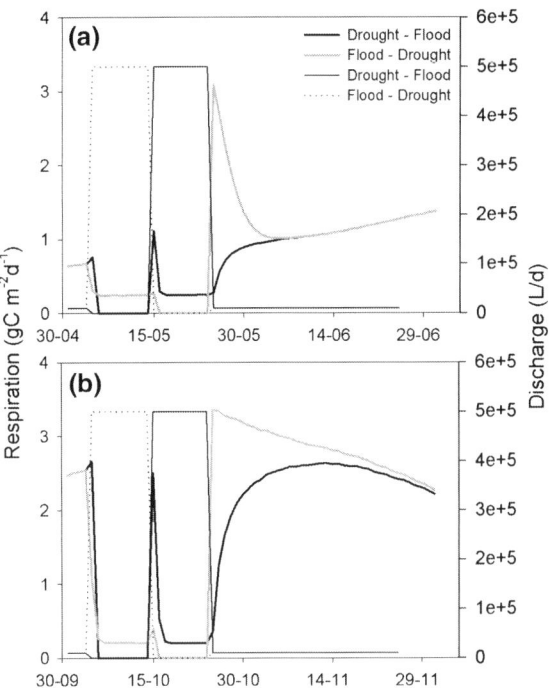

**Fig. 6** Effects of sequence of extreme flow events on respiration in May (**a**) and October (**b**). Floods and droughts were simulated as 10 days of discharge above the critical threshold of 500 $m^3$ $day^{-1}$ or 10 days of zero flow respectively

to the initial conditions after the sequence of extreme flow events. For example, the effects of extreme flow events in May lasted for 2–3 weeks (Fig. 6a), while more than 5 weeks were needed in October to return to original conditions (Fig. 6b).

## Discussion

Stream organic C dynamics responded very sensitively to both extreme flow events magnitude and timing, and showed that the specific sequence of extreme events played a distinct role on the stream ecosystem organic C dynamics. On the one hand, the amount of processed C respect the total amount of C entering the ecosystem was affected by extreme events such floods and droughts, but the relevance of those events differed along the year, with a maximal effect during the litterfall period. On the other hand, the amount of respired C in the ecosystem could be amplified or reduced depending on event sequence. Those findings indicate that the effects of extreme events sequences cannot be modeled by simply adding the separate effects of individual extreme events but, rather, that models should take into account event sequences.

Considerations about the modeling approach

The applied mechanistic biogeochemical model was a trade-off between the empirical and conceptual, rather than an empirical model. The used reach-scale model was based on a single mixed reactor unit, and provided an appropriate characterization of both water velocity and $C_{sed}$ of the Fuirosos stream despite its simplicity (Fig. 3).

The differences between observed and predicted $C_{sed}$ by the reach-scale model may rely on the input parameters of the model, rather than in the structure of the model itself. For example, C input in the model was assumed to be only through particulate organic C from terrestrial origin, while C in headwater reaches such as Fuirosos is influenced by both litterfall and the C entering the reach by drift from upstream reaches. The error introduced by ignoring particulate organic C entering the reach by drift is however apparently not crucial, given that 79% of the particulate organic C entering the simulated stream enters from the riparian forest (85 Kg C year$^{-1}$ from the riparian forest respect 22 Kg C year$^{-1}$ from upstream reaches; Acuña et al., 2007).

The use of simple and flexible models to simulate the effects of different components of climate change on different characteristics of freshwater ecosystems has been used by others (e.g., Moore et al., 2008). The calibration of those models with existing data is a fundamental step to obtain realistic model parameters. The main shortcut of these models is that it is not possible to validate their predictions based on scenarios.

Effects of extreme flow events timing and magnitude

The stream organic C dynamics reacted very sensitively to extreme events such floods and droughts. The extreme flow event magnitude was relevant in terms of effect on the stream organic C dynamics, but the timing of the extreme events was apparently more relevant than the magnitude, as the relevance of those events differed along the year. Temporal differences in the relevance of extreme flow events along the year was mainly related with the temporal variability in the C input to the simulated ecosystem, as litterfall is not temporarily homogenous along the year but concentrated in 3–4 months (from September to November). In headwater forested streams in most temperate areas, litterfall is indeed seasonal, as it is ecosystem respiration, which ranged from 1–2 g $O_2$ m$^{-2}$ day$^{-1}$ during late Spring to 30–35 g $O_2$ m$^{-2}$ day$^{-1}$ during late autumn (Acuña et al., 2004). A direct consequence of this seasonality is that any flow alteration during the litterfall period might have a more profound effect on the system. The relevance of flow during this period was reflected in the stronger effect of extreme flow events on the ratios of processed to total C (Fig. 4), as well as in the longer effect of extreme flow events on respiration (Fig. 6). Thus, floods shortly after litterfall peak input might stream $C_{sed}$ for a long period, whereas floods shortly before litterfall peak input have almost no consequences on $C_{sed}$. Previous studies in the simulated stream have shown higher heterotrophic activity during the litterfall period, with ecosystem respiration values up to $\sim$35 g $O_2$ m$^{-2}$ day$^{-1}$ (Acuña et al., 2004), and the highest microbial enzyme activities (Artigas et al., 2009). The results shown by those empirical studies support the simulation results indicating a major sensitivity of the ecosystem to extreme flow events during the litterfall period.

The impacts of shifts in the timing of extreme events owed to the alteration of the precipitation regime by climate change might have different consequences depending on the season, as a shift in the timing of floods from early winter to fall will have more important implications than a shift from early spring to early summer. Other modeling studies have also stressed the relevance of changes in the seasonal distribution of river flows on, for example nutrient loads (Moore et al., 2008). On the other hand, changes in the litterfall dynamics caused by longer droughts might also affect the organic C dynamics, as the sensitivity to floods might slightly decrease if the litterfall period is stretched. Thus, more intense droughts might advance the litterfall of the most drought sensitive deciduous trees from November to August, temporally extending the litterfall input (Acuña et al., 2007). A temporally extended supply of organic C might in turn allow faster recoveries of $C_{sed}$ after floods during this period.

### Effects of the sequence of extreme flow events

As stated in the previous section, the simulated stream ecosystem responded to single events of drought or flood as would be more or less expected based on the existing literature. However, obtained results in this modeling study highlighted two potentially unappreciated issues. First, that response to multiple extreme events depends on event sequence, so that sequence order drought–flood depressed respiration much more than the alternative sequence flood–drought. Second, that the relevance of the sequence of extreme flow events had a timing effect, with maximal differences between the two explored sequences during late summer and fall (Fig. 5), that is, the period of maximum litterfall.

Research on extreme events ("event-focused" in contrast to "trend-focused") has increased in recent years. Numerous examples, ranging from microbiology and soil science to biogeography, demonstrate how extreme weather events can accelerate shifts in species composition and distribution, thereby facilitating changes in ecosystem functioning (Jentsch et al., 2007). A more recent study by Miao et al. (2009) has experimentally pointed out the relevance of extreme events sequence. Those authors documented, on the basis of an experiment with seedlings of three types of subtropical wetland tree species, that mortality can be amplified and growth can be stimulated depending on event sequence. There are also some empirical studies in stream ecosystems showing the effects of the sequence of extreme events (Acuña et al., 2007; Artigas et al., 2009). For example, Acuña et al. (2007) reported that floods occurring after the litterfall involved massive organic C transport to downstream system, therefore depleting headwater reaches of available organic C, while floods occurring before the litterfall showed lesser influence on the organic C dynamics. In the same system of the previous study, Artigas et al. (2009) reported that the autumnal peaks of microbial biomass and activity were two to three times lower in years when there was permanent flow over the dry season compared to years when the flow was interrupted during the dry season. Despite of the differences in the nature and approach between these studies, their conclusions pointed out that the sequence of extreme events is relevant, as the effects of extreme events on ecosystems cannot be properly understood unless the sequence is considered.

### Implications for the stream organic C dynamics

The results of this study have noteworthy implications for understanding and predicting stream organic C dynamics in response to the ongoing alterations of the flow regime. The Mediterranean region is experiencing increasingly extreme flow events and is among the most vulnerable ecosystems to climate change (Hirabayashi et al., 2008). Results from this study highlight that the relevance of extreme events for the stream organic C dynamics strongly depends on the season, so that, the C cycle could be more or less affected by the ongoing alteration of the precipitation patterns if, for example, floods increase during fall (more influence) or during winter (less influence). Changes in the timing of extreme flow events can therefore modify the role of headwater forested streams in the C cycle, as the ratio of processed to total C is particularly sensitive to extreme flow events. In this direction, Acuña & Tockner (2009) stressed that increasing intermittency extent (number of days with zero flow) might have a profound effect on the stream organic C dynamics, with an effect even more relevant than warming water temperatures. Overall, the relevance of flow regime alterations on the C dynamics plus the large amounts of organic C processed in inland waters stress the need to further explore the coupling between the flow regime and the C cycle.

**Acknowledgments** Constructive comments by Urs Uehlinger improved the manuscript. The research was supported by the Catalan Institute for Water Research (ICRA).

### References

Acuña, V. & K. Tockner, 2009. The effects of alterations in temperature and flow regime on organic carbon dynamics in Mediterranean river networks. Global Change Biology (in press).

Acuña, V., A. Giorgi, I. Munoz, U. Uehlinger & S. Sabater, 2004. Flow extremes and benthic organic matter shape the metabolism of a headwater Mediterranean stream. Freshwater Biology 49: 960–971.

Acuña, V., I. Munoz, A. Giorgi, M. Omella, F. Sabater & S. Sabater, 2005. Drought and postdrought recovery cycles in an intermittent Mediterranean stream: structural and functional aspects. Journal of the North American Benthological Society 24: 919–933.

Acuña, V., A. Giorgi, I. Munoz, F. Sabater & S. Sabater, 2007. Meteorological and riparian influences on organic matter dynamics in a forested Mediterranean stream. Journal of the North American Benthological Society 26: 54–69.

Algesten, G., S. Sobek, A. K. Bergstrom, A. Agren, L. J. Tranvik & M. Jansson, 2004. Role of lakes for organic carbon cycling in the boreal zone. Global Change Biology 10: 141–147.

Artigas, J., A. M. Romaní, A. Gaudes, I. Muñoz & S. Sabater, 2009. Organic matter availability structures microbial biomass and activity in a Mediterranean stream. Freshwater Biology 54: 2025–2036.

Boulton, A. J. & P. S. Lake, 1992. The ecology of 2 intermittent streams in Victoria, Australia. III. Temporal changes in faunal composition. Freshwater Biology 27: 123–138.

Chatfield, C., 1995. Model uncertainty, data mining and statistical-inference. Journal of the Royal Statistical Society Series A, Statistics in Society 158: 419–466.

Chow, V. T., 1959. Open Channel Hydraulics. McGraw-Hill, New York.

Cole, J. J., Y. T. Prairie, N. F. Caraco, W. H. McDowell, L. J. Tranvik, R. G. Striegl, C. M. Duarte, P. Kortelainen, J. A. Downing, J. J. Middelburg & J. M. Melack, 2007. Plumbing the global carbon cycle: integrating inland waters into the terrestrial carbon budget. Ecosystems 10: 172–185.

Gillooly, J. F., J. H. Brown, G. B. West, V. M. Savage & E. L. Charnov, 2001. Effects of size and temperature on metabolic rate. Science 293: 2248–2251.

Giorgi, F. & P. Liolello, 2008. Climate change projections for the Mediterranean region. Global and Planetary Change 63: 90–104.

Grimm, N. B. & S. G. Fisher, 1989. Stability of periphyton and macroinvertebrates to disturbance by flash floods in a Desert stream. Journal of the North American Benthological Society 8: 293–307.

Henderson, F. M., 1966. Open Channel Flow. MacMillan, New York.

Hirabayashi, Y., S. Kanae, S. Emori, T. Oki & M. Kimoto, 2008. Global projections of changing risks of floods and droughts in a changing climate. Hydrological Sciences Journal 53: 754–772.

Huntington, T. G., 2006. Evidence for intensification of the global water cycle: review and synthesis. Journal of Hydrology 319: 83–95.

IPCC, 2007. Summary for policymakers. In: Salomon S., M. Qin, M. Manning, Z. Chen, M. Marquis, K. B. Avergt, M. Tignor, H. L. Miller (eds), Climate Change 2007: The Physical Science Basis. Contribution of Working Group I to the Fourth Assessment Report of the Intergovernmental Panel on Climate Change. Cambridge University Press, Cambridge, UK.

Jentsch, A., J. Kreyling & C. Beierkuhnlein, 2007. A new generation of climate-change experiments: events, not trends. Frontiers in Ecology and the Environment 5: 365–374.

Johnson, J. B. & K. S. Omland, 2004. Model selection in ecology and evolution. Trends in Ecology & Evolution 19: 101–108.

Miao, S. L., C. B. Zou & D. D. Breshears, 2009. Vegetation responses to extreme hydrological events: sequence matters. American Naturalist 173: 113–118.

Moore, K., D. Pierson, K. Pettersson, E. Schneiderman & P. Samuelsson, 2008. Effects of warmer world scenarios on hydrologic inputs to Lake Malaren, Sweden and implications for nutrient loads. Hydrobiologia 599: 191–199.

New, M., M. Todd, M. Hulme & P. Jones, 2001. Precipitation measurements and trends in the twentieth century. International Journal of Climatology 21: 1899–1922.

Reichert, P., 1994. Aquasim—a tool for simulation and data-analysis of aquatic systems. Water Science and Technology 30: 21–30.

Reichert, P., 1998. AQUASIM 2.0—User Manual. Swiss Federal Institute for Environmental Science and Technology (EAWAG), Dubendorf, Switzerland.

Sanchez, E., C. Gallardo, M. A. Gaertner, A. Arribas & M. Castro, 2004. Future climate events in the Mediterranean simulated by a regional climate model: a first approach. Global and Planetary Change 44: 163–180.

Schindler, D. W., 2001. The cumulative effects of climate warming and other human stresses on Canadian freshwaters in the new millennium. Canadian Journal of Fisheries and Aquatic Sciences 58: 18–29.

Sillmann, J. & E. Roeckner, 2008. Indices for extreme events in projections of anthropogenic climate change. Climatic Change 86: 83–104.

Stanley, E. H., D. L. Buschman, A. J. Boulton, N. B. Grimm & S. G. Fisher, 1994. Invertebrate resistance and resilience to intermittency in a Desert stream. American Midland Naturalist 131: 288–300.

Uehlinger, U., H. Buhrer & P. Reichert, 1996. Periphyton dynamics in a floodprone prealpine river: evaluation of significant processes by modelling. Freshwater Biology 36: 249–263.

GLOBAL CHANGE AND RIVER ECOSYSTEMS

# A multi-modeling approach to evaluating climate and land use change impacts in a Great Lakes River Basin

M. J. Wiley · D. W. Hyndman · B. C. Pijanowski · A. D. Kendall · C. Riseng · E. S. Rutherford · S. T. Cheng · M. L. Carlson · J. A. Tyler · R. J. Stevenson · P. J. Steen · P. L. Richards · P. W. Seelbach · J. M. Koches · R. R. Rediske

Received: 22 September 2009 / Accepted: 15 March 2010 / Published online: 12 April 2010
© Springer Science+Business Media B.V. 2010

**Abstract** River ecosystems are driven by linked physical, chemical, and biological subsystems, which operate over different temporal and spatial domains. This complexity increases uncertainty in ecological forecasts, and impedes preparation for the ecological consequences of climate change. We describe a recently developed "multi-modeling" system for ecological forecasting in a 7600 km² watershed in the North American Great Lakes Basin. Using a series of linked land cover, climate, hydrologic, hydraulic, thermal, loading, and biological response models, we examined how changes in both land cover and climate may interact to shape the habitat suitability of river segments for common sport fishes and alter patterns of biological integrity. In scenario-based modeling, both climate and land use change altered multiple ecosystem properties. Because water temperature has a controlling influence on species distributions, sport fishes were overall more sensitive to climate change than to land cover change. However, community-based biological integrity metrics were more sensitive to land use change than climate change; as were nutrient export rates. We discuss the implications of this result for regional preparations for climate change adaptation, and the extent to which the result may be constrained by our modeling methodology.

**Keywords** Climate change · Land use · Land cover · Rivers · Modeling · Fisheries · Salmon · Trout

Guest editors: R. J. Stevenson & S. Sabater / Global Change and River Ecosystems – Implications for Structure, Function and Ecosystem Services

M. J. Wiley (✉) · C. Riseng · S. T. Cheng
University of Michigan, Ann Arbor, MI, USA
e-mail: mjwiley@umich.edu

D. W. Hyndman · A. D. Kendall · R. J. Stevenson
Michigan State University, East Lansing, MI, USA

B. C. Pijanowski
Purdue University, West Lafayette, IN, USA

E. S. Rutherford
NOAA-Great Lakes Environmental Research Laboratory, Ann Arbor, MI, USA

M. L. Carlson · P. W. Seelbach
USGS Great Lakes Science Center, Ann Arbor, MI, USA

P. J. Steen
Huron River Watershed Council, Ann Arbor, MI, USA

J. A. Tyler
Fisheries Projections Ltd., Farmington, CT, USA

P. L. Richards
SUNY Brockport, Brockport, NY, USA

J. M. Koches · R. R. Rediske
Annis Water Resources Center, Grand Valley State University, Muskegon, MI, USA

## Introduction

River ecosystems consist of complex linkages between dynamic physical, chemical, and biological subsystems; each operating at different characteristic spatial scales and frequencies (Maxwell et al., 1995). This complexity makes ecological forecasting difficult; the ensuing methodological uncertainty being one of the obstacles slowing regional preparation for anticipated ecological impacts of climate change (NRC, 2007). Climate forecasters frequently employ "ensemble" modeling approaches (e.g., Murphy et al., 2004) in which multiple models of the same endpoint, each with its own characteristic strengths and weaknesses, are used together to generate more robust forecasts and to quantitatively evaluate model specification-related errors. In control systems engineering "multi-modeling" systems link a series of separately optimized models that describe distinct aspects (parameter domains) of a single problem in order to represent and control complex dynamic, nonlinear system processes (Johansen & Murray-Smith, 1997). In contrast, most water quality, hydrologic and fisheries forecasting typically involves single model simulations that are designed for the detailed representation of a single endpoint or suite of related parameters of interest. Examples of this "one-at-a time" approach in the river management field are legion. Even widely used water quality models which conceptually link hydrologic, hydraulic, and load generation processes (e.g., SPARROW Schwarz et al., 2001; SWAT Santhi et al., 2005) still are largely single purpose constructions which significantly abstract hydrologic/hydraulic process detail in the service of efficient water quality endpoint predictions.

In real-world management settings groups of stakeholders representing diverse interests (e.g., water quality, fisheries, farming, and forestry) are forced to understand and make decisions about whole systems and not constituent parts. For watershed stakeholders single focus, stand alone modeling often leads to large collections of partial and sometimes competing analyses; leaving unanswered the critical question of how models of different components should be integrated to evaluate the overall impact of alternate management strategies and choices. Nor do they typically provide much sense of forecast uncertainty with respect to the parameters that are predicted. The utility of a "multi-modeling" approach is that it can provide integrated forecasts over a wide range of ecological components and ecosystem services. The need for such forecast capacity is growing with the urgency of linking large-scale climate change modeling to local hydrologic, water quality, and ultimately biological consequences (e.g., Christensen & Lettenmaier, 2006; Moore et al., 2009, Nelson et al., 2009). Furthermore, ensemble modeling capability (multiple representations of single process domains) is quite easily implemented inside of a multi-modeling system (multiple models across multiple process domains), and this seems a promising approach to evaluating uncertainty in complex ecological forecasting. Explicit multi-modeling applications in ecological forecasting are at present rare; but they are now emerging in the context of global change preparation (e.g., see Nelson et al., 2009). Here we describe an analysis employing this approach to explore land use and climate change impacts in the Muskegon River watershed, a major tributary system of Lake Michigan in the USA.

The North American Laurentian Great Lakes (GL) and tributary watersheds support a $4.3 billion per year fishery, and the water needs of over 40 million inhabitants including five class one urban centers (Chicago, Detroit, Toledo, Cleveland, Buffalo, and Toronto). The water rich GL basin is likely to be particularly vulnerable ecologically to anticipated climate change (Kling et al., 2003), but like many parts of the U.S. has been the subject of only limited climate change planning to date (NRC, 2007; Dinse et al., 2009). In this paper, we briefly describe an ecological "multi-modeling" system developed for integrated assessment and ecological forecasting on the Muskegon River, a major watershed in the GL basin (Stevenson et al., 2008; Wiley et al., 2008). Using a series of linked land cover, climate, hydrologic, hydraulic, loading, and biological response models we examined the potential influence of land management practices and climate change on the ecological future of this important tributary of Lake Michigan. Specifically, we discuss the potential impacts of changes in climate and land use on water quality, channel stability, fisheries-relevant habitat, and biological integrity as reflected in its fish and macroinvertebrate communities.

# Methods

## Study area

The Muskegon River drains 7600 km² of Michigan's Lower Peninsula (Fig. 1). It is the second longest (~353 km) river in the Lake Michigan watershed and provides key spawning habitat for the region's economically important anadromous (adfluvial) fisheries. Included in its headwaters are Higgins (3,885 ha), Houghton (8,112 ha), and Mitchell-Cadillac (1,510 ha) lakes, and along its lower main stem a series of hydropower reservoirs. The river terminates in an extensive freshwater wetland that drains to Muskegon Lake, a drowned river mouth basin connected to Lake Michigan. The river drops a total of 175 m from its headwaters to Lake Michigan and has approximately 94 perennial tributaries comprised of over 2500 km of stream channel (at a scale of 1:100,000).

Over 47% of the watershed is forest. Between 1978 and 1998, the amount of urban in the watershed nearly doubled from 4% total coverage to over 7%. Approximately 22% of the agriculture in 1978 was lost by 1998, largely replaced by forest, which increased from 44% to 47% during the same 20 year period. According to the US Census Bureau, 358,184 people live in cities, villages, and townships that were wholly or partially located within the watershed; this represents approximately 3.6% of Michigan's population. The largest cities (and year 2000 population) include Muskegon (40,105), North Shores (22,527), and Cadillac (10,000). The watershed is located within a humid, temperature climate zone, receiving approximately 83 cm of precipitation annually from 1980 to 2005. The basin is dominated by coarse textured soils atop largely glacial drift deposits. This combination of climate, topography, and sediment types produce relatively stable flows in the Muskegon River that are primarily derived from groundwater sources (Kendall & Hyndman, in review). At these latitudes (43–45 N) groundwater dominated rivers support primarily cold and cool-water fish assemblages (Wiley et al., 1997; Zorn et al., 2002; Wehrly et al., 2003). The Muskegon River supports regionally important trout, salmon, and walleye sport fisheries (O'Neal, 1997).

## Modeling approach

The Muskegon River Ecological Modeling System (MREMS) is a "multi-model" in the sense of Johansen & Murray-Smith (1997). It consists of a set of independent component models targeted toward various aspects of the Muskegon River ecosystem, using different spatial and temporal domains as appropriate. Models are synchronized by shared inputs from climate, land cover, and GIS river network models, and also by inter-model data exchange as required. Data geo-referencing protocols help models operating at different spatial scales to communicate and integrate outputs. Model execution mimics the hierarchical organization of real river ecosystems (Fig. 2) by routing all model calculations

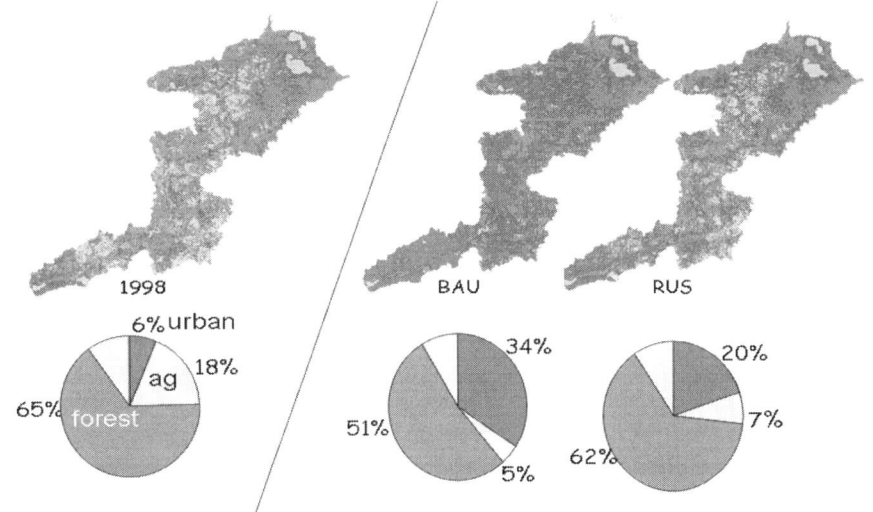

Fig. 1 Muskegon River watershed. Current and future forecast land cover maps used in the land management scenarios. Pie diagrams report corresponding land cover proportions for entire basin. Current scenario (1998 baseline) was derived from hand digitized aerial photography. BAU and RUS scenario land covers generated by LTM2 (Pijanowski et al. 2000, 2002a, b, 2005). The Muskegon River is tributary to the Eastern shore of Lake Michigan (inset)

from upstream to downstream elements and by organizing data flows from climate and landscape structure (land use) models to reach-referenced distributed hydrologic and physical models. The output of these in turn are used as input to a series of biological response models. We developed MREMS as an open modeling system to which any type of model could in theory be added. For the simulations described below, however, the specific suite of hydrologic, loading, and biological models are detailed in Table 1. Models were run by researchers at separate collaborating universities, and assembled and shared on a central web-accessible server and data execution directory (http://mwrp.net).

In the simulations reported below, we used the MREMS system to explore the potential impacts on the Muskegon River of two climate change scenarios along with two future land management trajectories. This study was a part of a larger, long-term, integrated planning exercise involving 16 watershed stakeholder organizations, a collaborating group of university and agency scientists, and a consortium of regional funders (Wiley et al., 2004, 2008; Stevenson et al., 2008).

The MREMS framework

The spatial framework of MREMS was adapted from the MRI-VSEC v1.0 system of Seelbach et al. (2006) by correcting local mapping errors and transferring it to the 1:24,000 scale. The VSEC system is a GIS representation of the drainage net itself, with longitudinal units defined "ecologically"; each VSEC unit is a contiguous channel segment, delimited to represent a relatively homogenous environment in terms of parameters meaningful to biological organisms (e.g., temperature, hydraulics, chemistry; Seelbach et al., 2002; Seelbach & Wiley, 2005). Higgins et al. (1999) referred to units of this type as fish macrohabitats. Ecological valley segments combine elements of local valley and channel geomorphology with catchment hydrology, the two dominant forces shaping riverine habitats. This approach is conceptually similar to the hydrogeomorphic 'HGM' concept used in wetland assessment (Hauer & Smith, 1998). The MREMS system identifies 138 distinct channel units in the Muskegon River, ranging from first to fifth order. Major reservoirs and Muskegon, Houghton, Mitchell-Cadillac, and Higgins lakes are included as separate VSEC units. In MREMS simulations, while basic computational resolution of constituent models varies, each directly or indirectly (via a post-processing) provides output for all 138 segments. The VSEC unit map then serves as the underlying skeleton on which model input and output is organized. Models communicate by placing spatially referenced outputs into a structured directory system that is organized by specific timeframe (simulation year) and problem context (scenario). A complete MREMS run for a specific scenario involves the serial execution of the set of component models (Table 1, Fig. 2) for each time frame, using links to scenario-specific inputs and outputs. In many cases, the output written by one model may be used as input by the next. Model execution order is determined by data dependency, thus execution order would typically start with the generation of a land cover map, followed by hydrologic, chemical and sediment loading, reach hydraulics (for key fisheries habitats) and finally biological models.

Overview of component models

Two climate scenarios were used in the simulations described below. A *standard climate scenario* was based on observed weather across the Muskegon watershed from 1980 to 2005 (Andresen, 2007; MAWN, 2008; NCDC, 2008) along with NEXRAD distributed precipitation (Fulton et al., 1998), along with a solar radiation model prior to the availability of such data (Yang & Koike, 2005). Hourly measured and simulated values from each gauge were distributed across the basin into 425 m grids using an inverse distance weighted interpolation. Beginning in 1996, NEXRAD data, at 4 km resolution, replaced interpolated gauge values provided that air temperatures are not below 0°C due to calibration issues with the radar-derived frozen precipitation data (Jayawickreme & Hyndman, 2007). A *climate change scenario* for the end of this century was then constructed from the standard climate scenario using A1B scenario model results from the Fourth IPCC Assessment (Meehl et al., 2007). A1B is a "conservative" forecast and assumes a peak of greenhouse gas emissions near mid-century, followed by modest reduction in emissions through 2100 (Nakicenovic, 2000). The IPCC regional model provided predicted anomalies by month for average daily temperature

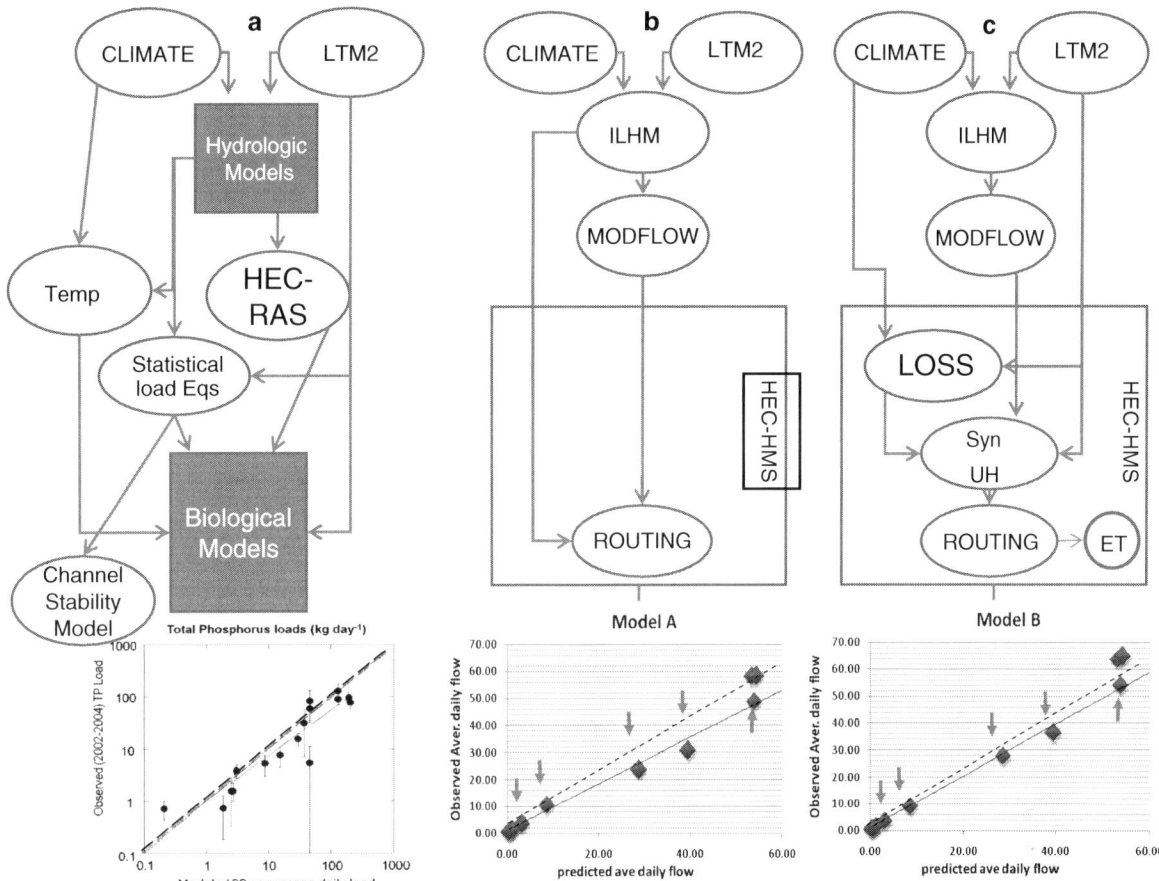

**Fig. 2** MREMS multi-model organization. **a** Overall model structure. Climate and Land cover models drove multiple (shaded box) distributed hydrologic model variants to produce a series of 20-year basin-wide flow estimates. These in turn were used to model temperature, hydraulics, material loading, channel stability, and a suite of biological responses (shaded box) using models listed in Table 1. Data below show predicted and observed annual average TP loads for 14 Muskegon River sites sampled during 2002–2004; dashed line is the one-to-one ratio, verticals represent sample variance as ±2 standard errors. Predicted values are from model runs using hydrologic variant 2c (below) and are computed from linked outputs of seven sub component models. **b** Example of one of the simpler hydrologic simulation variants ("cr0") based on a linked ILHM and MODFLOW implementation with channel routing managed by HEC-HMS. Of the four variants, this was the most mechanistically tractable, carefully preserving hydrologic mass-balance in every computational step. Data plotted below it show predicted and observed annual average daily discharge (cms) at five permanent (arrows) and several other short-term gauging stations; dashed line is the one-to-one ratio. **c** Hydrologic simulation variant ("m×4") which linked ILHM recharge and MODFLOW groundwater estimates with surface water models managed by HEC-HMS. Data below show predicted and observed annual average daily discharge (cms) at five permanent (arrows) and short-term gauging stations; dashed line is the one-to-one ratio. This was the best performing of the variants and was used in all of the physical and biological response estimates reported here

and total daily precipitation. The regional model was downscaled for our runs by using its' predicted anomalies to "offset" our higher resolution "standard climate model" values. The resulting climate change scenario model is on average both warmer and wetter than our standard climate model. It retains, however, the substantial east–west spatial variability of the standard model reflecting strong attenuation of lake-effect thermal buffering and precipitation.

In MREMS modeling, land use/land cover estimates are based on 1:24,000 air photo mapping for 1978 and 1998, and on projections of future land cover developed for this study using the neural-net based Land Transformation Model v.3 (LTM:

**Table 1** Component models linked in MREMS and used in analyses reported here

| Model | Predicts | Type | References |
|---|---|---|---|
| LTM v.3 | Land use change over time | Suite of linked neural net and linear models | Pijanowski et al. (2002a, 2005) |
| ILHM | Evapo-transpiration, recharge, runoff, soil moisture storage, snow dynamics | Process-based distributed, high resolution (120 m$^2$ grids) simulation | Hyndman et al. (2007) and Kendall & Hyndman (in review) |
| MODFLOW | Water table elevation, Groundwater flows | Standard FORTRAN codes | Harbaugh et al. (2000) |
| MRI_VSEC 1.0 | Basic channel segment and contributing basin physical attributes | National hydrography-based GIS layer with empirical and model-based attribution | Seelbach et al. (1997, 2006), Seelbach & Wiley (2005) |
| HEC-HMS | Surface water routing | Hydrologic simulation system | USACE (2009a, b) |
| HEC-RAS | Surface water hydraulics | 1D hydraulic simulation system | USACE (2009a, b) |
| MRI_LOADS, DOMQ | Flow-dependent dissolved, suspended, and bed loads and channel stability | Regional regression models | Baker et al. (2001), Ladewig (2006), Benson & Thomas (1966) |
| RPSTM | Daily water temperature statistics by reach | Reduced –parameter energy balance with channel routing | Cheng (unpublished) |
| GLGAP-SFM | Reach suitability for key fish species | Regional CART classification model for river segments | Steen et al. (2006, 2010) |
| MRI bioassessment models | Probability of ecological impairment by reach | CART reach classification model based on regionally normalized assessment | Wiley et al. (2002) and Riseng et al. (2006) |
| DWUA | Life-stage specific weighted useable area analyses | HSI analysis from HEC-RAS output | Raleigh et al. (1984) and Tyler (unpublished) |

Pijanowski et al., 2000, 2002a, b, 2005). LTM has been widely used in the United States, Africa, and Europe for forecasting and backcasting land cover distributions in ecological planning contexts (Pijanowski et al., 2006, 2007). LTM spatially distributes prescribed county-level rates of land use transitions using neural networks to train on the relationship of surrogates to drivers of land use change. The model was calibrated using standard land change goodness of fit statistics (Pontius et al., 2008). Working with a group of regional stakeholders, we explored a series of different land management scenarios as a part of the Muskegon watershed integrated assessment (Stevenson et al., 2008). Joint stakeholder-modeler workshops developed a series of land management scenarios to evaluate potential management strategies of particular interest to the watershed stakeholders (including urban sprawl containment, riparian setback rules, and agricultural land preservation). For each scenario, a series of land cover projections was developed covering the period 2000 to 2090 by constraining LTM runs with scenario-appropriate land transformation rules. Of the 13 primary management scenarios produced, four have been evaluated for sensitivity to climate change to date.

Here we report some of the results from the first two land management scenario analyses (Fig. 1). In our Business as Usual (BAU) scenario, a baseline future landscape reflects a continuation of the average rates of urban and forest growth observed from 1978 to 1998. In the BAU Scenario urban and forest land expansions occur at the expense of agriculture, and farming in this version of the future continues to decline steadily across the watershed, a pattern observed in this basin since the 1920s. In our Reduced Urban Sprawl (RUS) scenario, the LTM halved the region's historic rate of urban sprawl, and allowed forest re-growth to continue at a relatively high rate. The reduced rate of urban sprawl led to reduced extent of urbanization which in turn left more agriculture in place at the end of the 21st century.

Hydrologic forecasting was based on an ensemble of four variant simulations that linked a series of existing simulation modules including ILHM, MODFLOW, a regional synthetic hydrograph model, and HEC-HMS channel routing (examples in Fig. 2b and c). The Integrated Landscape Hydrology Model (ILHM; Hyndman et al., 2007; Kendall & Hyndman, in review), is a high resolution, distributed model developed to evaluate influences of both land use and climate on hydrology at scales pertinent to land managers. ILHM simulates all major surface and near-surface hydrologic processes including evapotranspiration (ET), snowmelt, groundwater recharge, overland flow, and stream discharge. Moisture is redistributed from precipitation to various subsurface and surface pathways, including canopy interception, snowmelt, surface depression storage, infiltration, evapotranspiration, throughflow, recharge, and stream routing. Input for the model consists of gridded topographic, climate, land cover, leaf area index (LAI), and other available information about the distribution of soils and glacial sediments. In the simulations, ILHM recharge estimates were linked to MODFLOW codes (Harbaugh et al., 2000) to estimate groundwater flux and accruals to tributary sub-basins. In two ensemble variants runoff estimates from ILHM were combined with MODFLOW estimates for each model sub-basin and then routed through the channel system. These variants represented the most mechanistically realistic of the four hydrologic simulations and included a version (Fig. 2b) which maintained a strict hydrologic mass-balance throughout all calculations. In the other two ensemble variants MODFLOW groundwater flux was combined with calibrated estimates of runoff from a regional synthetic unit hydrograph model (Fig. 2c) for downstream routing after corrections for calibrated estimates of riparian wetland and reservoir ET losses. Channel routing, data summary, and output management in all four model variants was handled in HEC-HMS v3.1.0 (USACE, 1998). Each model variant provided daily channel flows at each of the 138 VSEC river segments throughout the basin. Validation tests indicated that the most accurate ensemble variant ("mx4", Fig. 2c) predicted historical water balances for the five USGS gauged locations in the Muskegon basin within 5% of measured values over the 10-year simulation period (1996–2005). All four of the model variants frequently over-predicted flows in the lowest zone of freshwater estuary where backwater effects were common. In this paper, we used the most accurate of the hydrologic ensemble simulations to drive all subsequent loading and biological models as reported here. We used the entire ensemble of hydrologic forecasts to explore model variance and error bounds around modeled hydrologic and loading responses (see "Discussion" section).

Sediment and nutrient loads were estimated by linking hydrologic and land cover simulation outputs to a series of empirical "instantaneous" load regression models developed from state-wide data sets as part of the Michigan Rivers Inventory program (Seelbach & Wiley, 1997). Similar in general approach to the USGS SPARROW model (Schwarz et al., 2001), the loading regressions typically explained 85–90% of the N and P flux from sample catchments; and simulated load estimates for the Muskegon were well correlated (e.g., Fig. 2a) with observed fluxes measured during 2002–2004. Daily average concentrations were estimated as the ratio of daily loads to daily discharges. Changes in river channel stability were assessed for each river segment in terms of relative deviations from baseline (1980–2005) dominant discharge (Knighton, 1984) magnitudes. When deviations above or below baseline values exceeded 20%, stream segments were flagged as unstable and therefore likely to respond geomorphically. Increasing dominant discharge is associated with more channel and bank erosion, and sediment transport. Decreasing dominant discharge is typically associated with channel aggradation, and local bed sediment accumulation. Both conditions can initiate complex responses in terms of meander re-configuration, slope adjustment, and changes in bed material composition and structure (Schumm, 1977; Knighton, 1984).

Daily water temperatures for each VSEC unit were estimated using a new reduced parameter energy balance and channel routing model (RPSTM; Cheng & Wiley, 2007; Cheng, unpublished). Linked to HEC-HMS the model uses sub-basin groundwater flux from ILHM and runoff routing from HMS to estimate daily minimum, maximum, and mean temperature at the beginning and end of each HMS hydrologic routing unit.

We modeled the responses of the Muskegon River invertebrate and fish community to changing climate and land management trajectories by linking an ensemble of biological models to output from the physical modeling portions of MREMS. Building on earlier work in which simple summer air temperature offsets were added to regional habitat models (Steen et al., 2010) we linked these and other models with spatially explicit predictions of water temperature, stream flow, and water quality driven by statistically downscaled changes in precipitation and temperature. Biological response models used in these simulations include: multiple linear regression-based normalized assessment models for fish and macroinvertebrate communities (Wiley et al., 2002; Riseng et al., 2006) forecast for individual river segments using Bayesian probabilities from a CART (De'ath and Fabricius, 2000) analyses of a large sample of observed assessment scores ($n = 2000$; Riseng et al., 2006), Classification and Regression Tree (CART) models of sport fish habitat (Steen et al., 2006, 2010), dynamic weighted usable area models (DWUA) of sport fish habitat (e.g., Gouraud et al., 2001) and an agent-based model of steelhead (Oncorhyncus mykiss) recruitment dynamics (e.g., Tyler and Rutherford, 2007).

For the DWUA model and individual-based steelhead modeling, we focused on the approximately 40 km section of the lower Muskegon River immediately downstream of Croton Dam, a key fishery resource, and divided the section into cells for hydrological analysis (depth, flow velocity) by a one-dimensional channel routing model (HEC-RAS). A GIS model of river substrate was paired to the HEC-RAS grid of the river segments so that each cell of the river model included water depth, water velocity, and substrate characteristics that could be used to determine cell-by-cell habitat preferences for life stages of selected sport fishes. For the DWUA model, fish life stage habitat preferences were computed for each river cell based on habitat suitability indices (HSI) (e.g., Raleigh et al., 1984), which use known information on habitat preferences for a particular species-life stage combination to predict the amount of suitable habitat available in an aquatic environment for that life stage (USFWS, 1980). Species preference values for habitat variables range 0.0–1.0 for each variable. To compute the WUA for a cell, we multiplied the preference value ($P$) assigned for each environmental characteristic in the cell times the area of the cell. In results reported here integration with MREMS climate change scenario outputs for these two models included mean monthly temperature but not the hydrologic responses. Fully integrated runs of the DWUA and the steelhead IBM are currently underway.

## Results

Both climate change and land management scenarios altered modeled hydrologic, water quality, geomorphic, and biological conditions across the Muskegon River watershed compared to current conditions. Pairwise contrasts (alternate climate change scenarios × alternate land management scenarios) suggest that the impact of climate change on this river system will not only be large, but also that response will vary significantly depending upon future land use trajectory.

Land use change scenarios under current climate

Projected changes in land cover had significant effects on the distribution of predicted flows in the watershed (Tables 2 and 3) and as a result over many other ecologically important characteristics of the modeled river ecosystem. Under the current climate regime, the BAU scenario led to future reductions in agriculture and forested land cover, reduced ET losses and increased both rates of groundwater recharge and storm runoff. As a result base flow, storm flow, and median discharge all increased in the lower main stem by 15–20% (Table 2) and across most of the rest of the watershed as well (Table 3); flow variability across the basin increased by 30%. Increases in the dominant discharge exceeded 20% in the lower main stem and in many other VSEC units suggesting wide spread channel destabilization would occur. Consistent with this, the loading model projected large increases in sediment loads for almost all sites (26% and 57% for mean and standard error, respectively). Increases in urban cover and associated flows, appear to cause an even larger increase in nutrient loads. Daily mean total phosphorus (TP) loads increased on average by 53% and daily total inorganic nitrogen (TIN) loads increased by 31% (Table 3). Nutrient concentrations (Table 4), however, appear to be somewhat buffered by the increasing flows and on average increased by only 32% and 12% for TP and TIN, respectively. Water temperature changes were minimal, tending to slightly lower driven by increases in groundwater flux to the river, but this assumes constant a groundwater temperature.

The RUS scenario also resulted in major losses of agricultural land cover, but in contrast to the BAU scenario it also had increases in forest cover

**Table 2** Modeled flow statistics for the lower main stem Muskegon River (terminal VSEC unit) at its confluence with Muskegon Lake

| Climate scenario: | Current | Current | | | | IPCC A1B adjusted | | | |
|---|---|---|---|---|---|---|---|---|---|
| Land use scenario: | Current | BAU | | RUS | | BAU | | RUS | |
| Parameter: | (cms) | (cms) | (%Δ) | (cms) | (%Δ) | (cms) | (%Δ) | (cms) | (%Δ) |
| Q05 | 105 | 124 | 18.1% | 105 | 0.0% | 128 | 21.9% | 112 | 6.7% |
| Q10 | 88 | 101 | 14.8% | 87 | −1.1% | 108 | 22.7% | 95 | 8.0% |
| Q50 | 53 | 63 | 18.9% | 58 | 9.4% | 70 | 32.1% | 65 | 22.6% |
| Q90 | 37 | 45 | 21.6% | 41 | 10.8% | 48 | 29.7% | 46 | 24.3% |
| Qmin | 23 | 26 | 13.0% | 25 | 8.7% | 29 | 26.1% | 31 | 34.8% |
| Qmax | 808 | 851 | 5.3% | 793 | −1.9% | 957 | 18.4% | 896 | 10.9% |
| DomQ | 52 | 63 | 21.2% | 60 | 15.4% | 71 | 36.5% | 61 | 17.3% |
| Basin land cover (%) | | | | | | | | | |
| Urban | 6 | 34 | 466% | 23 | 283% | 34 | 466% | 34 | 283% |
| Agricultural | 18 | 5 | −72% | 5 | −72% | 5 | −72% | 5 | −72% |
| Forested | 56 | 52 | −7% | 61 | 9% | 52 | −7% | 52 | 9% |
| Water and wetlands | 8 | 8 | 0% | 9 | 12% | 8 | 0% | 8 | 12% |
| Other | 12 | 1 | −92% | 2 | −83% | 1 | −92% | 1 | −83% |

Summaries are over the modeled 25 year period (nominally 1980–2005). Current climate scenario coupled with current land use scenario provide baseline conditions. %Δ = percent change from baseline. Land use scenarios as described in the text and Fig. 1 (Current = 1998, BAU = end of century with business as usual land management; RUS = end of century with reduced urban sprawl rate). Q05-Q90 are daily flow exceeded values for indicated percentiles. DomQ = estimated dominant discharge for the period

(Table 2). Reductions in agricultural land cover again led to increases in recharge and base flow, but these were less than half the magnitude observed in the BAU scenario. However, high flows changed little and were even reduced relative to simulations based on current land cover; resulting on average in a more seasonally stable flow regime (Tables 2 and 3). Despite this general increase in flow stability, future dominant discharge still increased for many stream segments because of increasing base flows; but, increments in sediment load transported were about half those estimated for the BAU scenario. TP loads and concentrations changed little from current values, and were much reduced relative to the BAU forecasts. However, the projections based on the RUS scenario show marginally larger TIN loads than in the in BAU scenario (Table 4) reflecting more agricultural land use in the basin.

Landuse change scenarios under the A1B climate change scenario

Adding climate change to the Muskegon ecosystem forecasts dramatically altered most modeling results.

Increasing air temperatures resulted in increased ET estimates across the basin. However, increasing precipitation, particularly during winter months led to even larger increases in recharge rates; the overall result being that water yields and flows in the river generally increased (Tables 2 and 3). The hydrologic impact was largest on the BAU land use scenario (roughly doubling % change in median and lower frequency water yields), but flow increases were also observed in the RUS case. Under the RUS × Climate Change scenario, average annual flows were essentially equivalent to the BAU scenario flows under the Current Climate scenario. For both land use scenarios, dominant discharge increased with respect to baseline, compared to land use change scenarios alone. Consistent with that result, sediment concentrations generally declined slightly (−7.8 ± 5.8%), but sediment loads and yields increased by about 50% for the BAU scenario and doubled for the RUS scenario. Nutrient concentrations changed little from the current climate scenario runs, but nutrient loads and yields increased for both TIN and TP. The relative impact of the A1B climate was higher on nutrients in the RUS scenario, with the change in

**Table 3** Summary of modeled average daily load statistics for 138 VSEC channel units of the Muskegon River ecosystem

| Climate scenario: | Current | Current | | | | IPCC A1B adjusted | | | |
|---|---|---|---|---|---|---|---|---|---|
| Land use scenario: | Current | BAU | | RUS | | BAU | | RUS | |
| Parameter: | | | (%Δ) | | (%Δ) | | (%Δ) | | (%Δ) |
| Channel flow, ave. daily flow (cms/km$^2$) | | | | | | | | | |
| Median | 0.0089 | 0.0108 | 21.3% | 0.0098 | 10.1% | 0.0116 | 30.3% | 0.0108 | 21.3% |
| Mean | 0.0094 | 0.0113 | 20.2% | 0.0102 | 8.5% | 0.0123 | 30.9% | 0.0111 | 18.1% |
| SE | 0.0002 | 0.0003 | 50.0% | 0.0002 | 0.0% | 0.0003 | 50.0% | 0.0002 | 0.0% |
| Min | 0.0006 | 0.0017 | 183.3% | 0.0016 | 166.7% | 0.0019 | 216.7% | 0.0017 | 183.3% |
| Max | 0.0519 | 0.0586 | 12.9% | 0.0496 | −4.4% | 0.0600 | 21.4% | 0.0530 | 2.1% |
| Total phosphorus flux, ave. daily load (kg/km$^2$) | | | | | | | | | |
| Median | 26 | 36 | 38.5% | 27 | 3.8% | 39 | 50.0% | 30 | 15.4% |
| Mean | 30 | 46 | 53.3% | 32 | 6.7% | 50 | 66.7% | 35 | 16.7% |
| SE | 2 | 1 | −50.0% | 1 | −50.0% | 2 | 0.0% | 1 | −50.0% |
| Min | 1 | 4 | 300.0% | 3 | 200.0% | 4 | 300.0% | 4 | 300.0% |
| Max | 457 | 197 | −56.9% | 152 | −66.7% | 214 | −53.2% | 164 | −64.1% |
| Total inorg. nitrogen flux, ave. daily load (kg/km$^2$) | | | | | | | | | |
| Median | 350 | 568 | 62.3% | 447 | 27.7% | 613 | 75.1% | 532 | 52.0% |
| Mean | 409 | 534 | 30.6% | 491 | 20.0% | 575 | 40.6% | 529 | 29.3% |
| SE | 14 | 11 | −21.4% | 11 | −21.4% | 12 | −14.3% | 12 | −14.3% |
| Min | 17 | 86 | 405.9% | 77 | 352.9% | 93 | 447.1% | 84 | 394.1% |
| Max | 2762 | 2175 | −21.3% | 1901 | −31.2% | 2324 | −15.9% | 2024 | −26.7% |
| Total sediment flux, ave. daily load [kg/km$^2$] | | | | | | | | | |
| Median | 72 | 88 | 22.2% | 81 | 12.5% | 98 | 36.1% | 86 | 19.4% |
| Mean | 183 | 230 | 25.7% | 200 | 9.3% | 251 | 37.2% | 219 | 19.7% |
| SE | 23 | 36 | 56.5% | 30 | 30.4% | 40 | 73.9% | 33 | 43.5% |
| Min | 4 | 5 | 25.0% | 4 | 0.0% | 5 | 25.0% | 5 | 25.0% |
| Max | 4974 | 8060 | 62.0% | 6030 | 21.2% | 9064 | 82.2% | 6864 | 38.0% |

Current climate scenario coupled with current land use scenario provide baseline conditions. %Δ = percent change from baseline. Summaries are over a modeled 25 year period (nominally 1980–2005), land use scenarios as described in the text and Fig. 1 (Current = 1998, BAU = end of century with business as usual land management; RUS = end of century with reduced urban sprawl rate)

median TP flux increasing from 3.5% to 15.4%, and TIN from 28% to 52%. This contrasts with the BUA scenario results that changed from 38% to 50%, and from 62% to 75% for TP and TIN, respectively (Table 3). Impacts of the climate change scenario on water temperature were large and pervasive. Although increased recharge drove somewhat higher rates of groundwater accrual and baseflow in the river (normally associated with lower temperature; Wiley et al., 1997), groundwater temperatures at this latitude strongly reflect annual average air temperature. The net effect was a 15–16% and 13–14% increase in mid-summer (July) daily average and daily maximum water temperatures, respectively. Perhaps most importantly, these changes occurred across the range of temperatures, 21–25°C, that generally define habitat transitions for Michigan's cold, cool, and warm-water fish guilds (Wehrly et al., 2003).

Modeled impacts of the climate change scenario on biological parameters varied substantially (Table 5). Community-based (fish and macroinvertebrate) assessment metrics, which responded strongly to land use scenarios, were relatively insensitive to the climate scenario. Under the BAU scenario, the invertebrate metric results suggested a slight improvement basin-wide. Under the RUS scenario, the reach impairment rate for invertebrates worsened

**Table 4** Summary of modeled water quality statistics (mean annual concentration values) for 138 VSEC channel units in the Muskegon River ecosystem

| Climate scenario: | Current | Current | | | | IPCC A1B adjusted | | | |
|---|---|---|---|---|---|---|---|---|---|
| Land use scenario: | Current | BAU | | RUS | | BAU | | RUS | |
| Parameter: | | | (%Δ) | | (%Δ) | | (%Δ) | | (%Δ) |
| July water temp (°C) | | | | | | | | | |
| Mean | 22.6 | 22.2 | −1.8% | 22.1 | −2.2% | 26.2 | 15.9% | 26.1 | 15.5% |
| Max | 25.7 | 25 | −2.7% | 24.9 | −3.1% | 29.2 | 13.6% | 29 | 12.8% |
| SE | 0.193 | 0.190 | −1.6% | 0.186 | −3.6% | 0.200 | 3.6% | 0.193 | 0.0% |
| Total phosphorus ave. daily concentration (ppm) | | | | | | | | | |
| Median | 0.025 | 0.036 | 44.0% | 0.027 | 8.0% | 0.034 | 36.0% | 0.027 | 8.0% |
| Mean | 0.028 | 0.037 | 32.1% | 0.029 | 3.6% | 0.037 | 32.1% | 0.029 | 3.6% |
| SE | 0.000574 | 0.000829 | 44.4% | 0.000639 | 11.3% | 0.000828 | 44.3% | 0.000645 | 12.4% |
| Min | 0.002 | 0.005 | 150.0% | 0.001 | 50.0% | 0.005 | 150.0% | 0.004 | 100.0% |
| Max | 0.083 | 0.094 | 13.3% | 0.063 | 24.1% | 0.094 | 13.3% | 0.062 | −25.3% |
| Total inorg. nitrogen ave. daily concentration (ppm) | | | | | | | | | |
| Median | 0.396 | 0.454 | 14.6% | 0.457 | 15.4% | 0.43 | 8.6% | 0.443 | 11.9% |
| Mean | 0.401 | 0.448 | 11.7% | 0.463 | 15.5% | 0.438 | 9.2% | 0.454 | 13.2% |
| SE | 0.007489 | 0.003295 | −56.0% | 0.000761 | 89.8% | 0.00268 | −64.2% | 0.005071 | −32.3% |
| Min | 0.09 | 0.195 | 116.7% | 0.17 | 88.9% | 0.192 | 113.3% | 0.163 | 81.1% |
| Max | 1.174 | 0.651 | −44.5% | 0.879 | 25.1% | 0.633 | −46.1% | 0.863 | −26.5% |
| Total sediment ave. daily concentration (ppm) | | | | | | | | | |
| Median | 96 | 95 | −1.0% | 95 | −1.0% | 96 | 0.0% | 95 | −1.0% |
| Mean | 218 | 202 | −7.3% | 201 | −7.8% | 203 | −6.9% | 201 | −7.8% |
| SE | 34 | 34 | 0.0% | 33 | −2.9% | 34 | 0.0% | 33 | −2.9% |
| Min | 8 | 8 | 0.0% | 8 | 0.0% | 8 | 0.0% | 8 | 0.0% |
| Max | 9029 | 8957 | −0.8% | 8763 | −2.9% | 8933 | −1.1% | 8747 | −3.1% |

Current climate scenario coupled with current land use scenario provide baseline conditions. %Δ = percent change from baseline. Summaries are over a modeled 25 year period (nominally 1980–2005). Land use scenarios as described in the text and Fig. 1 (Current = 1998, BAU = end of century with business as usual land management; RUS = end of century with reduced urban sprawl rate)

slightly. The fish community metric likewise indicated a small worsening of basin conditions for both land use scenarios (increasing by only about 4% in both cases). Impacts on individual fish taxa, however, were substantial and variable (Table 5, Fig. 3). Brook trout (*Salvelnis fontinalis*) were most negatively affected losing >60% of their currently available channel habitat. Resident (nonanadromous "steelhead") rainbow trout (*Orchorynchus mikus*) lost more than 30% of their habitat and brown tout (*Salmo trutta*) more than 15%. Lower river habitat supporting Lake Michigan and Muskegon-run salmonines were substantially impacted, although individually the extent varied depending largely on life cycle timing. Coho salmon were most negatively affected (climate-related reductions were >70%). Chinook salmon (*Onchorynchus tshawytscha*) and steelhead were relatively less sensitive but still were forecast to experience greater than 50% reductions in habitat availability. The species most sensitive to this climate change scenario was the northern pike (*Esox lucius*) which the CART models projected would experience close to a doubling of habitat in the Muskegon river. Smallmouth bass (*Micropterus dolomeiui*) also were predicted to benefit substantially (+24%). Walleye pike (*Sander vitreus*), an important sport fish in the larger lakes and lower main stem, were relatively insensitive to the climate change scenario. In almost

Table 5 Summary of GLGAP-SFM modeled sport fish habitat responses (km of useable habitat) in the entire Muskegon River basin

| Climate scenario: | Current | Current | | | | IPCC A1B adjusted | | | | Average climate sensitivity |
|---|---|---|---|---|---|---|---|---|---|---|
| Land use scenario: | Current | BAU | | RUS | | BAU | | RUS | | |
| Taxa | (km) | (km) | (%Δ) | (km) | (%Δ) | (km) | (%Δ) | (km) | (%Δ) | (%Δ) |
| Smallmouth | 501 | 515 | 3% | 497 | −1% | 627 | 25% | 622 | 24% | 24 |
| Brook_trout | 1030 | 984 | −4% | 1038 | 1% | 348 | −66% | 410 | −60% | −61 |
| Brown_trout | 1516 | 1586 | 5% | 1549 | 2% | 1263 | −17% | 1277 | −16% | −20 |
| Rainbow_trout | 1656 | 1612 | −3% | 1630 | −2% | 1067 | −36% | 1080 | −35% | −33 |
| Steelhead | 315 | 438 | 39% | 438 | 39% | 270 | −14% | 261 | −17% | −55 |
| Chinook | 112 | 223 | 99% | 228 | 103% | 165 | 47% | 171 | 52% | −51 |
| Coho | 120 | 143 | 19% | 141 | 18% | 55 | −54% | 53 | −55% | −74 |
| Walleye | 216 | 167 | −23% | 218 | 1% | 179 | −17% | 201 | −7% | −1 |
| GLwalleye | 55 | 35 | −36% | 64 | 18% | 41 | −24% | 51 | −6% | −6 |
| Northern_pike | 548 | 578 | 5% | 551 | 1% | 1026 | 87% | 995 | 81% | 27 |
| | (proportion) | (proportion) | | (proportion) | | (proportion) | | (proportion) | | |
| Adfl.salmonines | 0.064 | 0.094 | 47% | 0.094 | 48% | 0.057 | −10% | 0.057 | −0.11 | −58 |
| All coldwater | 0.278 | 0.292 | 5% | 0.294 | 6% | 0.185 | −33% | 0.190 | −0.32 | −38 |
| All warmwater | 0.116 | 0.114 | −2% | 0.117 | 1% | 0.164 | 42% | 0.164 | 0.416 | 42 |
| All trouts* | 0.491 | 0.489 | 0% | 0.493 | 0% | 0.313 | −36% | 0.324 | −0.34 | −35 |
| All walleye | 0.106 | 0.035 | −67% | 0.049 | −53% | 0.039 | −63% | 0.044 | −0.58 | −1 |
| Fish comm'ty impaired | 0.185 | 0.416 | 125% | 0.324 | 75% | 0.423 | 129% | 0.330 | 0.784 | 4 |
| inv. comm'ty impaired | 0.111 | 0.293 | 163% | 0.189 | 70% | 0.291 | 161% | 0.196 | 0.758 | 2 |

Current climate scenario coupled with current land use scenario provides baseline conditions. %Δ = percent change from baseline. Summaries are over a modeled 25-year period (nominally 1980–2005), Land use scenarios as described in the text and Fig. 1 (Current = 1998, BAU = end of century with business as usual land management; RUS = end of century with reduced urban sprawl rate). Smallmouth bass = *Micropterus dolomieui*; brook trout = *Salvelinus fontinalus*; brown trout = *Salmo trutta*; rainbow trout = stream-resident *Oncorhynchus mykiss*; steelhead = adfluvial *O.mykiss*; Chinook = *O. tshawytscha*; coho = *Oncorynchus kisutch*; walleye = all *Sander vitreus*; GLwalleye = Lake Michigan/Muskegon Lake resident *Sander vitreus*; northern pike = *Esox lucious*; adfluv. Salmonines = all spp of *Oncorhynchus* spp.

all cases, biological responses to the climate change scenario were less severe under the RUS than the BAU land management scenario which was associated with more severely reduced forest and agriculture. The BUA land use provided some thermal buffering due to increases in groundwater flows. This effect was, however, quite modest (preserving 6–9 km of otherwise lost main stem habitat).

Results of the dynamic weighted useable area modeling (DWUA) in the lower mainstem river were similar to the CART modeling results but did show some striking differences between juvenile and adult habitat. Useable habitat for juvenile and adults life stages of smallmouth bass, walleye, Chinook salmon and steelhead declined from 10% to 34% under the BAU scenario, and slightly less (0–33%) under the RUS (Table 6). In contrast, under the climate change scenarios habitat for warmwater and coolwater fishes was dramatically increased (+124 to 252% for smallmouth bass; 41–427% for walleye) and decreased for adult stages of coldwater species (Chinook salmon, steelhead) and juvenile steelhead. Useable habitat for juvenile Chinook salmon was predicted to increase under climate change due to an increase in spring temperatures (Table 6).

## Discussion

MREMS scenario simulations indicated that climate change impacts on the Muskegon river ecosystem

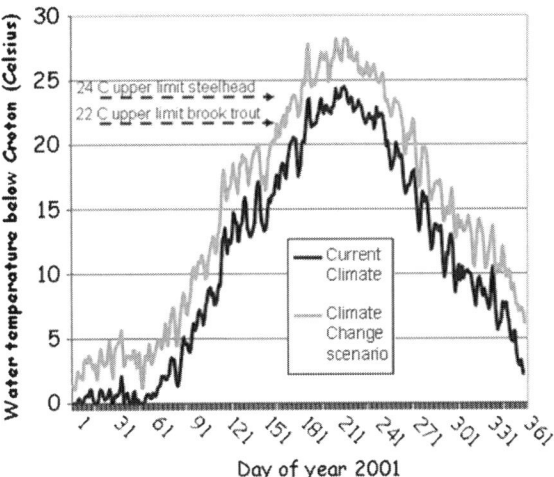

**Fig. 3** The IPCC A1B-modified scenario raised groundwater and temperatures in our models by several degrees shifting water temperatures in many channel segments across critical thermal thresholds for coldwater species. The example here is for channel unit VSEC 18, a critical spawning reach for adfluvial steelhead trout below a major hydropower dam in the lower part of the Muskegon main stem. With late summer temperatures currently too high for brook and marginal for steelhead trout due to reservoir warming, climate change scenarios pushed summer temperatures above levels tolerated by either species and removed much of the lower mainstem as useable summer habitat

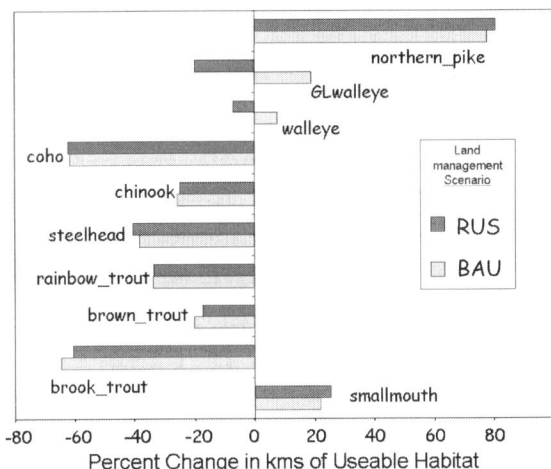

**Fig. 4** Basin-wide changes in habitat availability (kms of channel) for key sport fishes based on the BAU and RUS × IPCC A1B adjusted climate scenarios

will likely be pervasive, but also variable depending upon future land use trajectories. Significant reductions in negative impacts on hydrology, water quality,

and biological communities were associated with the highly forested future landscape produced in the RUS management scenario. This result suggests that traditional watershed management tools like land use planning could play an important role in developing climate adaptation strategies of the kind being promoted by IPCC and national governments (NRC, 2007; IPCC, 2007). However, even with the clearly preferable outcomes associated with the RUS scenario, the scale of ecological change projected was dramatic. Under the best-case RUS scenario, 57% of the Muskegon channel system would be destabilized by the end of this century; the BAU scenario modeling estimate was 76%. The same increase in river flows that lead to large-scale destabilization also drove 20–30% increases in sediment and nutrient loading (40–60% for the BAU scenario). The magnitude of these responses reflects the potency of climate impacts on river hydrology. Discharge rate is the principal organizing variable in fluvial systems (Schumm, 1977; Knighton, 1984); in rivers, ecological responses to future climate change will be necessarily linked to hydrologic response, and to the local details of basin geography, land use, and surficial geology that control hydrologic routing.

The Muskegon River, and most of the upper Great Lakes Basin, lies just south of the snow fall dominated northern latitudinal zone (50–80 N) where most GCMs predict increasing rainfall and net gains in annual river runoff (Palmer et al., 2008; Arnell, 2005; Nohara et al., 2006). The Great Lakes themselves complicate climate forecasting in this region and local anomalies associated with lake-effect dynamics are poorly represented in currently available GCMs (Lofgren et al., 2002; Croley, 2005). What is already clear is that the Muskegon basin has historically been strongly influenced by lake-effect precipitation (Kendall & Hyndman, in review), and that long-term gauging records for the region indicate increasing trends in rainfall and river discharge since the turn of the last century (Arnet, 2005; Dore, 2005). Relatively high annual rates of precipitation (up to 93 cm) on a landscape dominated by highly permeable glacial drift supports efficient recharge dynamics (Holtschlag, 1994; Boutt et al., 2001; Hyndman et al., 2007), high base flows and a characteristically cold and cool-water ecology for which the rivers of the northern lower Michigan Peninsula are well known (Wiley et al., 1997; Seelbach et al., 2006).

Table 6 Summary of dynamic weighted usable area habitat model (DWUA) and individual-based model (IBM) of sportfish in the lower Muskegon River, below Croton Dam

| Climate scenario: | Current | Current | | | | IPCC A1B adjusted | | | |
|---|---|---|---|---|---|---|---|---|---|
| Land use scenario: | Current | BAU | | RUS | | BAU | | RUS | |
| Taxa/life stage: | (ha) | (ha) | (%Δ) | (ha) | (%Δ) | (ha) | (%Δ) | (ha) | (%Δ) |
| Juvenile WUA | | | | | | | | | |
| Smallmouth | 4.10 | 3.55 | −14% | 3.88 | −5% | 7.96 | 94% | 8.88 | 116% |
| Chinook | 1.44 | 0.95 | −34% | 0.97 | −33% | 1.34 | −07% | 1.43 | −01% |
| Steelhead | 6.97 | 7.69 | 10% | 6.93 | −1% | 3.80 | −45% | 3.27 | −53% |
| Adult WUA | | | | | | | | | |
| Smallmouth | 1.09 | 0.89 | −18% | 1.00 | −9% | 3.14 | 188% | 3.62 | 232% |
| Chinook | 4.74 | 5.36 | 13% | 4.35 | −8% | 1.35 | −72% | 1.16 | −76% |
| Steelhead | 51.93 | 48.95 | −6% | 52.07 | 0% | 45.70 | −12% | 44.4 | −15% |
| Walleye | 0.33 | 0.28 | −17% | 0.27 | −18% | 1.31 | 336% | 1.44 | 336% |

Current climate scenario coupled with current land use scenario provides baseline conditions. %Δ = percent change from baseline. Summaries are over a modeled 25-year period (nominally 1980–2005). Land use scenarios as described in the text and Fig. 1 (Current = 1998, BAU = end of century with business as usual land management; RUS = end of century with reduced urban sprawl rate). Smallmouth bass = *Micropterus dolomieui*; steelhead *Oncorhynchus mykiss*; Chinook salmon = *O. tshawytscha*; walleye = Lake Michigan/Muskegon Lake resident *Sander vitreus*. The steelhead IBM outputs are in number (No.) and mean length (mm TL) of age-0 individuals surviving to Oct 31 in a model year

In this largely groundwater driven river, our hydrologic models predicted both increasing groundwater and increasing runoff deliveries to the channel system under the IPCC A1B scenario. Other modeling studies for the region have variably reported annual mean flow increases (Lofgren, 2004, Palmer et al., 2008), decreases (Croley, 2005), or both depending on specifics of the climate change modeling (Lofgren et al., 2002). Uncertainties associated with variously designed GCMs, parameterizations particularly in the GL region remain large at the present time (Kling, et al., 2003; Palmer et al., 2008; Lofgren, 2006). Our purpose in this analysis was not to explore implications of that climate uncertainty but instead to examine in more local detail the implications of a typical climate change scenario on the local river ecosystem characteristics. The hydrologic responses we observed from our models (~ 20–30% increases) were roughly consistent with the regionally comparable analyses of Palmer et al., (2008) and Lofgren (2004) who estimated river flow increases in this region of up to 0-40%, and 15%, respectively. What our analysis adds is the clear implication that (1) land use patterns can modulate hydrologic and related loading responses to climate change, and (2) that, as a result, the ecological consequences of climate change in river systems will be inextricably linked to ongoing decisions about the management of landscapes and fisheries, as well as dams (Palmer et al., 2008) and consumptive water use (Croley & Luukkonen, 2003).

### Sources of variation and uncertainty

Forecasting ecosystem responses to future change is necessarily risky business. Uncertainty about the magnitude and even direction of modeled system response to land use or climate change could arise from errors in model representation (specification error, parameterization error, or computational error) and from patterns in propagation of those errors including their statistical inflation through the series of sequential estimations inherent in a multi-modeling system. Furthermore, there is substantial "real" spatial and temporal variability in the dynamic responses of river systems which reflects underlying geographic differences in geology, land cover, drainage history, antecedent moisture conditions, etc.; these too add substantial uncertainty to any overall interpretation of system response. We have not to date carried out any formal study of error propagation characteristics in the MREMS system. Validation studies comparing predicted and observed values from the 2002–2004 field seasons generally (as in Fig. 2) indicate the system does an adequate job (correlations typically > 0.9 for physical parameters, >0.6 for biological parameters) representing spatial variability over the watershed. A detailed comparison of MREMS hydrologic variants with a Muskegon River implementation NOAA's DLBRM (Kao et al. unpublished; for model description see Croley, 2005) showed that the MREM' "mixed" model variants more accurately captured long-term average flows and basin water-balance characteristics, but were more prone to peak flow timing errors than the NOAA model which was heavily calibrated to the downstream-most gauge. Peak arrival delays in the lower river of 1–2 days are not uncommon in our hydrology models; particularly during winter months when the different snow-melt modules in our simulation variants greatly affect flow timing. We know also that all of the Muskegon River hydrologic models have periodic large errors at the downstream-most sites associated with Lake Michigan seiche activity, and with flooding due to high flow constraints imposed by bridges and channel engineering. These result in occasional large but temporary backwater effects near the mouth that reduce river discharge rates and increase local wetland storage. Inclusion of a hydrodynamic model at the estuary mouth will be required to correct this problem.

The ensemble of hydrologic model variants does provide some sense of the forecast variability in MREMS although we are still in the process of completing more detailed analyses. Ensemble mean and variance plots for the lower river (Fig. 5) suggest that overall the hydrologic models are in good agreement that the impact of climate on flow will be consistently larger than that of land use. Inter-ensemble variation in predicted magnitude of the land use response reflects the fact that the "mixed" variant models (including the $m \times 4$ variant used throughout this paper) were consistently more responsive to land use change than the variants with ILHM generated runoff (Fig. 2b in contrast to Fig. 2c). There was also a large spatial variability (see Table 3) across the watershed in the direction and magnitude of response, so site-specific behavior can deviate substantially

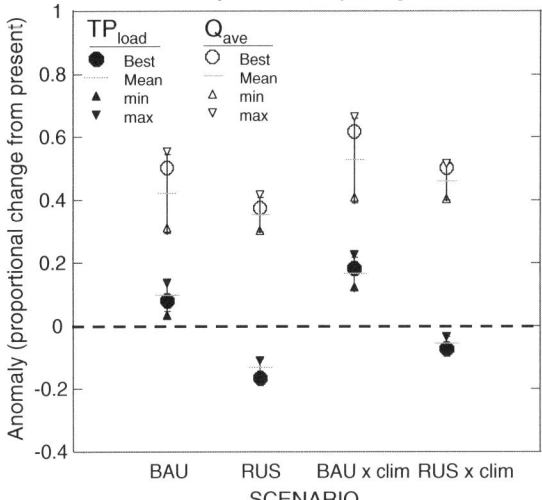

**Fig. 5** Hydrologic ensemble variation in anomaly forecasts for river reach VSEC 18. Results from four hydrologic simulation variants (see "Methods" section) were pooled to estimate mean (±)1 standard deviation responses for MREMS forecasts of average daily discharge (*open symbols*) and average total phosphorus concentration (*closed symbols*). Range in response represented as minimum and maximum values. The large circle symbols represent the forecast from our "m×4" variant (see Fig. 2c) which had the best overall performance in hydrologic validation tests and was used to estimate ecological responses reported in Tables 2–6

from downstream integrated response, and from system average response.

So far in our analysis it appears that variability in flow estimates due to (hydrologic) model choice is much smaller than spatial variability across the watershed; coefficients of variation between models ranging from 4% to 9% (across scenarios) compared to 227–288% between sites. When comparing uncertainty in estimates due to hydrologic model choice further down the computational chain of the multi-model (e.g., TP load in Fig. 5) coefficients of variation change little; ranging from 4% to 11% and showing no evidence of problematic error propagation. As this multi-modeling system matures, we hope to add ensemble modeling capability to more levels of the multi-modeling system. We currently do have multiple model representations for certain sport fishes in the biological ensemble but have not yet tried to analyze their coherence or variability. In the case of these biological models, differences in life history representation and habitat scale substantially complicate comparisons.

## Climate change and biological assessment

Linking hydrologic and loading model output to CART and MLR-based biological models led to forecasts of significant changes for a number of regionally important fish species. The Muskegon River is the single largest natural source of Chinook salmon reproduction in Lake Michigan and supports a regionally well known and economically valuable multi-species riverine sport fishery (O'Neal, 1997). In our simulations, cold water fishes in general were particularly sensitive to the climate change scenario and adfluvial salmonines most notably so. These model responses were driven principally by changes in summer water temperature (Fig. 2) which in this river system are controlled by largely by groundwater temperatures. Fish community composition is particularly sensitive to temperature changes in the 18–23°C range (July mean temperature; Wehrly et al., 2003) with a critical upper thresholds for many cold water species laying between 22 and 24°C (Eaton & Scheller, 1996), While the climate change scenario produced an average 3°C rise in annual air temperature and an average 4°C rise in mean July water temperature, these changes occurred across an extremely sensitive physiological range for local fishes (Fig. 2). The result was that modeled habitat availability dramatically and disproportionately declined for some important species, and expanded for others (Fig. 3, Tables 5 and 6). It is interesting to note, however, the independently derived assessment models for both invertebrate and fish communities were much more conservative in response to climate change than to land use change. Their pattern of sensitivity was almost the reverse of the observed in the individual fish models. Taken together, these results could be interpreted as implying that species distributions and community composition may be dramatically altered, but the underlying biological integrity of the community (*sensu* Karr, 1991) may be relatively unchanged (Fig. 4).

Indeed, in our forecasts the biological assessment metrics were impacted much more severely by land use scenarios than by the climate change. This response is in general consistent with the water quality modeling results which forecast relatively stable and decreasing concentrations of TP and sediment (respectively) under the climate change forecasts, but large increases in concentrations of both with land use change. This occurred despite significant increases in total load transport under the climate change scenario, which in the modeling was offset by dilution from higher flows. The assessment models we used were based on regionally normalized assessments of community composition data which estimate reference condition statistically (Wiley et al., 2002; Riseng et al., 2006). It is possible that the underlying empirical models were insufficiently sensitive to proximate physical variables forecast by MREMS and overly influenced by basin land use metrics. The fish assessment model did, however, explicitly evaluate base flow yield changes. The invertebrate model evaluated both changes in base flow yields and in summer water temperature. Furthermore, same kind of statistical over-fitting concern could also apply to the CART habitat models' relatively high sensitivity to the climate change scenario. But in this case, completed runs of the more mechanistic simulation models (DWUA and IBM recruitment modeling) to date agree with and confirm their predictions.

Whether or not wholesale shifts in species distributions related to global warming should be interpreted as indicating ecological impairment, is in itself, an interesting question (cf. Parmesan, 2006). We should hope that assessment metrics designed to reflect proximal insults on ecological integrity might indeed be relatively insensitive to regional shifts in climate. Ecological assessments are traditionally referenced in terms of either historical condition or by minimally disturbed regional baseline conditions. In either case, wholesale change in climate which affects underlying basin hydrology and temperature regimes are likely to disrupt current empirical relationships between community composition and traditional watershed stressors associated with land use, pollution, and local hydraulic modification. Even if baseline conditions themselves are changing, we will still need assessment protocols and metrics to manage and protect river ecosystems. If our current assessment methodologies are not in fact relatively neutral to direct climate impacts then we risk a growing probability that river assessment studies will be swamped by the effects changing climate. If so, then they will become relatively useless for evaluating the effects of the more common (and less "global") ecological insults that originate in the watershed.

## Land use as an adaptation strategy

Comparisons of our modeling results from land use management scenarios with and without the climate change scenario clearly illustrate the extent to which the impacts of climate on rivers can be modified and filtered by landscape condition. Reductions in flow, sediment loading, nutrient loading, and even biological impacts were achieved in the RUS scenario relative to the BAU scenario, with the difference between the two often magnified under the climate change scenario. This occurred because land cover related differences in hydrologic routing and ET losses modified the consequences of changing climate signals. At a time when many are concerned over the lack of institutional preparation for the ecological impacts of climate change, linking climate change preparation to land use change may be a way to engage more traditional watershed managers and institutions. Land use planning is already a central, widely accepted, and in many areas institutionalized feature of both academic and practical watershed management. Linking land use management with climate change may provide both a new rationale and a new opportunity to take action on what we already know is often an effective river management strategy.

**Acknowledgments** This work was supported by the Great Lakes Fishery Trust (2001.71), the National Science Foundation (EAR-0233648), and the USEPA STAR Program (G2M104070). Any opinions, findings, and conclusions or recommendations expressed in this publication are those of the authors and do not necessarily reflect the views of our granting agencies. Our sincere thanks to all of the many folks who contributed work in the field, on the computer, and in meetings; especially: Gary Noble (Muskegon Watershed Assembly), Richard O'Neal (Michigan Department of Natural Resources), Julie Metty (Great Lakes Fishery Trust), Jack Bails (Great Lakes Fishery Trust), Matt Ladewig, Jonah Duckles, Deepak Ray, Shaw Lacy, Solomon David, Yu-Chun Koa, Lori Ivan, Beth Sparks-Jackson, Kyung Seo Park, and Kurt Thompson.

## References

Andresen, J. S., 2007. Automated Weather Observing System (AWOS) Climate Data. Archived from the National Weather Service.

Arnell, N. W., 2005. Implications of climate change for freshwater inflows to the Arctic Ocean. Journal of Geophysical Research - Atmospheres, 110, (D07 105). doi:10.1029/2004 JD005348.

Baker, M. E., M. J. Wiley & P. W. Seelbach, 2001. GIS-based hydrologic modeling of riparian areas: implications for stream water quality. Journal of the American Water Resources Association 37(6): 1615–1628.

Benson, M. A. & D. M. Thomas, 1966. A Definition of Dominant Discharge. Bulletin of the International Association of Science and Hydrology. U.S. Geological Survey, Washington, D.C. [available on Internet at http://iahs.info/hsj/112/112007.pdf].

Boutt, D. F., D. W. Hyndman, B. C. Pijanowski & D. T. Long, 2001. Modeling impacts of land use on groundwater and surface water quality. Ground Water 39: 24–34.

Cheng, S. & M. J. Wiley, 2007. An Assessment of the Potential Thermal Impacts of Flow Reduction in Cedar Creek, Michigan. A Report to the Michigan Groundwater Conservation Advisory Council. University of Michigan, Ann Arbor, MI.

Christensen, N. & D. P. Lettenmaier, 2006. A multimodel ensemble approach to assessment of climate change impacts on the hydrology and water resources of the Colorado River basin. Hydrology and Earth System Sciences Discussions 3: 3727–3770.

Croley II, T. E., 2005. Using climate predictions in Great Lakes hydrologic forecasts. In Garbrecht, J. (ed.), Climate Variability, Climate Change, and Water Resources Management. American Society of Civil Engineers, Arlington, VA: 166–187

Croley II, T. E. & C. L. Luukkonen, 2003. Potential effects of climate change on ground water in Lansing, Michigan. Journal of the American Water Resources Association 39(1): 149–163.

De'ath, G. & K. E. Fabricius, 2000. Classification and regression trees: a powerful yet simple technique for the analysis of complex ecological data. Ecology 81: 3178–3192.

Dinse, K., J. Read & D. Scavia, 2009. Preparing for Climate Change in the Great Lakes Region. Michigan Sea Grant, Ann Arbor, MI.

Eaton, J. G. & R. M. Scheller, 1996. Effects of climate warming on fish thermal habitat in streams of the United States. Limnology and Oceanography 41(5): 1109–1115.

Fulton, R., J. Breidenbach, D.-J. Seo & D. Miller, 1998. WSR-88D rainfall algorithm. Weather Forecasting 13: 377–395.

Gouraud, V., J. L. Bagliniere, P. Baran, C. Sabaton, P. Lim & D. Ombredane, 2001. Factors regulating brown trout populations in two French rivers: application of a dynamic population model. Regulated Rivers: Research & Management 17: 557–569.

Harbaugh, A. W., E. R. Banta, M. C. Hill & M. G. McDonald, 2000. MODFLOW-2000. The U.S. Geological Survey Modular Ground-Water Model—User Guide to Modularization Concepts and the Ground-Water Flow Process. U.S. Geological Survey Open-File Report 00-92: 121 p.

Hauer, F. R. & R. D. Smith, 1998. The hydrogeomorphic approach to functional assessment of riparian wetland: evaluating impacts and mitigation on river floodplains in the U.S.A. Freshwater Biology 40: 517–530.

Higgins, J., M. Lammert & M. Bryer, 1999. Including Aquatic Targets in Ecoregional Portfolios: Guidance for Ecoregional Planning Teams. Designing a Geography of Hope Update #6. The Nature Conservancy, Arlington, VA.

Hyndman, D. W., A. D. Kendall & N. R. H. Welty, 2007. Evaluating Temporal and Spatial Variations in Recharge and Streamflow Using the Integrated Landscape Hydrology Model (ILHM), AGU Monograph Data Integration in Subsurface Hydrology. American Geophysical Union: 121–142. doi:10.1029/170GM01.

IPCC, 2007. Climate change 2007: synthesis report. In Pachauri, R. K. & A. Reisinger (eds), Contribution of Working Groups I, II and III to the Fourth Assessment Report of the Intergovernmental Panel on Climate Change. Core Writing Team, Geneva: 104 pp.

Jayawickreme, D. H. & D. W. Hyndman, 2007. Evaluating the influence of land cover on seasonal water budgets using next generation radar (NEXRAD) rainfall and streamflow data. Water Resources Research 43: W02408.

Johansen, T. A. & R. Murray-Smith, 1997. The operating regime approach to non-linear modelling and control. Chapter one. In Murray-Smith, R. & T. A. Johansen (eds), Multiple Model Approaches to Modeling and Control. Taylor Francis, Bristol, PA.

Kao, Y., S. Adlerstein, C. Delucchi & M. Wiley, 2008. Unpublished. DLBRM and MREMS: a comparison and evaluation. Report to the University of Michigan.

Karr, J. R., 1991. Biological integrity: a long-neglected aspect of water resource management. Ecological Applications 1(1): 66–84.

Kendall, A. D. & D. W. Hyndman, in review. Simulating the spatial and temporal variability of regional evapotranspiration and groundwater recharge: influences of land use, soils, and lake-effect climate. Water Resources Research.

Kling, G. W., K. Hayhoe, L. B. Johnson, J. J. Magnuson, S. Polasky, S. K. Robinson, B. J. Shuter, M. M. Wander, D. J. Wuebbles, D. R. Zak, R. L. Lindroth, S. C. Moser & M. L. Wilson, 2003. Confronting Climate Change in the Great Lakes Region: Impacts on Our Communities and Ecosystems. Union of Concerned Scientists, Cambridge, MA.

Knighton, D., 1984. Fluvial Forms and Processes. Edward Arnold, Baltimore, MD.

Ladewig, M. D., 2006. Sediment Transport Rates in the Lower Muskegon River and Its Tributaries. M.Sc. Thesis. University of Michigan, Ann Arbor, MI [available on Internet at http://deepblue.lib.umich.edu/].

Lofgren, B. M., 2004. Global warming effects on Great Lakes water: more precipitation but less water? In Proceedings 18th Conference on Hydrology, 84th Annual Meeting of the American Meteorological Society, Seattle, WA, January 11–15, 2004. American Meteorological Society: 3 pp.

Lofgren, B. M., 2006. Land Surface Roughness Effects on Lake Effect Precipitation. J. Great Lakes Res. 32: 839–851.

Lofgren, B. M., F. H. Quinn, A. H. Cites, R. A. Assel, A. J. Eberhardt & C. L. Luukkonen, 2002. Evaluation of potential impacts on Great Lakes water resources based on climate scenarios of two GCMs. Journal of Great Lakes Research 28(4): 537–554.

MAWN, 2008. Michigan Automated Weather Network Hourly Climate Data [available on Internet at http://www.agweather.geo.msu.edu/mawn].

Maxwell, J. R., C. J. Edwards, M. E. Jensen, S. J. Paustian, H. Parrott & D. M. Hill, 1995. A hierarchical framework of aquatic ecological units in North America (Nearctic Zone). USDA Forest Service, North-Central Forest Experiment Station, General Technical Report NC-176, St. Paul, MN.

Meehl, G. A., C. Covey, T. Delworth, M. Latif, B. McAvaney, J. F. B. Mitchell, R. J. Stouffer & K. E. Taylor, 2007. The WCRP CMIP3 multi-model dataset: a new era in climate change research. Bulletin of the American Meteorological Society 88: 1383–1394.

Moore, N., N. Torbick, B. Pijanowski, B. Lofgren, J. Wang, J. Andresen, D. Kim & J. Olson, 2009. Adapting MODIS-derived LAI and fractional cover into the Regional Atmospheric Modeling System (RAMS) in East Africa. International Journal of Climatology [available on Internet at http://www.interscience.wiley.com]. doi: 10.1002/joc.2011.

Murphy, J., D. Sexton, D. Barnett, G. Jones, M. Webb, M. Collins & D. Stainforth, 2004. Quantification of modeling uncertainties in a large ensemble of climate change simulations. Nature 430: 768–772.

Nakicenovic, N., et al., 2000. Special Report on Emissions Scenarios: A Special Report of Working Group III of the Intergovernmental Panel on Climate Change. Cambridge University Press, Cambridge, UK: 599 pp.

NCDC Hourly, 2008. National Climatic Data Center hourly precipitation gauge data [available on Internet at http://www.ncdc.noaa.gov/].

Nelson, K. C., M. A. Palmer, J. E. Pizzuto, G. E. Moglen, P. L. Angermeier, R. H. Hilderbrand, M. Dettinger & K. Hayhoe, 2009. Forecasting the combined effects of urbanization and climate change on stream ecosystems: From impacts to management options. Journal of Applied Ecology 46: 154–163.

Nohara, D., A. Kitoh, M. Hosaka, T. Oki, 2006. Impact of climate change on river discharge projected by multi-model ensemble. Journal of Hydrometeorology 7(5): 1076–1089.

NRC, 2007. Analysis of Global Change Assessments: Lessons Learned. Committee on the Analysis of Global Change Assessments. National Academy Press, Washington, D.C.

O'Neal, R. P., 1997. Muskegon River Watershed Assessment. Michigan Department of Natural Resources Fisheries Division, Lansing, MI. Fisheries Special Report Number 19.

Palmer, M. A., C. A. Reidy Liermann, C. Nilsson, M. Flörke, J. Alcamo, P. S. Lake & N. Bond, 2008. Climate change and the world's river basins: Anticipating management options. Frontiers in Ecology and the Environment: Vol. 6: 81–89.

Parmesan, C., 2006. Evolutionary and ecological responses to climate change. Annual Review of Ecology, Evolution and Systematics 37: 637–669.

Pijanowski, B. C., S. H. Gage, D. T. Long & W. C. Cooper, 2000. A land transformation model: integrating policy, socioeconomics and environmental drivers using a geographic information system. In Harris L. & J. Sanderson (eds.), Landscape Ecology: A Top Down Approach. Lewis Publishers, Boca Raton.

Pijanowski, B. C., D. G. Brown, B. Manik & B. Shellito, 2002a. Using artificial neural networks and GIS to forecast land use changes: A land transformation model. Computers, Environment and Urban Systems. 26: 553–575.

Pijanowski, B. C., B. Shellito & S. Pithadia, 2002b. Using artificial neural networks, geographic information systems and remote sensing to model urban sprawl in coastal watersheds along eastern Lake Michigan. Lakes and Reservoirs 7: 271–285.

Pijanowski, B., S. Pithadia, K. Alexandridis & B. Shellito, 2005. Forecasting large-scale land use change with GIS and neural networks. International Journal of Geographic Information Science 19: 197–215.

Pijanowski, B., K. Alexandridis & D. Mueller, 2006. Modeling urbanization in two diverse regions of the world. Journal of Land Use Science 1: 83–108.

Pijanowski, B., D. K. Ray, A. D. Kendall, J. M. Duckles & D. W. Hyndman, 2007. Using backcast land-use change and groundwater travel-time models to generate land-use legacy maps for watershed management. Ecology and Society 12: 25.

Pontius, R. Jr., W. Boersma, J. Castella, K. Clarke, T. de Nijs, C. Dietzel, Z. Duan, E. Fotsing, N. Goldstein, K. Kok, E. Koomen, C. Lippitt, W. McConnell, B. Pijanowski, S. Pithadia, A. Sood, S. Sweeney, T. Trung & P. Verburg, 2008. Comparing input, output and validation maps for several models of land change. Annals of Regional Science 42: 11–47.

Raleigh, R. F., T. Hickman, R. C. Solomon & P. C. Nelson, 1984. Habitat Suitability Index Models and Instream Flow Suitability Curves: Rainbow Trout. U.S. Fish and Wildlife Service Bervices Program FWS/OBS-82/10.60.

Riseng, C. M., M. J. Wiley, R. J. Stevenson, T. Zorn & P. W. Seelbach, 2006. Comparison of coarse versus fine scale sampling on statistical modeling of landscape effects and assessment of fish assemblages of the Muskegon River, Michigan In Landscape Influences on Stream Habitats and Biological Communities. American Fisheries Society Symposium 48. Bethesda, Maryland.

Santhi, C., R. Srinivasan, J. G. Arnold & J. R. Williams, 2005. A modeling approach to evaluate the impacts of water quality management plans implemented in a watershed in Texas. Environmental Modelling & Software. 21(2006): 1141–1157.

Schumm, S. A., 1977. The Fluvial System. Wiley, New York. 338 p.

Schwarz, G. E., R. A. Smith, R. B. Alexander, & J. R Gray, 2001. A spatially referenced regression model (SPARROW) for suspended sediment in streams of the conterminous U.S., in U.S. Subcommittee on Sedimentation. In Proceedings of the Seventh Federal Interagency Sedimentation Conference, March 25–29, 2001, Reno, Nevada, USA: VII-80-7.

Seelbach, P. W. & M. J. Wiley, 1997. Overview of the Michigan Rivers Inventory (MRI) Project. Fisheries Technical Report No. 97–3. Michigan Department of Natural Resources, Ann Arbor, MI: 31 pp.

Seelbach, P. W., M. J. Wiley, P. A. Soranno & M. T. Bremigan, 2002. Aquatic conservation planning: using landscape maps to predict ecological reference conditions for specific waters. Chapter 24. In Gutzwiller, K. (ed.), Concepts and Applications of Landscape Ecology in Biological Conservation. Springer-Verlag, New York, NY.

Seelbach, P. W. & M. J. Wiley, 2005. Landscape-based modeling as the basis for a prototype information system for ecological assessment of Lake Michigan tributaries. In Edsall, T. & M. Munawar (eds), State of Lake Michigan: Ecology, Health, and Management. Ecovision World Monograph Series. Aquatic Ecosystem Health and Management Society.

Seelbach, P. W., M. J. Wiley, J. C. Kotanchik & M. E. Baker, 1997. A Landscape-Based Ecological Classification System for River Valley Segments in Lower Michigan (MI-VSEC version 1.0). Fisheries Research Report 2036, Department of Natural Resources, Ann Arbor, MI.

Seelbach, P. W., M. J. Wiley, M. E. Baker, & K. E. Werhly, 2006. Initial classification of river valley segments across Michigan's Lower Peninsula. In Landscape Influences on Stream Habitats and Biological Communities. American Fisheries Society Symposium 48, Bethesda, MD: 25–48

Steen, P. J., D. R. Passino-Reader, & M. J. Wiley, 2006. Modeling brook trout presence and absence from landscape variables using four different analytical methods. In Landscape Influences on Stream Habitats and Biological Communities. American Fisheries Society Symposium 48, Bethesda, MD.

Steen P. J., M. J. Wiley & J. S. Schaeffer, 2010. Predicting future change in Muskegon River watershed game fish distributions under future land cover alteration and climate change scenarios. Transactions of the American Fisheries Society 139: 396–412.

Stevenson, R. J., S. T. Rier, C. M. Riseng, R. E. Schultz & M. J. Wiley, 2006. Comparing effects of nutrients on algal biomass in streams in two regions with different disturbance regimes and with applications for developing nutrient criteria. Hydrobiologia 561: 149–165.

Stevenson, R. J., M. J. Wiley, V. L. Lougheed, C. Riseng, S. H. Gage, J. Qi, D. T. Long, D. W. Hyndman, B. C. Pijanowski & R. A. Hough, 2008. Chapter 19: watershed science: essential, complex, multidisciplinary and collaboratory. In Ji, W. (ed.), Wetland and Water Resource Modeling and Assessment: A Watershed Perspective. Taylor & Francis, London.

Tyler, J. A. & E. S. Rutherford, 2007. River restoration effects on steelhead populations in the Manistee River, Michigan: analysis using an individual-based model. Transactions of the American Fisheries Society 136: 1654–1673.

U.S. Army Corps of Engineers Hydrologic Engineering Center (USACE-HEC), 1998. HEC-HMS Hydrologic Modeling System User's Manual. USACE-HEC, Davis, CA.

USACE. 2009a. [available on Internet at http://www.hec.usace.army.mil/software/hec-hms/].

USACE, 2009b. [available on Internet at http://www.hec.usace.army.mil/software/hec-ras/].

USFWS [US Fish and Wildlife Service], 1980. Habitat Evaluation Procedures: Using Habitat as a Basis for Environmental Impact Assessment. Washington (DC): Division of Ecological Services, USFWS. (11 Nov 2008; [available on internet at http://www.fws.gov/policy/870FW1.html]).

Wehrly, K. E., M. J. Wiley & P. W. Seelbach, 2003. Classifying Regional Variation in thermal regime based on stream fish community patterns. Transactions of the American Fisheries Society 132: 18–32.

Wiley, M. J., S. L. Kohler & P. W. Seelbach, 1997. Reconciling landscape and site based views of aquatic stream communities. Freshwater Biology 37: 133–148.

Wiley, M. J., P. W. Seelbach, K. Wehrly & J. Martin, 2002. Regional ecological normalization using linear models: A meta-method for scaling stream assessment indicators. Chapter 12. In Simon, T. P. (ed.), Biological Response signatures: indicator patterns using aquatic communities. CRC Press, Lewis Publishers Inc., Boca Raton.

Wiley, M. J., B. C. Pijanowski, P. Richards, C. Riseng, D. Hyndman, P. Seelbach & R. J. Stevenson, 2004. Combining valley segment classification with neural net modeling of landscape change: A new approach to integrated risk assessment for river ecosystems. In Proceedings of WEF 2004 Specialty Conference Series: Watershed 2004. Water Environment Federation, Dearborn, MI.

Wiley, M., B. Pijanowski, R. J. Stevenson, P. Seelbach, P. Richards, C. Riseng, D. Hyndman & J. Koches, 2008. Integrated modeling of the Muskegon River: Tools for ecological risk assessment in a Great Lakes watershed. Chapter 20. In Ji, W. (ed.), Wetland and Water Resource Modeling and Assessment: A Watershed Perspective. Taylor & Francis, London.

Yang, K. & T. Koike, 2005. A general model to estimate hourly and daily solar radiation for hydrological studies. Water Resources Research 41: W10403.

Zorn, T. G., P. W. Seelbach & M. J. Wiley, 2002. Distributions of stream fishes and their relationship to stream size and hydrology in Michigan's Lower Peninsula. Transactions of the American Fisheries Society 131: 70–85.

GLOBAL CHANGE AND RIVER ECOSYSTEMS

# Implications of global change for the maintenance of water quality and ecological integrity in the context of current water laws and environmental policies

Anna T. Hamilton · Michael T. Barbour · Britta G. Bierwagen

Received: 22 September 2009 / Accepted: 12 June 2010 / Published online: 11 July 2010
© US Government: US Environmental Protection Agency 2010

Guest editors: R. J. Stevenson, S. Sabater / Global Change and River ecosystems – Implications for Structure, Function and Ecosystem Services

A. T. Hamilton (✉) · M. T. Barbour
Center for Ecological Sciences, Tetra Tech, Inc.,
Santa Fe, NM, USA
e-mail: Anna.Hamilton@tetratech.com

M. T. Barbour
e-mail: Michael.Barbour@tetratech.com

B. G. Bierwagen
Global Change Research Program, National Center for Environmental Assessment, Office of Research and Development, U.S. Environmental Protection Agency, Washington, DC, USA
e-mail: Bierwagen.Britta@epa.gov

**Abstract** There is both a fundamental and applied need to define expectations of changes in aquatic ecosystems due to global changes. It is clear that programs using biological indicators and reference-based comparisons as the foundation for assessments are likely to make increasingly erroneous decisions if the impacts of global change are ignored. Global changes influence all aspects of water resource management decisions based on comparisons to reference conditions with impacts making it increasingly problematic to find an "undisturbed" water body to define acceptable conditions of ecological integrity. Using a more objective scale for characterizing reference conditions that is anchored in expectations for what would be attainable under undisturbed conditions, such as the Biological Condition Gradient (BCG) is one approach that maintains consistent definitions for ecosystem conditions. In addition, protection of reference stations and of unique or undisturbed aquatic resources is imperative, though the scope of protection options is limited. Projections indicate that encroaching land use will affect 36–48% of current reference surface waters by the year 2100. The interpretation of biological indicators is also at risk from global changes. Distinguishing taxonomic attributes based on temperature or hydrologic preferences can be used to enhance the ability to make inferences about global change effects compared to other stressors. Difficulties arise in categorizing unique indicators of global changes, because of similarities in some of the temperature and hydrologic effects resulting from climate change, land use changes, and water removal. In the quest for biological indicators that might be uniquely sensitive to one global stressor as an aid in recognizing probable causes of ecosystem damage, the potential similarities in indicator responses among global and landscape-scale changes needs to be recognized as a limiting factor. Many aspects of global changes are not tractable at the local to regional scales at which water quality regulations are typically managed. Our ability to implement water policies through bioassessment will require a shift in the scale of assessment, planning, and adaptations in order to fulfill our ultimate regulatory goals of preserving good water quality and ecological integrity. Providing clear expectations of effects due to global change for key species and communities in freshwater ecosystems will help water quality programs achieve their goals under changing environmental conditions.

**Keywords** Global change · Climate change · Water law · Environmental policy · Bioassessment

## The underpinnings of environmental protection of aquatic resources

The expression of water laws and environmental policies is to protect the health of aquatic ecosystems. The maintenance of natural ecological processes and ecosystem function is paramount to addressing this goal. This is reflected in the U.S. Clean Water Act (CWA; USGPO, 1972a, b) long-term goal of maintaining and restoring the ecological integrity of aquatic systems, defined as comprising physical, chemical and biological integrity (CWA). Similarly, the goal of the European Union Water Framework Directive (WFD), Directive, 2000), is to achieve "good status" of aquatic systems, including both "good ecological status" and "good chemical status" (see Noges et al., 2009).

In principle, assessment of status is based on a combination of biological, physical, and chemical indicators; however, there is recognition that biological indicators are an integrating response to all environmental conditions (Barbour et al., 2000; Moog & Chovanec, 2000). Based on this, in the U.S., biological assessment plays a central role in numerous water quality programs that are components of the CWA. Bioassessment data is used to assess water quality, identify biologically impaired waters, and develop National Water Quality Inventory reports. It is used to develop biocriteria and set aquatic life use categories, which represent different protection standards. Bioassessment data are used to determine whether conditions of the waterbody support designated uses, and if not, to develop total maximum daily load (TMDL) limitations for the pollutant(s) contributing to the impairment. Bioassessment results are used to help identify causes of observed impairments, based on the assumption that various components of aquatic communities will respond differently to different types of stressors. Bioassessment is used to determine the impacts of point-source discharges as well as of episodic spills, defining the extent of damage, responses to remediations, and supporting enforcement actions. Other CWA programs that depend on bioassessment data include permit evaluation and issuance, tracking responses to restoration actions, and other components of watershed management. In Europe, the WFD emphasizes biological assessment methods supported by the evaluation of hydromorphological and chemical parameters, definition of strategies against pollution (combined approach for point and diffuse sources), and establishment of river basin management plans (Chovanec et al., 2000).

Assessing the ecological health of rivers and streams is a fundamental and increasingly important water management issue worldwide (Bunn & Davies, 2000; Norris & Barbour, 2009) and relates directly to the preservation of ecosystem services of aquatic resources. The concept of ecosystem services embraces those processes by which the environment produces resources that we often take for granted, such as clean water, and other services, such as mitigation of drought and floods, the cycling and transport of nutrients, the maintenance of biodiversity, and detoxification and decomposition of waste (Daily, 1997; Postel & Carpenter, 1997). Efforts such as the UN Millennium Ecosystem Assessment (MEA, 2005) are raising public awareness that high quality aquatic resources provide society with ecological benefits and should be valued.

On a global scale, different countries are at varying levels of success in the endeavor of protection and restoration of their water resources (Jungwirth et al., 2000). Those that have had programs in place for many years have accomplished much in determining the status of their resources and what must be done to restore degrading waters (Barbour and Paul, 2010). However, few have considered how to effectively account for large scale, global changes, such as progressive climate change, increasing encroachment of developed and agricultural land uses, and increasing human demands for freshwater, all with ramifications to aquatic resources.

Water laws have historically focused more on point-source impacts than on landscape or larger scales of pollution or habitat degradation. During the twentieth century, the focus of environmental problems has changed from local or regional issues like sewage discharge in the first decennia toward global issues like climate change today (Verdonschot, 2000). This means that now, water resource management requires consideration of stressors that operate at very different scales, such as climate change, which operates on a global scale. It will alter air and water temperatures; alter flow and other hydrological parameters; reduce ice and snow cover; alter the timing of snowmelt; alter stratification regimes in lakes; increase sea levels; and increase salinity in some coastal areas (IPCC, 2007). Global changes in land use and water abstraction are landscape-scale stressors that are directly related to human population increases.

An increase in the spatial scale of stressors implies a need to increase the spatial and temporal scales of management and assessment (Verdonschot, 2000). This change in scale complicates the already complex management and restoration of impaired water resources. While ecosystem management will be most effective when all stressors impacting a resource are considered, certain aspects of global change impacts are not tractable for control at a local, watershed, or state agency scale.

The environmental research community is engaged in understanding the effects of global changes at a variety of spatial scales. Much of the research in the US on global change and its effects on aquatic ecosystems are being supported by federal agencies, such as the National Science Foundation (NSF), US Environmental Protection Agency (USEPA) and the National Oceanic and Atmospheric Administration (NOAA). The European Union (EU) supports a series of research projects being conducted by several institutions throughout Europe. The Australian government, through the Department of Climate Change the Australian Greenhouse Office, and the Australian Climate Change Science Program, also supports substantial research on climate change effects and adaptations. As the effects of global changes become more evident and attention is drawn to this complicating factor in maintaining good ecological status, scientists and environmental managers should work together to ensure that the goals of environmental protection are met. Research partnerships and information sharing may lead to more effective resource management and decision-making tools that account for global change effects.

## Global changes and ecosystem health

The concept of ecological health can be vague, but is generally borrowed from the human health/medical paradigm, such that a "healthy" ecosystem, i.e., one with "integrity", is conceptually one that is functioning within "normal" ranges. We assume this is true for pristine systems, but emphasize the question of whether an acceptable level of ecological integrity exists in systems with some degradation, often termed "minimally disturbed" or "leased disturbed" (Bailey et al., 2004; Stoddard et al., 2006). Conditions observed in "minimally" or "least" disturbed streams or rivers often become the basis for defining restoration or remediation goals, or for determining impairment, in similar systems that are impacted by human uses and development.

The effects of land use alterations on river ecosystem condition have been reasonably well studied (see for instance Helms et al., 2009; Allan 2004, Paul and Meyer 2001). However, taking a global view of the future impacts of progressive land use changes on ecosystem integrity is a more recent concern. There have been many studies of a variety of water use and hydrological alteration impacts on stream ecological condition, including flow regulation (e.g., Nilsson et al., 2005), increasing human uses (e.g., Bunn & Arthington, 2002), and hydrologic alteration in general (e.g., Konrad et al., 2008; Richter et al., 2003). Now there is recognition that global changes in climate combined with a growing world population and changes in global patterns of land use, socio-economics, and technical changes in water resource utilization and management will combine to increase water stress (i.e., decrease water availability and increase demand, Bates et al., 2008). Increased water stress will be particularly notable in regions where river runoff is projected to decrease in the future, as well as in regions where water resources are dependent on snow pack (e.g., the western U.S., the Mediterranean Basin (including southern Europe, Northern Africa, and southwestern Asia), southeastern Australia, southern Africa, and the west coast of South America) (Bates et al., 2008). Progressive increases in regional water stress will put increasing demands on freshwater in likely competition with ecological uses. This will bring conflicts in water use to a head, and put greater pressure on management to consider all sources of impacts to water resources (Brekke et al., 2009).

Increasing attention has been given to whether ecosystem responses to climate change are sufficient to cause concern and whether that concern should extend to our ability to preserve and restore ecological integrity. Freshwater ecosystems are considered sensitive to climate change impacts, owing to their fundamental dependence on hydrology and thermal regimes, their dominance by poikilotherms, and the risks of interactions with other stressors (Durance & Ormerod, 2007). However, documentation of aquatic biological responses to climate change on a basis that is meaningful to water quality and resource managers has been slow in coming, with much early attention focused on terrestrial ecosystems (e.g., Root et al., 2003; Thuiller, 2004; Walther et al., 2002, 2005; Parmesan, 2006; Tobin et al., 2008; Zuckerberg et al., 2009). Climate change effects in aquatic systems will manifest at all ecological levels (Fig. 1). Examples of shifts in aquatic community structure that are relevant in a bioassessment framework, documented in 10 relevant references, are presented in Table 1. Many of these effects will be species and region specific; therefore, there is substantial variation in aquatic biological responses that may impact the evaluation of ecological status. There will be potentially major consequences both for ecosystem function and for the interpretation of biomonitoring results relative to assessment of ecosystem health.

There are some similar and confounding effects among global changes that have ramifications to the implementation of water policy. For example, stream temperatures can be increased by climate change, and also by land uses (e.g., urbanization, deforestation) (Table 2). Base flows can be reduced by climate change, and also by water abstraction. Though the scale of these effects can be different, their similarities can make it difficult to separate climate change impacts from those of other landscape-scale stressors within management-relevant time frames. Climate change effects are also compounded by other large-scale climate drivers such as the North Atlantic

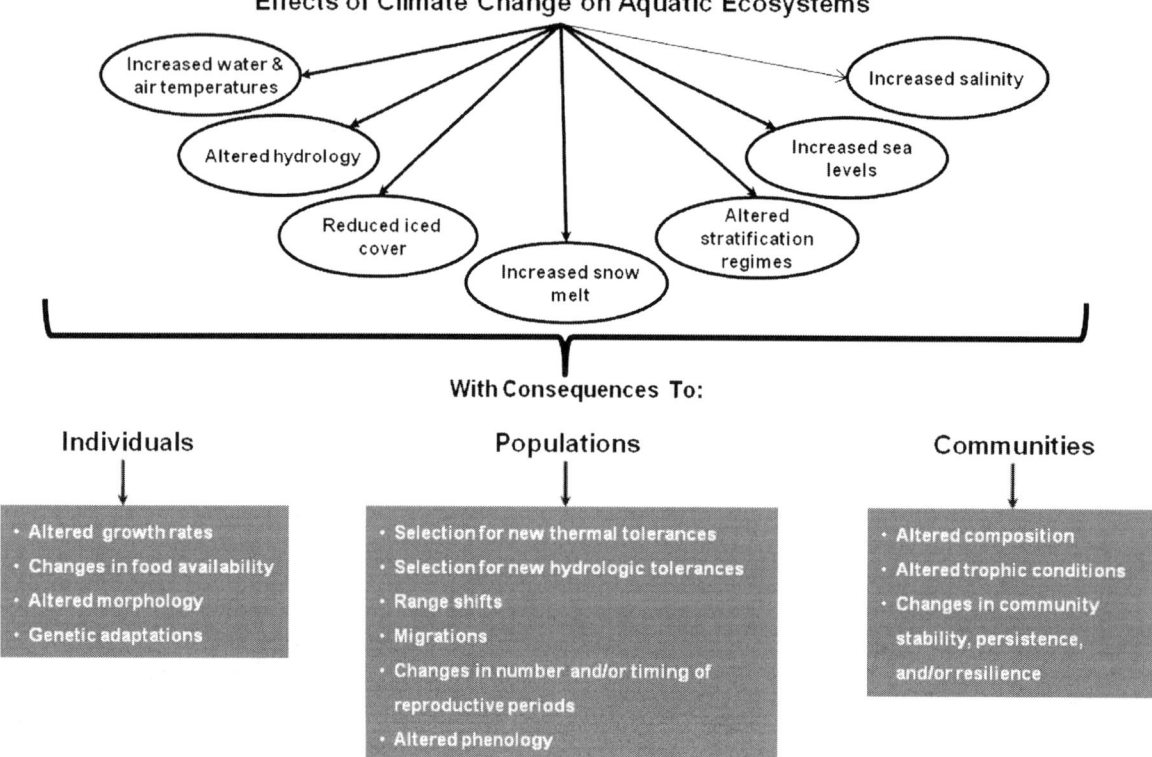

**Fig. 1** Illustration of climate change effects on aquatic ecosystems and the subsequent individual-, population- and community-level responses to these changes. Effective indicators are those that respond to the consequences of climate change

Table 1 Documented examples of shifts in aquatic community structure due to climate changes

| Examples of aquatic community changes | Reference |
| --- | --- |
| Increases in abundance, species richness, and proportion of southern and of warm-water species of fish in large rivers | Daufresne & Boet (2007) |
| Loss of cold-water fishes from headwater streams, but also extension of more tolerant, thermophilic fishes from larger streams and rivers into newly suitable habitat | Buisson et al. (2008) |
| Increases in fish species richness with increasing temperatures at higher latitudes | Hiddink & ter Hofstede (2008) |
| Displacement of upstream, cold-water invertebrate taxa with downsteam, warm-water taxa | Daufresne et al. (2004) |
| An increase in lentic and thermophilic invertebrates with increasing temperature | Doledec et al. (1996) |
| Reductions of spring abundance of dominant taxa, shifts in invertebrate assemblage composition from cooler to warmer water taxa, and possible losses (local extinctions) of more scarce taxa with increasing temperatures | Durance & Ormerod (2007) |
| Significant long-term trends related to the thermophily and rheophily of benthic taxa, with groups preferring cold waters and higher flows declining | Chessman (2009) |
| Changes in stability and persistence | Collier (2008) |
| Changes in species composition in lakes | Burgmer et al. (2007) |
| Changes in structure and diversity of riverine mollusk communities with reduction in community resilience during hot years | Mouthon & Daufresne (2006) |

Table 2 Examples of temperature and hydrologic effects resulting from climate change, land use changes, and water removal

| Effects on | Climate change | Land use change | Water abstraction/diversion |
| --- | --- | --- | --- |
| Water temperature | Increases on a global scale with regional differences, modified to some extent by local to regional variations in vulnerability; superimposed on year-to-year variations | Local to watershed scale increases due to altered land cover (e.g., runoff from impervious surfaces) and altered channel morphology; possible decreases due to reforestation, restoration of riparian cover | Local to watershed scale increases due to reduced flows, altered channel morphology; variable temperature effects related to dam releases |
| Hydrology | Regionally variable, often including reduced annual discharge, reduced summer flows and baseflow, increased flow variability, increased flooding episodes, increased flashiness; some streams changing from perennial to intermittent flows | Decreased infiltration, increased surface runoff, altered flood runoff patterns, higher peak flood flows, reduced groundwater recharge and reduced baseflow | Reduced discharge, flow displacement, altered timing and magnitude of peak and base flows, altered flow variability |

Oscillation (NAO) (Bradley & Ormerod, 2001). The presence of development and water abstraction or diversion infrastructure can be documented independently, e.g., by quantification of watershed land cover or water withdrawal data, which can help define contributing factors in a weight of evidence framework. However, climate change effects are pervasive, and if other global change influences are present, quantitative partitioning of biological responses among these factors will be problematic in many circumstances. In the quest for biological indicators that might be uniquely sensitive to one global stressor as an aid in recognizing probable causes of ecosystem damage, the potential similarities in indicator responses among global and landscape-scale changes should be recognized as a limiting factor.

## Susceptible aspects of assessment programs to global changes

In the framework of water laws and environmental policy aimed at protecting and restoring the condition of aquatic ecosystems, bioassessments result in information on the ecological health of a water body (Norris & Barbour, 2009). Ecological indicators serve as easily interpretable surrogates to gauge condition of an aquatic ecosystem (Niemi & McDonald, 2009). They can also be early warnings of degradation that link appropriately to the stressors expected in the system, which now include global changes in climate, land use, and water use.

The effects of global change on bioassessment programs will vary regionally. Land and water use effects are largely driven by locations of and projected future changes in major population and agricultural centers. Differences in climate change threats and responsiveness are instead driven by regional variability in climate, as well as regional differences in the vulnerability of aquatic ecosystems. Differences in regional climate and disturbance regimes are important contributors to species sensitivities to environmental changes (Helmuth et al., 2006). Many factors can influence susceptibility to changing water temperature or hydrologic regime due to climate change, such as elevation (Cereghino et al., 2003; Diaz et al., 2008; Chessman, 2009), stream order (Minshall et al., 1985; Cereghino et al., 2003), degree of ground water influence, or factors that affect water depth and flow rate, such as water withdrawals (Poff, 1997; Poff et al., 2006a; Chessman, 2009).

Essential components of the assessment and management of ecological health include the reference condition paradigm and the use of biological indicators. In a regulatory context, impairment represents a level of departure from defined reference conditions considered unacceptable for maintenance of ecological integrity. However, the continued good status of reference locations, and therefore, their use as a basis for comparison, is significantly threatened by global change. Both climate change effects and encroachment of development can be expected to impact streams that were previously categorized as minimally altered, probably more so than previously conceived or documented. We have found that in some examples from state biomonitoring data in the U.S. losses of cold-preference taxa and replacement by warm-preference taxa had the potential to degrade the condition classification of reference stations by a full level (e.g., from "excellent" to "good", or "good" to "fair"), as defined by the particular state bioassessment schema (Fig. 2) (USEPA, 2010).

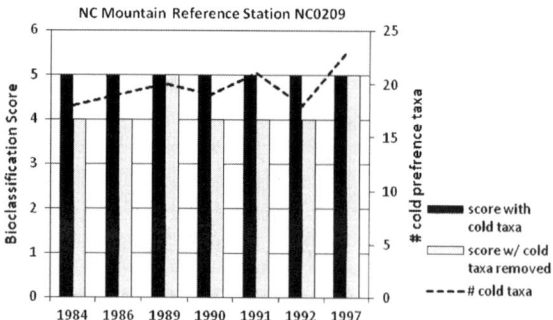

**Fig. 2** Final bioclassification (station quality) scores at a North Carolina reference site in the Blue Ridge Mountain ecoregion (NC0209) before and after all cold-preference taxa are eliminated from the observed benthic invertebrate community

Land use impacts on reference stations have long been considered a factor that can be controlled in the process of selecting sampling stations. However, finding undisturbed reference conditions is already a challenge in the U.S. and Europe (e.g., Herlihy et al., 2008; Noges et al., 2009), and often the only comparisons for assessment are to the "best of what is left", or least disturbed conditions (Stoddard et al., 2006). For state biomonitoring programs in the U.S., it is often the case that the extent of developed land uses affecting established reference locations is not quantitatively documented, nor updated over time. We have found that the influence of urban/suburban land uses on established reference locations can be greater than previously assumed. In one example, 20–25% of reference sites classified as high quality in Florida were actually surrounded by >20% urban/suburban land uses (Fig. 3). Furthermore, using spatially explicit projections of population growth, it was estimated that by the year 2100, from 36 to 48% of current reference locations could be compromised by encroachment of developed land uses (Fig. 3) (USEPA, 2010).

The combined effects of climate change and land use encroachment are thus reducing availability of reference locations for comparative analyses, and definition of the reference baseline will continue to shift toward more degraded conditions ("reference station drift", Fig. 4). Noges et al. (2007) suggest that in Europe, climate change will alter reference conditions enough to potentially alter typologies (classifications) and impact restoration targets. They point to the need for periodic re-evaluation and adjustment

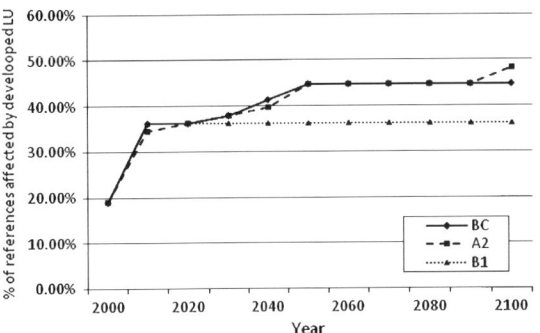

**Fig. 3** Percent of existing Florida reference stations ($N = 58$, classified as "exceptional"), that have >20% developed land use (with 25 houses per square mile or more) within a 1-km buffer surrounding the station; for current land use conditions, and for projected developed land use distributions consistent with IPCC (2000) base case, A2 and B1 scenarios, projected for decadal time periods through 2100 (from USEPA 2010)

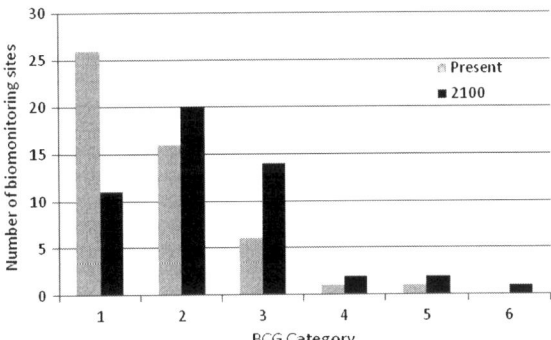

**Fig. 4** Hypothetical example of reference station drift over time as climate change degrades the condition of sites, based on classification of stations along a Biological Condition Gradient (BCG)

of reference conditions and associated standards to accommodate climate change effects that cannot be mitigated. With regard to the Clean Water Act (CWA), this might be interpreted as a lowering of standards in conflict with anti-degradation policies.

The influence of global change thus poses a serious risk to management decisions regarding environmental protection and restoration. Unaccounted climate change impacts, especially when further compromised by increasing influence of developed or agricultural land uses, are likely to lead to fewer determinations of impairment and listings of impaired stream reaches. This is expected because biological responses to climate change, such as decreases in mean abundances or species richness of cold-preference or other sensitive taxa and trait groups, increases in warm-preference or other tolerant taxa and groups (USEPA, 2010), and increases in the variability of these indicators, drive reference sites to greater similarity with non-reference areas, as well as greater difficulty in establishing statistical differentiation (USEPA, 2008). Overall, progressive under-protection of water resources can be expected.

The interpretation of biological indicators and metrics is also at risk from global changes. Several traditional, taxonomically based biological metrics, including total species richness; Ephemeroptera (mayflies)–Plecoptera (stoneflies)–Trichoptera (caddisflies) (EPT) abundance and richness metrics; and Hilsenhoff's biotic index (HBI), are shown to be composed of both cold- and warm-water preference taxa (USEPA, 2010). The relative contribution from cold- and warm-preference taxa varies regionally and among metrics (USEPA, 2010). This mixed composition of temperature traits leads to differential responses to climate-associated increases in water temperature over time, and therefore, the vulnerability of that metric to climate change effects. If a metric like EPT richness is evaluated in a vulnerable location where cold-preference taxa are declining over time, the EPT metric would decline with the potential to alter the ecological status rating of that location. This could happen due to climate change in the absence of any other impairment to the water resource conditions at that location. In another example, we have found a widespread, moderate but significant relationship between temperature sensitivity and sensitivity to organic pollution (USEPA, 2010). Thus, metrics such as Hilsenhoff's Biotic Index (HBI), originally adopted to represent responses to organic pollution, are susceptible to changes in index values due only to climate change effects. The magnitude of biological responses due to climate change-mediated increases in temperature may be sufficient to cross thresholds of impairment and alter management decisions (USEPA, 2010). Without accounting for these climate responses, the traditional interpretation of changes in EPT, HBI, or other metrics could be in error and increase the uncertainty of currently employed processes of impairment characterization and interpretation of causes.

Tracking long-term responses in trait groups defined by temperature preferences is a means to determine biological responses to global changes in temperature (Poff, 1997). Categorization by traits reduces variation across geographical areas that can be associated with taxonomic composition, and thus supports analyses of regional or larger-scale data sets. Such regional-scale analyses are consistent with the larger scale of global change effects. Evaluation of temperature trait groups that are subsets of traditional taxonomically based metrics also illustrates the vulnerabilities of current bioassessment metrics.

Predictive models such as the River Invertebrate Prediction and Assessment Classification System (RIVPACS) used to assess condition and make compliance or other regulatory decisions are susceptible to responses to climate change, predominantly through the relative composition of cold- and warm-preference trait groups in the "observed communities" ($O$ in the $O/E$ [observed/expected] index) (USEPA, 2010). The predictive models for expected communities may be more robust to climate change than individual indices, given periodic model recalibration, due to the relative stability of using long-term averages in estimating predictor variables (USEPA, 2010). The process of periodic recalibration may still have the undesirable effect of altering the reference baseline of comparison over the long term.

## Integration of monitoring and assessment for global change into environmental policy

It is clear that programs using biological indicators and reference-based comparisons as the foundation for assessments are likely to make increasingly erroneous decisions, especially in the most climatically vulnerable or in increasingly populated regions and watersheds, if the impacts of global change are ignored. Conversely, programs that adapt their biological assessment framework by characterizing global change vulnerabilities and sensitive ecological traits will be in a better position to make informed decisions regarding the synergism of multiple stressors in the context of global changes. We have highlighted several critical components of the typical water quality management process that are demonstrably vulnerable to global change impacts. These include impacts to biological indicators that dissociate their responses from conventionally interpreted causes; degradation of reference conditions that will rapidly make it difficult or impossible to define desirable levels of ecological integrity; and the synergism of global changes that will alter water distribution and availability and increase the uncertainty under which flows needed to maintain ecological functioning will be maintained. If assessment approaches are not modified, the increasing loss of aquatic habitat to global shifts in abundance and distribution of flowing waters will become institutionalized because of lack of action to deal with the implications of global change. This will introduce additional uncertainty into a system that requires knowledge of relatively predictable biological indicator responses to different types of "conventional" stressors.

The framework for a new approach needs to address these major vulnerabilities in a manner that accommodates the inherent scale differences between "global change" stressors and many conventional stressors. We recognize that many aspects of on-the-ground implementation of water policy and water quality management will continue to be at a local (e.g., stream reach) level. Local actions that are augmented with a regional (or larger) scale focus will allow bioassessment and implementation of water law and policy to incorporate considerations of global change stressors. Any supplementary monitoring or analyses at a regional scale should also provide inputs that are meaningful at the local scale.

*What do we do about loss of reference conditions?* Global changes influence all aspects of making water resource manage decisions that are based on comparisons to reference conditions. Clearly it will become increasingly problematic to find an "undisturbed" water body of a particular type as a way to define acceptable conditions of ecological integrity. The baseline described from existing reference locations will increasingly reflect only the best available or least disturbed among available locations, and so will reflect increasingly degraded conditions. As stated earlier, the option of periodic recalibration of reference conditions based on what is left leads directly to acceptance of widespread and institutionalized degradation of water quality conditions and calls into question how anti-degradation policies, typically included in water quality regulations in the U.S., can be managed considering the additional influences of climate change. There needs to be a basic understanding that recalibration of reference in a downward direction should never be done. Use of a biological condition gradient approach will aid in anchoring the reference condition founded on an achievable schema (Davies & Jackson, 2006). Special attention of management and regulatory agencies will be needed to consider implementation of anti-degradation policies with respect to global changes.

On the other hand, there will have to be consideration given in some circumstances to the possibility that global changes may lead to irreversible changes in habitat conditions over the long term that may irretrievably alter attainable ecological conditions and uses. For example, some cold-water streams could take on cool water characteristics, with declining abundances or richness of sensitive cold-water taxa and possible increases in warm-water taxa. Regulated parameters such as temperature, dissolved oxygen, and ammonia, may also be sensitive to climate change effects, and their values may need to be adjusted relative to revised designated uses (Table 3). Refinement of aquatic life uses can be applied to guard against lowering of water quality protective standards. More refined aquatic uses could create more narrowly defined categories, which could accommodate potentially "irreversible" changes, but with sufficient scope to maintain protection and support anti-degradation from regulated causes. In addition, the application of use attainability analyses (UAA) on vulnerable water bodies may be pertinent for characterizing climate impacts.

Detecting and monitoring shifting conditions in the reference population should be part of any water quality program. As described earlier, a more objective scale for defining ecological condition, and thus for characterizing reference conditions, that is anchored in expectations for what would be attainable under undisturbed conditions will be integral to such a program. While seemingly a tall order, developing a scale of condition that reflects the full range of biological potential for a region would have two values as an adaptation of the process of bioassessment in the face of global changes. First, it would allow existing reference locations to be ranked in terms of ecological status, where such ranking could be corroborated with documentation of existing levels of land use encroachment, water withdrawals and flow alterations. Second, it would define the scale against which future reference station degradation

Table 3 Variables addressed in criteria and pathways through which they may be affected by climate change (from Hamilton et al., 2009)

| Criteria | Climate change impacts |
| --- | --- |
| Pathogens | Increased heavy precipitation and warming water temperatures may require the evaluation of potential pathogen viability, growth, and migration |
| Sediments | Changing runoff patterns and more intense precipitation events will alter sediment transport by potentially increasing erosion and runoff |
| Temperature | Warming water temperatures from warming air temperatures may directly threaten the thermal tolerances of temperature-sensitive aquatic life and result in the emergence of harmful algal blooms (HABs), invasion of exotic species, and habitat alteration |
| Nutrients | Warming temperatures may enhance the deleterious effects of nutrients by decreasing oxygen levels (hypoxia) through eutrophication, intensified stratification, and extended growing seasons |
| Chemical | Some pollutants (e.g., ammonia) are made more toxic by higher temperatures |
| Biological | Climate changes such as temperature increases may impact species distribution and population abundance, especially of sensitive and cold-water species in favor of warm-tolerant species including invasive species. This could have cascading effects throughout the ecosystem |
| Flow | Changing flow patterns from altered precipitation regimes is projected to increase erosion, sediment and nutrient loads, pathogen transport, and stress infrastructure. Depending on region it is also projected to change flood patterns and/or drought and associated habitat disturbance |
| Salinity | Sea level rise will inundate natural and manmade systems resulting in alteration and/or loss of coastal and estuarine wetland, decreased storm buffering capacity, greater shoreline erosion, and loss of habitat of high value aquatic resources such as coral reefs and barrier islands. Salt water intrusion may also affect groundwater |
| pH | Ocean pH levels have risen from increased atmospheric $CO_2$, resulting in deleterious effects on calcium formation of marine organisms and dependent communities and may also reverse calcification of coral skeletons |

from combined global change impacts could be tracked and quantified. Characterization of conditions at test locations would utilize the same scale, and relative changes between reference and test conditions would provide one piece of evidence in the evaluation of contributions of global and conventional stressors to impairment.

An example of an objective scale of condition is the Biological Condition Gradient (BCG), which has been under research and development by the USEPA (Davies & Jackson, 2006). Its application in general is justified as supporting a more uniform interpretation of condition. This approach would also establish a best estimate of undisturbed biological potential, and therefore, a reference a baseline against which to track future global changes. This would be a valuable adaptation of the existing bioassessment approach, and in the long term would help meet the objectives of maintaining and restoring good levels of water quality and ecological integrity. It is highly recommended that an objective scaling of biological condition, such as the BCG, be given much greater focus and support at the national and regional scale.

Development of an objective scale for characterizing ecological status and condition would have to be regional and be supported by classification of river system types (development of typologies). This follows the need to define reference conditions regionally within types, an approach used in the US (e.g., Gerritsen et al., 2000; Barbour & Gerritsen, 2006), Europe (e.g., Verdonschot, 2006; Noges et al., 2007), and Australia (e.g., Kennard et al., 2006). BCGs have been developed for some regions in the US (e.g., Gerritsen & Leppo, 2005). However, there is still a legacy of conducting site-specific condition assessments based on comparison to an "upstream" reference site or using a paired-watershed approach. Site-specific comparisons cannot capture or account for effects of large scale, pervasive global changes outside of a regional context that defines the range of variation of conditions that exist for that water body type.

Implicit in tracking and accounting for global changes as a part of bioassessment is a time component to sampling and analyses. Despite the relatively large number of reference stations that may be sampled within a jurisdiction or ecoregion, there are typically few stations that are sampled repeatedly and have long-term data. Preservation of adequate long-term data records, with ongoing, regular (at least annual) monitoring is desirable to increase the robustness of water program assessments to the confounding effects of climate change. Adequate long-term monitoring can be a burden for local to regional management agencies. Considering the desirability of repeated temporal monitoring with the recommendation that reference conditions be established on a regional basis, it is recommended that a network of comprehensive monitoring locations be established and maintained. A regional level of implementation would be consistent with the scale of controlling factors in climate change patterns and vulnerabilities, such as climatic type, geology, topography, elevation, ground water influence, latitude, vegetation, etc. Such conditions often cross state, tribal or other jurisdictional boundaries. Collaboration would be best suited in a modest initial focus of monitoring vulnerable areas and watershed types.

Environment agencies are charged with embracing the concept of protecting remaining high quality stream reaches that define reference conditions, mostly at the watershed management scale (e.g., Palmer et al., 2009). However, the range of possible protective actions is limited, and for the most part, only addresses climate change indirectly. Protection options encompass minimization, mitigation, and/or buffering from land use impacts of non-point source runoff, erosion, and hydrologic changes, as well as consideration of water withdrawal impacts. Categories can include socio-political action at the municipal to state scale, such as zoning restrictions, green building incentives, riparian buffer zones, designation of conservation zones or protected areas (e.g., preserves, national parks or forests, wilderness areas), implementation of environmental flow regulations, or other limitations on water withdrawals or diversions from particular stream reaches. Protection actions can be valuable on a local to regional scale, and can be targeted at high quality or unique aquatic resources (Palmer et al., 2009). However, there are often disconnects between the agency with interest in protection of high quality reference locations (e.g., state environment or natural resource departments) and the agencies with the authority to implement such actions. In addition, implementation of comprehensive protection measures takes time, which is in short supply relative to impacts from global changes which are happening now. Given the existing and future projected levels of global change impacts to aquatic resources, targeted protection cannot be solely relied onto maintain the future viability of bioassessment and implementation of water policy.

*How do we resolve mixed messages from existing biological indicators?* Appropriate indicators for assessing ecological status and integrity depend on the vulnerability of potential indicators to the range of stressors being tracked, their ease of measurement and interpretability, and the applicability of the information they provide at the spatial or temporal scale of interest. Indicators can be drawn from a wide range of ecosystem properties, including functional and structural components, production and metabolism, nutrient use and cycling, energy supply, and species composition and feeding types represented in its biotic assemblages (Allan & Castillo, 2007). There is increasing emphasis on using measures that reflect functional processes (Paul, 1997; Fellows et al., 2006), or that incorporate both ecosystem structure and function (Udy et al., 2006). Moss (2008) argues that true properties of ecological quality include efficiency of nutrient use, habitat connectivity, mechanisms of resilience, and characteristic biological structures and functions, but not necessarily a characteristic species list; and that biomonitoring should not rely mainly on taxonomic indices. Still, it is widely accepted and applied in bioassessment that characteristics of biological assemblages, including both taxonomic and trait groups, strongly reflect ecosystem status and health (Bonada et al., 2007a, b; Johnson & Hering, 2009; Norris & Barbour, 2009).

Difficulties arise in categorizing unique indicators of global changes, because of similarities in some of the temperature and hydrologic effects resulting from climate change, land use changes, and water removal (Table 2). In addition, climate change is a global influence in the environment, which differentiates it from most conventional stressors, including land use and water demand changes, which are landscape-scale stressors with global implications. Biomonitoring tools should be tailored to the types and scales of stressors expected. If it is a fundamentally sound concept that larger scale (global change) stressors require a comparably large scale of assessment and management, then both the selection of potential indicators and the way they are used in a bioassessment framework must be appropriate for regional (or larger) scale application. This will require a conscientious refocusing of water policy implementation from the local scale fostered by the current stream reach-specific listing process, to a watershed scale or larger scale, integrated approach to evaluation and management.

With increasing knowledge of the types of biological responses to climate change evident around the US, Europe, and elsewhere, as well as the categories of organisms that are showing relatively predictable responses, it may become possible to adjust bioassessment metrics to enable a clearer interpretation of stressor identification and causal analysis. In a management or regulatory context, biological indicators likely to express the strongest, most regionally consistent and interpretable responses to climate variables will be most effective to document effects early and establish credible links between causes (i.e., the stressor of climate change) and observed biological responses.

Groupings of macroinvertebrates (or other organisms used in bioassessment) based on ecological traits related to temperature or hydrologic preferences provide a link to temperature- and flow-related global change effects. They are interpretable with regard to causal relationships, though with limitations, and offer predictive ability and transferability among regions (Lamouroux et al., 2004; Poff et al., 2006b; Horrigan and Baird, 2008; Verberk et al., 2008a, b). Trait-based analysis may increase the diagnostic power of taxon-based bioindicators (Dolédec et al., 2000) and contribute to a broader understanding of environmental stressor responses (Dolédec & Statzner, 2008). We have already indicated that complete separation of climate change and other global changes will often be difficult and is unlikely to be accomplished through a single biological indicator unique to a specific global stressor. Nevertheless, we propose that separating taxonomic metrics (e.g., EPT taxa, HBI) into sub-categories based on temperature preferences can be used to enhance the ability to make inferences about global change effects compared to other stressors. Tracking metrics by temperature preference would provide a mechanism for documenting climate change-related taxa losses or replacements; this could be used in conjunction with other stressor data and trends at reference locations to help differentiate among global and conventional stressor contributions to observed responses.

Preliminary evaluations have been undertaken of the efficacy of separating taxonomic metrics, such as metrics related to EPT taxa, and the HBI, into new metrics that account for temperature preferences of the component taxa (e.g., a "cold-EPT richness" metric, etc.) (USEPA, 2010). Additional testing of potential climate-revised metrics would be needed on a regional basis.

Continued success in associating biological responses with probable causes of impairment will require redefinition and recalibration of many traditional metrics. In terms of policy implications, it is likely that such efforts would be viewed by local resource managers as extensive. "Federal" agencies are already providing support in the form of substantial trait data bases (e.g., Schmidt-Kloiber et al., 2006; USEPA, 2010). Additional regional efforts to define temperature traits will be needed to provide an important component in the foundation for evaluating climate change affects on aquatic ecosystems. If the relative contributions from the major global changes can not be reliably separated, this weakens specific impairment or other management decisions with the potential effect of losing the confidence of the regulated community.

*Why embrace a management paradigm shift?* The legacy of focusing on point sources of pollution carries an additional legacy of justifying remedial action based on cause. "Punishment of the guilty" describes the assessment and management of point sources of pollution, in which the imposition of some regulatory action (e.g., permit limits, mitigations, restoration requirements) is justified based on specific attribution of the cause of impairment. This approach gives a narrow focus to the process of water resource quality protection and restoration and does not encompass adaption strategies to address and manage global change impacts. We recognize there will always be local pollution problems that require local solutions; thus, examination of local causes will continue to be applicable in the context of defining impairment from point source dischargers, supporting issuance and re-issuance of discharge permits, tracking responses to specific restoration actions, and developing total maximum daily load (TMDL) limits. However, local management for preservation of ecological integrity also is impacted by stressors at non-local (global) scales; hence, the need to utilize multiple scales of assessment.

Recognize the management implications of the discontinuities between the scale of global stressors, the scale of the indicators employed in the assessment process (local to regional), and the scale of management options (local to regional) is a first step in developing an appropriate response. The causes of climate change cannot be altered at the scale of most water resource management activities. Even if maximum reductions of greenhouse gas emissions were achieved immediately on a national to global scale, "committed" climate responses would continue for decades. Similarly, population growth along with increases in developed and agricultural land uses will continue, as will demands for human uses of fresh water, outside of the purview of local water quality agencies. But water resource managers are still faced with protection or restoration of good water quality and ecological integrity.

In the context of climate change it is increasingly important to consider adaptation strategies or other management actions for remediation or restoration that are not necessarily directly related to the proximal cause of the problem. This is a diverse concept that has been touched on by the IPCC ("no regrets" adaptations, IPCC, 2001) and the U.S. Climate Change Science Programs (CCSP, 2008). We suggest that managers be encouraged to consider actions that can improve or ameliorate an impaired condition in an aquatic system, considering broad factors such as feasibility of implementation, possibility of multiple environmental benefits, and direct or indirect impacts on conditions of concern. The actions may not be directly related to the causes of concern, but given the increasing likelihood that multiple causes may be contributing to a particular condition at multiple scales, this is inevitable. Adaptations can be selected to have recognizable environmental benefits, whether projected global changes ultimately impact the site or not ("no regrets" adaptations). The need to take whatever management actions that may be feasible is starting to be recommended in response to climate change in particular (Palmer et al., 2009).

In the US, the USEPA develops the scientific basis for water law and environmental policy actions and provides recommendations and guidance to state water quality agencies for implementation. In Europe, the guiding principal is the Water Framework Directive and the oversight organization that includes representatives of the partner countries. Each country is then charged with the mandate to implement the Directive. Regardless, the agency with the responsibility of assessing and protecting its water resources should help advance a revised assessment process, including inputs on:

- Methods and criteria for evaluating the relative vulnerabilities of regions or watersheds to global change effects, focusing on alterations to thermal and hydrologic regimes, and future projections of land use patterns. Assessments should use objective criteria to define vulnerabilities. Providing clear expectations of effects due to global change for key species and communities in freshwater ecosystems will help these programs achieve their goals under changing environmental conditions.
- Support of additional research needs through agency and academic channels, including focus on how to implement program modifications, BCG implementation, traits and biotic response data base expansion, and metrics modifications.
- Development and implementation of more sophisticated watershed modeling to bridge the scale gap between General Circulation Models (GCMs) and watershed responses. The EuroLimpacs project of the EU has done much in this regard (e.g., Whitehead et al., 1998a, b, 2002; Wade et al., 2002). In the US, EPA's Global Change Research Program is supporting research to evaluate the impacts of climate change and urbanization on hydrology and water quality in 20 major river basins throughout the US.

**Acknowledgments** The authors would like to thank L. Yuan, R. Novak, K. Metchis, and several anonymous reviewers for their comments, which greatly improved this article. The authors would also like to thank R. Cantilli of the U.S. EPA for supporting some of our initial thinking on climate change implications to water policy. The Global Change Research Program in the National Center for Environmental Assessment in the U.S. Environmental Protection Agency's Office of Research and Development provided financial support for the analyses contributing to this article through contract # GS-10F-0268K, DO 1107 to Tetra Tech, Inc. The views expressed in this article are those of the authors and they do not necessarily reflect the views or policies of the U.S. Environmental Protection Agency.

## References

Allan, J. D., 2004. Landscapes and riverscapes: the influence of land use on stream ecosystems. Annual Review of Ecology, Evolution, and Systematics 35: 257–284.

Allan, J. D. & M. M. Castillo, 2007. Stream Ecology: Structure and Function of Running Waters, 2nd edn. Springer, the Netherlands. ISBN 978-1-4020-5582-9: 436 pp.

Bailey, R. C., R. H. Norris & T. B. Reynoldson, 2004. Bioassessment of Freshwater Ecosystems: Using the Reference Condition Approach. Kluwer Academic Publishers, Norwell, MA, USA.

Barbour, M. T. & J. Gerritsen, 2006. Key features of bioassessment development in the United States of America. In Ziglio, G., M. Siligardi, & G. Flaim (eds), Biological Monitoring of Rivers: Applications and Perspectives. Wiley, Chichester, England: 351–366 (469 pp).

Barbour, M. T., W. F. Swietlik, S. K. Jackson, D. L. Courtemanch, S. P. Davies & C. O. Yoder, 2000. Measuring the attainment of biological integrity in the USA: a ciritial element of ecological integrity. Hydrobiologia 422(423): 453–464.

Barbour, M. T. & M. J. Paul, 2010. Adding value to water resource management through biological assessment of rivers. In Moog, O., S. Sharma, & D. Hering (eds), Rivers in the Hindu Kush-Himalaya – Ecology and Environmental Assessment. Hydrobiologia 651: 17–24. doi:10.1007/s10750-010-0287-7.

Bates, B. C., Z. W. Kundzewicz, S. Wu & J. P. Palutikof (eds) (2008). Climate Change and Water. Technical Paper of the Intergovernmental Panel on Climate Change, IPCC Secretariat, Geneva: 210 pp.

Bonada, N., M. Rieradevall & N. Prat, 2007a. Macroinvertebrate community structure and biological traits related to flow permanence in a Mediterranean river network. Hydrobiologia 589: 91–106.

Bonada, N., S. Dolédec & B. Statzner, 2007b. Taxonomic and biological trait differences of stream macroinvertebrate communities between Mediterranean and temperate regions: implications for future climate scenarios. Global Change Biology 13: 1658–1671.

Bradley, D. C. & S. J. Ormerod, 2001. Community persistence among stream invertebrates tracks the North Atlantic Oscillation. Journal of Animal Ecology 70: 987–996.

Brekke, L. D., J. E. Kiang, J. R. Olsen, R. S. Pulwarty, D. A. Raff, D. P. Turnipseed, R. S. Webb & K. D. White, 2009. Climate Change and Water Resources Management—A Federal Perspective: U.S. Geological Survey Circular 1331: 65 p. [Also available online at http://pubs.usgs.gov/circ/1331/].

Buisson, L., W. Thuiller, S. Lek, P. Lim & G. Grenouillet, 2008. Climate change hastens the turnover of stream fish assemblages. Global Change Biology 14: 2232–2248.

Bunn, S. E. & A. H. Arthington, 2002. Basic principles and ecological consequences of altered flow regimes for aquatic biodiversity. Environmental Management 30: 4492–4507.

Bunn, S. E. & P. M. Davies, 2000. Biological processes in running waters and their implications for the assessment of ecological integrity. Hydrobiologia 422(423): 61–70.

Burgmer, T., H. Hillebrand & M. Pfenninger, 2007. Effects of climate-driven temperature changes on the diversity of freshwater macroinvertebrates. Oecologia 151: 93–103.

CCSP, 2008. Preliminary review of adaptation options for climate-sensitive ecosystems and resources. A Report by the U.S. Climate Change Science Program and the

Subcommittee on Global Change Research (Julius, S. H. & J. M. West (eds); Baron, J. S., B. Griffith, L. A. Joyce, P. Kareiva, B. D. Keller, M. A. Palmer, C. H. Peterson & J. M. Scott (Authors)). U.S. Environmental Protection Agency, Washington, DC, USA.

Cereghino, R., Y.-S. Park, A. Compin & S. Lek, 2003. Predicting the species richness of aquatic insects in streams using a limited number of environmental variables. Journal of the North American Benthological Society 22(3): 442–456.

Chessman, B., 2009. Climatic changes and 13-year trends in stream macroinvertebrate assemblages in New South Wales, Australia. Global Change Biology. doi: 10.1111/j.1365-2486.2008.01840.x.

Chovanec, A., P. Jäger, M. Jungwirth, V. Koller-Kreimel, O. Moog, S. Muhar & St. Schmutz, 2000. The Austrian way of assessing the ecological integrity of running waters: a contribution to the EU Water Framework Directive. Hydrobiologia 422(423): 445–452.

Collier, K. J., 2008. Temporal patterns in the stability, persistence and condition of stream macroinvertebrate communities: relationships with catchment land-use and regional climate. Freshwater Biology 53: 603–616.

Daily, G.C., 1997. Introduction: what are ecosystem services? In Daily, G. (ed.), Nature's Services: Societal Dependence on Natural Ecosystems. Island Press, Washington, DC: 1–10.

Daufresne, M. & P. Boet, 2007. Climate change impacts on structure and diversity of fish communities in rivers. Global Change Biology 13: 2467–2478.

Davies, S. P. & S. K. Jackson, 2006. The biological condition gradient: a descriptive model for interpreting change in aquatic ecosystems. Ecological Applications 16: 1251–1266.

Diaz, A. M., M. L. Suarez Alonso & M. R. Vidal-Abarca Gutierrez, 2008. Biological traits of stream macroinvertebrates from a semi-arid catchment: patterns along complex environmental gradients. Freshwater Biology 53: 1–21.

Directive, 2000. Directive 2000/60/EC of the European Parliament and of the council of 23 October 2000 establishing a framework for community action in the field of water policy. Official Journal of the European Communities L 327:1–72.

Dolédec, S. & B. Statzner, 2008. Invertebrate traits for the biomonitoring of large European rivers: an assessment of specific types of human impact. Freshwater Biology 53: 617–634.

Doledec, S., J. Dessaix & H. Tachet, 1996. Changes within the Upper Rhone River macro-benthic communities after the completion of three hydroelectric schemes: anthropogenic effects or natural change? Archiv Fur Hydrobiologie 136: 19–40.

Dolédec, S., J. M. Olivier & B. Statzner, 2000. Accurate description of the abundance of taxa and their biological traits in stream invertebrate communities – effects of taxonomic and spatial resolution. Archiv für Hydrobiologie 148: 25–43.

Durance, I. & S. J. Ormerod, 2007. Climate change effects on upland stream macroinvertebrates over a 25-year period. Global Change Biology 13: 942–957.

Fellows, C. S., J. E. Clapcott, J. W. Udy, S. E. Bunn, B. D. Harch, M. J. Smith & P. M. Davies, 2006. Benthic metabolism as an indicator of stream ecosystem health. Hydrobiologia 572: 71–87.

Gerritsen, J. & E. W. Leppo, 2005. Biological Condition Gradient for Tiered Aquatic Life Use in New Jersey. Prepared for the U.S. Environmental Protection Agency (USEPA), Office of Science and Technology, USEPA Region 2, and the New Jersey Department of Environmental Protection, by Tetra Tech, Inc., Center for Ecological Sciences.

Gerritsen, J., M. T. Barbour & K. King, 2000. Apples, oranges, and ecoregions: on determining pattern in aquatic assemblages. Journal of the North American Benthological Society 19(3): 487–496.

Helms, B. S., J. E. Schoonover & J. W. Feminella, 2009. Seasonal variability of landuse impacts on macroinvertebrate assemblages in streams of western Georgia, USA. Journal of the North American Benthological Society 28(4): 991–1006.

Helmuth, B., B. R. Broitman, C. A. Blanchette, S. Gilman, P. Halpin, C. D. G. Harley, M. J. O'Donnell, G. E. Hofmann, B. Menge & D. Strickland, 2006. Mosaic patterns of thermal stress in the rocky intertidal zone: implications for climate change. Ecological Monographs 76(4): 461–479.

Herlihy, A. T., S. G. Paulsen, J. Van Sickle, & J. L. Stoddard, 2008. Striving for consistency in a national assessment: the challenges of applying a reference-condition approach at a continental scale. Journal of the North American Benthological Society 27(4): 860–877.

Hiddink, J. G. & R. ter Hofstede, 2008. Climate induced increases in species richness of marine fishes. Global Change Biology 14: 453–460.

Horrigan, N. & D. J. Baird, 2008. Trait patterns of aquatic insects across gradients of flow-related factors: a multivariate analysis of Canadian national data. Canadian Journal of Fisheries and Aquatic Sciences 65: 670–680.

IPCC (Intergovernmental Panel on Climate Change), 2000. Emissions Scenarios. A Special Report of Working Group III of the Intergovernmental Panel on Climate Change. Cambridge University Press, UK. http://www.grida.no/climate/ipcc/emission/091.htm.

IPCC (Intergovernmental Panel on Climate Change), 2001. Climate Change 2001: Impacts, Adaptation, and Vulnerability. Contribution of Working Group II to the Third Assessment Report of the Intergovernmental Panel on Climate Change, Cambridge University Press, Cambridge, United Kingdom.

IPCC (Intergovernmental Panel on Climate Change), 2007. Climate change 2007. The physical science basis. Summary for policy makers. Contribution of Working Group I to the Fourth Assessment Report of the Intergovernmental Panel on Climate Change. IPCC Secretariat, Geneva, Switzerland.

Johnson, R. K. & D. Hering, 2009. Response of taxonomic groups in streams to gradients in resource and habitat characteristics. Journal of Applied Ecology 46: 175–186.

Jungwirth, M., S. Muhar & S. Schmutz, 2000. Assessing the ecological integrity of running waters. Hydrobiologia 422/423. Kluwer Academic Publishers, Dordrecht: 487 pp.

Kennard, M. J., B. D. Harch, B. J. Pusey & A. H. Arthington, 2006. Accurately defining the reference condition for summary biotic metrics: a comparison of four approaches. Hydrobiologia 572: 151–170.

Konrad, C. P., A. M. D. Brasher, & J. T. May, 2008. Assessing streamflow characteristics as limiting factors on benthic invertebrate assemblages in streams across the western United States. Freshwater Biology 53: 1983–1998.

Lamouroux, N., S. Doledec & S. Gayraud, 2004. Biological traits of stream macroinvertebrate communities: effects of microhabitat, reach, and basin filters. Journal of the North American Benthological Society 23(3): 449–466.

Millennium Ecosystem Assessment (MEA), 2005. Ecosystems and Human Well-being: Synthesis. Island Press, Washington, DC.

Minshall, G. W., R. C. Petersen & C. F. Nimz, 1985. Species richness in streams of different size from the same drainage basin. American Naturalist 125: 16–38.

Moog, O. & A. Chovanec, 2000. Assessing the ecological integrity of rivers: walking the line among ecological, political and administrative interests. Hydrobiologia 422(423): 99–109.

Moss, B., 2008. The water framework directive: total environment or political compromise? Science of the Total Environment 400: 32–41.

Mouthon, J. & M. Daufresne, 2006. Effects of the 2003 heatwave and climatic warming on mollusc communities of the Saone: a large lowland river and of its two main tributaries (France). Global Change Biology 12: 441–449.

Niemi, G. J. & M. E. McDonald, 2009. Applications of ecological indicators. Annual Review of Ecology Evolution and Systematics 35: 89–111.

Nilsson, C., C. A. Reidy, M. Dynesius & C. Revenga, 2005. Fragmentation and flow regulation of the world's large river systems. Science 308: 405–408.

Noges, P., W. Van de Bund, A. C. Cardoso & A.-S. Heiskanen, 2007. Impact of climatic variability on parameters used in typology and ecological quality assessment of surface waters – implications on the Water Framework Directive. Hydrobiologia 584: 373–379.

Noges, P., W. van de Bund, A. C. Cardoso, A. G. Solimini & A. Stiina Heiskanen, 2009. Assessment of the ecological status of European surface waters: a work in progress. Hydrobiologia 633: 197–211.

Norris, R. H. & M. T. Barbour, 2009. Bioassessment of aquatic ecosystems. Encyclopedia of Inland Waters 3:21–28.

Palmer, M. A., D. P. Lettenmaier, N. L. Poff, S. L. Postel. B. Richter & R. Warner, 2009. Climate change and river ecosystems: protection and adaptation options. Environmental Management. doi:10.1007/s00267-009-9329-1.

Parmesan, C., 2006. Ecological and evolutionary responses to recent climate change. Annual Review of Ecology Evolution and Systematics 37: 637–669.

Paul, M. J., 1997. Back to Odum: using ecosystem functional measures in stream ecosystem management. In Hatcher, K. J. (ed.), Proceedings of the 1997 Georgia Water Resources Conference. March 20–22, 1997, University of Georgia.

Paul, M. J. & J. L. Meyers, 2001. Streams in the urban landscape. Annual Review of Ecology, Evolution, and Systematics 32:333–365.

Poff, N. L., 1997. Landscape filters and species traits: towards mechanistic understanding and prediction in stream ecology. Journal of the North American Benthological Society 16: 391–409.

Poff, N. L., B. P. Bledsoe & C. O. Cuhaciyan, 2006a. Hydrologic variation with land use across the contiguous United States: geomorphic and ecological consequences for stream ecosystems. Geomorphology 79: 264–285.

Poff, N. L., J. D. Olden, N. K. M. Vieira, D. S. Finn, M. P. Simmons & B. C. Kondratieff, 2006b. Functional trait niches of North American lotic insects: traits-based ecological applications in light of phylogenetic relationships. Journal of the North American Benthological Society 25(4): 730–755.

Postel, S. & S. R. Carpenter, 1997. Freshwater ecosystem services. In Daily, G. (ed.), Nature's Services. Island Press, Washington, DC: 195–214.

Richter, B. D., R. Matthews, D. L. Harrison & R. Wigington, 2003. Ecologically sustainable river flows for river integrity. Ecological Applications 13: 206–224.

Root, T. L., J. T. Price, K. R. Hall, S. H. Schneider, C. Rosenzweig & J. A. Pounds, 2003. Fingerprints of global warming on wild animals and plants. Nature 421: 57–60.

Schmidt-Kloiber, A., W. Graf, A. Lorenz & O. Moog, 2006. The AQUEM-STAR taxa list – a pan-European macroinvertebrate ecological database and taxa inventory. Hydrobiologia 566: 325–342.

Stoddard, J. L., D. P. Larsen, C. P. Hawkins, R. K. Johnson & R. H. Norris, 2006. Setting expectations for the ecological condition of streams: the concept of reference conditions. Ecological Applications 16: 1267–1276.

Thuiller, W., 2004. Patterns and uncertainties of species' range shifts under climate change. Global Change Biology 10: 2020–2027.

Tobin, P., N. S. Nagarkatti, G. Loeb & M. C. Saunders, 2008. Historical and projected interactions between climate change and insect voltinism in a multivoltine species. Global Change Biology 14: 951–957.

Udy, J. W., C. S. Fellows, M. E. Bartkow, S. E. Bunn, J. E. Clapcott & B. D. Harch, 2006. Measures of nutrient processes as indicators of stream ecosystem health. Hydrobiologia 572: 89–102.

USEPA (2008) Climate Change Effects on Stream and River Biological Indicators: A Preliminary Analysis. U.S. Environmental Protection Agency, Washington, DC. EPA/600/R-07085.

USEPA (2010) Implications of Climate Change for Bioassessment Programs and Approaches to Account for Effects. U.S. Environmental Protection Agency, Washington, DC. EPA/600/R-09/xxx.

USGPO (U.S. Government Printing Office), 1972a. Report of the Committee on Public Works – United States House of Representatives with Additional and Supplemental Views on H.R. 11896 to Amend the Federal Water Pollution Control Act. House Report 92–911. 92nd congress, 2d session, March 11, 1972: 149 p.

USGPO (U.S. Government Printing Office), 1972b. Report of the Senate Public Works Committee – United States Senate. Federal Water Pollution Control Act Amendments of 1972. P.L. 92–500. Senate Report 92–414: 1468 p.

Verberk, W. C. E. P., H. Siepel & H. Esselink, 2008a. Life-history strategies in freshwater macroinvertebrates. Freshwater Biology 53: 1722–1738.

Verberk, W. C. E. P., H. Siepel & H. Esselink, 2008b. Applying life-history strategies for freshwater macroinvertebrates to lentic water. Freshwater Biology 53: 1739–1753.

Verdonschot, P. F. M., 2000. Integrated ecological assessment methods as a basis for sustainable catchment management. Hydrobiologia 422(423): 389–412.

Verdonschot, P. F. M., 2006. Evaluation of the use of Water Framework Directive typology descriptors, reference sites and spatial scale in macroinvertebrate stream typology. Hydrobiologia 566: 39–58.

Wade, A. J., P. G. Whitehead & D. Butterfield, 2002. The Integrated Catchments model of Phosphorus dynamics (INCA-P), a new approach for multiple source assessment in heterogeneous river systems: model structure and equations. Hydrology and Earth System Sciences 6: 583–606.

Walther, G.-R., E. Post, P. Convey, A. Menzel, C. Parmesan, T. J. C. Beebee, J.-M. Fromentin, O. Hoegh-Guldberg & F. Bairlein, 2002. Ecological responses to recent climate change. Nature 416: 389–395.

Walther, G. R., S. Berger & M. T. Sykes, 2005. An ecological 'footprint' of climate change. Proceedings of the Royal Society B-Biological Sciences 272: 1427–1432.

Whitehead, P. G., E. J. Wilson & D. Butterfield, 1998a. A semi-distributed integrated nitrogen model for multiple source assessment in catchments (INCA). Part I – model structure and process equations. The Science of the Total Environment 210(211): 547–558.

Whitehead, P. G., E. J. Wilson, D. Butterfield & K. Seed, 1998b. A semi-distributed integrated flow and nitrogen model for multiple source assessment in catchments (INCA). Part II – application to large river basins in south Wales and eastern England. The Science of the Total Environment 210(211): 559–583.

Whitehead, P. G., D. J. Lapworth, R. A. Skeffington & A. Wade, 2002. Excess nitrogen leaching and C/N decline in the Tillingbourne catchment, southern England: INCA process modelling for current and historic time series. Hydrology and Earth System Sciences 6: 455–466.

Zuckerberg, B., A. M. Woods & W. F. Porter, 2009. Poleward shifts in breeding bird distributions in New York State. Global Change Biology 15: 1866–1883.